PROBABILITY RANDOM PROCESSES AND QUEUEING THEORY

PROBABILITY RANDOM PROCESSES AND QUEUEING THEORY

A.M. Natarajan
Principal
Kongu Engineering College
Erode, Tamil Nadu

A. Tamilarasi
Dept. of CSE-PG
Kongu Engineering College
Erode, Tamil Nadu

PUBLISHING FOR ONE WORLD

NEW AGE INTERNATIONAL (P) LIMITED, PUBLISHERS
New Delhi • Bangalore • Chennai • Cochin • Guwahati • Hyderabad
Jalandhar • Kolkata • Lucknow • Mumbai • Ranchi
Visit us at **www.newagepublishers.com**

Published by New Age International (P) Ltd., Publishers
First Edition: 2003
Second Edition: 2005
Reprint: 2009

Branches:

- 36, Malikarjuna Temple Street, Opp. ICWA, Basavanagudi, **Bangalore**. © (080) 26677815
- 26, Damodaran Street, T. Nagar, **Chennai**. © (044) 24353401
- CE–39/1016, Carrier Station Road, Near Paulson Park Hotel, Ernakulam South, **Cochin** © (0484) 4051303
- Hemsen Complex, Mohd. Shah Road, Paltan Bazar, Near Starline Hotel, **Guwahati**. © (0361) 2543669
- No. 105, 1st Floor, Madhiray Kaveri Tower, 3-2-19, Azam Jahi Road, Nimboliadda, **Hyderabad**. © (040) 24652456
- RDB Chambers (Formerly Lotus Cinema) 106A, Ist Floor, S.N. Banerjee Road, **Kolkata**. © (033) 22275247
- 16-A, Jopling Road, **Lucknow**. © (0522) 2209578
- 142C, Victor House, Ground Floor, N.M. Joshi Marg, Lower Parel, **Mumbai**. © (022) 24927869
- 22, Golden House, Daryaganj, **New Delhi**. © (011) 23262370, 23262368

ISBN (10): 81-224-1670-5

ISBN (13): 978-81-224-1670-1

Rs. 225.00

C-09-08-3975

Printed in India at A. H. Print Pack, Delhi.
Typeset at Goswami Associates, Delhi.

PUBLISHING FOR ONE WORLD
NEW AGE INTERNATIONAL (P) LIMITED, PUBLISHERS
4835/24, Ansari Road, Daryaganj, New Delhi-110002
Visit us at **www.newagepublishers.com**

Preface to the Second Edition

We take great pride and pleasure in bringing out this revised, enlarged and improved second edition of this book entitled ''Probability, Random Processes and Queueing Theory'' for the benefit of all the readers. All engineering college students, in particular B.E. (CSE, IT, ECE) of Anna University have found this to be extremely useful. This book discusses the theoretical aspects, which have been refined and updated. Two more chapters 8 and 9 have been added in this revised edition, which discusses Statistical Inference, and Design of Experiments, Quality Control. Further, chapter 6 is splitted into two chapters, one is on Markov Process, and another chapter is Reliability Engineering. Many misprints have been corrected.

We hope that the book will prove to be of much greater utility to the students as well as teachers of the subject. Corrections, suggestions and comments for the improvement of the book will be appreciated and duly acknowledged.

We wish to thank our Secretary, correspondent and all other members of KVIT trust for their encouragement and moral support for bringing out this edition successfully. Also we would like to thank New Age International Publishers for their untiring efforts they have put in, in order to bring out this edition elegantly, efficiently and in quick time.

Authors

Preface to the First Edition

This book aims to develop the subject of Probability, Random Process and Queueing Theory which play important roles in nearly all phases of human endeavour. Theoretical concepts developed here have been used to solve a large number of practical problems.

The book covers the entire syllabus framed by Anna University for second semester B.Tech (I.T.) and fourth semester B.E (CSE, ECE) courses of all engineering colleges in Tamilnadu. Many examples have been included here in order to provide an insight into the concept of probability. It can also be used as a reference book. Further it will stimulate an interest in the readers on this topic.

The salient features of this book include a simple and concise presentation; self-contained details of each topic and more than three hundred solved problems. In each chapter number of exercises are included for the readers to practise and use it. Though the book is presented in the form of a textbook, it can help the readers to learn on self-study basis. The prerequisites needed for using this book are a course on Laplace and Fourier transform.

First chapter provides basic concepts on probability theory. Random variables, expectations, moment-generating function are also defined and explained in chapter 2. Also emphasis is given to discrete as well as continuous distributions. Binomial, Poisson, Gamma, Normal and Weibull distributions are also given due weightage. Third chapter deals with random process and their applications. Chapter four and five discuss correlation function and power spectral density. Markovian chain and Reliability theory which are extensively used in several disciplines are studied in chapter 6. Markovian queueing models are dealt with separately in chapter 7.

It is our duty to express our sincere gratitute to our secretary Mr. D. Venkateshwaran B.E., M.S. (USA), KVIT, the correspondent Mr. M.N. Ramasamy and KVIT Trust members for their kind encouragement and support during this project. Also we would like to thank our colleagues and students for giving valuable comments. We would like to thank Mr. P. Shiv Kumar, of New Age International Publishers for bringing out this book in time.

Suggestions for the improvement of the book will be gratefully acknowledged and executed.

Authors

Contents

1
Probability

Most individuals have an intuitive knowledge of probability as a measure of uncertainty regarding an event. A more precise understanding is necessary for an engineer or any scientist to construct and analyse probabilistic models. The theory has its origin in the unusual interest in gambling that prevaded France in the 17th century. Two mathematician Pascal and Fermat (1601–1665) laid the foundation for the theory of probability. Later, Huyghens (1629–1695) started working on problems in games of chances and published in 1654 the first book on "The theory of probability". Following this, Jacob Bernoulli (1654–1705) finds variety of applications of this theory. Important contributions were made by Laplace (1749–1827), Abraham De Moivre (1667–1754), Markov (1856–1922) etc. Now-a-days, interest in the theory of probability is worldwide but the outstanding contributions are from Bernstein, Khintchine, Kolmogorov, Feller, etc.

In this chapter we consider only those experiments for which the sample space contains a finite number of elements. The likelihood of the occurrence of an event resulting from a statistical experiment is evaluated by means of a set of real numbers called weights or probabilities ranging from 0 to 1. To every point in the sample space we assign a probability such that the sum of all probabilities is 1. If certain sample point is quite likely to occur when the experiment is conducted, then the probability assigned should be close to 1. On the other hand, a probability close to zero is assigned to a sample point that is not likely to occur. In many experiments such as tossing a coin, all sample points have the same chance of occurring and are assigned equal probabilities. For points outside the sample space, the probability assigned is zero.

Now we define terms like events, sample space, etc.

1.1 TERMINOLOGY

1.1.1 Random Experiment

It is an experiment whose outcome is not known in advance. For example, if a uniform unbiased coin is tossed, then the outcome may be tail or head.

1.1.2 Trial and Event

The performance of a random experiment is called a **trial** and the outcome is called an **event** or **case.** For example, throwing of a die is a trial and getting 1 or 2 or 3 . . . or 6 is an event.

1.1.3 Exhaustive Events

All possible outcome of an experiment is called **exhaustive events** or **exhaustive cases**. In tossing a coin, either the head or tail turns up. There is no other possibility and therefore these are the exhaustive events.

1.1.4 Favourable Events

The number of outcomes which result in the happening of a desired events are called **favourable events (or cases)**. Thus in drawing a card from a deck of cards, the number of favourable cases of getting a spade is 13 (as there are 13 spade cards in the deck of 52 cards)

1.1.5 Mutually Exclusive Events

Two or more events are said to be **mutually exclusive** if the happening of any one of them excludes the happening of all others in a single experiment. In the throw of a single dice, the events of getting 1,2, ... 6 are mutually exclusive since when one side is on top, the other sides cannot be on top.

1.1.6 Equally Likely Events

Two or more events are said to be **equally likely** if the chance of their happening is equal. There is no preference of any one event over the other. For example, in the throw of a coin, getting a head or a tail is equally likely.

1.1.7 Compound Events

When two or more events occur together, their joint occurrence is called a **compound event.**

Two or more events occurring together may be either independent or dependent.

1.1.8 Independent Events

Two or more events are said to be **independent** if the occurrence of one event will not affect the occurrence of the remaining events. In the throw of a dice repeatedly coming up of any side (say 4) on the first throw is independent of coming up of the same side (4) again in the second throw.

1.1.9 Permutation

The word permutation refers to **arrangements** and combination refers to **'grouping'**. These terms are used in the calculation of probability. Some simple rules of permutation are:

(i) The permutation of n distinct things taken all at a time is factorial n $(1 \times 2 \times 3 \times \ldots (n-1) \times n = n!)$. For example total number of ways of arranging 3 letters x, y, z is xyz, xzy, yxz, yzx, zxy, zyx. That is, 3 letters x, y, z can be arranged in $1 \times 2 \times 3 = 6 = 3!$ ways.

(ii) The permutation of n distinct things taken r at a time is given by $n\,P_r$ or $\dfrac{n!}{(n-r)!}$. For example, the arrangement of any two letters out of 3 letters x, y, z is: xy, yx, xz, zx, yz, zy. Hence the permutation of 3 letters taken 2 at a time is $3\,P_2 = \dfrac{3!}{1!} = 3!$

1.1.10 Permutation with Repetition

(i) The number of permutations of n things when n_1 of them are of one kind and n_2 of another kind is $\dfrac{n!}{n_1!\,n_2!}$. For example, the number of permutations of the word 'ENGINEERING', is $\dfrac{11!}{3!2!3!2!}$

(E occurs 3 times, N occurs 3 times, I occurs 2 times, G occurs 2 times).

1.1.11 Rule of Sum

If an event can occur in m ways and another event can occur in n ways there are $m + n$ ways in which exactly one event can occur. For example, consider a box containing 7 red balls and 4 white balls, there are 7 + 4 ways to choose either a red ball or a white ball.

1.1.12 Rule of Product

If there are m outcomes for event E_1 and n possible outcomes for event E_2, then there are mn outcomes for the composite event E_1E_2. For example, when a pair of dice is thrown once, the number of possible outcomes are $6 \times 6 = 36$ since each dice have 6 possible outcomes.

1.1.13 Combination

The number of combinations of n different things taken r at a time is $\dfrac{n!}{r!(n-r)!}$. It is represented as nC_r or $C(n, r)$

Thus in a set of 3 different letters (say A, B, C), if any two letters at a time are picked up, the different combination is $3C_2$. This can be computed as $3C_2 = \dfrac{3!}{2!(3-2)!} = \dfrac{(1\times 2\times 3)}{(1\times 2)\times(1)} = 3.$

1.1.14 Mathematical Probability

If a trial results in n exhaustive, mutually exclusive and equally likely cases of which m are favourable to an event E, then the probability P of the happening of the event E is

$$P(E) = \frac{\text{Favourable number of cases}}{\text{Exhaustive number of cases}} = \frac{m}{n}.$$

The probability of not happening of an event E is denoted by $Q(E) = 1 - P(E)$. Since m events are favourable to the event E, $n - m$ events are not favourable to the event E. Therefore, the probability Q for not happening of the event E is $Q = \dfrac{n-m}{n} = 1 - \dfrac{m}{n} = 1 - P$. That is $P + Q = 1$.

Note that P as well as Q are non-negative and cannot exceed 1. If $P(E) = 1$, then E is called *certain event* and if $P(E) = 0$, E is called an *impossible event.*

1.1.15 Statistical or Empirical Probability

If in n trials an event E happens m times, then the probability P of the happening of E is given by

$$P = P(E) = \underset{n\to\infty}{\text{Lt}} \frac{m}{n}.$$

1.2 AXIOMS OF PROBABILITY

The set of all possible outcomes of an experiment is known as the **sample space** of the experiment and is denoted by S. Each outcome in a sample space is called a **sample point**.

1.2.1 Example

If the experiment consists of flipping two coins, then the sample space consists of the following four elements:

$$S = \{HH, HT, TH, TT\}$$

For any two events E and F of a sample space S, we define the new event $E \cup F$ called the **union** of E and F to consists of all sample points which occur either in E or in F or in both. For example, if $E = \{HH, HT\}$, $F = \{TH\}$ then $E \cup F = \{HH, HT, TH\}$.

Similarly for any two events E and F we define the new event $E \cap F$ called the **intersection** of E and F consists of all outcomes which occur both in E and in F. We extend the definitions to the events $E_1, E_2, \ldots E_n \ldots$ For any event E we define the new event E^c as the complement of E to consist of all points in S that are not in E. Also note that $S^c = \phi$, empty set.

Axiom 1. The probability of an event E, $P(E)$, is the sum of the probabilities of all sample points in E. Therefore,

$$0 \leq P(E) \leq 1, \; P(\phi) = 0.$$

Axiom 2. Since all the n possible outcomes of a trial are favourable to S,

$$P(S) = \frac{\text{Favourable number of cases}}{\text{Exhaustive number of cases}} = \frac{n}{n} = 1.$$

Axiom 3. For any sequence of mutually exclusive events $E_1, E_2, \ldots$ (that is events for which $E_i \cap E_j = 0$ when $i \neq j$),

$$P\left(\bigcup_{i=1}^{\infty} E_i\right) = \sum_{i=1}^{\infty} P(E_i)$$

1.2.2 Example

If a die is rolled and suppose that all six sides are equally likely to appear, then $P(\{1\}) = P(\{2\}) = P(\{3\}) = P(\{4\}) = P(\{5\}) = P(\{6\}) = \frac{1}{6}$. The probability of rolling an even number 2, 4, 6 is given by $P(\{2,4,6\}) = P(\{2\}) + P(\{4\}) + P(\{6\}) = \frac{3}{6} = \frac{1}{2}$.

1.2.3 Proposition

If $E \subset F$, then $P(E) \leq P(F)$

Proof: Since $E \subset F$, it follows that

$$F = E \cup (E^c \cap F).$$

Since E and $(E^c \cap F)$ are mutually exclusive events, by Axiom 3,

$$P(F) = P(E) + P(E^c \cap F)$$

which proves the result, since $P(E^c \cap F) \geq 0$.

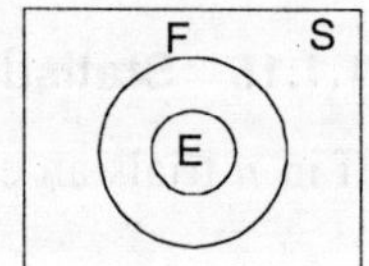

Fig. 1.1

1.3 ADDITION AND MULTIPLICATION THEOREMS

1.3.1 Addition Theorem

If E and F are not mutually exclusive events, then

$$P(E \cup F) = P(E) + P(F) - P(E \cap F)$$

Proof: $E \cup F$ can be written as the union of two disjoint events E and $E^c \cap F$. Thus from Axiom 3,

$$P(E \cup F) = P(E \cup (E^c \cap F)) = P(E) + P(E^c \cap F)$$

Further, since $F = (E \cap F) \cup (E^c \cap F)$ (see Fig 1.2), we obtain

$$P(F) = P(E \cap F) + P(E^c \cap F)$$

Hence

$$P(E \cup F) = P(E) + P(F) - P(E \cap F)$$

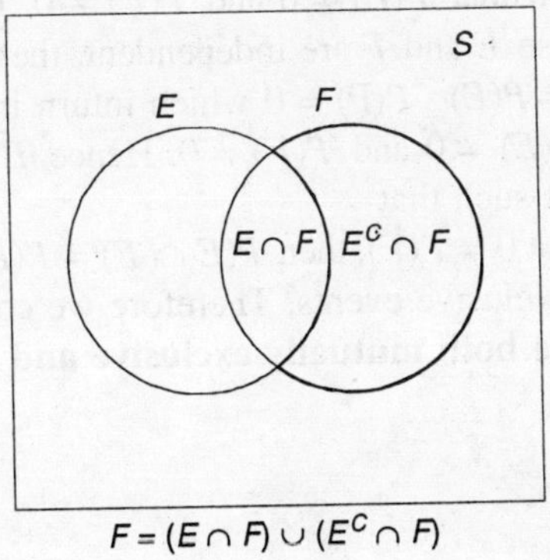

Fig. 1.2

1.3.2. Multiplication Theorem

If two events E and F are **independent** and can happen simultaneously, the probability of their joint occurrence $P(E \cap F) = P(E).P(F)$

The theorem can be extended to three or more events.

1.4 MARGINAL AND CONDITIONAL PROBABILITY

1.4.1 Marginal Probability

A probability of only one event that takes place is called a **marginal probability**.

1.4.2 Joint Probability

The probability of occurrence of both events E and F together, denoted by $P(E \cap F)$, is known as **joint probability** of E and F.

1.4.3 Conditional Probability

Two events E and F are said to be dependent when F can occur given that E has already occurred or vice versa. The probabilities associated with such events are called **conditional probabilities.** It is denoted in the following manner:

$P(A/B)$ would mean probability of A given that B has occurred.

$P(B/A)$ would mean probability of B given that A has occurred.

Now

$$P(A/B) = \frac{P(A \cap B)}{P(B)}$$

$$P(B/A) = \frac{P(A \cap B)}{P(A)}$$

Hence the multiplication theorem in terms of conditional probabilities would be

$$P(A \cap B) = P(B) \cdot P(A/B)$$
$$P(A \cap B) = P(A) \cdot P(B/A)$$

If E and F are mutually independent, then

$$P(E/F) = P(E) \text{ and } P(F/E) = P(F). \text{ Hence}$$
$$P(E \cap F) = P(E/F)\, P(F) = P(E) \cdot P(F)$$

Let E and F be two events such that $P(E) \neq 0$ and $P(F) \neq 0$. If E and F are mutually exclusive events, then $P(E \cap F) = 0$. Suppose E and F are independent, then $P(E \cap F) = P(E) \cdot P(F)$. Hence $P(E \cap F) = 0$ implies $P(E \cap F) = P(E) \cdot P(F) = 0$ which inturn implies $P(E) = 0$ or $P(F) = 0$. This contradicts our assumption that $P(E) \neq 0$ and $P(F) \neq 0$. Hence E and F are not independent. Conversely if E and F are independent such that

$$P(E) \neq 0 \neq P(F), \text{ then } P(E \cap F) = P(E) \cdot P(F) \neq 0$$

Hence E and F are not mutually exclusive events, Therefore we conclude that

Two events E and F cannot be both mutually exclusive and independent.

1.5 BAYE'S THEOREM

1.5.1 Theorem

Let $B_1, B_2, \ldots B_k$ be a partition of the sample space S. Let A be an event associated with S. Let $P(A/B_i)$ and $P(B_i/A)$ be the conditional probabilities for $i = 1$ to k, then

$$P(B_i/A) = \frac{P(A/B_i)\,P(B_i)}{\sum_{j=1}^{k} P(A/B_j)\,P(B_j)} \qquad i = 1, 2, \ldots k$$

Proof Given $B_1, B_2, \ldots B_k$ be a partition of the sample space S. That is (i) $B_i \cap B_j = \phi$ for all $i \neq j$, (ii) $\bigcup_{i=1}^{k} B_i = S$, (iii) $P(B_i) > 0$ for all i. Let A be an event associated with S. Then

$$(A \cap B_1) \cup (A \cap B_2) \cup \ldots \cup (A \cap B_k) = A \cap (B_1 \cup B_2) \cup (A \cap B_3) \ldots \cup (A \cap B_k)$$

(by distributive law)

$$= A \cap (B_1 \cup B_2 \cup B_3) \cup \ldots \cup (A \cap B_k)$$

$\ldots \quad \ldots \quad \ldots$ (by distributive law)

$$= A \cap (B_1 \cup B_2 \cup \ldots \cup B_k) = A \cap S = A$$

Also all the events $A \cap B_1, A \cap B_2, \ldots A \cap B_k$ are pairwise mutually exclusive. For,

$$(A \cap B_i) \cap (A \cap B_j) = A \cap (B_i \cap B_j) \qquad i \neq j$$
$$= A \cap \phi = \phi$$

Now we apply addition theorem for mutually exclusive events $A \cap B_1, A \cap B_2, \ldots A \cap B_k$ and write

$$P(A) = P(A \cap B_1) + P(A \cap B_2) + \ldots + P(A \cap B_k)$$

However each term $P(A \cap B_j)$ may be expressed as $P(A / B_j)\, P(B_j)$ and hence we obtain the **theorem on total probability.**

$$P(A) = P(A/B_1)\,P(B_1) + P(A/B_2)\,P(B_2) + \ldots + P(A/B_k)\,P(B_k). = \sum_{i=1}^{k} P(A/B_j)\,P(B_j) \qquad (1.1)$$

Again by the definition of condition probability and (1.1) we have

$$P(B_i/A) = \frac{P(A \cap B_i)}{P(A)} = \frac{P(A/B_i)\,P(B_i)}{\sum_{j=1}^{k} P(A/B_j)\,P(B_j)}$$

Note: Since B_i's are partition of the sample space S, only one of events B_i occurs. Hence the above formula gives us the probability of a particular B_i given that the event A has occurred. In order to apply this theorem we must know the values of $P(B_i)$.

Relationship between Conditional, Joint and Marginal Probabilities.

The conditional probability of event B given that A has already happened is given by $P(B/A)$ $= \dfrac{P(A \cap B)}{P(A)}$, where $P(A \cap B)$ is the joint probability of events A and B happening together and $P(A)$, the marginal probability of the happening of event A.

1.6 SOLVED PROBLEMS

1. Out of 50 students in a class, what is the probability of a single student to opt for a picnic.

Solution. P (event of a single student opt for a picnic) $= \dfrac{1}{50} = .02$

2. What is the probability of obtaining two heads in two throws of a single coin?

Solution. The probability of obtaining a head in the first throw is $\dfrac{1}{2}$. The probability of obtaining a head in the second throw is also $\dfrac{1}{2}$ (it is not affected by the first throw of the coin). Two throws being independent, the probability of obtaining head in both throws is the product of probability of obtaining a head in the first throw and the probability of obtaining a head in the second throw. Hence the required probability is $\dfrac{1}{2} \times \dfrac{1}{2} = \dfrac{1}{4}$.

3. Find the probability of drawing two red balls in succession from a bag containing 3 red and 6 black balls when

(i) the ball that is drawn first is replaced,

(ii) it is not replaced.

Solution. Let A_1 be the event that the first ball drawn is red and A_2 be the event that the second ball drawn is red.

(i) If the first ball drawn is replaced, the events are independent. Then

$$P(A_1 \cap A_2) = P(A_1)\,P(A_2) = \frac{3}{9} \times \frac{3}{9} = \frac{1}{9}.$$

(ii) If the first ball is not replaced after taking a red ball the bag will contain only 8 balls out of which 2 are red. The events are not independent. Therefore

$$P(A_1 \cap A_2) = P(A_1) \cdot P(A_2/A_1) = \frac{3}{9} \times \frac{2}{8} = \frac{6}{72}$$

4. What is the probability of picking an ace and a king from a deck of 52 cards.
Solution. Let A be the event of picking an ace and B be the event of picking a king. Probability of picking ace and king
= Probability of picking an ace first and a king second + probability of picking a king first and an ace second

$$= P(A) \cdot P(B/A) + P(B) \cdot P(A/B)$$

$$= \frac{4}{52} \times \frac{4}{51} + \frac{4}{52} \times \frac{4}{51} = \frac{8}{663}$$

5. A fair coin is tossed 5 times. What is the probability of having atleast one head?

Solution. The probability of getting no head in 5 tosses of a coin is given by $P(A) = \left(\frac{1}{2}\right)^5 = \frac{1}{32}$.

Hence, the probability of getting atleast one head is given by $P(A^c) = P(S) - P(A) = 1 - \frac{1}{32} = \frac{31}{32}$.

6. Prove that if $P(A) > P(B)$, then $P(A/B) > P(B/A)$

Solution. $P(A/B) = \dfrac{P(A \cap B)}{P(B)}$ implies $P(B) = \dfrac{P(A \cap B)}{P(A/B)}$

$P(B/A) = \dfrac{P(A \cap B)}{P(A)}$ implies $P(A) = \dfrac{P(A \cap B)}{P(B/A)}$

If $P(A) > P(B)$, then $\dfrac{P(A \cap B)}{P(B/A)} > \dfrac{P(A \cap B)}{P(A/B)}$ and hence $P(A/B) > P(B/A)$.

7. One card is drawn from a deck of 52 cards. What is the probability of the card being either red or a king?
Solution. Let A represents an event that the card drawn is red and B represents that the card drawn is king. For a card to be either red or a king, it is required to find the probability of $A \cup B$. Now

$$P(A) = \frac{26}{52} = \frac{1}{2}; \; P(B) = \frac{4}{52} = \frac{1}{13}.$$

There are two red coloured king cards. So

$$P(A \cap B) = \frac{2}{52} = \frac{1}{26}$$

Therefore, $\quad P(A \cup B) = P(A) + P(B) - P(A \cap B) = \dfrac{1}{2} + \dfrac{1}{13} - \dfrac{1}{26} = \dfrac{7}{13}$

8. A bag contains 12 balls numbered from 1 to 12. If a ball is taken at random, what is the probability of having a ball with a number which is a multiple of either 2 or 3.
Solution. Let A be an event that the ball number is a multiple of 2 and B be an event that the ball number is a multiple of 3. Then

$$A = \{2, 4, 6, 8, 10, 12\}, \; B = \{3, 6, 9, 12\}, \; A \cap B = \{6, 12\}$$

$$P(A) = \frac{6}{12} = \frac{1}{2};\ P(B) = \frac{4}{12} = \frac{1}{3},\ P(A \cap B) = \frac{1}{12} = \frac{1}{6}.$$

Hence

$$P(A \cup B) = P(A) + P(B) - P(A \cap B) = \frac{1}{2} + \frac{1}{3} - \frac{1}{6} = \frac{4}{6} = \frac{2}{3}.$$

9. If two events A and B are independent, show that
(i) A^C and B^C are independent
(ii) A^C and B are independent
(iii) A and B^C are independent

Solution. Given A and B are independent. So
(i) $P(A \cap B) = P(A) \cdot P(B)$

Now $P(A^C \cap B^C) = P((A \cup B)^C) = 1 - P(A \cup B)$
$= 1 - [P(A) + P(B) - P(A \cap B)] = 1 - [P(A) + P(B) - P(A)\,P(B)]$
$= [1 - P(B)] - P(A)\,[1 - P(B)] = (1 - P(B))\,(1 - P(A)) = P(B^C)\,P(A^C)$

Hence A^C and B^C are independent.

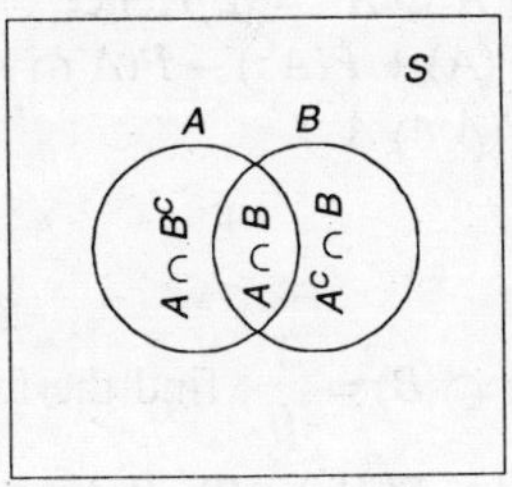

Fig. 1.3

(ii) We know $(A^C \cap B) \cup (A \cap B) = B$. Hence by addition theorem

$$P(A^C \cap B) = P(B) - P(A \cap B), \text{ as } (A^C \cap B) \text{ and } (A \cap B)$$

are mutually exclusive. Now

$$P(A^C \cap B) = P(B) - P(A \cap B) = P(B) - P(A)\,P(B)$$
$$= P(B)\,(1 - P(A)) = P(B) \cdot P(A^C)$$

Hence A^C and B are independent.

(iii) $P(A \cap B^C) = P(A) - P(A \cap B) = P(A) - P(A)\,P(B)$
$= P(A)\,(1 - P(B)) = P(A)\,P(B^C)$

Hence A and B^C are independent.

10. If $B \subset A$, prove that $P(A \cap B^C) = P(A) - P(B)$.

Solution. $B \cup (A \cap B^C) = A$. Here B and $A \cap B^C$ are mutually exclusive. Hence by addition theorem of probability we have

$P(B) + P(A \cap B^C) = P(A)$

Hence $P(A \cap B^C) = P(A) - P(B)$.

Fig. 1.4

11. For any two events A and B show that

$$P(A \cap B) \le P(A) \le P(A \cup B) \le P(A) + P(B)$$

Solution. By the Venn diagram it is clear that $A \cap B^C$ and $A \cap B$ are two disjoint event such that

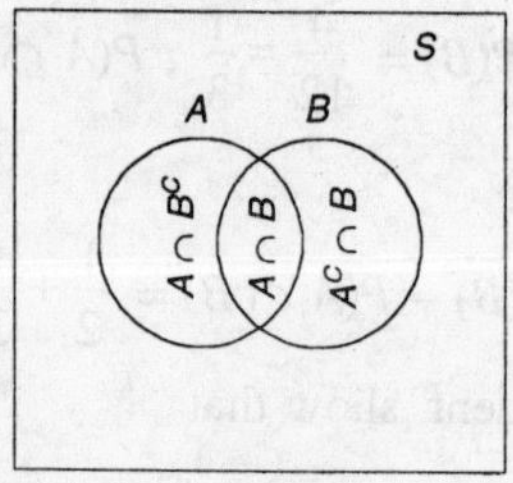

Fig. 1.5

$(A \cap B^C) \cup (A \cap B) = A$. Hence $P(A \cap B^C) + P(A \cap B) = P(A)$. Since $P(A \cap B^C) \geq 0$, $P(A) \geq P(A \cap B)$.

We know that $P(A \cup B) = P(A) + P(B) - P(A \cap B)$. Since $P(B) \geq P(A \cap B)$, $P(B) - P(A \cap B) \geq 0$. Hence $P(A \cup B) \geq P(A)$. Also since $P(A \cap B) \geq 0$, $P(A \cup B) \leq P(A) + P(B)$. Hence $P(A \cap B) \leq P(A) \leq P(A \cup B) \leq P(A) + P(B)$.

12. Prove that for any event A in S,

$$P(A \cap A^C) = 0$$

Solution. We know $P(S) = 1$ and $A \cup A^C = S$. Hence

$$1 = P(S) = P(A \cup A^C) = P(A) + P(A^C) - P(A \cap A^C)$$

$$\Rightarrow \quad 1 = P(A) + (1 - P(A)) - P(A \cap A^C)$$

$$\Rightarrow \quad 1 = 1 - P(A \cap A^C)$$

$$\Rightarrow \quad P(A \cap A^C) = 0$$

13. Given $P(A) = \frac{1}{3}$, $P(B) = \frac{1}{4}$, $P(A \cap B) = \frac{1}{6}$, find the following probabilities

$$P(A^C),\ P(A^C \cup B),\ P(A^C \cap B^C)$$

Solution. $$P(A^C) = 1 - P(A) = 1 - \frac{1}{3} = \frac{2}{3}$$

$$P(A^C \cup B) = P(A^C) + P(B) - P(A^C \cap B) = P(A^C) + P(B) - [P(B) - P(A \cap B)]$$

$$= P(A^C) + P(A \cap B) = \frac{2}{3} + \frac{1}{6} = \frac{5}{6}.$$

$$P(A^C \cap B^C) = P((A \cup B)^C) = 1 - P(A \cup B) = 1 - [P(A) + P(B) - P(A \cap B)]$$

$$= 1 - \left[\frac{1}{3} + \frac{1}{4} - \frac{1}{6}\right] = \frac{7}{12}$$

14. Suppose from a pack of 52 cards one card is drawn at random what is the probability that it is either a king or a queen.

Solution. Since the events are mutually exclusive (if a card is a king it cannot be a queen and vice versa) the probability of drawing a king is $\frac{4}{52}$ and similarly drawing a queen is $\frac{4}{52}$. Hence the probability of drawing a king K or a queen Q is

$$P(K \cup Q) = P(K) + P(Q) = \frac{4}{52} + \frac{4}{52} = \frac{2}{13}.$$

15. One ticket is drawn at random from a bag containing 30 tickets numbered from 1 to 30. Find the probability that it is a multiple of 5 or 7.

Solution. One ticket can be drawn out of 30 in $30C_1$ ways. The multiples of 5 are 5, 10, 15, 20, 25, 30 and multiples of 7 are 7, 14, 21, 28. Hence there are 6 multiples of 5 and 4 multiples of 7. None of these are common. So the events are mutually exclusive. The probability of having a multiple of 5 or 7 is

$$\frac{6}{30}+\frac{4}{30}=\frac{10}{30}=\frac{1}{3}.$$

16. The probability that A will live upto 60 years is $\frac{3}{4}$ and probability that B will live upto 60 years is $\frac{2}{3}$. What is the probability that both A and B will live upto sixty years.

Solution. $P(A \cap B) = P(A) \cdot P(B) = \frac{3}{4}\times\frac{2}{3}=\frac{1}{2}.$

Hence the probability that both A and B will live upto sixty year is $\frac{1}{2}$.

17. Find the probability of drawing two kings from a pack of cards in two successive draws, the card drawn not being replaced.

Solution. The probability of drawing a king in the first draw is $\frac{4}{52}$. The probability of drawing a king in the second draw, given that the first draw has already given a king is 3/51. The combined probability of two events is $\frac{4}{52}\times\frac{3}{51}=\frac{1}{221}$.

18. A bag contains 3 red and 4 white balls. Two draws are made without replacement; what is the probability that both the balls are red.

Solution. Probability of drawing a red ball in the first draw is $P(A)=\frac{3}{7}$. Probability of drawing a red ball in the second draw given that the first ball drawn is red is $P(B/A)=\frac{2}{6}$ (since only six balls are left and only two out of them are red). The combined probability of the two events are

$$P(A \cap B) = P(A) \cdot P(B/A) = \frac{3}{7}\times\frac{2}{6}=\frac{1}{7}$$

19. Three coins are tossed simultaneously. What is the probability that they will fall two heads and one tail.

Solution. The probability P of getting a head is $\frac{1}{2}$. Now, the probability that out of 3 coins getting exactly 2 heads and one tail is

$$3C_2\, P^2Q = 3\times\frac{1}{2}\times\frac{1}{2}\times\left(1-\frac{1}{2}\right)=\frac{3}{8},$$

where Q denotes the probability of getting a tail and is $Q = 1 - P$.

20. If A and B are mutually exclusive event, prove that $P(A/B^C) = \dfrac{P(A)}{1-P(B)}$

Solution. $P(A/B^C) = \dfrac{P(A \cap B^C)}{P(B^C)} = \dfrac{P(A \cap B^C)}{1-P(B)} = \dfrac{P(A)-P(A \cap B)}{1-P(B)}$

Since A and B are mutually exclusive, $P(A \cap B) = 0$.

Hence $P(A/B^C) = \dfrac{P(A)}{1-P(B)}$.

21. If A and B are two mutually exclusive events $P(A \cup B) \neq 0$, prove that $P(A/A \cup B) = \dfrac{P(A)}{P(A)+P(B)}$.

Solution. $$P(A/A \cup B) = \frac{P(A \cap (A \cup B))}{P(A \cup B)} = \frac{P(A)}{P(A)+P(B)-P(A \cap B)}$$

$$= \frac{P(A)}{P(A)+P(B)} \text{ since } P(A \cap B) = 0.$$

22. If A and B are two independent events, show that
$$P(A \cup B) = 1 - P(A^C) \cdot P(B^C).$$

Solution. $$P(A \cup B) = 1 - P(A \cup B)^C = 1 - P(A^C \cap B^C)$$
$$= 1 - P(A^C) \cdot P(B^C),$$

since if A and B are independent, A^C and and B^C are also independent.

23. Box 1 contains 1 white and 999 red balls. Box 2 contains 1 red and 999 white balls. A ball is picked from a randomly selected box. If the ball is red, what is the probability that it came from box 1?

Solution. Let A be the event that a ball picked is red and A_i be the event that i^{th} box is selected $(i = 1, 2)$

$$P(A/A_1) = \frac{999}{1000} \text{ and } P(A/A_2) = \frac{1}{1000} \text{ and}$$

$$P(A_i) = \frac{1}{2}. \text{ Hence by Baye's theorem}$$

$$P(A_1/A) = \frac{P(A_1)\cdot P(A/A_1)}{\sum_{i=1}^{2} P(A/A_i)\,P(A_i)} = \frac{\frac{1}{2}\times\frac{999}{1000}}{\left(\frac{999}{1000}+\frac{1}{1000}\right)\frac{1}{2}} = \frac{999}{1000}.$$

Hence the required probability is $\dfrac{999}{1000}$.

24. Let A and B be boxes that contain 5 black, 6 white; 8 black, 4 white balls respectively. Two balls are transferred from B to A and then a ball is drawn from A.

(a) What is the probability that this ball is white?

(b) Given that the ball drawn is white, what is the probability that atleast one white ball was transferred to A?

Solution. (a) Let E_1 be the event that a white ball is drawn from A.

E_2 be the event that two white balls are transferred from B.

E_3 be the event that one white ball and one black ball are transferred from B.

E_4 be the event that two black balls are transferred from B. Then

$$P(E_1) = P(E_2) \cdot P(E_1/E_2) + P(E_3) \cdot P(E_1/E_3) + P(E_4) \cdot P(E_1/E_4)$$

$$= \frac{4C_2}{12C_2} \times \frac{8}{13} + \frac{8C_1 \times 4C_1}{12C_2} \times \frac{7}{13} + \frac{8C_2}{12C_2} \times \frac{6}{13}$$

$$= \frac{6}{66} \times \frac{8}{13} + \frac{32}{66} \times \frac{7}{13} + \frac{28}{66} \times \frac{6}{13} = \frac{136}{429}.$$

25. A box contains 2000 components of which 5% are defective. Second box contains 500 components of which 40% are defective. Two other boxes contain 1000 components each with 10% defective components. We select at random one of the above boxes and remove from it at random a single component.

(i) What is the probability that this component is defective?

(ii) Finding that the selected component is defective, what is the probability that it was drawn from box 2?

Solution. Let A_i be the event of selecting i^{th} box. B be the event consisting of all defective components. Therefore, we have

$$P(A_1) = P(A_2) = P(A_3) = P(A_4) = \frac{1}{4}$$

$$P(B/A_1) = \frac{100}{2000} = 0.05$$

$$P(B/A_2) = \frac{200}{500} = 0.4$$

$$P(B/A_3) = \frac{100}{1000} = 0.1$$

$$P(B/A_4) = \frac{100}{1000} = 0.1$$

Probability that the component is defective is given by

$$P(B) = P(A_1)\, P(B/A_1) + P(A_2)\, P(B/A_2) + P(A_3)\, P(B/A_3) + P(A_4)\, P(B/A_4)$$

$$= \frac{1}{4} \times 0.05 + \frac{1}{4} \times 0.4 + \frac{1}{4} \times 0.1 + \frac{1}{4} \times 0.1 = 0.1625.$$

26. Of three independent events, the probability of first only happening is $\frac{1}{4}$, the probability of second only happening is $\frac{1}{8}$, and the probability that only third happens is $\frac{1}{12}$. Find the marginal probability of each event.

Solution. Let A, B, C be events. Then

$$P(A \cap B^C \cap C^C) = P(A) \cdot P(B^C)\, P(C^C) = \frac{1}{4}$$

$$P(A^C \cap B \cap C^C) = P(A^C)\, P(B)\, P(C^C) = \frac{1}{8}$$

$$P(A^C \cap B^C \cap C) = P(A^C)\, P(B^C)\, P(C) = \frac{1}{12}$$

Let $\quad P(A^C) = x,\ P(B^C) = y,\ P(C^C) = z.$ Then

$$(1 - x)\, yz = \frac{1}{4};\ xz\,(1 - y) = \frac{1}{8};\ xy\,(1 - z) = \frac{1}{12}$$

By remainder theorem, we get the solution of x as $\frac{1}{2}$. Therefore $yz = \frac{1}{2}$ and $z\,(1 - y) = \frac{1}{4}$ implies $z = \frac{3}{4}$, $y = \frac{2}{3}$. Hence

$$P(A) = 1 - P(A^C) = \frac{1}{2}$$

$$P(B) = 1 - P(B^C) = 1 - \frac{2}{3} = \frac{1}{3}$$

$$P(C) = 1 - P(C^C) = 1 - \frac{3}{4} = \frac{1}{4}.$$

27. It is given that $P(A \cup B) = \frac{5}{6}$, $P(A \cap B) = \frac{1}{3}$ and $P(B^C) = \frac{1}{2}$. Show that the events A and B are independent.

Solution. $\quad P(B) = 1 - P(B^C) = 1 - \frac{1}{2} = \frac{1}{2}.$

Now $\quad P(A \cup B) = P(A) + P(B) - P(A \cap B)$

$$\Rightarrow \quad \frac{5}{6} = P(A) + \frac{1}{2} - \frac{1}{3}$$

$$\Rightarrow \quad P(A) = \frac{5}{6} + \frac{1}{3} - \frac{1}{2} = \frac{2}{3}$$

Hence $P(A) \cdot P(B) = \frac{2}{3} \cdot \frac{1}{2} = \frac{1}{3} = P(A \cap B)$. So A and B are independent.

28. A box contains four tickets with numbers 112, 121, 211, 222 and one ticket is drawn. Let A_i (i = 1, 2, 3) be the event that the ith digit of the number of tickets drawn is 1. Discuss the independent of the events A_1, A_2, A_3.

Solution.

$$P(A_1) = \frac{1}{2};\ P(A_3) = \frac{1}{2};\ P(A_1 \cap A_2 \cap A_3) = 0$$

$$P(A_2) = \frac{1}{2}; \; P(A_1 \cap A_2) = \frac{1}{4}$$

$$P(A_1 \cap A_3) = \frac{1}{4}; \; P(A_2 \cap A_3) = \frac{1}{4}$$

$$P(A_1)\, P(A_2) = \frac{1}{4}; \; P(A_1) \cdot P(A_3) = \frac{1}{4}; \; P(A_2)\, P(A_3) = \frac{1}{4}.$$

Hence events are pairwise independent. Now $P(A_1)\, P(A_2)\, P(A_3) = \frac{1}{8}$. But $P(A_1 \cap A_2 \cap A_3) = 0$.

Hence $P(A_1)\, P(A_2)\, P(A_3) \neq P(A_1 \cap A_2 \cap A_3)$ and so A_1, A_2, A_3 cannot be simultaneously independent.

29. A man has three coins A, B and C, A is unbiased; the probability that a head will result when B is tossed is $\frac{2}{3}$; the probability that a head will result when C is tossed is $\frac{1}{3}$. If one of the coins chosen at random is tossed three times giving a total of two heads and one tail, find

(i) the probability that the chosen coin was A

(ii) the probability that a fourth toss of the same coin will give a head.

Solution. Let D denote the probability of giving two heads and one tail. Then

$$P(D/A) = 3C_2 \left(\frac{1}{2}\right)^2 \frac{1}{2} = \frac{3}{8}$$

$$P(D/B) = 3C_2 \left(\frac{2}{3}\right)^2 \frac{1}{3} = \frac{4}{9}$$

$$P(D/C) = 3C_2 \left(\frac{1}{3}\right)^2 \frac{2}{3} = \frac{2}{9}$$

(i) $$P(A/D) = \frac{P(A) \cdot P(D/A)}{P(A) \cdot P(D/A) + P(B) \cdot P(D/B) + P(C) \cdot P(D/C)}$$

$$= \frac{\frac{1}{3} \times \frac{3}{8}}{\frac{1}{3} \times \frac{3}{8} + \frac{1}{3} \times \frac{4}{9} + \frac{1}{3} \times \frac{2}{9}} = \frac{9}{25}$$

(ii) The fourth toss of A will give head is $\frac{1}{2}\, P(A)\, P(D/A)$. Similarly for B and C the probabilities will be $\frac{2}{3}\, P(B)\, P(D/B)$, $\frac{1}{3}\, P(C)\, P(D/C)$ respectively. Hence the required probabilities

$$= \left(\frac{1}{3} \times \frac{3}{8} \times \frac{1}{2}\right) + \left(\frac{2}{3} \times \frac{1}{3} \times \frac{4}{9}\right) + \frac{1}{3} \times \frac{1}{3} \times \frac{2}{9} = \frac{241}{1296}.$$

30. Three candidates A, B, C are selected for the position of a general manager in a company whose chances of getting the appointment are in the proportion 4 : 2 : 3 respectively. The probability that A is selected will improve the office canteen is 0.3. The probability of B and C doing the same are respectively 0.5 and 0.8. What is the probability that the office canteen will be improved.

Solution. Let H_1, H_2, H_3, be the events that the candidates are selected for the position of general manager. Let D be the event that the office canteen being improved. Then from Baye's Theorem,

$$P(D) = P(H_1) \cdot P(D/H_1) + P(H_2) \cdot P(D/H_2) + P(H_3) \cdot P(D/H_3)$$

$$= \frac{4}{9} \times 0.3 + \frac{2}{9} \times 0.5 + \frac{3}{9} \times 0.8 = .13 + .11 + .27 = .51$$

31. A box contains 10 white, 5 yellow and 10 black balls. A ball is chosen at random from the box and it is noted that it is not one of the black balls. What is the probability that it is yellow.

Solution. Let Y denote the event that the ball selected is yellow, and let B^C denote the event that it is not black. Now

$$P(Y/B^C) = \frac{P(Y \cap B^C)}{P(B^C)}$$

But $P(Y \cap B^C) = P(Y)$ since the ball will be both yellow and not black if and only if it is yellow. Hence, assuming that each of the 25 balls is equally likely to be chosen, we obtain that

$$P(Y/B^C) = \frac{\frac{5}{25}}{\frac{15}{25}} = \frac{1}{3}.$$

32. For any three events A, B, C

$$P((A \cap B^C)/C) + P((A \cap B)/C) = P(A/C)$$

Solution.

$$P((A \cap B^C)/C) + P((A \cap B)/C)$$

$$= \frac{P(A \cap B^C \cap C)}{P(C)} + \frac{P(A \cap B \cap C)}{P(C)}$$

$$= \frac{P[(A \cap B^C \cap C) \cup (A \cap B \cap C)]}{P(C)}$$

$$= \frac{P\big[[(A \cap C) \cap B^C] \cup [(A \cap C) \cap B]\big]}{P(C)}$$

$$= \frac{P\big[(A \cap C) \cap (B^C \cup B)\big]}{P(C)} \quad \text{by distributive law}$$

$$= \frac{P(A \cap C)}{P(C)}$$

Note.

$P((A_2 \cup A_3 \cup \ldots \cup A_n)/A_1) = P(A_2/A_1) + P(A_3/A_1) + \ldots + P(A_n/A_1)$

provided $A_2, A_3, \ldots A_n$ are pairwise disjoint sets.

33. If $A \cap B = \phi$, then show that $P(A) \le P(B^C)$

Solution. $$A = (A \cap B) \cup (A \cap B^C) = \phi \cup (A \cap B^C) = A \cap B^C$$

Hence $A \subseteq B^C$ which implies $P(A) \le P(B^C)$.

34. Two urn contains 4 white and 6 black balls and 4 white and 8 black balls. One urn is selected at random and a ball is taken out. It turns out to be white. Find the probability that it is from the first urn.

Solution. Let E_1 and E_2 be events that the first and the second urn respectively were selected. Since the urn was selected at random

$$P(E_1) = \frac{1}{2} = P(E_2).$$

Let A be the event that the ball taken out is white. Then

$A = (E_1 \cap A) \cup (E_2 \cap A)$ and $(E_1 \cap A) \cap (E_2 \cap A) = \phi$.

Now

$$P(A) = P(E_1 \cap A) + P(E_2 \cap A)$$

$$= P(E_1) \cdot P(A/E_1) + P(E_2) \cdot P(A/E_2) = \frac{1}{2} \times \frac{4}{10} + \frac{1}{2} \times \frac{4}{12} = \frac{11}{30}$$

By Baye's theorem

$$P(E_1/A) = \frac{P(E_1)\,P(A/E_1)}{P(A)} = \frac{\frac{1}{2} \times \frac{4}{10}}{\frac{11}{30}} = \frac{6}{11}.$$

35. The contents of urns I, II, III are as follows.

1 white,	2 black,	3 red balls
2 white,	1 black,	1 red ball
4 white,	5 black,	3 red balls

One urn is chosen at random and two balls are drawn. They happen to be white and red. What is the probability that they come from urns I, II or III?

Solution. Let E_1, E_2, E_3 denote the events that the urns I, II, III are chosen respectively and let A be the event that the two balls taken from the selected urn are white and red. Then

$$P(E_1) = P(E_2) = P(E_3) = \frac{1}{3}$$

$$P(A/E_1) = \frac{1 \times 3}{6C_2} = \frac{1}{5}$$

$$P(A/E_2) = \frac{2 \times 1}{4C_2} = \frac{1}{3}$$

$$P(A/E_3) = \frac{4 \times 3}{12C_2} = \frac{2}{11}$$

Hence

$$P(E_2/A) = \frac{P(E_2)\,P(A/E_2)}{\sum_{i=1}^{3} P(E_i)\,P(A/E_i)} = \frac{\frac{1}{3}\times\frac{1}{3}}{\frac{1}{3}\times\frac{1}{5}+\frac{1}{3}\times\frac{1}{3}+\frac{1}{3}\times\frac{2}{11}} = \frac{55}{118}$$

$$P(E_3/A) = \frac{\frac{1}{3}\times\frac{2}{11}}{\frac{1}{3}\times\frac{1}{5}+\frac{1}{3}\times\frac{1}{3}+\frac{1}{3}\times\frac{1}{11}} = \frac{30}{118}$$

$$P(E_1/A) = 1 - \frac{55}{118} - \frac{30}{118} = \frac{33}{118}$$

36. A box contains five balls. Two balls are drawn and found to be white. What is the probability that all of the balls being white?

Solution. Let B be the probability that two balls drawn are white. Let A_1, A_2, A_3, A_4 be the probability that the box contains 2, 3, 4 or 5 white balls, since these are the only possibilities after the knowledge that two white balls are drawn. Hence

$$P(A_1) = \frac{1}{4} = P(A_2) = P(A_3) = P(A_4)$$

$P(B/A_1)$ is the probability that the box contains 2 white balls and both been drawn is $\frac{1}{4}\times\frac{2}{5} = \frac{1}{10}$. Similarly $P(B/A_2) = \frac{1}{4}\times\frac{3}{5} = \frac{3}{20}$, $P(B/A_3) = \frac{1}{4}\times\frac{4}{5} = \frac{1}{5}$, $P(B/A_5) = \frac{1}{4}\times\frac{5}{5} = \frac{1}{4}$.

37. In a bolt factory machines *A, B, C* manufacture respectively 25%, 35%, and 40% of the total. Of their output 5, 4, 2 percent are defective bolts. A bolt is drawn at random from the product and is found to be defective. What are the probabilities that it was manufactured by machines *A, B, C*?

Solution. Let E_1, E_2, E_3 be the events that a bolt selected at random is manufactured by the machines *A, B, C* respectively and let E denote the event of its being defective. Now

$$P(E_1) = 0.25,\ P(E_2) = 0.35,\ P(E_3) = 0.40$$

The probability of drawing a defective bolt manufactured by machine A is $P(E/E_1) = 0.05$. Similarly we have

$$P(E/E_2) = 0.04,\ P(E/E_3) = 0.02$$

Hence the probability that a defective bolt selected at random is manufactured by machine A is given by

$$P(E_1/E) = \frac{P(E_1)\,P(E/E_1)}{\sum_{i=1}^{3} P(E_1)\,P(E/E_1)}$$

$$= \frac{0.25\times 0.05}{0.25\times 0.05 + 0.35\times 0.04 + 0.40\times 0.02} = \frac{25}{69}$$

Similarly we get

$$P(E_2/E) = \frac{28}{69}, P(E_3/E) = \frac{16}{69}.$$

Baye's Theorem for future events.

The probability of materialization of another event C given $P(C/A \cap E_1)$, $P(C/A \cap E_2)$, . . . is

$$P(C/A) = \frac{\sum_{i=1}^{n} P(E_i)\, P(A/E_i)\, P(C/A \cap E_i)}{\sum_{i=1}^{n} P(E_i)\, P(A/E_i)}$$

38. Three boxes of the same appearance have the following proportion of balls

I	2 black	1 white
II	1 black	2 white
III	2 black	2 white

One of the urn is selected and one ball is drawn. It turns out to be white. What is the probability of drawing white ball again, if the first one drawn is not replaced.

Solution. Let E_1, E_2, E_3 be the event of drawing I, II, III urns respectively. Let A be the event of drawing white ball. Now

$$P(E_1) = P(E_2) = P(E_3) = \frac{1}{3}$$

$$P(A/E_1) = \frac{1}{3};\ P(A/E_2) = \frac{2}{3};\ P(A/E_3) = \frac{2}{4}$$

Let C denote the future event of drawing other white ball from the urns.

$$P(C/A \cap E_1) = 0;\ P(C/A \cap E_2) = \frac{1}{2};\ P(C/A \cap E_3) = \frac{1}{3}.$$

$$P(C/A) = \frac{\sum_{i=1}^{3} P(E_i)\, P(A/E_i)\, P(C/E_i \cap A)}{\sum_{i=1}^{3} P(E_i)\, P(A/E_i)}$$

$$= \frac{\frac{1}{3}\times\frac{1}{3}\times 0 + \frac{1}{3}\times\frac{2}{3}\times\frac{1}{2} + \frac{1}{3}\times\frac{1}{2}\times\frac{1}{3}}{\frac{1}{3}\times\frac{1}{3} + \frac{1}{3}\times\frac{2}{3} + \frac{1}{3}\times\frac{1}{2}} = \frac{1}{3}$$

39. A certain item is manufactured by three factories say 1, 2 and 3. It is known that 1 turns out twice as many items as 2, and that 2 and 3 turns out the same number of item (during a specified production period). It is also known that 2 percent of the items produced by 1 and 2 are defective, while 4 percent of those manufactured by 3 are defective. All the items produced are put into one stockpile, and then one item is chosen at random. What is the probability that this item is defective?

Solution. Let A be the event that the item produced is defective. Let B_1, B_2, B_3 be the event that the items come from factories 1, 2, 3 respectively. Let the factory 1 manufacture 50 number of items. Then 2 and 3 manufacture 25 number of items each. Hence

$$P(B_1) = \frac{50}{100} = \frac{1}{2};\ P(B_2) = P(B_3) = \frac{25}{100} = \frac{1}{4}$$

$$P(A/B_1) = P(A/B_2) = 0.02;\ P(A/B_3) = 0.04$$

Hence

$$P(A) = P(B_1)\,P(A/B_1) + P(B_2)\,P(A/B_2) + P(B_3)\,P(A/B_3)$$

$$= \frac{1}{2} \times 0.02 + \frac{1}{4} \times 0.02 + \frac{1}{4} \times 0.04 = 0.025$$

40. Suppose that among six bolts, two are shorter than a specified length. If two bolts are chosen at random, what is the probability that the two shorts bolts are picked?

Solution. Let A_1, A_2 be the events that the first and second bolt chosen are short, respectively. Now

$$P(A_1 \cap A_2) = P(A_2/A_1)\,P(A_1) = \frac{1}{5} \times \frac{2}{6} = \frac{1}{15}$$

41. A pair of dice is thrown twice. What is the probability of getting totals of 7 and 11?

Solution. Let A_1, A_2, B_1, and B_2, be respective independent events that a 7 occurs on the first throw, 7 occurs on the second throw, an 11 occurs on the first throw, and an 11 occurs on the second throw. $A_1 \cap B_2$, $B_1 \cap A_2$ are mutually exclusive events and

$$P[(A_1 \cap B_2) \cup (B_1 \cap A_2)] = P(A_1 \cap B_2) + P(B_1 \cap A_2)$$

$$= P(A_1)\,P(B_2) + P(B_1)\,P(A_2)$$

$$= \left(\frac{1}{6}\right)\left(\frac{1}{18}\right) + \left(\frac{1}{18}\right)\left(\frac{1}{6}\right) = \frac{1}{54}.$$

42. Three cards are drawn in succession without replacement from a deck of cards. Find the probability that the event $A_1 \cap A_2 \cap A_3$ occurs, where A_1 is the event that the first card is a red ace, A_2 is the event that a second card is a 10 or a Jack and A_3 is the event that the third card is greater than 3 but less than 7.

Solution. $P(A_1) = \frac{2}{52};\ P(A_2/A_1) = \frac{8}{51};\ P(A_3/A_1 \cap A_2) = \frac{12}{50}$

Hence

$$P(A_1 \cap A_2 \cap A_3) = P(A_1)\,P(A_2/A_1)\,P(A_3/A_1 \cap A_2)$$

$$= \frac{2}{52} \times \frac{8}{51} \times \frac{12}{50} = \frac{8}{5525}$$

43. A coin is biased so that a head is twice as likely to occur as a tail. If the coin is tossed 3 times, what is the probability of getting 2 tails and 1 head.

Solution. The sample space S = {HHH, HHT, HTH, THH, HTT, TTH, THT, TTH}. Since a coin is biased and a head is twice as likely to occur as a tail, $P(H) = \frac{2}{3}$ and $P(T) = \frac{1}{3}$. Let A be the event of getting 2 tails and 1 head in the 3 tosses of the coin. Then

$$A = \{TTH, THT, HTT\}$$

And since the outcomes of the 3 tosses are independent,

$$P(\text{TTH}) = P(\text{T} \cap \text{T} \cap \text{H}) = P(\text{T})\, P(\text{T})\, P(\text{H}) = \frac{1}{3} \times \frac{1}{3} \times \frac{2}{3} = \frac{2}{27}$$

$$P(\text{THT}) = P(\text{HTT}) = \frac{2}{27}$$

Hence $$P(A) = \frac{2}{27} + \frac{2}{27} + \frac{2}{27} = \frac{2}{9}$$

EXERCISES

1. Two students A and B work independently on a problem. The probability that the first one will solve it is $\frac{3}{4}$ and the probability that second one will solve it is $\frac{2}{3}$. What is the probability that the problem will be solved?
2. For given three events A, B, C, verify that
$$P((A \cup B)/C) = P(A/C) + P(B/C) - P((A \cap B)/C)$$
3. What is the chance that a leap year selected at random will contain 53 sundays?
4. A bag contains 3 red, 6 white, 7 blue balls. What is the probability that two balls drawn are white and blue.
5. If $P(A_1 \cup A_2) = \frac{2}{3}$, $P(A_1 \cap A_2) = \frac{1}{6}$. Find $P\left(A_1^C \cup A_2^C\right)$ and $P\left(A_1^C \cap A_2^C\right)$.
6. If $P(A) = \frac{1}{3}$, $P(B) = \frac{1}{4}$, $P(A \cap B) = \frac{1}{6}$. Find $P(A^C \cup B^C)$.
7. A problem in statistics is given to three students A, B, C whose chances of solving it are $\frac{1}{2}, \frac{3}{4}, \frac{1}{5}$ respectively. What is the probability that the problem will be solved if all of them try independently.
8. Suppose there are three boxes containing 2 white and 3 black; 3 white and two black, 4 white and 1 black balls respectively. There is equal probability of each box being chosen. One ball is drawn from a box at random. What is the probability that a ball drawn is white.
9. A bag contains 8 white and 6 red balls. Find the probability of drawing two balls of the same colour.
10. A box contains 4 bad and 6 good tubes. Two are drawn out together. One of them is tested and found to be good. What is the probability that the other one is also good.
11. Show that the multiplication theorem $P(A \cap B) = P(A/B)\, P(B)$ established for two events, may be generalized to three events as follows.
$$P(A \cap B \cap C) = P(A/B \cap C)\, P(B/C)\, P(C).$$
12. Prove: If $P(A/B) > P(A)$, then $P(B/A) > P(B)$.
13. Suppose that colored balls are distributed in three indistinguishable boxes as follows:

Box	Red	White	Blue
I	2	3	5
II	4	1	3
III	3	4	3

A box is selected at random from which a ball is selected at random and it is observed to be red what is the probability that box 3 was selected.

14. A factory has two machines A and B. Past records show that machine A produces 30% of the total output and machine B the remaining 70%. Machine A produces 5% defective articles and machine B produces 1% defective articles. An item is drawn at random and found to be defective. What is the probability that it was produced by machine A.

15. A coin is tossed. If it turns up head, two balls will be drawn from an urn A, otherwise two balls will be drawn from urn B. Urn A contains 3 black and 5 white balls. Urn B contains 7 black and 1 white ball. In both cases selections are to be made with replacements, what is the probability that urn A is used given that both the balls drawn are black.

ANSWERS

(1) $\frac{11}{12}$ (3) $\frac{2}{7}$ (4) $\frac{7}{20}$

(5) $P(A_1^C \cup A_2^C) = \frac{5}{6}$, $P(A_1^C \cap A_2^C) = \frac{1}{3}$ (6) $P(A^C \cup B^C) = \frac{5}{6}$

(7) $\frac{29}{32}$ (8) $\frac{2}{9}$ (9) $\frac{43}{91}$ (10) $\frac{5}{9}$

(13) $\frac{3}{10}$ (14) 0.682 (15) $\frac{1}{8}$.

2

Random Variables and Probability Distributions

2.1 DISTRIBUTION FUNCTION

2.1.1 Random Variable

A random variable is a function from the sample space S into R, the set of real numbers.

We represent a random variable by a bold face letters like X, Y etc. The real numbers $X(s)$, $Y(t)$ are called the *values* of the random variable. The values are represented by lower case letters like x, y, s, t etc.

2.1.2 Example

Suppose that we toss two coins and consider the sample space associated with this experiment. Then $S = \{$ HH, HT, TH, TT$\}$. Define the random variable X as follows: X is the number of heads obtained in the two tosses. Hence X (HH) = 2, X (HT) = 1 = X(TH) and X (TT) = 0.

Note that to every $s \in S$ there corresponds exactly one value $X(s)$. Different values of s may leads to the same value of X. For example X (HT) = X (TH) = 1.

2.1.3 Discrete and Continuous Random Variable

A random variable that can take at most a countable number of possible values is said to be **discrete.**

A random variable which can take any value between some intervals is called a **continuous** random variable.

2.1.4 Probability Mass Function

If X is a discrete random variable which assumes values $x_1, x_2, \ldots x_n$ with probabilities $p(x_1), p(x_2) \ldots p(x_n)$ respectively, then p is called **probability mass function** (*pmf*) of X provided thc following conditions are satisfied

(i) $p(x_i) \geq 0$

(ii) $\sum_{i=1}^{n} p(x_i) = 1.$

2.1.5 Probability Density Function

If X is a continuous random variable, then a function f is said to be the **probability density function** (*pdf*) of X, if it satisfies the following conditions:

(i) $f(x) \geq 0$ for all x

(ii) $\int_{-\infty}^{\infty} f(x)\,dx = 1.$

(iii) For any a, b with $-\infty < a < b < \infty$,

$$P(a \le X \le b) = \int_a^b f(x)\,dx.$$

2.1.6 Cumulative Distribution Function

Let X be a random variable, discrete or continuous. We define F to be the cumulative distribution function of X (abbreviated as *cdf*) if $F(x) = P(X \le x)$ and

$$P(X \le x) = \sum_{i=0}^{x} p(x_i) \text{ if } X \text{ is discrete}$$

$$= \int_{-\infty}^{x} f(x)\,dx \text{ if } X \text{ is continuous}$$

where $p(x_i)$ and $f(x)$ are probability mass function and probability density function respectively.

2.1.7 Properties of Distribution Function

1. In a random variable X, all the assigned real numbers vary from $-\infty$ to ∞. There are no real number less than $-\infty$. Hence
   $F_X(-\infty) = P(X \le -\infty) = 0$
2. $F_X(\infty) = P(X \le \infty) = P(S) = 1$.
3. Since $\{X \le a\}$ and $\{a < X \le b\}$ are disjoint sets,

$$P(X \le b) = P(X \le a) + P(a < X \le b)$$

$$\Rightarrow \quad F_X(b) = F_X(a) + P(a < X \le b)$$

$$\Rightarrow \quad P(a < X \le b) = F_X(b) - F_X(a)$$

$$\int_a^b f(x)\,dx = [F_X(x)]_a^b = F_x(b) - F_x(a) = P(a < X \le b)$$

4. Let F be the *cdf* of a continuous random variable with *pdf* f then
   $\frac{d}{dx}F(x) = f(x)$, if the derivative exists.

   Recall the definition of the derivative of the function F:

$$F'(x) = \underset{h \to 0+}{\text{Lt}} \frac{F(x+h) - F(x)}{h}$$

$$= \underset{h \to 0+}{\text{Lt}} \frac{P(X \le x+h) - P(X \le x)}{h}$$

$$= \underset{h \to 0+}{\text{Lt}} \frac{1}{h}[P(X \le x+h) - P(X \le x)]$$

Thus if h is small and positive

$$F'(x) = f(x) \simeq \frac{P(x < X \le x+h)}{h}$$

That is, $f(x)$ is approximately equal to the "amount of probability in the interval $(x, x+h)$ per length h". Hence the name probability density function.

2.1.8 Example

Suppose that continuous random variable has *cdf F* given by

$$F(x) = 0 \qquad x \le 0$$
$$= e^{-x}, \qquad x > 0$$

Then $F'(x) = e^{-x}$ for $x > 0$ and thus the *pdf f* is given by

$$f(x) = e^{-x} - e^{-x} \quad x \ge 0$$
$$= 0 \text{ elsewhere.}$$

Note. Finally we fix the terminology. When we speak of the probability distribution of a random variable X we mean its *pdf f* if X is continuous, or its *pmf p* defined for $x_1, x_2, \ldots$ if X is discrete. When we speak of the cumulative distribution function or just distribution function we always mean F, where $F(x) = P(X \le x)$.

2.1.9 Example

Check whether the function

$$f(x) = 4x^3, 0 \le x \le 1$$

is a probability density function or not.

Solution. Clearly $f(x) \ge 0$ for all $x \in [0, 1]$. Also

$$\int_0^1 f(x)\,dx = \int_0^1 4x^3\,dx = 4\left(\frac{x^4}{4}\right)_0^1 = 1.$$

Hence the given function $f(x)$ is a probability density function.

2.1.10 Example

The pmf. of a random variable X is given by $p(i) = \dfrac{c\lambda^i}{i!}$ $i = 0, 1, 2, \ldots$, where λ is some positive value. Find

(i) $P(X = 0)$ (ii) $P(X > 2)$

Solution. Since $\sum_{i=0}^{\infty} p(i) = 1$, we have

$$c\sum_{i=0}^{\infty} \frac{\lambda^i}{i!} = 1$$

As $e^{\lambda} = \sum_{i=0}^{\infty} \frac{\lambda^i}{i!}$, we have $ce^{\lambda} = 1$ or $c = e^{-\lambda}$.

Hence

(i) $P(X = 0) = \dfrac{e^{-\lambda}\lambda^0}{0!} = e^{-\lambda}$

(ii) $P(X > 2) = 1 - P(X \le 2)$

$$= 1 - \{P(X = 0) + P(X = 1) + P(X = 2)\} = 1 - \left\{e^{-\lambda} + \lambda e^{-\lambda} + \frac{\lambda^2 e^{-\lambda}}{2}\right\}$$

2.1.11 Example

A random variable X has the density function

$$f(x) = \begin{cases} 1/4 & -2 < x < 2 \\ 0 & \text{elsewhere} \end{cases}$$

Obtain

(i) $P(X < 1)$ (ii) $P(|X| > 1)$ (iii) $P(2X + 3 > 5)$

Solution.

(i) $P(X < 1) = \int_{-\infty}^{1} f(x)\,dx = \int_{-2}^{1} 1/4\,dx = \frac{1}{4}[x]_{-2}^{1} = 3/4$

(ii) $P(|X| > 1) = 1 - P(|X| \le 1) = 1 - \int_{-1}^{+1} f(x)\,dx = 1 - \int_{-1}^{1} \frac{1}{4}\,dx$

$$= 1 - 1/4\,[x]_{-1}^{1} = 1 - 1/2 = 1/2.$$

(iii) $P[(2X + 3) > 5] = P(X > 1 \text{ and } X < 2)$, since $2X + 3 > 5$ implies $X > 1$. Now

$$P(X > 1 \text{ and } X < 2) = \int_{1}^{2} f(x)\,dx = \int_{1}^{2} 1/4\,dx = 1/4\,[x]_{1}^{2} = 1/4.$$

2.1.12 Example

If the probability density of a random variable is given by

$$f(x) = \begin{cases} K(1 - x^2), & 0 < x < 1 \\ 0 & \text{otherwise} \end{cases}$$

Find (i) K, (ii) the cumulative distribution function of the random variable.

Solution.

(i) $\int_{-\infty}^{\infty} f(x)\,dx = 1$

$$\Rightarrow \int_{0}^{1} K(1 - x^2)\,dx = 1$$

$$\Rightarrow K\left[x - \frac{x^3}{3}\right]_0^1 = 1$$

$$\Rightarrow K[2/3] = 1 \Rightarrow K = 3/2.$$

(ii) The distribution function $F(x) = \int_{-\infty}^{\infty} f(x)\,dx = \int_{0}^{x} 3/2(1 - x^2)\,dx$

$$= \begin{cases} 3/2[x - x^3/3] & 0 < x < 1 \\ 0 & \text{otherwise} \end{cases}$$

2.1.13 Example

A random variable X has the density function

$$f(x) = \begin{cases} K \cdot \dfrac{1}{1+x^2} & \text{if } -\infty < x < \infty \\ 0 & \text{otherwise} \end{cases}$$

Determine K and the distribution function

Solution.

(i) $\int_{-\infty}^{\infty} f(x)\,dx = 1$

$$\Rightarrow \int_{-\infty}^{\infty} K \frac{1}{1+x^2}\,dx = 1$$

$$\Rightarrow K \cdot \left[\tan^{-1} x\right]_{-\infty}^{\infty} = 1$$

$$\Rightarrow K\,[\tan^{-1} \infty - \tan^{-1}(-\infty)] = 1$$

$$\Rightarrow K\,[\pi/2 - (-\pi/2)] = 1$$

$$\Rightarrow K \cdot \pi = 1 \quad \text{or} \quad K = \frac{1}{\pi}.$$

(ii) $F(x) = \int_{-\infty}^{\infty} f(x)\,dx = \int_{-\infty}^{\infty} \frac{1}{\pi} \cdot \frac{1}{1+x^2}\,dx = \frac{1}{\pi}\left[\tan^{-1} x\right]_{-\infty}^{\infty}$

$$= \begin{cases} 1/\pi(\tan^{-1} x), & -\infty < x < \infty \\ .0, & \text{otherwise} \end{cases}$$

2.2 MOMENTS AND MOMENT-GENERATING FUNCTIONS

Mean: Let X be a random variable with probability distribution $f(x)$. The **mean** or **expected value of** X is

$$E(X) = \sum_x x f(x) \quad \text{if } X \text{ is discrete}$$

$$E(X) = \int_{-\infty}^{\infty} x f(x)\,dx \quad \text{if } X \text{ is continuous}$$

We generalize the definition of $E(X)$ as:

2.2.1 Definition

Let X be a random variable with probability distribution $f(x)$. The mean or expected value of the random variable $g(X)$ is

$$\mu_{g(X)} = E(g(X)) = \sum_x g(x) f(x) \text{ if } X \text{ is discrete}$$

$$E(g(X)) = \int_{-\infty}^{\infty} g(x) f(x)\,dx \quad \text{if } X \text{ is continuous}$$

2.2.2 Definition

Let X be a random variable. The variance of X denoted by

$$\text{Var}(X) = E(X - E(X))^2$$

$$= \sum_x (x - E(X))^2 f(x) \quad \text{if } X \text{ is discrete}$$

and $$\text{Var}(X) = E(X - E(X))^2 = \int_{-\infty}^{\infty} (x - E(X))^2 f(x)\,dx \text{ if } X \text{ is continuous.}$$

The positive square root of the variance is called the **standard deviation** of X.
Generalizing, the variance of a random variable X to a random variable $g(X)$ as:

2.2.3 Definition

Let X be a random variable with probability distribution $f(x)$. The variance of the random variable $g(X)$ is

$$E(g(X) - \mu_{g(X)})^2 = \sum_x [g(x) - \mu_{g(x)}]^2 f(x) \quad \text{if } X \text{ is discrete}$$

and $$E(g(X) - \mu_{g(X)})^2 = \int_{-\infty}^{\infty} (g(x) - \mu_{g(x)})^2 f(x)\,dx \text{ if } X \text{ is continuous}$$

In particular, if $g(X) = X$, we obtain Definition 2.2.2.

2.2.4 Definition

Moments

If $g(X) = X^r$ for $r = 0, 1, 2, 3, \ldots$, then the expectation of X^r is called the rth **moment** about the origin of the random variable X, which we denote by μ_r^1.

$$\mu_r^1 = E(X^r) = \begin{cases} \sum_x x^r f(x) & \text{if } X \text{ is discrete} \\ \int_{-\infty}^{\infty} x^r f(x)\,dx & \text{if } X \text{ is continuous} \end{cases}$$

Since the first and second moments about the origin are given by $\mu_1^1 = E(X)$ and $\mu_2^1 = E(X^2)$, we write mean = first moment about the origin and $\text{Var}(X) = \mu_2^1 - (E(X))^2 = \mu_2^1 - \left(\mu_1^1\right)^2$.
The rth moment about a point c is defined by $E(X - c)^r$. Central moment of order r is equal to $E(X - E(X))^r = E(X - \mu_1^1)^r$ which is denoted by m_r.
Now the first and second central moment are given by:

$$m_1 = E(X - E(X)) = E(X) - E(X) = 0$$

$$m_2 = E(X - E(X))^2 = E(X^2 - 2X\,E(X) + (E(X))^2$$

$$= E(X^2) - 2(E(X))^2 + (E(X))^2$$

$$= E(X^2) - (E(X))^2 = \text{Var}(X)$$

Moment-Generating Function

An important technique/method that can be used to calculate the higher moments is the moment generating function.

2.2.5 Definition

The moment generating function (mgf) of a random variable X about the origin whose probability distribution function is $f(x)$ is given by $E(e^{tX})$ and is denoted by $M_X(t)$. Hence

$$M_X(t) = E(e^{tX}) = \begin{cases} \sum_x e^{tx} f(x) & \text{if } X \text{ is discrete} \\ \int_{-\infty}^{\infty} e^{tx} f(x) dx & \text{if } X \text{ is continuous} \end{cases}$$

Note. Moments generating functions will exist only if the sum or integral of above definition converges.

Moment generating function of a random variable X is used to generate all the moments of that variable.

2.2.6 Theorem

Let X be a random variable with mgf $M_X(t)$. Then

$$\mu_r^1 = \left(\frac{d^r M_X(t)}{dt^r}\right)_{t=0}$$

Proof. Assuming that term by term integration and differentiation is valid, we get

$$\frac{d^r M_X(t)}{dt^r} = \begin{cases} \sum_x x^r e^{tx} f(x) & \text{if } X \text{ is discrete} \\ \int_{-\infty}^{\infty} x^r e^{tx} f(x)\, dx & \text{if } X \text{ is continuous} \end{cases}$$

Setting $t = 0$, we see that both cases reduce to $E(X^r) = \mu_r^1$

The moment generating function of X about the point $X = a$ is $M_X(t) = E(e^{t(X-a)})$.

2.2.7 Theorem. (Uniqueness Theorem)

Let X and Y be two random variables with moment-generating functions $M_X(t)$, $M_Y(t)$ respectively. If $M_X(t) = M_Y(t)$ for all values of t, then X and Y have the same probability distribution.

2.2.8 Theorem

$$M_{X+a}(t) = e^{at} M_X(t)$$

Proof. $M_{X+a}(t) = E(e^{t(X+a)}) = E(e^{tX}).\ e^{at} = e^{at}.\ M_X(t)$

2.2.9 Theorem

$$M_{aX}(t) = M_X(at)$$

Proof. $M_{aX}(t) = E(e^{t(aX)}) = E(e^{(at)X}) = M_X(at)$

2.2.10 Theorem

If $X_1, X_2, \ldots X_n$ are independent random variables with moment-generating functions $M_{X_1}(t), M_{X_2}(t), \ldots M_{X_n}(t)$ respectively and $Y = X_1 + X_2 + \ldots + X_n$, then

$$M_Y(t) = M_{X_1}(t)\, M_{X_2}(t) \ldots M_{X_n}(t)$$

Proof. For the continuous case

$$M_Y(t) = E(e^{tY}) = E\left(e^{t(X_1 + X_2 + \ldots + X_n)}\right)$$

$$= \int_{-\infty}^{\infty} \ldots \int_{-\infty}^{\infty} e^{t(X_1 + X_2 + \ldots + X_n)} f(x_1, x_2, \ldots x_n)\, dx_1 dx_2 \ldots dx_n$$

Since the variables are independent, we have

$$f(x_1\, x_2 \ldots x_n) = f_1(x_1)\, f_2(x_2) \ldots f_n(x_n)$$

and then

$$M_Y(t) = \int_{-\infty}^{\infty} e^{tx_1} f_1(x_1)\, dx_1 \int_{-\infty}^{\infty} e^{tx_2} f_2(x_2)\, dx_2 \ldots \int_{-\infty}^{\infty} e^{tx_n} f_n(x_n)\, dx_n$$

$$= M_{X1}(t)\, M_{X2}(t) \ldots M_{Xn}(t)$$

Similarly we prove the result for discrete case.

In some cases, the moment-generating function does not exist. For example, the moment-generating function of the Cauchy random variable X with parameter 'a' and pdf $f(x) = \dfrac{a}{\pi(x^2 + a^2)}, -\infty < x < \infty$ does not exist, since $E(X^n) \to \infty$, for $n \geq 2$. For this type of situations, we look for alternative tool to generate moments.

2.2.11 Definition

The characteristic function of a random variable X is defined by

$$\phi_X(w) = E(e^{iwx})$$

$$= \begin{cases} \sum_x e^{iwx} f(x) & \text{if } X \text{ is discrete} \\ \int_{-\infty}^{\infty} e^{iwx} f(x)\, dx & \text{if } X \text{ is continuous} \end{cases}$$

Since

$$|e^{iwx}| = |\cos wx + i \sin wx|^{1/2}$$

$$= (\cos^2 wx + \sin^2 wx)^{1/2} = \sqrt{1} = 1,$$

$$|\phi_X(w)| = |E(e^{iwx})| = \left|\int_{-\infty}^{\infty} e^{iwx} f(x)\, dx\right|$$

$$\leq \int_{-\infty}^{\infty} |e^{iwx}|\, f(x)\, dx = \int_{-\infty}^{\infty} f(x)\, dx$$

$$= 1 \text{ since } X \text{ is a random variable.}$$

Hence the characteristic function $\phi_X(w)$ always exist even when moment- generating function may not exist.
Moreover, if the characteristic function of a distribution is given, the distribution can be uniquely determined by taking the inverse Fourier transform of $\phi_X(w)$. That is, *pdf* of X is

$$f(x) = \frac{1}{2\pi}\int_{-\infty}^{\infty} \phi_X(w)\, e^{-iwx}\, dw$$

2.2.12 Example

The characteristic function of a random variable X is given by

$$\phi_X(w) = \begin{cases} 1-|w|, & |w|<1 \\ 0, & |w|>1 \end{cases}$$

Find the pdf of X.
Solution. The pdf of X is

$$f(x) = \frac{1}{2\pi}\int_{-\infty}^{\infty} \phi_X(w)\, e^{-iwx}\, dw = \frac{1}{2\pi}\left[\int_{-1}^{0} (1-|w|)\, e^{-iwx}\, dw\right]$$

$$= \frac{1}{2\pi}\left[\int_{-1}^{0} (1+w)\, e^{-iwx}\, dw + \int_{0}^{1} (1-w)\, e^{-iwx}\, dw\right]$$

$$= \frac{1}{2\pi x^2}(2 - e^{ix} - e^{-ix}) = \frac{1}{\pi x^2}(1 - \cos x)$$

$$= \frac{1}{2\pi}\left[\frac{\sin(x/2)}{x/2}\right]^2 \qquad -\infty < x < \infty$$

2.2.13 Example

Find the characteristic function of a random variable X with parameter 'a' and its pdf given by $f(x) = \dfrac{a}{\pi(x^2+a^2)}$, $-\infty < x < \infty$.

Solution.

$$\phi_X(w) = \int_{-\infty}^{\infty} e^{iwx} f(x)\, dx = \int_{-\infty}^{\infty} e^{iwx} \cdot \frac{a}{\pi(x^2+a^2)}\, dx$$

$$= 2 \text{ real part of } \int_{0}^{\infty} \frac{a}{\pi(x^2+a^2)} \cos wx\, dx$$

$$= e^{-a|x|} \quad \text{by Example 2.2.14}$$

2.2.14 Example

Find the Fourier cosine transform of $e^{-a|x|}$, $x > 0$, 'a' constant and deduce that

$$\int_{0}^{\infty} \frac{\cos wx}{w^2+a^2}\, dw = \frac{\pi}{2a} e^{-a|x|}$$

Solution. Given $f(x) = e^{-a|x|}$. So inverse Fourier cosine transform of $f(x)$ is

$$\sqrt{\frac{2}{\pi}}\int_0^\infty e^{-ax}\cdot\cos wx\,dx = \sqrt{\frac{2}{\pi}}\left[\frac{e^{-ax}}{w^2+a^2}(-a\cos wx + w\sin wx)\right]_0^\infty$$

$$= \sqrt{\frac{2}{\pi}}\left(\frac{a}{w^2+a^2}\right) = F_c(f(w))\ \text{(say)}$$

Now $$f(x) = \sqrt{\frac{2}{\pi}}\int_0^\infty F_c(f(w))\cos wx\,dw$$

$$= \sqrt{\frac{2}{\pi}}\int_0^\infty \sqrt{\frac{2}{\pi}}\cdot\frac{a}{w^2+a^2}\cos wx\,dw = \frac{2a}{\pi}\int_0^\infty \frac{\cos wx}{w^2+a^2}\,dw$$

But $$f(x) = e^{-a|x|}$$

Hence $$e^{-a|x|} = \frac{2a}{\pi}\int_0^\infty \frac{\cos wx}{w^2+a^2}\,dw$$

$$\Rightarrow \quad \frac{\pi}{2a}\cdot e^{-a|x|} = \int_0^\infty \frac{\cos wx}{w^2+a^2}\,dw$$

Hence the result.

2.2.15 Definition

If X and Y are random variables the **covariance** between them is

$$\text{cov}(X, Y) = E(XY) - E(X)\,E(Y) = E((X - E(X))\,(Y - E(Y)))$$

2.2.16 Example

Prove that $\text{cov}(aX, bY) = ab\ \text{cov}(X, Y)$.

Solution.
$$\text{cov}(aX, bY) = E((aX - E(aX))\,(bY - E(bY)))$$
$$= E(a(X - E(X))\,b(Y - E(Y)))$$
$$= abE((X - E(X))\,(Y - E(Y))) = ab\ \text{cov}(X, Y).$$

2.3 BINOMIAL DISTRIBUTION

An experiment often consists of repeated trials each with possible outcomes that may be labeled success and failure. The most obvious application deals with the testing of items, where each test or trial may indicate a defective or a non-defective item. Consider the case of cards drawn in succession from an ordinary deck of cards and each trial is labeled a success or a failure depending whether the card is a spade or not a spade. If each card drawn is replaced and the deck is shuffled before the next draw, the two experiments just described have similar properties, in that the repeated trials are independent and the probability of a success remains constant from trial to trial. The process is referred to as a *Bernoulli process.* Each trial is called a *Bernoulli trial.* Observe that in the card-drawing example, the probabilities of a success for the repeated trials change if the cards are not replaced.

The probability of selecting a spade on the first draw is $\frac{13}{52}=\frac{1}{4}$, but on the second draw it is a conditional probability having a value of $\frac{13}{51}$ or $\frac{12}{51}$ depending on whether or not a spade occurred on the first draw. This, then, would no longer be considered a set of Bernoulli trials.

Special discrete distributions. The Bernoulli probability mass function is the density function of a discrete random variable X having 0 and 1 as its only possible values. That is, if we let $X = 1$ when the outcome of an experiment is a success and $X = 0$ when it is a failure, then the probability mass function of X is given by

$$\left.\begin{aligned} P(0) &= P\{X=0\} = 1-p \\ P(1) &= P\{X=1\} = p \end{aligned}\right\} \qquad (2.1)$$

where p, $0 \le p \le 1$ is the probability that the trial is a success.

A random variable X is said to be Bernoulli random variable if pmf satisfies (2.1) for some $p \in (0, 1)$.

To generate a Bernoulli pmf a single trial is performed. Suppose n independent trials are performed each of which results in a success with probability p and in a failure with probability $1 - p$. If X represents the number of successes that occur in n trials, then X is said to be a **binomial random variable** with parameters (n, p).

The pmf of a binomial random variable having parameters (n, p) is given by

$$P(i) = nC_i\, p^i\, (1-p)^{n-i},\ i = 0, 1, 2, \ldots n \qquad (2.2)$$

We refer X with the probability mass function (2.2) has a binomial distribution.

Note. The binomial distribution is applicable only when

(i) each trial has exactly two mutually exclusive outcomes namely success and failure

(ii) The outcomes of successive trials are mutually independent.

Also note,
$$\sum_{i=0}^{\infty} P(i) = \sum_{i=0}^{\infty} nC_i\, p^i\, (1-p)^{n-i}$$
$$= (p + (1-p))^n \qquad \text{by Binomial expansion}$$
$$= 1$$

2.3.1 Example

Five fair coins are tossed. If the outcomes are assumed independent, find the probability mass function of the number of heads obtained.

Solution. Let X = number of heads (success). Then X is a binomial random variable with parameters (n = 5, p = 1/2). Hence

$$P(X=0) = 5C_0\left(\frac{1}{2}\right)^0\left(\frac{1}{2}\right)^5 = \frac{1}{32}$$

$$P(X=1) = 5C_1\left(\frac{1}{2}\right)\left(\frac{1}{2}\right)^4 = \frac{5}{32}$$

$$P(X=2) = 5C_2\left(\frac{1}{2}\right)^2\left(\frac{1}{2}\right)^3 = \frac{10}{32}$$

$$P(X = 3) = 5C_3 \left(\frac{1}{2}\right)^3 \left(\frac{1}{2}\right)^2 = \frac{10}{32}$$

$$P(X = 4) = 5C_4 \left(\frac{1}{2}\right)^4 \left(\frac{1}{2}\right) = \frac{5}{32}$$

$$P(X = 5) = 5C_5 \left(\frac{1}{2}\right)^5 \left(\frac{1}{2}\right)^0 = \frac{1}{32}$$

2.3.2 Example

It is known that car produced by an automobile company will be defective with probability 0.01 independently of each other. The company sells the cars in packages of 10 and offers a money-back guarantee that almost 1 of the 10 cars is defective. What proportion of packages sold must the company replace?

Solution. If X is the number of defective cars in a package, then X is a binomial random variable with probability (0.01). Hence the probability that a package will have to be replaced is

$$\begin{aligned} P(X > 1) &= 1 - \{P(X \le 1)\} \\ &= 1 - \{P(X = 0) + P(X = 1)\} \\ &= 1 - \{10C_0 (.01)^0 (.99)^{10} + 10C_1 (.01)(.99)^9\} = .004 \end{aligned}$$

Hence only .4 percentage of the packages will have to be replaced.

Note. Let X be a binomial with parameters (n, p). The relationship between $P(X = k + 1)$ and $P(X = k)$ is given by

$$P(X = k + 1) = \left(\frac{p}{1-p}\right)\left(\frac{n-k}{k+1}\right) P(X = k)$$

2.3.3 Example

Let X be a binomial random variable with parameters $n = 6$, $p = .4$. Let $P(X = 0) = (.6)^6$. Then find $P(X = 3)$.

Solution.

$$P(X = 1) = \left(\frac{4}{6}\right)\left(\frac{6}{1}\right)(.6)^6 = .1866$$

$$P(X = 2) = \left(\frac{4}{6}\right)\left(\frac{5}{2}\right) P(X = 1) = .3110$$

$$P(X = 3) = \left(\frac{4}{6}\right)\left(\frac{4}{3}\right) P(X = 2) = .2765.$$

2.3.4 Example

Calculate the mean and variance of the binomial random variable X with pmf

$$P(X \le x) = \sum_{x=0}^{n} nC_x \, p^x \, (1-p)^{n-x}$$

Solution.

$$\text{mean} = E(X) = \sum_{x=0}^{n} x\, nC_x\, p^x\,(1-p)^{n-x} = \sum_{x=1}^{n} x\, nC_x\, p^x\,(1-p)^{n-x}$$

$$= \sum_{x=1}^{n} x \cdot \frac{n\,(n-1)!}{x!\,(n-x)!}\, p^x\,(1-p)^{n-x} = \sum_{x=1}^{n} \frac{xn(n-1)!\,p \cdot p^{x-1}\,(1-p)^{n-x}}{x(x-1)!\,[(n-1)-(x-1)]!}$$

$$= np \sum_{x=1}^{n} \frac{(n-1)!}{(x-1)!\,[(n-1)-(x-1)]!}\, p^{x-1}\,(1-p)^{n-x}$$

$$= np \sum_{x=1}^{n} {}^{n-1}C_{x-1}\, p^{x-1}\,(1-p)^{(n-1)-(x-1)}$$

$$= np\,(p + (1-p))^{n-1}, \text{ since } [p + (1-p)]^n = \sum_{x=0}^{n} nC_0\, p^x\,(1-p)^{n-x}$$

$$= np.$$

Second moment $\mu_2^1 = E(X^2) = \sum_{x=0}^{n} x^2\, nC_x\, p^x\,(1-p)^{n-x}$

Since the first term of the series is equal to zero

$$E(X^2) = \sum_{x=0}^{n} x^2\, nC_x\, p^x\,(1-p)^{n-x} = \sum_{x=0}^{n} [x(x-1)+x]\, nC_x\, p^x\,(1-p)^{n-x}$$

$$= \sum_{x=1}^{n} x(x-1)\, nC_x\, p^x\,(1-p)^{n-x} + \sum_{x=1}^{n} x\, nC_x\, p^x\,(1-p)^{n-x}$$

$$= \left[\sum_{x=2}^{n} x(x-1)\, nC_x\, p^x\,(1-p)^{n-x}\right] + np$$

when $x = 1$, the first term vanishes

$$= \left[\sum_{x=2}^{n} x(x-1)\frac{n!}{x!\,(n-x)!}\, p^x\,(1-p)^{n-x}\right] + np$$

$$= \left[\sum_{x=2}^{n} x(x-1)\frac{n\,(n-1)\,(n-2)!\, p^x\,(1-p)^{n-x}}{x(x-1)\,(x-2)!\,[(n-2)-(x-2)]!}\right] + np$$

$$= n(n-1)\,p^2 \left[\sum_{x=2}^{n} \frac{(n-2)!\, p^{x-2}\,(1-p)^{(n-2)-(x-2)}}{(x-2)!\,[(n-2)-(x-2)]!}\right] + np$$

$$= n\,(n-1)\,p^2 \left[\sum_{x=2}^{n} {}^{(n-2)}C_{(x-2)}\, p^{x-2}\,(1-p)^{(n-2)-(x-2)}\right] + np$$

$$= n\,(n-1)\,p^2\,(p + (1-p))^{n-2} + np = n\,(n-1)\,p^2 + np = n^2p^2 - np^2 + np$$

$$\text{Var}(X) = E(X^2) - (E(X))^2$$
$$= n^2p^2 + np - np^2 - n^2p^2 = np - np^2 = np(1-p)$$

2.3.5 Example

Suppose X is a binomial distribution with parameters n and p. Then the moment-generating function of X is given by

$$M_X(t) = E(e^{tX}) = \sum_{x=0}^{n} e^{tx}\, nC_x\, p^x (1-p)^{n-x} = \sum_{x=0}^{n} nC_x (pe^t)^x (1-p)^{n-x}$$
$$= [pe^t + (1-p)]^n \text{ in accordance with binomial theorem.}$$

2.3.6 Example

Suppose that X is binomially distributed with parameters n and p. Then the characteristic function is given by

$$\phi_X(w) = E(e^{iwX}) = \sum_{x=0}^{n} e^{iwx}\, nC_x\, p^x (1-p)^{n-x}$$

$$= \sum_{x=0}^{n} nC_x (pe^{iw})^x (1-p)^{n-x} = (pe^{iw} + (1-p))^n$$

2.3.7 Example

Suppose that X has a binomial distribution with parameters n and p. Given $M_X(t) = (pe^t + q)^n$. Then prove $\text{Var}(X) = np(1-p)$

Solution. $M_X(t) = [pe^t + (1-p)]^n$

$$\frac{d}{dt}(M_X(t)) = M'_X(t) = n(pe^t + (1-p))^{n-1} pe^t$$

$$\frac{d^2}{dt^2}(M_X(t)) = M''_X(t) = np\,[e^t(n-1)(pe^t + (1-p))^{n-2} pe^t + (pe^t + (1-p))^{n-1} e^t]$$

Since $\quad \text{Var}(X) = E(X)^2 - (E(X))^2 = M''_X(0) - (M'_X(0))^2$,

We have $\quad E(X) = M'_X(0) = np$ and

$\quad E(X^2) = M''_X(0) = np\,[(n-1)p + 1]$.

Hence $\quad \text{Var}(X) = M''_X(0) - (M'_X(0))^2 = np(1-p)$.

2.3.8 Example

For a binomial distribution with mean 6, standard deviation $\sqrt{a}$, find the first two terms of the distribution.

Solution. For a binomial distribution with parameters n and p, mean $= np$ and the standard deviation $=$ $\sqrt{np(1-p)}$. Given $np = 6$ and $\sqrt{np(1-p)} = \sqrt{2}$, which implies $1 - p = \frac{2}{np} = \frac{2}{6} = \frac{1}{3}$. Hence $p = \frac{2}{3}$ and $n = 9$. Now $f(x) = nC_x p^x (1-p)^{n-x}$ implies that the first term is :

$$f(0) = 9C_0 \left(\frac{2}{3}\right)^0 (1-p)^9 = \left(\frac{1}{3}\right)^9.$$

The second term is $f(1) = 9\left(\frac{2}{3}\right)\left(\frac{1}{3}\right)^8 = \frac{2}{2187}$.

2.4 POISSON DISTRIBUTION

The probability density function of the poisson variate can be obtained as a limiting case of the Binomial probability density function under the following assumption.

(i) The number of trials is increased indefinitely $(n \to \infty)$

(ii) The probability of success in a single trial is very small $(p \to 0)$

(iii) np is a finite constant say $np = \lambda$.

Consider the probability density function of a Binomial random variable X as

$$P(X = x) = nC_x p^x (1-p)^{n-x} = \frac{n!}{x!(n-x)!} p^x (1-p)^{n-x}$$

$$= \frac{n(n-1)\ldots(n-x+1)}{x!}\left(\frac{\lambda}{n}\right)^x \left(1-\frac{\lambda}{n}\right)^{n-x} \qquad \text{since } np = \lambda$$

$$= \frac{n(n-1)\ldots(n-x+1)}{x!\,n^x}\cdot\lambda^x \frac{(1-\lambda/n)^n}{(1-\lambda/n)^x}$$

$$= \frac{\left(\frac{n}{n}\right)\left(\frac{n-1}{n}\right)\ldots\left(\frac{n-x+1}{n}\right)\lambda^x}{x!}\frac{(1-\lambda/n)^n}{(1-\lambda/n)^x}$$

For given x, as $n \to \infty$, the terms $1 - 1/n, 1 - 2/n, \ldots 1 - \frac{x-1}{n}$, $(1 - \lambda/n)^x$ tends to 1. It is well known from the definition of e that $\underset{n\to\infty}{\text{Lt}} (1 - \lambda/n)^n = e^{-\lambda}$. Hence

$$\underset{n\to\infty}{\text{Lt}}\ P(X = x) = \frac{\lambda^x}{x!} e^{-\lambda}.$$

Poisson random variable

A random variable X taking on one of the values 0, 1, 2, . . . is said to be a **Poisson random variable** with parameter λ if for some $\lambda > 0$,

$$P(x) = P(X = x) = \frac{e^{-\lambda}\lambda^x}{x!} \quad x = 0, 1, 2, \ldots$$

Note that

$$\sum_{x=0}^{\infty} P(x) = \sum_{x=0}^{\infty} \frac{e^{-\lambda}\lambda^x}{x!} = e^{-\lambda}\sum_{x=0}^{\infty}\frac{\lambda^x}{x!} = e^{-\lambda}e^{\lambda} = 1$$

2.4.1 Example

Find the mean and variance of the Poisson random variable X with pmf

$$P(X \le x) = \sum_{x=0}^{\infty} \frac{e^{-\lambda}\lambda^x}{x!}$$

Solution.

$$E(X) = \sum_{x=0}^{\infty} x\,P(X=x) = \sum_{x=0}^{\infty} x\cdot\frac{e^{-\lambda}\lambda^x}{x!} = \sum_{x=0}^{\infty}\frac{e^{-\lambda}\lambda^x}{(x-1)!}$$

$$= \lambda e^{-\lambda}\sum_{x=1}^{\infty}\frac{\lambda^{x-1}}{(x-1)!} = \lambda e^{-\lambda}\left[1+\frac{\lambda}{1!}+\frac{\lambda^2}{2!}+\ldots\right]$$

$$= \lambda e^{-\lambda}\cdot e^{\lambda} = \lambda$$

$$\text{Second moment} = E(X^2) = \sum_{x=0}^{\infty} x^2\,\frac{e^{-\lambda}\lambda^x}{x!} = \sum_{x=0}^{\infty}[x(x-1)+x]\frac{e^{-\lambda}\lambda^x}{x!}$$

$$= \sum_{x=0}^{\infty} x(x-1)\cdot\frac{e^{-\lambda}\lambda^x}{x!} + \sum_{x=0}^{\infty}\frac{xe^{-\lambda}\lambda^x}{x!} = \sum_{x=2}^{\infty}\frac{e^{-\lambda}\lambda^x}{(x-2)!}+E(X)$$

since when $x = 0, 1$, the first two terms vanishes

$$= \lambda^2\cdot e^{-\lambda}\sum_{x=2}^{\infty}\frac{\lambda^{x-2}}{(x-2)!}+\lambda = \lambda^2 e^{-\lambda}\cdot e^{\lambda}+\lambda = \lambda^2+\lambda.$$

$$\text{Var}(X) = E(X^2) - (E(X))^2 = \lambda^2+\lambda-\lambda^2 = \lambda.$$

2.4.2 Example

If X is a Poisson random variable with probability density function $P(x) = \sum_{x=0}^{\infty}\frac{e^{-\lambda}\lambda^x}{x!}$ then find its moment generating function and characteristic function.

Solution. $$M_X(t) = E(e^{tX}) = \sum_{x=0}^{\infty} e^{tx}P(x) = \sum_{x=0}^{\infty} e^{tx}\,\frac{e^{-\lambda}\lambda^x}{x!} = e^{-\lambda}\sum_{x=0}^{\infty}\frac{(\lambda e^t)^x}{x!}$$

$$= e^{-\lambda}\left\{1+\frac{\lambda e^t}{1!}+\frac{(\lambda e^t)^2}{2!}+\ldots\right\} = e^{-\lambda}\cdot e^{\lambda e^t} = e^{\lambda(e^t-1)}$$

Characteristic function is

$$\phi_X(w) = \sum_{x=0}^{\infty} e^{iwx}\cdot\frac{e^{-\lambda}\lambda^x}{x!} = e^{-\lambda}\sum_{x=0}^{\infty}\frac{(e^{iw}\lambda)^x}{x!}$$

$$= e^{-\lambda}\left[1+\frac{\lambda e^{iw}}{1!}+\frac{(\lambda e^{iw})^2}{2!}+\ldots\right] = e^{-\lambda}e^{\lambda e^{iw}} = e^{\lambda(e^{iw}-1)}$$

2.4.3 Example

At a busy traffic intersection the probability p of an individual car having an accident is very small say $p = 0.0001$. However, during a certain peak hours of the day say between 4 p.m and 6 p.m, a large number of cars (say 1000) pass through the intersection. Under these conditions what is the probability of two or more accidents occurring during that period.

Solution. Here $p = 0.0001$, $n = 1000$,

$$\lambda = np = 1000 \times 0.0001 = 0.1$$

$$P(X = x) = \frac{e^{-\lambda}\lambda^x}{x!}$$

Now the probability of two or more accidents occurring is

$$P(X \geq 2) = 1 - [p(X = 0) + p(X = 1)]$$

$$= 1 - \left[\frac{e^{-0.1}(0.1)^0}{0!} + \frac{e^{-0.1}(0.1)}{1!}\right] = 1 - e^{-0.1}(1 + 0.1) = 0.0045.$$

2.4.4 Example

In a component manufacturing industry, there is a small probability of 1/500 for any component to be defective. The components are supplied in packets of 10. Use Poisson distribution to calculate the approximate number of packets containing

(i) no defective (ii) two defective components in a consignment of 10,000 packets.

Solution. Given $p = \frac{1}{500}$, $n = 10$, x denote the number of defective component in one packet.

$$\lambda = np = \frac{1}{50} = .02$$

$$P(X = x) = \frac{e^{-\lambda}\lambda^x}{x!} = \frac{(.02)^x e^{-0.02}}{x!}$$

(i) $P(X = 0) = \dfrac{(.02)^0 e^{-0.02}}{0!} = 0.9802$

Hence the number of packets free from defective is $0.9802 \times 10{,}000 = 9802$

(ii) $P(X = 1) = \dfrac{(.02)^1 e^{-0.02}}{1!} = 0.0196$

Hence the number of packets in which one may be defective is $0.0196 \times 10{,}000 = 19.6$

2.5 GEOMETRIC DISTRIBUTION

Suppose that independent trials, each having a probability p of success, are performed until a success occurs. If we let $X = n$, then

$$P(X = n) = (1 - p)^{n-1} p, \quad n = 1, 2, \ldots. \tag{2.3}$$

Since in order that $X = n$, it is necessary and sufficient that the first $(n - 1)$ trials are failures and the nth trial is a success. Equation (2.3) follows as the outcomes of the successive trials are assumed to be independent. Also

$$\sum_{n=1}^{\infty} P(X=n) = p\sum_{n=1}^{\infty}(1-p)^{n-1} = \frac{p}{1-(1-p)} = 1$$

Hence any random variable X whose probability mass function given by (2.3) is said to be **Geometric random variable** with parameter p.

2.5.1 Example

In a chemical engineering process industry it is known that, on the average, 1 in every 100 items is defective. What is the probability that the fifth item inspected is the first defective item found.

Solution. Given $n = 5$, $p = 0.01$, hence by Geometric distribution

$$P(X=5) = (1-p)^4 p = (0.01)(0.99)^4 = 0.0096.$$

2.5.2 Example

At busy time, a telephone exchange will be working busy with full capacity. So people cannot get a line to use immediately. It may be of interest to know the number of attempts necessary in order to get a connection. Suppose that $p = 0.05$, then find the probability that 5 attempts are necessary for a successful call connection.

Solution. Given $n = 5$, $p = 0.05$. Then

$$P(X=5) = (0.05)(0.95)^4 = 0.041$$

2.5.3 Example

The mean and variance of a random variable following the geometric distribution are

$$E(X) = \frac{1}{p},\ \text{Var}(X) = \frac{1-p}{p^2}.$$

Solution.

$$E(X) = \sum_{n=1}^{\infty} n(1-p)^{n-1} p = p\sum_{n=1}^{\infty} nt^{n-1},\ t = 1-p$$

$$= p\,[1 + 2t + 3t^2 + 4t^3 + \ldots]$$

$$= p\,[1-t]^{-2} \qquad \text{(since } t<1)$$

$$= p\,[1-(1-p)]^{-2} = \frac{p}{p^2} = \frac{1}{p}.$$

Second moment

$$E(X^2) = \sum_{n=0}^{\infty} n^2 (1-p)^{n-1} p = \sum_{n=1}^{\infty} n^2 pt^{n-1} \qquad (t = 1-p)$$

$$= \sum_{n=1}^{\infty} [n(n-1)+n]\,pt^{n-1} = \sum_{n=1}^{\infty} n(n-1)\,pt^{n-1} + \sum_{n=1}^{\infty} npt^{n-1}$$

$$= pt\sum_{n=1}^{\infty} n(n-1)t^{n-2} + \frac{1}{p}$$

$$= pt\,[2 + 3\times 2\times t + 4\times 3\times t^2 + 5\times 4\times t^3 + \ldots] + \frac{1}{p}$$

$$= 2pt\,[1 + 3t + 6t^2 + \ldots] + \frac{1}{p} = 2pt\,[1 - t]^{-3} + \frac{1}{p}$$

$$\left[\text{since } (1-x)^{-n} = 1 + \frac{nx}{1!} + \frac{n(n+1)}{2!}x^2 + \frac{n(n+1)(n+2)}{3!}x^3 + \ldots\right]$$

$$= \frac{2p(1-p)}{p^3} + \frac{1}{p} = \frac{2-2p+p}{p^2} = \frac{2-p}{p^2}$$

Now

$$\text{Var}(X) = E(X^2) - (E(X))^2 = \frac{2-p}{p^2} - \left(\frac{1}{p}\right)^2 = \frac{1-p}{p^2}$$

2.5.4 Example

Find the moment generating function of a Geometric random variable X.

Solution.

$$M_X(t) = E(e^{tX}) = \sum_{x=0}^{\infty} e^{tx} P(X = x) = \sum_{x=1}^{\infty} e^{tx} (1-p)^{x-1} \cdot p$$

$$= \frac{p}{1-p}\sum_{x=1}^{\infty} (e^t (1-p))^x = \frac{p}{1-p}\left\{(1-p)e^x + ((1-p)e^t)^2 + \ldots\right\}$$

$$= \frac{p}{1-p}\,\{[1-(1-p)e^t]^{-1}(1-p)e^t = \frac{p}{1-p}\left[\frac{(1-p)e^t}{1-(1-p)e^t}\right]$$

$$= \frac{pe^t}{1-(1-p)e^t}$$

The Geometric distribution has an important property known as **memory less** or **Markov property**; it is the only discrete distribution with this property. That is, suppose that any event A has not occurred during the first s repetitions. Then the probability that it will not occur during the next t repetitions is same as the probability that it will not occur during the first t repetitions. In other words, the information of no success is "forgotten" so far as subsequent calculations are concerned.

The following theorem explains the above concept.

2.5.5 Theorem

If the random variable X has the geometric distribution, then for any two positive integers s and t

$$P(X > s + t / X > s) = P(X > t)$$

Proof. Since X has the geometric distribution

$$P(X = n) = p(1-p)^{n-1}$$

Let $1 - p = q$. Now

$$P(X > s) = \sum_{n=s+1}^{\infty} pq^{n-1} = pq^s + pq^{s+1} + \ldots$$

$$= pq^s\,[1 + q + q^2 + \ldots]$$

$$= pq^s\,(1-q)^{-1} = \frac{pq^s}{1-q} = q^s$$

Consider the conditional probability,

$$P\,(X > s + t/X > s) = \frac{P(X > s+t \text{ and } X > s)}{P(X > s)}$$

$$= \frac{P(X > s+t)}{P(X > s)} = \frac{q^{s+t}}{q^s} = q^t = P\,(X > t)$$

2.6 NEGATIVE BINOMIAL DISTRIBUTION

Let us consider an experiment in which the properties are same as those listed for a binomial experiment with the exception that the trials will be repeated until a fixed number of success occur. Therefore, instead of finding the probability of r success in n trials, when n is fixed, we are now interested in the probability that the rth success occurs on the xth trial. Experiments of this type is called **negative binomial experiment.** The number of x trials to produce r success in a negative binomial experiment is called a **negative binomial random variable** and its probability distribution is called the **negative binomial distribution.**

Let the random variable Y denotes the number of failures before the occurrence of rth success. Then $y + r$ denotes the number of trials necessary to produce exactly r successes and y failures with the rth success occurring at the $(y + r)^{\text{th}}$ trial. By the multiplication rule of probability, the pmf of y say $g(y)$ equals the product of the probability

$$^{(y+r-1)}C_{(r-1)}\, p^{r-1}\, q^y,\ q = 1 - p$$

of obtaining exactly $r - 1$ success in the first $y + r - 1$ trials and the probability p of a success on the $(y + r)^{\text{th}}$ trial. Thus

$$g\,(y) = {}^{(y+r-1)}C_{r-1}\, p^r\, q^y,\ y = 0, 1, 2, \ldots$$

We say that Y has a negative binomial distribution.

We show that the sum of all probabilities for the negative binomial distribution is equal to one.

We know $\quad {}^{(y+r-1)}C_{r-1} = \dfrac{(y+r-1)!}{(r-1)![(y+r-1)-(r-1)]!} = \dfrac{(y+r-1)}{(r-1)!\,y!} = {}^{(y+r-1)}C_y,$

and

$$(1-a)^{-n} = 1 + \frac{na}{1!} + \frac{n(n+1)}{2!}a^2 + \frac{n(n+1)(n+2)}{3!}a^3 + \ldots$$

$$= 1 + \frac{(-1)(-n)}{1!}a + \frac{(-n)(-n-1)}{2!}a^2 - \frac{(-n)(-n-1)(-n-2)}{3!}a^3 + \ldots$$

$$= \sum_{k=0}^{\infty} (-1)^k \frac{(-n)(-n-1)\ldots(-n-k+1)}{k!}a^k$$

$$= \sum_{k=0}^{\infty} {}^{(n+k-1)}C_k\, a^k,\ |a| < 1, \tag{2.4}$$

We have $\sum_{y=0}^{\infty} g(y) = \sum_{y=0}^{\infty} {}^{(y+r-1)}C_{(r-1)}\, p^r q^y, \; q = 1 - p$

$$= \sum_{y=0}^{\infty} {}^{(y+r-1)}C_y\, p^r q^y = p^r \sum_{y=0}^{\infty} {}^{(y+r-1)}C_y\, q^y$$

$$= p^r [1 - q]^{-r} = p^r p^{-r} = 1.$$

2.6.1 Example

Find the moment-generating function of a negative binomial distribution.

Solution.

$$M_Y(t) = E(e^{tY})$$

$$= \sum_{y=0}^{\infty} e^{ty}\, {}^{(y+r-1)}C_{(r-1)}\, p^r q^y, \; q = 1 - p$$

$$= p^r \sum_{y=0}^{\infty} {}^{(y+r-1)}C_{(r-1)}\, [qe^t]^y$$

But, from (2.4) with $a = qe^t$, we have

$$M_Y(t) = p^r (1 - qe^t)^{-r}, \; t < -\log q \quad \text{or} \quad qe^t < 1.$$

2.6.2 Example

Find the mean and variance of the negative binomial variable y.

$$\text{Mean} = \mu'_1 = \left[\frac{d}{dt} M_Y(t)\right]_{t=0}$$

$$= [-rp^r (1 - qe^t)^{-r-1} (-qe^t)]_{t=0} = [rqp^r e^t (1 - qe^t)^{-(r+1)}]_{t=0}$$

$$= rqp^r (1 - q)^{-(r+1)} = rqp^r \cdot p^{-(r+1)} = \frac{rq}{p}$$

$$\text{Variance} = \mu'_2 - (\mu'_1)^2. \text{ Now}$$

$$\mu'_2 = \left[\frac{d^2}{dt^2} M_Y(t)\right]_{t=0}$$

$$= [rqp^r \{(1 - qe^t)^{-(r+1)} e^t + e^t \cdot (1 - qe^t)^{-(r+2)} (-(r+1)) (-qe^t)\}]_{t=0}$$

$$= rqp^r \{(1 - q)^{-(r+1)} + (1 - q)^{-(r+2)} (r + 1)\, q\}$$

$$= rqp^r \cdot p^{-(r+1)} \{1 + (r + 1)\, q\, p^{-1}\}$$

$$= \frac{rq}{p}\left\{1 + \frac{(r+1)q}{p}\right\}$$

and $$\mu'_2 - (\mu'_1)^2 = \frac{rq}{p}\left\{1 + \frac{(r+1)q}{p}\right\} - \frac{r^2 q^2}{p^2}$$

$$= \frac{rq}{p} + \frac{r^2q^2}{p^2} + \frac{rq^2}{p^2} - \frac{r^2q^2}{p^2} = \frac{rq}{p^2}[p+q]$$

$$= \frac{rq}{p^2} \text{ since } p + q = 1$$

2.6.3 Example

If $g(y)$ represents probability function of rth order negative binomial distribution, then

$$g(y+1) = \frac{y+r}{y+1}\, q\, g(y),\ q = 1 - p$$

Solution. Now

$$g(y) = {}^{(y+r-1)}C_{(r-1)}\, p^r q^y$$

$$g(y+1) = {}^{(y+r)}C_{(r-1)}\, p^r q^{y+1}$$

$$\frac{g(y+1)}{g(y)} = \frac{(y+r)!(r-1)!\,y!}{(r-1)!(y+r-1)!(y+1)!}\, q = \frac{y+r}{y+1}\, q$$

Hence $$g(y+1) = \frac{y+r}{y+1}\, q\, g(y)$$

2.6.4 Example

A fair die is cast on successive independent trials until the second six is observed. The probability of observing exactly ten non-sixes before the second six is cast is

$${}^{(10+2-1)}C_1 \left(\frac{1}{6}\right)^2 \left(\frac{5}{6}\right)^{10} = 0.049.$$

2.6.5 Example

Find the probability that a person tossing three coins will get either all heads or all tails for the second time on the fifth toss.

Solution. Using negative binomial distribution, we have the probability as

$${}^{(5-1)}C_{(2-1)} \left(\frac{1}{4}\right)^2 \left(\frac{3}{4}\right)^3 = \frac{27}{256}$$

2.6.6 Example

The probability that an experiment will succeed is 0.8. If the experiment is repeated until four successful outcomes have occurred, what is the expected number of repetitions required?

Solution. The number of repetitions is $E(X) = \frac{rq}{p} = \frac{4 \times 0.2}{0.8} = 1.$

Relationship between the binomial and negative binomial.

Let X have a binomial distribution with parameters n, p. Let Y have a negative binomial distribution with parameters r and p. (That is Y = number of trials required to obtain r successes with probability of success p). Then

(i) $P(Y \leq n) = P(X \geq r)$

(ii) $P(Y > n) = P(X < r)$

2.7 EXPONENTIAL DISTRIBUTION

We shall now direct our study to several important continuous probability distributions. They are the exponential, gamma, weibull, and normal distributions. Exponential distribution finds application in reliability and queueing theory. In those applications, the exponential distribution corresponds to the lifetime of an equipment or service time of the counter in queueing which we discuss in chapter 7.

In the process of Poisson type, we counted the number of changes occurring in a given interval. This number was a discrete-type random variable with a Poisson distribution. But not only is the number of changes is a random variable; the waiting time between successive changes are also random variable. But this one is of continuous type. In particular, let W denote the waiting time until the first change occurs when observing a Poisson process in which the mean number of changes in the unit interval is λ. Then W is a continuous type random variable and we find its distribution function. Clearly W is non-negative.

The distribution function $F(w) = 0,\ w < 0$ and

$$F(w) = P(W \leq w) = 1 - P(W > w)$$
$$= 1 - P[\text{no changes in } [0, w]] = 1 - e^{-\lambda w},\ w \geq 0 \tag{2.5}$$

This, when $w > 0$, the probability distribution function is $F'(w) = \lambda e^{-\lambda w}$. Hence

2.7.1 Definition

The random variable X has an exponential distribution with parameter $\lambda > 0$, if its probability distribution function is given by

$$f(x) = \begin{cases} \lambda e^{-\lambda x} & \text{if } x \geq 0 \\ 0 & \text{if } x < 0 \end{cases}$$

2.7.2 Example

Find the mean and variance of exponential distribution of a random variable X.

Solution.

$$E(X) = \int_0^\infty x\,\lambda\, e^{-\lambda x}\,dx$$
$$= \left[-xe^{-\lambda x}\right]_0^\infty + \int_0^\infty e^{-\lambda x}\,dx, \text{ by integration by parts taking}$$
$$u = x,\ dv = \lambda e^{-\lambda x}\,dx$$
$$= \frac{1}{\lambda}.$$

Thus the expected value equals the reciprocal of the parameter λ. We can find $E(X^2) = \frac{2}{\lambda^2}$.

Hence

$$\text{Var}(X) = E(X^2) - (E(X))^2 = \frac{2}{\lambda^2} - \frac{1}{\lambda^2} = \frac{1}{\lambda^2}$$

2.7.3 Example

Find the moment-generating function of a random variable X having exponential distribution.

Solution. $M_X(t) = E(e^{tX})$

$$= \int_0^\infty e^{tx} \lambda e^{-\lambda x} dx = \lambda \int_0^\infty e^{-(\lambda - t)x} dx$$

$$= \lambda \left[\frac{e^{-(\lambda - t)x}}{-(\lambda - t)} \right]_0^\infty = -\lambda \left[\frac{1}{-(\lambda - t)} \right] = \frac{\lambda}{\lambda - t}$$

2.7.4 Example

Let X have an exponential distribution with a mean of 100. Find the probability that $X < 90$.

Solution.

We know $E(X) = \frac{1}{\lambda} = 100$ (given). So $\lambda = \frac{1}{100}$

The pdf of X is

$$f(x) = \frac{1}{100} e^{-x/100} \quad 0 \le x < \infty$$

$$P(X < 90) = \int_0^{90} \frac{1}{100} e^{-x/100} dx = 1 - e^{-90/100} = 0.593$$

2.7.5 Example

Customers arrive in a certain shop according to an approximate Poisson process at a mean rate of 20 per hour. What is the probability that the shopkeeper will have to wait for more than 5 minutes for his first customer to arrive?

Solution. Let X denote the waiting time in minutes until the first customer arrives and $E(X) = 1/\lambda = 1/3$ is the expected number of arrivals per minute. Then

$$f(x) = 3e^{-3x} \quad 0 \le x < \infty$$

Hence

$$P(X > 5) = \int_5^\infty 3e^{-3x} dx = 3\left[\frac{e^{-3x}}{-3} \right]_5^\infty = e^{-5}$$

2.7.6 Theorem

For any $s, t > 0$ the memoryless or Markov property of exponential distribution is

$$P(X > s + t / X > s) = \frac{P((X > s + t) \cap (X > s))}{P(X > s)} = \frac{e^{-\lambda(s+t)}}{e^{-\lambda s}} = e^{-\lambda t} = P(X > t)$$

where

$$P(X \ge t) = \int_t^\infty \lambda e^{-\lambda t} dt = \lambda \left[\frac{e^{-\lambda t}}{-\lambda} \right]_t^\infty = e^{-\lambda t}$$

$$P(a \le X \le h) = \int_a^b \lambda e^{-\lambda t} dt = \lambda \left[\frac{e^{-\lambda t}}{-\lambda} \right]_a^b = e^{-\lambda a} - e^{-\lambda b}$$

2.7.7 Example

Suppose that the length of a phone call in minutes is an exponential random variable with parameter $\lambda = \frac{1}{10}$. If A arrives immediately ahead of B at a public telephone booth, find the probability that B will have to wait (i) more than 10 minutes, and (ii) between 10 and 20 minutes.

Solution. Let X denote the length of the call made by the person in the booth, we have

$$\text{(i) } P\{X > 10\} = \int_{10}^{\infty} \frac{1}{10} e^{-x/10} dx = \left[-e^{-x/10}\right]_{10}^{\infty} = e^{-1} = 0.368$$

$$\text{(ii) } P\{10 < X < 20\} = \int_{10}^{20} \frac{1}{10} e^{-x/10} dx = \left[-e^{-x/10}\right]_{10}^{20} = e^{-1} - e^{-2} = 0.233$$

2.8 GAMMA DISTRIBUTION

Before discussing about Gamma probability distribution, let us define Gamma function and illustrate some of its properties.

2.8.1 Definition

The gamma function, denoted by Γ is defined as

$$\Gamma_{(x)} = \int_0^{\infty} e^{-t} t^{x-1} dt, \; x > 0 \tag{2.6}$$

Properties:

1. Integrating (2.6) by parts, we have

$$\Gamma_{(x)} = [-e^{-t} t^{x-1}]_0^{\infty} + \int_0^{\infty} (x-1) t^{x-2} e^{-t} dt$$

$$= (x-1) \int_0^{\infty} e^{-t} t^{(x-1)-1} dt = (x-1)\, \Gamma_{(x-1)} \tag{2.7}$$

The relation (2.7) gives the recursive relation of gamma function Γ.

2. $\Gamma_{(1)} = \int_0^{\infty} e^{-t} dt = 1$

3. Put $x = n$. Then $\Gamma_{(n)} = (n-1)\, \Gamma_{(n-1)}$
$= (n-1)(n-2)\, \Gamma_{(n-2)} = \ldots$
$= (n-1)(n-2) \ldots 2.1 = (n-1)!$

and, for this reason, the gamma function is called the *generalized factorial.* $\Gamma_{(1)}$ corresponds to $0!$ and $\Gamma_{(1)} = 0! = 1$.

In the Poisson process with mean λ, we have seen that the waiting time until the first change has an exponential distribution. In this section let W denote the waiting time until the nth change occurs. Now we find the distribution of W. The distribution function of w, when $w > 0$, is given by

$$F(w) = P(W \le w) = 1 - P(W > w)$$
$$= 1 - P \text{ (less than } n \text{ changes)}$$

$$= 1 - \sum_{k=0}^{n-1} \frac{(\lambda w)^k \, e^{-\lambda w}}{k!}$$

since the number of changes in the inteval [0, w] has a Poisson distribution with mean λw. Because w is a continuous-type random variable, $F(w)$ is equal to pdf of w. Then, for $w > 0$,

$$f(w) = F'(w) = \lambda e^{-\lambda w} - e^{-\lambda w} \sum_{k=1}^{n-1} \left[\frac{k(\lambda w)^{k-1}\lambda}{k!} - \frac{(\lambda w)^k_\lambda}{k!} \right]$$

$$= \lambda e^{-\lambda w} - e^{-\lambda w} \left[\lambda - \frac{\lambda(\lambda w)^{n-1}}{(n-1)!} \right]$$

$$= \frac{\lambda(\lambda w)^{n-1}}{(n-1)!} e^{-\lambda w} = \frac{\lambda(\lambda w)^{n-1}}{\Gamma_{(n)}} e^{-\lambda w}$$

This distribution depends on two parameters n and λ, where $n \geq 1$ and $\lambda > 0$. The function defined above represents a probability distribution function, since

$$\int_0^\infty f(x)\, dx = \int_0^\infty \frac{\lambda(\lambda x)^{n-1}}{\Gamma_{(n)}} e^{-\lambda x}\, dx$$

$$= \frac{\lambda}{\Gamma_{(n)}} \int_0^\infty t^{n-1} e^{-t} \frac{dt}{\lambda}, \text{ put } \lambda x = t$$

$$= \frac{1}{\Gamma_{(n)}} \Gamma_{(n)} = 1$$

If $n = 1$, then $f(x) = \lambda e^{-\lambda x}$, which is an exponential distribution.

2.8.2 Example

Find the moment-generating function of Gamma distribution.

Solution. $M_X(t) = E(e^{tX})$

$$= \int_0^\infty e^{tx} \cdot \frac{\lambda(\lambda x)^{n-1}}{\Gamma_{(n)}} e^{-\lambda x}\, dx = \frac{\lambda}{\Gamma_{(n)}} \int_0^\infty (\lambda x)^{n-1} e^{-(\lambda - t)x}\, dx$$

Put $(\lambda - t)\, x = z$, $dx = \dfrac{dz}{\lambda - t}$,

$$= \frac{\lambda}{\Gamma_{(n)}} \left(\frac{\lambda}{\lambda - t} \right)^{n-1} \int_0^\infty z^{n-1} e^{z} \frac{dz}{\lambda - t}$$

$$= \left(\frac{\lambda}{\lambda - t} \right)^n \cdot \frac{1}{\Gamma_{(n)}} \cdot \int_0^\infty z^{n-1} e^{-z}\, dz$$

$$= \left(\frac{\lambda}{\lambda - t} \right)^n \frac{1}{\Gamma_{(n)}} \Gamma_{(n)} = (1 - t/\lambda)^{-n}$$

2.8.3 Example

Find the mean and variance of Gamma distribution

Solution. If μ'_r represents r^{th} order moment about origin, then

$$\mu'_r = \left[\frac{d^r}{dt^r} M_X(t)\right]_{t=0} = \left[\frac{d^r}{dt^r}\left(1-\frac{t}{\lambda}\right)^{-n}\right]_{t=0}$$

$$= \left[\frac{n(n+1)(n+2)\ldots(n+r-1)}{\lambda^r}(1-t/\lambda)^{-n-r}\right]_{t=0}$$

$$= \frac{n(n+1)\ldots(n+r-1)}{\lambda^r}$$

Hence $\qquad \text{mean} = E(X) = \mu'_1 = \dfrac{n}{\lambda}$

$$E(X^2) = \mu'_2 = \frac{n(n+1)}{\lambda^2}$$

$$\text{Var}(X) = E(X^2) - (E(X))^2$$

$$= \frac{n(n+1)}{\lambda^2} - \frac{n^2}{\lambda^2} = \frac{n}{\lambda^2}$$

If $\lambda = 1$, mean = n = variance.

2.8.4 Theorem

Sum of two independent gamma variables is a gamma variate.

Proof. Let X_1 and X_2 be two independent random variables following gamma distribution with parameters n_1, λ and n_2, λ respectively.

If $M_{X_1}(t)$ and $M_{X_2}(t)$ represents moment generating function of X_1 and X_2 respectively, then by the independence of X_1 and X_2, the moment-generating function of their sum $X_1 + X_2$ is

$$M_{X_1+X_2}(t) = M_{X_1}(t) \,.\, M_{X_2}(t)$$

$$= \left(1-\frac{t}{\lambda}\right)^{-n_1}\left(1-\frac{t}{\lambda}\right)^{-n_2} = \left(1-\frac{t}{\lambda}\right)^{-(n_1+n_2)} \qquad (2.8)$$

(2.8) represents the moment-generating function of a gamma variate with parameters $n_1 + n_2$, λ.

2.8.5 Example

The random variable X has the gamma distribution with density function given by

$$f(x) = \begin{cases} 0 & \text{for } x \le 0 \\ 2e^{-2x} & \text{for } x > 0 \end{cases}$$

Find the probability that X is not smaller than 3.

Solution. $P(X \geq 3) = 2\int_3^{\infty} e^{-2x}\,dx = 2\left[\frac{e^{-2x}}{-2}\right]_3^{\infty} = e^{-6}$

2.8.6 Example

Suppose that an average of 30 customers per hour arrive at a shop in accordance with a Poisson process. That is, if a minute is our unit, then $\lambda = 1/2$. What is the probability that the shopkeeper will wait more than 5 minutes before both of the first two customers arrive?

Solution. If X denotes the waiting time in minutes until the second customer arrives, then X has a gamma distribution with $n = 2$, $\lambda = \frac{1}{2}$

$$P(X > 5) = \int_5^{\infty} \frac{\lambda(\lambda w)^{n-1}}{\Gamma_{(n)}} e^{-\lambda w}\,dw$$

$$= \int_5^{\infty} \frac{\frac{1}{4} w.e^{-w/2}}{\Gamma_{(2)}}\,dw = \int_5^{\infty} \frac{1}{4} we^{-w/2}\,dw \qquad \text{since } \Gamma_{(2)} = 1\ \Gamma_{(1)} = 1$$

$$= \frac{1}{4}\left[-2we^{-w/2} - 4e^{-w/2}\right]_5^{\infty} = \frac{7}{2}e^{-5/2} = 0.287$$

2.8.7 Example

Telephone calls arrive at a switch board at a mean rate of $\lambda = 2$ per minute according to a Poisson process. Let X denote the waiting time in minutes until the fifth call arrives. The probability density function of X, $n = 5$ and $\lambda = 2$ is

$$f(w) = \frac{\lambda^n w^{n-1}}{\Gamma_{(n)}} e^{-\lambda w} \qquad 0 \leq w < \infty$$

$$= \frac{2^5 w^4}{5!} e^{-2w}$$

The mean and variance of X are respectively $E(X) = 5/2$ and Var $(X) = 5/4$.

Note. Let X have a gamma distribution with $\lambda = 1/2$ and $n = r/2$, where r is a positive integer. The pdf of X is

$$f(x) = \begin{cases} \dfrac{w^{\frac{r}{2}-1}}{\Gamma(r/2)\,2^{r/2}} e^{-w/2}, & 0 \leq w < \infty \\ 0 & w < 0 \end{cases}$$

We say that X has a **chi-square distribution** $\chi^2(r)$ **with** r **degrees of freedom**. The mean and variance of this chi-square distribution are

$$E(X) = \frac{n}{\lambda} = \frac{(r/2)}{(1/2)} = r$$

$$\text{Var}(X) = \frac{n}{\lambda^2} = \frac{(r/2)}{1/4} = 2r$$

Hence mean equals the number of degrees of freedom and the variance equals twice the number of degrees of freedom. The moment-generating function is

$$M_X(t) = (1 - 2t)^{-r/2}, \quad t < 1/2$$

2.9 WEIBULL DISTRIBUTION

Suppose that we are considering a component (or entire components assembled into a system) which is put under some sort of "stress". This situation may be a steel beam under a load, a fuse inserted into a circuit, or an electronic device put into service. Suppose that for any such component (or system) a state which we may denote as "failure" can be defined. Thus the steel beam may crack/break, or the fuse may blow or the electronic device may fail to function as the case may be. If such a component is put under stress conditions at some specified time starting with $t = 0$, and observed until it fails, the **time to failure** or **life length** say T may be considered as a continuous random variable with pdf f. Note that the value of T in any case, cannot be predicated from deterministic model and it can be predicted only by means of probabilistic model with T considered as a random variable. We now introduce the following concept used in sequel.

2.9.1 Definition

The reliability of a component (or system) at time t, say $R(t)$ is defined as $R(t) = P(T > t)$, where T is the life length of the component. R is called the **reliability function**. That is, the reliability of a component equals the probability that the component does not fail during the interval $[0, t]$.

For example, if a particular item $R(t_1) = 0.90$ means approximately 90% of such items, used under certain conditions will still be functioning at time t_1.

In terms of the probability density function of T, say f, we say

$$R(t) = \int_t^{\infty} f(s)\,ds$$

In terms of cumulative distribution function of T, say F we have

$$R(t) = 1 - P(T \le t) = 1 - F(t)$$

In addition to the reliability function R, another function plays an important role in describing the failure characteristics of an item is:

2.9.2 Definition

The **failure rate** Z (some times **hazard function**) associated with the random variable T is given by

$$Z(t) = \frac{f(t)}{1 - F(t)} = \frac{f(t)}{R(t)}$$

To interpret $Z(t)$, suppose that the item has survived for t hours and we desire the probability that it will not survive for an additional time Δt units. Applying the definition of conditional probability, we have

$$P\{T \in (t, t + \Delta t)/T > t\} = \frac{P\{T \in (t, t+\Delta t), T > t\}}{P\{T > t\}} = \frac{P\{T \in (t, t+\Delta t)\}}{P(T > t)}$$

$$= \frac{\int_t^{t+\Delta t} f(x)\,dx}{R(t)} \approx \frac{f(t)\,\Delta t}{R(t)} \approx \Delta t\, Z(t)$$

The above argument says that the pdf of T uniquely determines the failure rate Z. The converse is also true. That is, the failure rate Z uniquely determines the pdf of T.

2.9.3 Theorem

If T, the time to failure, is a continuous random variable with pdf f and if $F(0) = 0$, where F is the cdf of T then f may be expressed in terms of Z as

$$f(t) = Z(t) \cdot e^{-\int_0^t Z(s)\,ds} \tag{2.9}$$

Proof. Since $R(t) = 1 - F(t)$, $R'(t) = -F'(t) = -f(t)$

Hence $$Z(t) = \frac{f(t)}{R(t)} = -\frac{R'(t)}{R(t)}$$

Integrating on both sides from 0 to t,

$$\int_0^t z(s)\,ds = -\int_0^t \frac{R'(s)}{R(s)}\,ds = -[\log R(s)]_0^t$$
$$= -\log R(t) + \log R(0)$$
$$= -\log R(t)$$

provided $\log R(0) = 0$ which holds if and only if $R(0) = 1$ [since $F(0) = 0$].

Hence $$R(t) = e^{-\int_0^t z(s)\,ds}$$

Thus $$f(t) = F'(t) = \frac{d}{dt}[1 - R(t)] = Z(t)\, e^{-\int_0^t z(s)\,ds}$$

An interesting relationship exists between the reliability function R and the mean time to failure $E(T)$.

2.9.4 Theorem

If $E(T)$ is finite, then

$$E(T) = \int_0^\infty R(t)\,dt$$

Proof.

Consider $$\int_0^\infty R(t)\,dt = \int_0^\infty \left[\int_t^\infty f(s)\,ds\right] dt = \left[t \int_t^\infty f(s)\,ds\right]_0^\infty + \int_0^\infty t f(t)\,dt$$

by integration by parts, taking $\int_t^\infty f(s)\,ds = u$, $dt = dv$

$$= \left[t \int_t^{\infty} f(s)\, ds \right]_0^{\infty} + E(T)$$

The vanishing of first integral at $t = 0$ is immediate. Since $E(T)$ is finite, the vanishing at $t = \infty$ follows.

Suppose that the life time distribution is exponential. Then by the memory- less property, it follows that the distribution of remaining life for a t year old item is the same as for a new item. Hence $Z(t)$ should be constant i.e.,

$$Z(t) = \frac{f(t)}{R(t)} = \frac{\lambda e^{-\lambda t}}{e^{-\lambda t}} = \lambda$$

Thus the failure rate function for the exponential distribution is constant. The parameter λ is called the **rate of the distribution**.

Consider for $\Delta t > 0$, $P(T \in [t, t + \Delta t]/T > t)$

This represents the probability that the item will fail during the next Δt units, given that it has not failed at time t.

$$P(T \in (t, t + \Delta t)/T > t) = \frac{e^{-\lambda t} - e^{-\lambda(t+\Delta t)}}{e^{-\lambda t}} = 1 - e^{-\lambda \Delta t}$$

Hence the conditional probability is independent of t, depending only on Δt. That is the exponential failure law implies that the probability of failure is independent of past history.

Weibull Failure Law

Let us modify the notion of constant failure rate which leads to the exponential failure law. Suppose the failure rate Z associated with T, the life length of an item has the form

$$Z(t) = (\alpha\, \beta)\, t^{\beta-1} \tag{2.9}$$

where α and β are positive constants. From (2.9) we obtain the probability density function (pdf) of T as

$$f(t) = (\alpha\beta) t^{\beta-1} \cdot e^{-\alpha t^{\beta}}, \quad t > 0,\ \alpha,\ \beta > 0 \tag{2.10}$$

The random variable having pdf given by (2.10) is said to have a **Weibull distribution**

Figure (2.9.1) shows the pdf for $\alpha = 1$, $\beta = 1, 2, 3$. The reliability function R is given by $R(t) = e^{-\alpha t^{\beta}}$ which is a decreasing function of t.

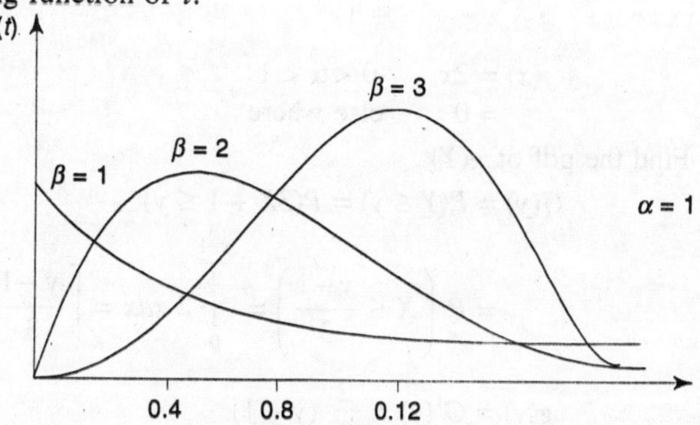

Fig. (2.9.1)

2.9.5 Theorem

The mean and variance of the Weibull distribution are

$$E(T) = \alpha^{-1/\beta}\,\Gamma\left(\frac{1}{\beta}+1\right)$$

$$\text{Var }(T) = \alpha^{-2/\beta}\left\{\Gamma\left(\frac{2}{\beta}+1\right)-\left(\Gamma\left(\frac{1}{\beta}+1\right)\right)^2\right\}$$

Using Weibull distribution, we may obtain an increasing and a decreasing failure rate by simply making the appropriate choice for the parameter β.

2.10 FUNCTIONS OF RANDOM VARIABLE

If we know the probability distribution of a random variable X, then we can also determine the distribution of a function of X say $g(X)$. To do so, it is necessary to express the event that $g(X) \le y$ in terms of X being in some set.

If X is a discrete random variable and $Y = g(X)$, then Y is also discrete random variable. For, suppose the possible values of X are $x_1, x_2, \ldots$, then certainly the possible values of Y are $g(x_1), g(x_2), \ldots$ If X is a continuous random variable with pdf 'f' and Y is a continuous function, then, $Y= g(X)$ is a continuous random variable with pdf 'h'. That is, obtain the cumulative distribution function of Y as $P(Y \le y)$. Then differentiate $P(Y \le y)$ with respect to y in order to get $h(y)$.

2.10.1 Example

Suppose that a random variable X assumes three values –1, 0, 1 with probabilities $\frac{1}{3}, \frac{1}{2}, \frac{1}{6}$ respectively. Let $Y = 3X + 1$. Find the possible values of Y with the same probability as of X.

Solution. If x_1, x_2, x_3 are the possible values of X with probabilities $p(x_1), p(x_2), p(x_3)$ and H is a function such that to each value y there corresponds exactly one value x and the probability distribution is $q(y_i) = P(Y = y_i) = p(x_i)$. Hence the answer follows.

2.10.2 Example

Let X has pdf

$$f(x) = 2x \qquad 0 < x < 1$$
$$= 0 \qquad \text{else where}$$

Let $g(Y)= 3X + 1$. Find the pdf of $g(Y)$

Solution.

$$G(y) = P(Y \le y) = P(3X + 1 \le y)$$

$$= P\left(X \le \frac{y-1}{3}\right) = \int_0^{\frac{y-1}{3}} 2\,x dx = \left[\frac{y-1}{3}\right]^2.$$

Thus

$$g(y) = G^1(y) = \frac{2}{9}(y-1)$$

2.10.3 Example

Let X be uniformly distributed over (0, 1). The distribution of the random variable Y, defined by $Y = X^n$ is

$$\begin{aligned} F(y) &= P\{Y \le y\}, \qquad 0 \le y \le 1 \\ &= P\{X^n \le y\} = P(X \le y^{1/n}) \\ &= F(y^{1/n}) = y^{1/n} \end{aligned}$$

Then density function of Y is

$$f(y) = \begin{cases} \dfrac{1}{n} y^{-\left(\frac{n-1}{n}\right)} & 0 \le y \le 1 \\ 0 & \text{otherwise} \end{cases}$$

2.10.4 Example

If X is a continuous random variable with probability density function f, then the distribution of $Y = X^2$ is

$$\begin{aligned} F(y) &= P(Y \le y) = P(X^2 \le y) \\ &= P\left(-\sqrt{y} \le X \le \sqrt{y}\right) = F\left(\sqrt{y}\right) - F\left(-\sqrt{y}\right) \end{aligned}$$

Hence

$$f(y) = F'(y) = \frac{1}{2\sqrt{y}}\left[f\left(\sqrt{y}\right) + f\left(-\sqrt{y}\right)\right]$$

2.10.5 Example

Suppose that a continuous random variable has pdf

$$\begin{aligned} f(x) &= 2x \qquad 0 < x < 1 \\ &= 0 \qquad \text{otherwise} \end{aligned}$$

Find the pdf of $Y = e^{-X}$.

Solution.

$$\begin{aligned} G(y) &= P(Y \le y) = P(e^{-X} \le y) \\ &= P(X \ge -\log y) = \int_{-\log y}^{1} 2x\,dx = 1 - (-\log y)^2 \end{aligned}$$

Hence $g(y) = G^1(y) = -\dfrac{2\log y}{y}$. Since $f(x) > 0$, for $0 < x < 1$, we find that $g(y) > 0$, for $\dfrac{1}{e} < y < 1$.

The method used in the above examples may be generalized as follows.

2.10.6 Theorem

Let X be a continuous random variable with pdf f, where $f(x) > 0$ for $a < x < b$. Suppose that $y = H(x)$ is strictly monotone (increasing or decreasing) function of x. Assume this function is differentiable (and hence continuous) for all x. Then the random variable Y defined as $Y = H(X)$ has a pdf 'h' given by

$$h(y) = f(x) \left| \frac{dx}{dy} \right|$$

where x is expressed in terms of y. If 'H' is increasing, then 'h' is non-zero for those values of y satisfying $H(a) < y < H(b)$. If H is decreasing, then 'h' is non-zero for those values of y satisfying $H(b) < y < H(a)$.

Proof. (a) Assume g is strictly increasing function. Hence

$$G(y) = P(Y \leq y) = P(H(X) \leq y)$$
$$= P(X \leq H^{-1}(y)) = F(H^{-1}(y))$$

Differentiating $G(y)$ with respect to y we obtain

$$\frac{d}{dy} G(y) = \frac{dG(y)}{dx} \cdot \frac{dx}{dy}, \qquad x = H^{-1}(y)$$

$$\Rightarrow \quad h(y) = G'(y) = \frac{dF(x)}{dx} \cdot \frac{dx}{dy}, \qquad F(x) = F(H^{-1}(y)) = G(y)$$

$$= f(x) \frac{dx}{dy} \tag{2.11}$$

(b) Assume H is a decreasing function. Then

$$G(y) = P(Y \leq y) = P(H(X) \leq y) = P(X \geq H^{-1}(y))$$
$$= 1 - P(X < H^{-1}(y)) = 1 - F(H^{-1}(y))$$

$$h(y) = G'(y) = \frac{dG(y)}{dx} \cdot \frac{dx}{dy} = \frac{d}{dx}[1 - F(x)]. \frac{dx}{dy}$$

$$= -f(x) \frac{dx}{dy} \tag{2.12}$$

Combining (2.11) & (2.12) we get,

$$h(y) = f(x) \left| \frac{dx}{dy} \right|$$

Applying Theorem 2.10.6 to Example 2.10.2, we have

$$f(x) = 2x,\ 0 < x < 1,\ Y = 3X + 1,\ x = \frac{y-1}{3},$$

$\left| \frac{dx}{dy} \right| = \frac{1}{3}$. Thus pdf of Y is

$$h(y) = f(x) \left| \frac{dx}{dy} \right| = 2\left(\frac{y-1}{3}\right) \cdot \frac{1}{3} = \frac{2}{9}(y-1)$$

Note. If $y = H(x)$ is not a monotone function of x we cannot apply the above method directly. In this case we use the general method.

2.11 CHEBYSHEV'S INEQUALITY

2.11.1 Theorem

If X is a random variable with mean μ and variance σ_x^2, then for any positive number k,

$$P\{|X - \mu| \geq k\sigma_x\} \leq \frac{1}{k^2}$$

or

$$P\{|X - \mu| < k\sigma_x\} \geq 1 - \frac{1}{k^2}.$$

Proof. $$\sigma_x^2 = E(X - E(X))^2 = E(X - \mu)^2 = \int_{-\infty}^{\infty} (x-\mu)^2 f(x)\, dx$$

$$= \int_{-\infty}^{\mu-k\sigma_x} (x-\mu)^2 f(x)\, dx + \int_{\mu-k\sigma_x}^{\mu+k\sigma_x} (x-\mu)^2 f(x)\, dx + \int_{\mu+k\sigma_x}^{\infty} (x-\mu)^2 f(x)\, dx$$

Since $\int_{\mu-k\sigma_x}^{\mu+k\sigma_x} (x-\mu)^2 f(x)\, dx \geq 0$, $\sigma_x^2 \geq \int_{-\infty}^{\mu-k\sigma_x} (x-\mu)^2 f(x)\, dx + \int_{\mu+k\sigma_x}^{\infty} (x-\mu)^2 f(x)\, dx$

In the above relation, for the first integral, the upper bound for x is given by

$$x \leq \mu - k\sigma_x \Rightarrow \mu - x \geq k\sigma_x$$

For the second integral, the lower bound for x is

$x \geq \mu + k\sigma_x \Rightarrow x - \mu \geq k\sigma_x$. Hence we have

$(\mu - x)^2 = (x - \mu)^2 \geq k^2\sigma_x^2$. Therefore

$$\sigma_x^2 \geq k^2\sigma_x^2 \left[\int_{-\infty}^{\mu-k\sigma_x} f(x)\, dx + \int_{\mu+k\sigma_x}^{\infty} f(x)\, dx\right]$$

$$= k^2\sigma_x^2\, [P(-\infty < X \leq \mu - k\sigma_x) + P(\mu + k\sigma_x \leq X < \infty]$$

$$\geq k^2\sigma_x^2\, [P(X - \mu \leq -k\sigma_x) + P(X - \mu \geq k\sigma_x]$$

$$\geq k^2\sigma_x^2\, [P(|X - \mu| \geq k\sigma_x]$$

$$1 \geq k^2\, [P(|X - \mu| \geq k\sigma_x]$$

$$\Rightarrow \quad P\{|X - \mu| \geq k\sigma_x\} \leq \frac{1}{k^2}, \ k \text{ is a constant}$$

We know that

$$P\{|X - \mu| \geq k\sigma_x\} + P\{|X - \mu| < k\sigma_x\} = 1$$

$$\Rightarrow \quad P\{|X - \mu| < k\sigma_x\} = 1 - P\{|X - \mu| \geq k\sigma_x \geq 1 - \frac{1}{k^2}.$$

2.11.2 Example

If X is the number scored in a throw of a fair die, show that the Chebyshev's inequality gives

$$P\{|X - \mu| > 2.5\} < 0.47$$

where μ is the mean of X, while the actual probability is zero.

Solution. Let X be a random variable which takes the values 1, 2, 3, 4, 5, 6 each with probability $\frac{1}{6}$.

$$\mu = E(X) = \Sigma x p(x) = \frac{1}{6} + \frac{2}{6} + \frac{3}{6} + \frac{4}{6} + \frac{5}{6} + \frac{6}{6} = \frac{7}{2}.$$

$$E(X^2) = \frac{1}{6}\,[1^2 + 2^2 + 3^2 + 4^2 + 5^2 + 6^2] = \frac{1}{6}\left[\frac{6 \times 7 \times 13}{6}\right] = \frac{91}{6}$$

$$\text{Var}(X) = E(X^2) - (E(X))^2 = \frac{91}{6} - \left(\frac{7}{2}\right)^2 = 2.9167$$

we know

$$P\{|X - \mu| > \varepsilon\} < \frac{\sigma_x^2}{\varepsilon^2},\ \varepsilon > 0 \text{ and } \varepsilon = k\sigma_x$$

If $\varepsilon = 2.5$

$$P\{|X - 3.5| > 2.5\} < \frac{2.9167}{6.25} = 0.47$$

The actual probability is given by

$$P\{|X - 3.5| > 2.5\} = P\{1 > X > 6\} = 0$$

since when a fair die is thrown there is no probability that a number scored is less than 1 or greater than 6.

2.11.3 Example

A random variable X has the density function e^{-x} for $x \geq 0$. Show that Chebyshev's inequality gives $P\{|X - 1| > 2\} < \frac{1}{4}$ and also show that the actual probability is e^{-3}.

Solution.

$$\text{Mean} = E(X) = \int_{-\infty}^{\infty} x\,f(x)\,dx = \int_{0}^{\infty} x e^{-x}\,dx$$

$$= \left[-x\,e^{-x}\right]_0^{\infty} + \int_0^{\infty} e^{-x}\,dx = 0 + \left[-e^{-x}\right]_0^{\infty} = 1$$

$$E(X^2) = \int_{-\infty}^{\infty} x^2 f(x)\,dx = \int_{-\infty}^{\infty} x^2\,e^{-x}\,dx$$

$$= [-x^2e^{-x}]_0 + \int_0^{\infty} 2x\, e^{-x}\, dx = 0 + 2\left[\int_0^{\infty} xe^{-x}\, dx\right] = 2$$

$$\text{Var}(X) = E(X^2) - (E(X))^2 = 2 - 1 = 1$$

By Chebyshev's inequality $P\{|X - \mu| > \varepsilon\} < \dfrac{\text{Var}(x)}{\varepsilon^2}$,

If $\varepsilon = 2$, we have

$$P\{|X - \mu| > 2\} < \frac{1}{4}$$

Actual probability:

$$P\{|X - 1| \le 2\} = P\{-1 \le x \le 3\} = \int_{-1}^{3} f(x)\, dx = \int_0^3 e^{-x} dx = -e^{-3} + 1$$

So $$P\{|X - 1| > 2\} = 1 - P\{|X - 1| \le 2\}$$
$$= 1 - [-e^{-3} + 1] = e^{-3}$$

Schwartz Inequality.

2.11.4 Theorem

If X and Y are two discrete random variables taking real values, with finite σ_X^2 and σ_Y^2, then

$$E(XY)^2 \le E(X^2)\, E(Y^2)$$

Proof: If X and Y are two discrete random variables taking real values, with finite σ_X^2 and σ_Y^2. Define a real valued function $H(\lambda)$, for any real λ as

$$H(\lambda) = (X + \lambda Y)^2$$

Taking expectations on both sides, we get

$$E(H(\lambda)) = E(X + \lambda Y)^2 = E(X^2 + 2XY\lambda + \lambda^2 Y^2)$$
$$= E(X^2) + 2\lambda E(XY) + \lambda^2 E(Y^2)$$

since $H(\lambda) \ge 0$, for any λ, $E(H\lambda) \ge 0$ for all real λ. The R.H.S. expression is a quadratic expression in λ and is of the form

$$\phi(\lambda) = A\lambda^2 + B\lambda + C \ge 0$$

This implies that the graph of the function $\phi(\lambda)$ touches λ–axes at only one point or not at all. This is equivalent to saying the discriminant of the function $\phi(\lambda)$ viz $B^2 - 4AC \le 0$, since the condition $B^2 - 4AC > 0$ implies that $\phi(\lambda)$ has two distinct real roots. Hence

$$B^2 - 4AC \le 0$$

$\Rightarrow$ $$4[E(XY)]^2 - 4E(X^2)E(Y^2) \le 0$$

$\Rightarrow$ $$E(XY)^2 \le E(X^2)\, E(Y^2)$$

2.11.5 Example

Two unbiased dice are thrown. Let X be a random variable that represents the numbers showing up. Prove by Chebyshev's inequality

$$P\{|X-7| \geq 2\} \leq \frac{35}{24}$$

Compare this result with the actual probability.

Solution. Let X be the random variable which represents the sum of numbers of points on throwing two unbiased dice.

If the point shown in each die is equal to one, the minimum sum of numbers is equal to two. If both dice show six, then the sum of numbers at the maximum is 12. Thus the variables takes values from 0 to 12. From the sample space of throwing two dice, we get

x	2	3	4	5	6	7	8	9	10	11	12
$p(x)$	$\frac{1}{36}$	$\frac{2}{36}$	$\frac{3}{36}$	$\frac{4}{36}$	$\frac{5}{36}$	$\frac{6}{36}$	$\frac{5}{36}$	$\frac{4}{36}$	$\frac{3}{36}$	$\frac{2}{36}$	$\frac{1}{36}$

$$E(X) = \sum_{i=2}^{12} x_i \, p(x_i)$$

$$= \left[\left(2\times\frac{1}{36}\right)+\left(3\times\frac{2}{36}\right)+\left(4\times\frac{3}{36}\right)+\left(5\times\frac{4}{36}\right)+\left(6\times\frac{5}{36}\right)+\left(7\times\frac{6}{36}\right)\right.$$

$$\left.+\left(8\times\frac{5}{36}\right)+\left(9\times\frac{4}{36}\right)+\left(10\times\frac{3}{36}\right)+\left(11\times\frac{2}{36}\right)+\left(12\times\frac{1}{36}\right)\right] = 7.$$

$$E(X^2) = \sum_{i=2}^{12} x_i^2 \, \phi(x_i)$$

$$= \frac{4}{36}+\frac{18}{36}+\frac{48}{36}+\frac{100}{36}+\frac{180}{36}+\frac{294}{36}+\frac{320}{36}$$

$$+\frac{324}{36}+\frac{300}{36}+\frac{242}{36}+\frac{144}{36} = \frac{329}{6}$$

$$\text{Var}(X) = \frac{329}{6} - 49 = \frac{35}{6}$$

By Chebyshev's inequality, $P\{|X-\mu| > \varepsilon\} \leq \frac{\text{Var}(X)}{\varepsilon^2}$, taking $\varepsilon = 2$, we have

$$P\{|X-7| \geq 2\} \leq \frac{35/6}{4} = \frac{35}{24}$$

The actual probability is given by

$$P\{|X-7| \geq 2\} = 1 - P\{|X-7| < 2\}$$

$$= 1 - P\{5 < X < 9\}$$

$$= 1 - \{P(X=6) + P(X=7) + P(X=8)\}$$

$$= 1 - \left\{\frac{5}{36}+\frac{6}{36}+\frac{5}{36}\right\} = 1 - \frac{16}{36} = \frac{20}{36} = \frac{5}{9}.$$

2.12 NORMAL DISTRIBUTION

The normal distribution is one of the most important distributions of continuous random variable. A large number of random variables occurring in many applications have a distribution closely resembling the normal distribution.

2.12.1 Definition

A random variable X, assuming all real values $(-\infty < x < \infty)$ has a normal (or Gaussian) distribution if its probability density function is of the form

$$f(x) = \frac{1}{\sigma\sqrt{2\pi}} e^{-\frac{(x-\mu)^2}{2\sigma^2}} \qquad -\infty < x < \infty \tag{2.13}$$

where μ and σ are called parameters.

The density function is a bell-shaped curve that is symmetric about μ. The values μ and σ^2 represent the average value (mean) and the variance of X.

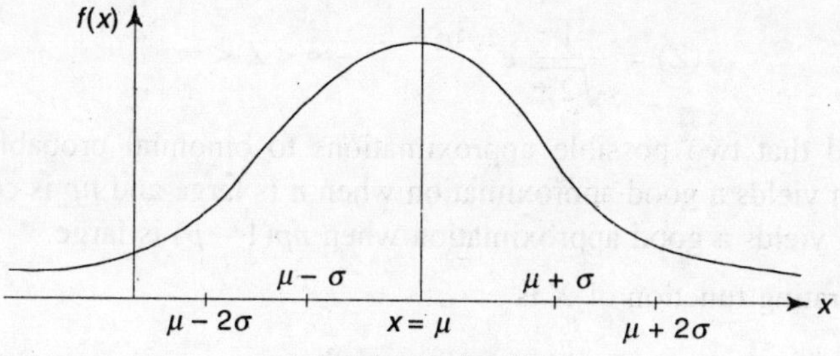

normal density function

First we verify $f(x)$ is a density function. That is, to prove $\int_{-\infty}^{\infty} f(x)\,dx = 1$. For, let $I = \int_{-\infty}^{\infty} f(x)\,dx$.

Then
$$I = \frac{1}{\sigma\sqrt{2\pi}} \int_{-\infty}^{\infty} e^{-\frac{(x-\mu)^2}{2\sigma^2}}\,dx$$

Let
$$\frac{x-\mu}{\sigma} = z\text{. Then}$$

$$I = \frac{1}{\sqrt{2\pi}} \int_{-\infty}^{\infty} e^{-z^2/2}\,dz$$

and
$$I^2 = \frac{1}{2\pi} \int_{-\infty}^{\infty} e^{-z^2/2}\,dz \int_{-\infty}^{\infty} e^{-w^2/2}\,dw$$

$$= \frac{1}{2\pi} \int_{-\infty}^{\infty} \int_{-\infty}^{\infty} e^{-(z^2+w^2)/2}\,dz\,dw$$

We now evaluate the double integral by means of a change of variables to polar coordinates. That is, let

$$w = r\cos\alpha,\ z = r\sin\alpha,\ dz\,dw = r\,d\alpha\,dr$$

Thus
$$I^2 = \frac{1}{2\pi}\int_0^{2\pi}\int_0^{\infty} r\,e^{-r^2/2}\,d\alpha\,dr$$

$$= \frac{1}{2\pi}\int_0^{2\pi}\left[-e^{-r^2/2}\right]_0^{\infty} d\alpha = \frac{1}{2\pi}\left[\int_0^{2\pi} d\alpha\right] = 1.$$

Hence $I = 1$. Thus (2.13) is a pdf of the random variable X which is normally distributed. We use $N(\mu, \sigma^2)$ to mean the normal distribution with parameter μ and σ^2.

If X is normally distributed with parameters μ and σ^2, then $z = \dfrac{x-\mu}{\sigma}$ is normally distributed with parameters 0 and 1. Such a random variable Z is said to have standard normal distribution with pdf ϕ as

$$\phi(Z) = \frac{1}{\sqrt{2\pi}}\,e^{-Z^2/2} \qquad -\infty < Z < \infty$$

It should be noted that two possible approximations to binomial probabilities are: the Poisson approximation, which yields a good approximation when n is large and np is constant and the normal approximation which yields a good approximation when $np(1-p)$ is large.

The moment-generating function of X is

$$M_X(t) = \int_{-\infty}^{\infty}\frac{e^{tx}}{\sigma\sqrt{2\pi}}\,e^{-\frac{(x-\mu)^2}{2\sigma^2}}\,dx$$

$$= \int_{-\infty}^{\infty}\frac{1}{\sigma\sqrt{2\pi}}\,e^{-\frac{1}{2\sigma^2}(x^2-2(\mu+\sigma^2 t)x+\mu^2)}\,dx$$

To evaluate this integral, write

$$x^2 - 2(\mu+\sigma^2 t)x + \mu^2 = [x-(\mu+\sigma^2 t)]^2 - 2\mu\sigma^2 t - \sigma^4 t^2$$

Thus

$$M_X(t) = e^{\left(\frac{2\mu\sigma^2 t+\sigma^4 t^2}{2\sigma^2}\right)}\int_{-\infty}^{\infty}\frac{1}{\sigma\sqrt{2\pi}}\,e^{-\frac{1}{2\sigma^2}(x-(\mu+\sigma^2 t))^2}\,dx$$

Note that the integrand in the last integral is like the pdf of a normal distribution with μ replaced by $\mu + \sigma^2 t$. However, the normal pdf integrates to one for all real μ, is particular to $\mu + \sigma^2 t$. Hence

$$M_X(t) = e^{\left(\mu t+\frac{\sigma^2 t^2}{2}\right)}$$

Taking logarithm on both sides, we get

$$\log M_X(t) = \mu t + \frac{\sigma^2 t^2}{2}$$

Differentiating with respect to t, on both sides

$$\frac{M'_X(t)}{M_X(t)} = \mu + \sigma^2 t$$

$$E(X) = \left[\frac{M'_X(t)}{M_X(t)}\right]_{t=0} = \mu$$

and $$\frac{d}{dt}\left[\frac{M'_X(t)}{M_X(t)}\right] = \frac{M_X(t)\,M''_x(t) - (M'_x(t))^2}{(M'_x(t))^2} = \sigma^2$$

Hence $$\text{Var}(X) = \left[\frac{d}{dt}\left[\frac{M'_x(t)}{M_X(t)}\right]\right]_{t=0} = \sigma^2.$$

Hence the parameters μ and σ^2 are the mean and variance of X.

2.12.2 Example

If the pdf of X is

$$f(x) = \frac{1}{\sqrt{32\pi}} e^{-\frac{(x+7)^2}{32}}, \quad -\infty < x < \infty$$

then Y is $N(-7, 16)$. That is X is a normal distribution with mean $\mu = -7$, Var $(X) = 16$ and $M_X(t) = e^{(-7t + 8t^2)}$.

2.12.3 Example

If X is $N(\mu, \sigma^2)$, then $Z = \left(\frac{X-\mu}{\sigma}\right)$ is $N(0, 1)$

Solution. $$P(Z \le z) = P\left(\frac{X-\mu}{\sigma}\right) = P(X \le z\sigma + \mu)$$

$$= \int_{-\infty}^{z\sigma+\mu} \frac{1}{\sigma\sqrt{2\pi}} e^{-\frac{(x-\mu)^2}{2\sigma^2}} dx$$

$$= \int_{-\infty}^{z} \frac{1}{\sqrt{2\pi}} e^{-w^2/2} dw, \quad w = \frac{x-\mu}{\sigma}$$

$$= \phi(z)$$

Hence Z is $N(0, 1)$. Hence we write, if X is $N(\mu, \sigma^2)$, then

$$P(a \le X \le b) = P\left(\frac{a-\mu}{\sigma} \le \frac{X-\mu}{\sigma} \le \frac{b-\mu}{\sigma}\right) = \phi\left(\frac{b-\mu}{\sigma}\right) - \phi\left(\frac{a-\mu}{\sigma}\right)$$

since $\frac{X-\mu}{\sigma}$ is $N(0, 1)$

2.12.4 Example

If X is $N(3, 16)$, then

$$P(4 \le X \le 8) = P\left(\frac{4-3}{4} \le \frac{X-3}{4} \le \frac{8-3}{4}\right)$$
$$= \phi(1.25) - \phi(0.25)$$
$$= 0.8944 - 0.5987 = 0.2957$$

(using normal table)

$$P(-2 \le X \le 1) = P\left(\frac{-2-3}{4} \le Z \le \frac{1-3}{4}\right)$$
$$= \phi(-0.5) - \phi(-1.25) = 0.2029$$

2.12.5 Example

Let X be the number of times that a fair coin, flipped 40 times, land heads. Find $P(X = 20)$. Use normal approximation and compare it to the exact solution.

Solution. $P(X = 20) = P(19.5 < X < 20.5)$

$$= P\left(\frac{19.5-20}{\sqrt{10}} < \frac{X-20}{\sqrt{10}} < \frac{20.5-20}{\sqrt{10}}\right)$$
$$= \phi(.16) - \phi(-.16) = .1272$$

The exact solution is

$$P(X = 20) = {}^{40}C_{20}\left(\frac{1}{2}\right)^{40} = .1268$$

2.13 SOLVED PROBLEMS

1. The probability density function of a random variable X is

$$f(x) = \begin{cases} x & 0 < x \le 1 \\ 2-x & 1 \le x < 2 \\ 0 & x \ge 2 \end{cases}$$

Compute the cumulative distribution function $F(x)$.

Solution. $F(x) = \int_{-\infty}^{x} f(x)\,dx = \int_{-\infty}^{x} o dx = 0$ for any $x \le 0$

$$F(x) = \int_{-\infty}^{0} f(x)\,dx + \int_{0}^{x} f(x)\,dx = \int_{-\infty}^{0} o dx + \int_{0}^{x} x dx$$
$$= x^2/2, \quad 0 < x < 1$$

$$F(x) = \int_{-\infty}^{0} f(x)\,dx + \int_{0}^{1} f(x)\,dx + \int_{1}^{x} f(x)\,dx$$
$$= \int_{-\infty}^{0} o dx + \int_{0}^{1} x dx + \int_{1}^{x} (2-x)\,dx = 2x - \frac{x^2}{2} - 1 \qquad 1 < x < 2$$

$$F(x) = \int_{-\infty}^{0} f(x)\,dx + \int_{0}^{1} f(x)\,dx + \int_{1}^{2} f(x)\,dx + \int_{2}^{\infty} f(x)\,dx$$

$$= \int_{-\infty}^{0} o\,dx + \int_{0}^{1} x\,dx + \int_{1}^{2} (2-x)\,dx + \int_{2}^{\infty} o\,dx = \frac{1}{2} + 2 - \frac{3}{2} = 1 \quad x \geq 2.$$

2. A continuous random variable X has a distribution function given by

$$f(x) = \begin{cases} a\,(1+x^2) & 2 \leq x \leq 5 \\ 0 & \text{otherwise} \end{cases}$$

Find $P(X < 4)$.

Solution. We know $\int_{-\infty}^{\infty} f(x)\,dx = 1$

$$\Rightarrow \quad \int_{2}^{5} a(1+x^2)\,dx = 1$$

$$\Rightarrow \quad a\left[x + \frac{x^3}{3}\right]_2^5 = 1$$

$$\Rightarrow \quad \left[5 + \frac{5^3}{3} - 2 - \frac{2^3}{3}\right] = 1$$

$$\Rightarrow \quad a = \frac{1}{42}$$

$$P(X < 4) = \int_{-\infty}^{4} f(x)\,dx = \int_{2}^{4} \frac{1}{42}(1+x^2)\,dx$$

$$= \frac{1}{42}\left[x + \frac{x^3}{3}\right]_2^4 = \frac{62}{126}.$$

3. Verify whether the following is a distribution function or not

$$F(x) = \begin{cases} 0 & x < -k \\ \dfrac{x}{2k} + 2 & -k \leq x \leq k \\ 1 & x > k \end{cases}$$

Solution. $F(x)$ is a distribution function only if $f(x)$ is a density function.

$$f(x) = \frac{dF(x)}{dx} = \frac{1}{2k} \qquad -k \leq x \leq k.$$

$$\int_{-\infty}^{\infty} f(x)\,dx = \int_{-k}^{k} \frac{1}{2k}\,dx = \frac{1}{2k}[x]_{-k}^{k} = \frac{1}{2k}[k+k] = 1$$

Therefore $F(x)$ is a distribution function.

4. A continuous random variable has the probability density function defined by

$$f(x) = \begin{cases} kxe^{-\lambda x} & x \geq 0, \lambda > 0 \\ 0 & \text{otherwise} \end{cases}$$

Determine the constant k and find mean and variance.

Solution. $\int_{-\infty}^{\infty} f(x)\, dx = 1$

$$\Rightarrow \quad \int_{0}^{\infty} kxe^{-\lambda x}\, dx = 1$$

$$\Rightarrow \quad k\left[x\left(\frac{e^{-\lambda x}}{-\lambda}\right) - \left(\frac{e^{-\lambda x}}{\lambda^2}\right)\right]_0^{\infty} = 1$$

$$\Rightarrow \quad k\left[\frac{1}{\lambda^2}\right] = 1 \quad \Rightarrow \quad k = \lambda^2.$$

Mean:

$$E(x) = \int_{-\infty}^{\infty} x\, f(x)\, dx = \int_{0}^{\infty} kx^2 e^{-\lambda x}\, dx$$

$$= \lambda^2\left[x^2\left(\frac{e^{-\lambda x}}{-\lambda}\right)_0^{\infty} - 2x\left(\frac{e^{-\lambda x}}{\lambda^2}\right)_0^{\infty} + 2\left(\frac{e^{-\lambda x}}{-\lambda^3}\right)_0^{\infty}\right] = \lambda^2 \times \frac{2}{\lambda^3} = 2/\lambda$$

Variance:

$$E(x^2) = \int_{-\infty}^{\infty} x^2\, f(x)\, dx$$

$$= \int_{0}^{\infty} x^2\, kx\, e^{-\lambda x}\, dx = \lambda^2 \int_{0}^{\infty} x^3\, e^{-\lambda x}\, dx$$

$$= \lambda^2\left[x^3\left(\frac{e^{-\lambda x}}{-\lambda}\right)_0^{\infty} - 3x^2\left(\frac{e^{-\lambda x}}{\lambda^2}\right)_0^{\infty} + 6x\left(\frac{e^{-\lambda x}}{-\lambda^3}\right)_0^{\infty} - 6\left(\frac{e^{-\lambda x}}{\lambda^4}\right)_0^{\infty}\right]$$

$$= \lambda^2 \times \frac{6}{\lambda^4} = 6/\lambda^2.$$

5. A random variable X has the following probability function

X:	−2	−1	0	1	2	3
$P(X)$:	0.1	k	0.2	$2k$	0.3	$3k$

(a) Evaluate k

(b) Evaluate $P(X < 2)$ and $P(-2 < X \leq 2)$

(c) Evaluate the mean of X.

Solution.

(a) $\sum_{i=0}^{n} P(x) = 1$

$\Rightarrow \quad 0.1 + k + 0.2 + 2k + 0.3 + 3k = 1$

$\Rightarrow \quad 6k + 0.6 = 1$

$\Rightarrow \quad k = 0.066$

(b) $P(x < 2) = P(x = -2) + P(x = -1) + P(x = 0) + P(x = 1)$

$= 0.1 + 0.066 + 0.2 + 0.132$

$= 0.498$

$P(-2 < X \le 2) = P(X = -1) + P(X = 0) + P(X = 1) + P(X = 2)$

$= 0.066 + 0.2 + 0.132 + 0.3 = 0.698$

(c) Mean of $X = E(X) = \sum_{i=1}^{n} x_i\, P(x_i)$

$= (-2)(0.1) + (-1)(0.066) + (0)(0.2) + (1)(2)(0.066)$

$+ (2)(0.3) + (3)(3)(0.066) = 1.06.$

6. Show that if X is a discrete random variable and 'a' is a constant

(i) $E(aX) = a\, E(X)$

(ii) $\text{Var}(aX) = a^2\, \text{Var}(X)$

Solution.

(i) $E(aX) = \sum_x ax\, P(x) = a\, \Sigma x P(x) = a\, E(X)$

(ii) $Y = aX,\ E(Y) = a\, E(X)$

$Y - E(Y) = a(X - E(X))$

Squaring and taking expectation on both sides we get

$$E[Y - E(Y)]^2 = a^2\, E(X - E(X))^2$$

$$\Rightarrow \quad V(Y) = a^2 V(X)$$

$$\Rightarrow \quad V(aX) = a^2\, V(X)$$

7. Prove that $\text{Var}(X) = E(X^2) - [E(X)]^2$ where X is a random variable

Solution. $\text{Var}(X) = E(X - E(X))^2$

$= E[X^2 - 2XE(X) + (E(X))^2]$

$= E(X^2) - 2[E(X)]^2 + [E(X)]^2 = E(X)^2 - [E(X)]^2.$

8. For what values of α is the function

$$f(x) = \alpha\, x^{-2}\, u(x - \alpha)$$

(a) Probability density function, where $u(x)$ is the unit step function

(b) Find the corresponding cumulative distribution function

(c) Compute $P(X \ge 10)$

Solution.

(a) As $f(x)$ is a probability density function, $\int_{-\infty}^{\infty} f(x)\, dx = 1$

We know

$$u(x) = \begin{cases} 1 & \text{if } x > 0 \\ 0 & \text{if } x < 0 \end{cases}$$

Therefore $\int_{-\infty}^{\infty} f(x)\, dx = \int_{-\infty}^{\infty} \alpha\, x^{-2}\, u(x - \alpha)\, dx = 1$

$\Rightarrow \quad \int_{\alpha}^{\infty} \alpha\, x^{-2}\, dx = 1 \qquad$ since $u(x - \alpha) = 1,\ x > \alpha$

$$\Rightarrow \qquad \alpha\left[\frac{x^{-1}}{-1}\right]_{\alpha}^{\infty} = 1$$

$$\Rightarrow \qquad \alpha\left[\frac{1}{\alpha}-\frac{1}{\infty}\right] = 1$$

$$\Rightarrow \qquad \alpha\cdot\frac{1}{\alpha} = 1$$

$\Rightarrow$ α can take value between $-\infty$ to ∞.

(b) The cumulative distribution function is

$$F(x) = \int_{-\infty}^{x} f(x)\,dx = \int_{-\infty}^{x} \alpha\, x^{-2}\, u(x-\alpha)\,dx$$

$$= \int_{\alpha}^{x} \alpha\, x^{-2}\,dx = \alpha\left[-\frac{1}{x}\right]_{\alpha}^{x} = 1 - \alpha/x$$

(c) $P(X \geq 10) = 1 - P(x < 10)$

$= 1 - [1 - \alpha/10] = \alpha/10.$

9. A continuous random variable X has a pdf $f(x) = 3x^2$, $0 \leq x \leq 1$. Find 'a' and 'b' such that

(i) $P[X \leq a] = P[X > a]$

(ii) $P[X > b] = 0.05$

Solution.

(i) Since $P(X \leq a) = P(X > a)$, each must be equal to $\frac{1}{2}$, because $P(X \leq a) + P(X > a) = 1$.

Therefore

$$P(X \leq a) = \frac{1}{2}$$

$$\Rightarrow \qquad \int_0^a f(x)\,dx = \frac{1}{2}$$

$$\Rightarrow \qquad \int_0^a 3x^2\,dx = \frac{1}{2}$$

$$\Rightarrow \qquad 3\left[\frac{x^3}{3}\right]_0^a = \frac{1}{2}$$

$$\Rightarrow \qquad (x^3)_0^a = \frac{1}{2} \quad \Rightarrow \quad a^3 = \frac{1}{2} \quad \Rightarrow \quad a = \frac{1}{2^{1/3}}.$$

(ii) $P(X > b) = 0.05$

$$\Rightarrow \qquad \int_b^1 f(x)\,dx = 0.05$$

$$\Rightarrow \quad \left[3\cdot\frac{x^3}{3}\right]_b^1 = 0.05$$

$$\Rightarrow \quad 1 - b^3 = 0.05$$

$$\Rightarrow \quad b^3 = 0.95$$

$$\Rightarrow \quad b = (0.95)^{1/3}.$$

10. Find the mean value of the random variable whose probability density function is given by

$$f(x) = \frac{3}{5}\,10^{-5}\,x\,(100 - x), \qquad 0 \le x \le 100$$

Solution.
$$E(X) = \int_{-\infty}^{\infty} xf(x)\,dx = \int_0^{100} \frac{3}{5}\,10^{-5}\,x^2\,(100 - x)\,dx$$

$$= \frac{3}{5}\,10^{-5}\int_0^{100}(100\,x^2 - x^3)\,dx$$

$$= \frac{3\times10^{-5}}{5}\left[\frac{100\,x^3}{3} - \frac{x^4}{4}\right]_0^{100} = 50.$$

11. Find the expected value of the number on a die when thrown.

Solution. Let X be the random variable which represents the number on a die when thrown i.e. X takes the value 1, 2, 3, 4, 5, 6, with equal probability $\frac{1}{6}$. Therefore

$$E(x) = \sum_{i=1}^{6} x_i p(x_i)$$

$$= 1\left(\frac{1}{6}\right)+2\left(\frac{1}{6}\right)+3\left(\frac{1}{6}\right)+4\left(\frac{1}{6}\right)+5\left(\frac{1}{6}\right)+6\left(\frac{1}{6}\right)$$

$$= \frac{1}{6}[1 + 2 + 3 + 4 + 5 + 6] = \frac{1}{6}\left[\frac{6\times7}{2}\right] = \frac{7}{2}.$$

12. A fair coin is tossed three times. Let X be the number of tails appearing. Find the probability distribution of X. Calcualte $E(X)$.

Solution. When a coin is tossed three times the sample space S = {HHH, HHT, HTH, THH, HTT, THT, TTH, TTT}. Let X be the number of tails appearing. Then the probability distribution is given by

$$P(X = 0) = \frac{1}{8}; \qquad P(X = 2) = \frac{3}{8}$$

$$P(X = 1) = \frac{3}{8}; \qquad P(X = 3) = \frac{1}{8}$$

We show table as:

X	0	1	2	3
$p(x)$	$\frac{1}{8}$	$\frac{3}{8}$	$\frac{3}{8}$	$\frac{1}{8}$

So $$E(x) = \sum_x xp(x)$$

$$= \left[\left(0\times\frac{1}{8}\right)+\left(1\times\frac{3}{8}\right)+\left(2\times\frac{3}{8}\right)+\left(3\times\frac{1}{8}\right)\right] = \frac{3}{2}$$

13. A coin is tossed until a tail appears. What is the expectation of the number of tosses.

Solution. For the given problem, the probability density function is a geometric distribution given by $P(X = n) = (1 - p)^{n-1} . p$, with expected value of X as $\frac{1}{p}$. We know the probability of appearance of tail is $\frac{1}{2}$.

Hence $$E(X) = \frac{1}{p} = \frac{1}{1/2} = 2$$

14. Obtain the moment-generating function of the random variable X having probability density function

$$f(x) = \begin{cases} x & 0 \le x < 1 \\ 2 - x & 1 \le x < 2 \\ 0 & \text{otherwise} \end{cases}$$

Solution. $$M_x(t) = E(e^{tX})$$

$$= \int_{-\infty}^{\infty} e^{tx} f(x)\, dx = \int_0^1 xe^{tx}\, dx + \int_1^2 (2 - x)e^{tx}\, dx$$

$$= \left[\frac{xe^{tx}}{t}\right]_0^1 - \int_0^1 \frac{e^{tx}}{t}\, dx + \int_1^2 2e^{tx}\, dx - \int_1^2 xe^{tx} dx$$

$$= \frac{e^t}{t} - \frac{1}{t}\left[\frac{e^t - 1}{t}\right] + 2\left[\frac{e^{2t} - e^t}{t}\right] - \left[\frac{2e^{2t} - e^t}{t} - \frac{e^{2t} - e^t}{t^2}\right]$$

$$= \frac{e^{2t} - 2e^t + 1}{t^2} = \frac{(e^t - 1)^2}{t^2}.$$

15. Find the moment generating function of the random variable whose moments are $\mu_r^1 = (r + 1)!\, 2^r$.

Solution.

$$M_X(t) = E(e^{tX})$$

$$= E\left(1 + \frac{tX}{1!} + \frac{t^2X^2}{2!} + \frac{t^3X^3}{3!} + \ldots \frac{t^rX^r}{r!} + \ldots\right)$$

$$= 1 + E(tX) + \frac{1}{2!}\, E(t^2X^2) + \frac{1}{3!}\, E(t^3X^3) + \ldots \frac{1}{r!}\, E(t^rX^r) + \ldots$$

$$= 1 + t\, E(X) + \frac{1}{2!}\, t^2 E(X^2) + \ldots \frac{1}{r!}\, t^r E(X^r) + \ldots$$

$$= 1 + t\mu_1^1 + \frac{t^2}{2!}\mu_2^1 + \frac{t^3}{3!}\mu_3^1 + \ldots = \sum_{r=0}^{\infty} \frac{t^r}{r!}\mu_r^1$$

$$= \sum_{r=0}^{\infty} \frac{t^r}{r!}(r+1)!\,2^r = \sum_{r=0}^{\infty}(r+1)(2t)^r$$

$$= 1 + 2(2t) + 3(2t)^2 + \ldots = (1-2t)^{-2} = \frac{1}{(1-2t)^2}$$

16. The mean and variance of binomial distribution are 4 and 3 respectively. Find $P(X \geq 1)$.

Solution. A binomial random variable X with parameters n and p whose mean is np and variance is $np(1-p)$. Now

$$\text{mean} = E(X) = np = 4$$

$$\text{Var}(X) = E(X^2) - (E(X))^2 = np(1-p) = 3 \Rightarrow 1 - p = \frac{3}{np} = \frac{3}{4}$$

The probability density function is

$$P(X \geq 1) = 1 - P(X < 1) = 1 - P(X = 0)$$

$$= 1 - nC_0\, p^0 (1-p)^{n-0} = 1 - (1-p)^n.$$

We have $1 - p = \frac{3}{4}$ or $p = \frac{1}{4}$. Since $np = 4$, $n = \frac{4}{p} = 16$.

So $P(X \geq 1) = 1 - (1-p)^n$

$$= 1 - \left(\frac{3}{4}\right)^{16} = 1 - 0.10023 = 0.9899.$$

17. When a fair coin is tossed 200 times, determine the mean and standard deviation.

Solution. In tossing a coin, let the probability of getting head (probability of success) is p and the probability of getting tail (probability of failure) is $1 - p = q$. Then $p = q = \frac{1}{2}$. Therefore

$$\text{mean} = np = 200 \times \frac{1}{2} = 100$$

$$\text{Var}(X) = npq = 200 \times \frac{1}{2} \times \frac{1}{2} = 50$$

Standard deviation $= \sqrt{npq} = \sqrt{50}$

18. The probability that a blade is defective is $\frac{1}{500}$. If the blades are supplied in packets of 10, calculate the number of packets containing no defective and two defective blades in a consignment of 10,000 packets.

Solution. Let p denote the probability of defective blade. Then

$$p = \frac{1}{500} = 0.002,\ q = 1 - p = 0.998$$

Let X be the number of defective blades in a packet of ten blades. Then

$$P(X = 0) = {}^{10}C_0\, p^0 q^{10} = (0.998)^{10} = 0.98$$

$$P(X = 2) = {}^{10}C_2\, p^2 q^8 = 45 \times 0.000004 \times 0.984 = 0.0001771$$

Hence

(i) number of packets containing no defective blades is $0.98 \times 10{,}000 = 9800$.

(ii) number of packets containing two defective blades is $0.0001771 \times 10{,}000 = 1.771 \approx 2$.

19. Five percent of the products of a machine are defective. How many items should be produced in order for the probability of atleast one defective to be $\frac{1}{2}$ or more.

Solution. Probability of getting defective is

$$p = \frac{5}{100} = 0.05$$

Let S_n be the random variable denoting the lot size n contains defective items. Then

$$P(S_n = k) = {}^nC_k\, p^k\, (1-p)^{n-k},$$

Now we want to find the smallest integer n such that

$$P(S_n \geq 1) \geq \frac{1}{2}$$

$$\Rightarrow \quad P(S_n \geq 1) = 1 - P(S_n = 0)$$

$$= 1 - {}^nC_0\, (0.05)^0\, (0.95)^n = 1 - (0.95)^n$$

Thus $1 - (0.95)^n \geq \frac{1}{2} \quad \Rightarrow \quad (0.95)^n \leq \frac{1}{2}$

By using logarithmic table we get $n \geq 13.5$. Hence $n = 14$ is the smallest lot size that should give atleast one defective item with probability $\frac{1}{2}$ or more.

20. The probability that a patient recovers from a rare blood disease is 0.4. If 15 people are known to have contracted this disease, what is the probability that (i) atleast 10 survive (ii) from 3 to 8 survive (iii) exactly 5 survive.

Solution. Let X be the number of people that survive.

$$n = 15, p = 0.4$$

(i) $P(X \geq 10) = 1 - P(X < 10)$

$$= 1 - [P(X=0) + P(X=1) + P(X=2) + P(X=3) + P(X=4) + P(X=5) + P(X=6) + P(X=7) + P(X=8) + P(X=9)]$$

$$= 1 - [({}^{15}C_0\, (0.4)^0\, (0.6)^{15} + {}^{15}C_1\, (0.4)^1\, (0.6)^{14} + \ldots + {}^{15}C_9\, (0.4)^9\, (0.6)^6]$$

$$= 1 - 0.9662 = 0.0338$$

(ii) $P(3 \leq X \leq 8) = P(X=3) + P(X=4) + P(X=5) + P(X=6) + P(X=7) + P(X=8)$

$$= {}^{15}C_3\, (0.4)^3\, (0.6)^{12} + {}^{15}C_4\, (0.4)^4\, (0.6)^{11} + {}^{15}C_5\, (0.4)^5\, (0.6)^{10} + {}^{15}C_6\, (0.4)^6\, (0.6)^9 + {}^{15}C_7\, (0.4)^7\, (0.6)^8 + {}^{15}C_8\, (0.4)^8\, (0.6)^7$$

$$= 0.1859.$$

21. The average number of radioactive particles passing through a counter during 1 millisecond is in a laboratory experiment is 4. What is the probability that 6 particles enter the counter in a given millisecond?

Solution. Using Poisson distribution with $x = 6$ and mean $= \lambda = 4$, we have

$$P(X = 6) = \frac{e^{-\lambda}\lambda^x}{x!} = \frac{e^{-4}\,4^6}{6!}$$

22. If a random variable has a Poisson distribution such that $P(1) = P(2)$, find $P(4)$.

Solution. We know $P(X = x) = \dfrac{e^{-\lambda}\lambda^x}{x!}$. Then

$$P(X = 1) = e^{-\lambda}\lambda$$

$$P(X = 2) = \frac{e^{-\lambda}\lambda^2}{2}$$

$$P(1) = P(2) \;\Rightarrow\; e^{-\lambda}\lambda = \frac{e^{-\lambda}\lambda^2}{2} \;\Rightarrow\; \lambda = 2.$$

Therefore

$$P(X = 4) = \frac{e^{-\lambda}\,2^4}{4!} = \frac{0.1353 \times 2}{3} = 0.0902.$$

23. If the probability of a defective fuse from a manufacturing unit is 2%, in a box of 200 fuses, find the probability that exactly 4 fuses are defective.

Solution. Let p denote the probability of a defective fuse. Then

$p = 2\% = \dfrac{2}{100} = 0.02$, $n = 200$, $\lambda =$ mean $= np = 200 \times 0.02 = 4$. Let X be set of all fuses which are defective. Then

$$P[X = x] = \frac{e^{-\lambda}\lambda^x}{x!}$$

implies $$P(X = 4) = \frac{4^4 \cdot e^{-4}}{4!} = 0.1952.$$

24. Suppose the cost of performing an experiment is Rs. 100. If the experiment fails, an additional cost of Rs. 20 occurs because of certain changes to be made before the next trial. If the probability of success in given trial is 0.2, if the individual trial is independent and if the experiment is continued until the first successful result is achieved, what is the expected cost of the entire procedure.

Solution. If C is the cost and X is the number of trials required to achieve success, we have

$$C = 100X + 20(X - 1) = 120X - 20.$$

Since X follows geometric distribution, $E(X) = \dfrac{1}{p} = \dfrac{1}{0.2}$. Hence

$$E(C) = 120E(X) - 20 = 120 \times \frac{1}{0.2} - 20 = 580$$

Hence the expected cost is 580.

25. If a biased coin falls heads r times in succession, find the probability that it will fall heads within next two trials.

Solution. Let X be the random variable denoting the number of tosses required to get, first head. Since the coin is biased, let p be the probability of getting head in each toss and $1 - p = q$ is the probability of getting tail. Now X follows geometric distribution.

Then $\quad P(X = n) = p(1 - p)^{n-1} = pq^{n-1}$

Probability of getting heads in r or $(r + 1)$ tosses is

$$P(X \leq r + 1/X \geq r) = \frac{P(X \leq r+1, X \geq r)}{P(X \geq r)} = \frac{P(r \leq X \leq r+1)}{P(X \geq r)}$$

$$P(X \geq r) = \sum_{n=r}^{\infty} q^{n-1} p = q^{r-1}\, p + q^r p + \ldots$$

$$= q^{r-1} p[1 + q + q^2 + \ldots] = \frac{q^{r-1}\, p}{1-q} = \frac{q^{r-1}\, p}{p} = q^{r-1}$$

$$P(r \leq X \leq r + 1) = P(X = r) + P(X = r + 1) = q^{r-1}\, p + q^r p$$

Thus

$$P(X \leq r + 1/X \geq r) = \frac{q^{r-1}\, p + q^r\, p}{q^{r-1}} = p + qp$$

which is independent of r.

26. A random variable X has a mean $\mu = 8$ and variance $\sigma_X^2 = 9$ and an unknown probability distribution. Find (i) $P(-4 < X < 20)$ and (ii) $P(|X - 8| \geq 6)$.

Solution. By Chebyshev's inequality

$$P(|X - \mu| \geq k\, \sigma_X) \leq \frac{1}{k^2}$$

or $$P(|X - \mu| < k\, \sigma_X) \geq 1 - \frac{1}{k^2},$$

we have

(i) $P(\mu - k\sigma_X < X < \mu + k\sigma_X) \geq 1 - \dfrac{1}{k^2}$

$\Rightarrow \quad P(8 - (4)\,3 < X < 8 + (4)\,3) \geq 1 - \dfrac{1}{4^2}$

$\Rightarrow \quad P(-4 < X < 20) \geq \dfrac{15}{16}$

(ii) $P(|X - 8| \geq 6) \leq \dfrac{1}{4}$, $k\sigma_X = 3k = 6$

$\Rightarrow \quad k = 2.$

27. Suppose that the number of miles that a car run before its battery wears out is exponentially distributed with an average value of 10,000 miles. If a person desires to take a 5000– mile trip what is the probability that he will be able to complete the trip without having to replace the car battery? What can be said when the distribution is not exponential.

Solution. By memoryless property of the exponential distribution, the remaining life time (in thousands of miles) of the battery is exponential with parameter $\lambda = \frac{1}{10}$. Hence

$$P(\text{remaining life time} > 5) = 1 - P(X \le 5)$$
$$= 1 - (1 - e^{-5\lambda}) = e^{-5\lambda} = e^{-1/2} = 0.604.$$

However, if the lifetime is not exponential, then

$$P(\text{life time} > (t+5)/\text{life time} > t) = \frac{1 - P(\text{lifetime} \le t+5)}{1 - P(\text{lifetime} \le t)}.$$

where t is the number of miles that the battery had been in use prior to the start of the trip.

28. Suppose that X has an exponential distribution with parameter λ. Then $E(X) = \frac{1}{\lambda}$. Compute the probability that X exceeds its expected vlaue

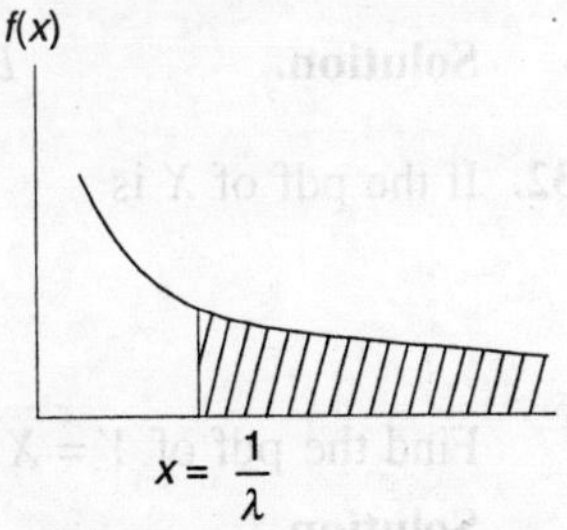

Solution. $P\left(X > \frac{1}{\lambda}\right) = e^{-\lambda \cdot \frac{1}{\lambda}}$

$$= e^{-1} < \frac{1}{2}$$

29. If X is a normal random variable with parameter $\mu = 3$ and $\sigma_X^2 = 9$. Find (i) $P(2 < X < 5)$, (ii) $P(X > 0)$, (iii) $P\{|X - 3| > 6\}$.

Solution.

(i) $P\{2 < X < 5\} = P\left\{\frac{2-3}{3} < \frac{X-3}{3} < \frac{5-3}{3}\right\}$

$$= P\left\{-\frac{1}{3} < Z < \frac{2}{3}\right\} = \phi\left(\frac{2}{3}\right) - \phi\left(-\frac{1}{3}\right)$$

$$= \phi\left(\frac{2}{3}\right) - \phi\left[1 - \phi\left(\frac{1}{3}\right)\right] = .3779 \quad \text{(see normal table)}$$

(ii) $P[X > 0] = P\left[\frac{X-3}{3} > \frac{0-3}{3}\right]$

$$= P(Z > -1) = 1 - \phi(-1) = \phi(1) = .8413.$$

(iii) $P(|X - 3| > 6) = P(X > 9) + P(X < -3)$

$$= P\left(\frac{X-3}{3} > \frac{9-3}{3}\right) + P\left(\frac{X-3}{3} < \frac{-3-3}{3}\right)$$

$$= P(Z > 2) + P(Z < -2) = 1 - \phi(2) + \phi(-2)$$
$$= 2[1 - \phi(2)] = .0456$$

30. If the number of speeders a radar unit spots per hour at a certain location on Route *I*, is a Poisson random variable with $\lambda = 8.4$, what is the probability of a waiting time less than 10 minutes between successive speeders?

Solution. Here $\lambda = 8.4$. Treating the waiting time as a random variable having an exponential distribution is

$$P\left(X \le \frac{1}{6}\right) = \int_0^{1/6} (8.4)\, e^{-8.4\, x}\, dx = [-e^{-(8.4)x}]_0^{1/6} = 1 - e^{-1.4} \approx 0.75$$

31. If X has the pdf

$$f(x) = \begin{cases} e^{-x} & \text{for } x > 0 \\ 0 & \text{otherwise} \end{cases}$$

Find $E(g(x))$ if $\quad g(x) = e^{3x/4}$.

Solution. $$E(e^{3x/4}) = \int_0^{\infty} e^{3x/4}\, e^{-x} \cdot dx = 4.$$

32. If the pdf of X is

$$f(x) = \begin{cases} 6x\,(1-x) & \text{for } 0 < x < 1 \\ 0 & \text{otherwise} \end{cases}$$

Find the pdf of $Y = X^3$.

Solution. $$G(y) = P(Y \le y) = P(X^3 \le y)$$

$$= P(X \le y^{1/3}) = \int_0^{y^{1/3}} 6x\,(1-x)\, dx = 3y^{2/3} - 2y$$

and hence pdf of Y is

$$g(y) = \frac{d}{dy}\, G(y) = 2\,(y^{-1/3} - 1),\ 0 < y < 1.$$

33. If X is the number of heads obtained in four tosses of a coin, find the probability distribution of $Y = \dfrac{1}{1+X}$.

Solution. The probability distribution of X is given by

X	0	1	2	3	4
$f(x)$	$\frac{1}{16}$	$\frac{4}{16}$	$\frac{6}{16}$	$\frac{4}{16}$	$\frac{1}{16}$

Using the relationship $Y = \dfrac{1}{1+X}$, the probability distribution of Y is given by

y	1	$\frac{1}{2}$	$\frac{1}{3}$	$\frac{1}{4}$	$\frac{1}{5}$
$g(y)$	$\frac{1}{16}$	$\frac{4}{16}$	$\frac{6}{16}$	$\frac{4}{16}$	$\frac{1}{16}$

Hence for $\quad n = 4,\ p = \dfrac{1}{2}$, we have

$$f(x) = {}^nC_x\, p^x (1-p)^{n-x} = 4C_x \left(\frac{1}{2}\right)^4 \quad \text{for } x = 0, 1, 2, 3, 4.$$

$$\text{and } g(y) = f\left(\frac{1}{y}-1\right) = 4C\left(\frac{1}{y}-1\right)\left(\frac{1}{2}\right)^4, \quad \text{for } y = 1, \frac{1}{2}, \frac{1}{3}, \frac{1}{4}, \frac{1}{5}$$

EXERCISES

1. A random variable X has range $0 < x < \infty$ and probability density $f(x) = kx^{\alpha-1} e^{-\beta x}$, $0 < x < \infty$. Calculate the constant k. Also find the mean and variance
2. Find $E(X)$ and $E(X^2)$ for a random variable X with density given by
$$f(x) = \begin{cases} 2e^{-2x}, & x > 0 \\ 0 & x \le 0 \end{cases}$$
3. Find a constant $b > 0$, so that the function
$$f(x) = \begin{cases} \dfrac{e^{3x}}{4} & 0 \le x \le b \\ 0 & \text{otherwise} \end{cases}$$
is a valid density function
4. Let X be a random variable with the following distribution

X	-3	6	9
$P(X)$	$\frac{1}{6}$	$\frac{1}{2}$	$\frac{1}{3}$

Find $E(X)$ and $E(X^2)$.
5. Let X be a continuous random variable with *pdf* $f(x) = e^{-x}$, $x > 0$. Let $Y = 2X + 1$. Find $P(Y \ge 5)$.
6. Suppose that the continuous random variable X has pdf $f(x) = e^{-x}$, $x > 0$. Find the *pdf* of $Y = X^3$.
7. *Prove:* Var $(X + C) =$ Var (X), C is a constant.
8. If X is a continuous random variable with pdf
$$\begin{aligned} f(x) &= 1 + x && \text{if } -1 \le x \le 0 \\ &= 1 - x && \text{if } 0 \le x \le 1 \end{aligned}$$
find Var (X).
9. If X has a Poisson distribution with parameter λ, and if $P(X = 0) = 0.2$, evaluate $P(X > 2)$.
10. Let X have a Poisson distribution with parameter λ. Find the value of k for which $P(X = k)$ is largest.
[*Hint:* Compare $P(X = k)$ with $P(X = k - 1)$]
11. Suppose that the probability that an item produced by a particular machine is defective equals 0.2. If 10 items produced from this machine are selected at random, what is the probability that not more than one defective is found? Use the binomial and Poisson distributions and compare the answers.
12. Suppose that X has the following pdf
$$f(x) = \lambda e^{-\lambda(x-a)} \qquad x \ge a$$
Find the moment-generating function of X.

13. Let X be the outcome when a fair die is tossed.
   (i) Find the moment-generating function of X.
   (ii) Using it, find $E(X)$ and Var (X).
14. A random variable X has a mean $\mu = 12$, a variance 9 and an unknown probability distribution. Using Chebyshev's theorem, find
   (i) $P(6 < X < 18)$ (ii) $P(3 < X < 21)$
15. A fruit grower claims that $\frac{2}{3}$ of his peach crop has been contaminated by the medfly infestation. Find the probability that among 4 peaches inspected by this grower
   (i) all 4 have been contaminated
   (ii) anywhere from 1 to 3 have been contaminated.
16. In testing a certain kind of truck tire over a rugged terrain, it is found that 25% of the trucks fail to complete the test run without a blowout. Of the next 15 trucks tested, find the probability that
   (i) from 3 to 6 have blowouts
   (ii) more than 5 have blowouts
17. Let X be a binomial random variable with pdf

$$f(x) = \begin{cases} 3C_x \left(\frac{2}{5}\right)^x \left(\frac{3}{5}\right)^{3-x} & x = 0, 1, 2, 3 \\ 0 & \text{otherwise} \end{cases}$$

   Find the probability distribution of $Y = X^2$.
18. The speed of a molecule in a uniform gas at equilibrium is a random variable V whose pdf is

$$f(v) = \begin{cases} kv^2\, e^{-bv^2} & v > 0 \\ 0 & \text{otherwise} \end{cases}$$

   where k is a constant and b depends on the absolute temperature and mass of the molecule. Find the pdf of the kinetic energy of the molecule W, where $W = \frac{mV^2}{2}$.
19. Suppose that the random variable Y has the moment-generating function $M(t) = (1 - t)^{-1}$. Find the mean and variance.
20. Suppose that the life of a certain type of electron tube has an exponential distribution with a mean life of 500 hours. Find the probability that it will last for another 600 hours if the tube has been in operation for 300 hours.

ANSWERS

(1) $k = \beta^\alpha/\Gamma_{(\alpha)}$ (2) $E(X) = \frac{1}{2} = E(X^2)$ (3) $b = 0.855$ (5) $1/e^2$

(6) $\frac{1}{3} y^{-2/3} e^{-y^{1/3}}$, $y > 0$ (8) $\frac{1}{6}$ (9) 0.219

(11) 0.3758 (binomial), 0.4060 (Poisson) (12) $\frac{\lambda e^{ta}}{\lambda - t}$

(13) $M_X(t) = \frac{1}{6}[e^t + e^{2t} + e^{3t} + e^{4t} + e^{5t} + e^{6t}]$

(14) (i) Atleast $\frac{3}{4}$ (ii) Atleast $\frac{8}{9}$. (15) (i) $\frac{16}{81}$ (ii) $\frac{64}{81}$

(16) (i) 0.7073 (ii) 0.1484

(17) $3C_{\sqrt{y}}\left(\frac{2}{5}\right)^{\sqrt{y}}\left(\frac{3}{5}\right)^{3-\sqrt{y}}$, $y = 0, 1, 4, 9$

(18) Gamma distribution with $\alpha = \frac{3}{2}$, $\beta = m/2b$.

(19) Mean = 1, Variance = 1. (20) $e^{-6/5}$.

3

Two-Dimensional Random Variables

3.1 MARGINAL AND CONDITIONAL DISTRIBUTIONS

The study of random variables and their probability distributions in the preceding chapter was restricted to one dimensional sample spaces. The outcomes of an experiment were assumed as values by a single random variable. There will be situations, we may record the simultaneous outcomes of more than one random variable. For example, the hardness H and tensile strength T of a manufactured piece of steel resulting in the outcome (h, t) or one might be interested in the study of height (H) and weight (W) of some chosen population, giving rise to an outcome (h, w).

3.1.1 Definition

Let E be an experiment and S a sample space associated with E. Let $X = X(s)$, $Y = Y(s)$ be two functions each assigning a real number to each outcome $s \in S$. (Fig. 3.1). We call (X, Y) a **two-dimensional random variable** (sometimes called a **random vector**).

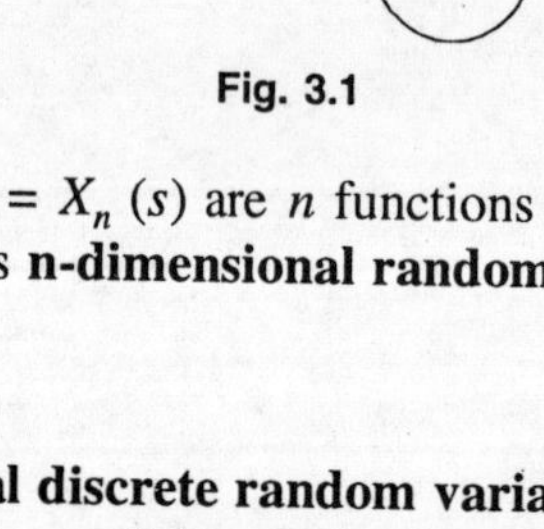

Fig. 3.1

If $X_1 = X_1(s)$, $X_2 = X_2(s)$, . . . $X_n = X_n(s)$ are n functions each assigning a real number to every outcome $s \in S$, we call $(X_1, \ldots X_n)$ as **n-dimensional random variable.**

3.1.2 Definition

(X, Y) is said to be a **two-dimensional discrete random variable** if the possible values of (X, Y) are finite or countably infinite. That is, the possible values of (X, Y) may be represented as (x_i, y_j) $i = 1, 2, \ldots n \ldots, j = 1, 2, \ldots m, \ldots$ (X, Y) is said to be a **two-dimensional continuous random variable** if (X, Y) can assume all values in some non countable set of R^2.

Note: (i) It may happen that one of the components of (X, Y), say X is discrete, while the other is continuous. However, in most applications we deal only with the cases, in which either both X and Y are discrete or both continuous.

3.1.3 Definition

Let (X, Y) be a two-dimensional discrete random variable. For each possible outcome (x_i, y_j) we associate a number $p(x_i, y_j)$ representing $P(X = x_i, Y = y_j)$ and satisfying the following conditions

(i) $p(x_i, y_j) \geq 0$ for all $(x_i, y_j) \in X \times Y$

(ii) $\sum_{j=1}^{\infty} \sum_{i=1}^{\infty} p(x_i, y_j) = 1$

The function p defined for all $(x_i, y_j) \in (X, Y)$ is called the **joint probability distribution** or **probability mass function** of (X, Y).

3.1.4 Definition

Let (X, Y) be a two-dimensional continuous random variable. The **joint probability density function** (pdf) f is a function satisfying the following conditions

(i) $f(x, y) \geq 0$ for all $(x, y) \in R \times R$

(ii) $\int\int_R f(x, y)\, dxdy = 1.$

3.1.5 Definition

Let (X, Y) be a two – dimensional random variable. The **cumulative distribution function** (cdf) F of the two-dimensional random variable (X, Y) is defined by

$$F(x, y) = P(X \leq x, Y \leq y).$$

Note: F is a function of two-dimensional variables and has a number of properties analogous to those discussed for the one–dimensional cdf. If F is the cdf of a two-dimensional random variable with joint pdf f, then

$$\frac{\partial^2 F(x, y)}{\partial x\, \partial y} = f(x, y)$$

wherever F is differentiable. This result is analogous to $\frac{d}{dx} F(x) = \frac{d}{dx} [P(X \leq x)] = \frac{d}{dx}\left[\int_{-\infty}^{x} f(s)\, ds\right]$ $= f(x)$, where f is the pdf of the one–dimensional random variable X.

Marginal and conditional probability distributions

With each two-dimensional random variable (X, Y) we associate two one-dimensional random variables, namely X and Y individually. That is, we find the probability distribution of X or the probability distribution of Y.

3.1.6 Definition

The marginal probability distribution of a discrete random variable X is

$$p(x_i) = P(X = x_i) = P(X = x_i, Y = y_1 \text{ or } X = x_i, Y = y_2, \text{ or } \ldots) = \sum_{j=1}^{\infty} p(x_i, y_j)$$

Analogously we define the marginal probability distribution of Y as

$$q(y_j) = P(Y = y_j) = \sum_{i=1}^{\infty} p(x_i, y_j)$$

For the continuous case we define as follows.

Let f be the joint pdf of the continuous two-dimensional random variable (X, Y). We define g and h, the marginal probability density functions of X and Y respectively as:

$$g(x) = \int_{-\infty}^{\infty} f(x, y)\, dy,\; h(y) = \int_{-\infty}^{\infty} f(x, y)\, dx$$

These pdf correspond to the basic pdf's of the one-dimensional random variables X and Y respectively. For example

$$P(a \le X \le b) = P[a \le X \le b, -\infty < Y < \infty]$$

$$= \int_a^b \int_{-\infty}^{\infty} f(x, y)\, dy\, dx = \int_a^b g(x)\, dx$$

The concept of conditional probability may be introduced in a very natural way.

3.1.7 Definition

Let (X, Y) be a discrete two-dimensional random variable with probability distribution $p(x_i, y_j)$. Then the conditional probability distribution of X given Y is

$$p(x_i/y_j) = P(X = x_i/Y = y_j) = \frac{p(x_i, y_j)}{q(y_j)} \text{ if } q(y_j) > 0$$

Analogously, the conditional probability distribution of Y given X is

$$q(y_j/x_i) = p(Y = y_j/X = x_i) = \frac{p(x_i, y_j)}{p(x_i)} \text{ if } p(x_i) > 0$$

Note: For given j, $p(x_i/y_j)$ satisfies all the conditions for a probability distribution. We have

(i) $p(x_i/y_j) \ge 0$

(ii) $\sum_{i=1}^{\infty} p(x_i/y_j) = \sum_{i=1}^{\infty} \frac{p(x_i, y_j)}{q(y_j)} = \frac{q(y_j)}{q(y_j)} = 1$

3.1.8 Definition

Let (X, Y) be a continuous two-dimensional random variable with joint pdf f. Let g and h be the marginal p.d.f's of X and Y, respectively.

The conditional pdf of X for given $Y = y$ is

$$g(x/y) = \frac{f(x, y)}{h(y)},\; h(y) > 0$$

The conditional pdf of Y for given $X = x$ is

$$h(y/x) = \frac{f(x, y)}{g(x)}\; g(x) > 0$$

Note The above conditional pdf's satisfies all the requirements for a one-dimensional pdf Thus for fixed y,

(i) $g(x/y) \ge 0$

(ii) $\int_{-\infty}^{\infty} g(x/y)\, dx = \int_{-\infty}^{\infty} \frac{f(x, y)}{h(y)}\, dx = \frac{1}{h(y)} \int_{-\infty}^{\infty} f(x, y)\, dx = \frac{h(y)}{h(y)} = 1$

Just as we defined the concept of independence between two events A, and B, we shall now define independent random variables.

3.1.9 Definition

Let (X, Y) be a two-dimensional discrete random variable. We say that X and Y are independent random variables if and only if $p(x_i, y_j) = p(x_i)\, q(y_j)$ for all i and j.

Let (X, Y) be a two-dimensional continuous random variable. We say X and Y are independent random variables if and only if $f(x, y) = g(x)\, h(y)$ for all (x, y), where f is the joint pdf, and g and h are the mariginal pdf's of X and Y respectively.

3.1.10 Theorem

Let (X, Y) be a two-dimensional discrete random variable. Then X and Y are independent if and only if $p(x_i/y_j) = p(x_i)$ for all i and j (equivalently, if and only if $q(y_j/x_i) = q(y_j)$ for all i and j).

Let (X, Y) be a two dimensional continuous random variable. Then X and Y are independent if and only if $g(x/y) = g(x)$ or equivalently, if and only if $h(y/x) = h(y)$ for all (x, y).

Proof: We know by definition, $p(x_i/y_j) = \dfrac{p(x_i, y_j)}{q(y_j)}$. Since X and Y are independent, $p(x_i, y_j) = p(x_i)\, q(y_j)$.

Hence

$$P(x_i/y_j) = \frac{p(x_i)\, q(y_j)}{q(y_j)} = p(x_i)$$

Similarly

$$p(y_j/x_i) = \frac{q(y_j)\, p(x_i)}{p(x_i)} = q(y_j)$$

$$g(x/y) = \frac{f(x, y)}{h(y)} = \frac{g(x)\, h(y)}{h(y)} = g(x)$$

$$h(y/x) = \frac{f(x, y)}{g(x)} = \frac{g(x)\, h(y)}{g(x)} = h(y)$$

3.1.11 Example

Let X and Y be the lifetimes of two electronic devices. Suppose that their joint pdf is given by

$$f(x, y) = e^{-(x+y)},\ x \geq 0,\ y \geq 0$$

then X and Y are independent, since

$$f(x, y) = e^{-(x+y)} = e^{-x} \cdot e^{-y} = g(x)\, h(y)$$

3.1.12 Example

Suppose that $f(x, y) = 8xy$, $0 \leq x \leq y \leq 1$. Check the independence of X and Y.

Solution. $$g(x) = \int_0^1 f(x, y)dy = \int_0^1 8xy\, dy = 8x\left[\frac{y^2}{2}\right]_0^1 = 4x$$

$$h(y) = \int_0^1 f(x, y)\, dx = \int_0^1 8xy\, dx = 8y\left[\frac{x^2}{2}\right]_0^1 = 4y.$$

$$f(x, y) = 8xy \neq g(x)\, h(y) = 16xy$$

Hence X and Y are not independent.

Note. From the definition of the marginal probability distribution (in either the discrete or the continuous case) it is clear that the joint probability distribution determines, uniquely the marginal probability distribution. But the converse is not true.

Note. Let (X, Y) be a two-dimensional random variable. Let A and B be events whose occurrence (or non-occurrence) depend only on X and Y respectively. Then if X and Y are independent random variables, then

$$P(A \cap B) = \iint_{A \cap B} f(x, y)\, dxdy = \iint_{A \cap B} g(x)\, h(y)\, dxdy$$

$$= \int_A g(x)\, dx \int_B h(y)\, dy = P(A)\, P(B).$$

3.1.13 Example

The joint density function of (X, Y) is given by

$$f(x, y) = \begin{cases} 2e^{-x}\, e^{-2y} & 0 < x < \infty, 0 < y < \infty \\ 0 & \text{otherwise} \end{cases}$$

Compute (i) $P(X > 1, Y < 1)$, (ii) $P(X < Y)$, (iii) $P(X < a)$

Solution.

(i) $$P(X > 1, Y < 1) = \int_0^1\left(\int_1^\infty 2e^{-x}\, e^{-2y}\, dx\right) dy = \int_0^1 2e^{-2y}\left(\frac{e^{-x}}{-1}\right)_1^\infty dy$$

$$= e^{-1}\int_0^1 2e^{-2y}\, dy = e^{-1}\,(1 - e^{-2})$$

(ii) $$P(X < Y) = \int_0^\infty\left(\int_0^y 2e^{-x}\, e^{-2y}\, dx\right) dy = \int_0^\infty 2e^{-2y}\,(1 - e^{-y})dy$$

$$= \int_0^\infty 2e^{-2y}\, dy - \int_0^\infty 2e^{-3y}\, dy = 1 - \frac{2}{3} = \frac{1}{3}$$

(iii) $$P(X < a) = \int_0^a\left(\int_0^\infty 2e^{-2y}\, e^{-x}\, dy\right) dx = \int_0^a e^{-x}\, dx = 1 - e^{-a}$$

3.1.14 Example

The joint density of X and Y is

$$f(x, y) = \begin{cases} e^{-(x+y)} & 0 < x < \infty, 0 < y < \infty \\ 0 & \text{otherwise} \end{cases}$$

Find the density function of the random variable $\dfrac{X}{Y}$.

Solution. We start by computing the distribution function of $\dfrac{X}{Y}$. For $a < 0$,

$$F(a) = P\left(\frac{X}{Y} \le a\right) = \iint_{x/y \le a} e^{-(x+y)}\, dxdy$$

$$= \int_0^\infty \int_0^{ay} e^{-(x+y)}\, dxdy = \int_0^\infty (1 - e^{-ay})\, e^{-y}\, dy$$

$$= \left[-e^{-y} + \frac{e^{-(a+1)y}}{a+1} \right]_0^\infty = 1 - \frac{1}{a+1}$$

$$\frac{d}{da} F(a) = f(a) = \frac{1}{(1+a)^2}, \; 0 < a < \infty$$

Sums of Independent Random Variables.

Let X and Y be two independent continuous random variables having probability density functions g and h. The cumulative distribution function of $X + Y$ is obtained as follows:

$$F(a) = P(X + Y \le a) = \iint_{x+y \le a} g(x)\, h(y)\, dx\, dy$$

$$= \int_{-\infty}^{\infty} \left(\int_{-\infty}^{a-y} g(x)\, dx \right) h(y)\, dy = \int_{-\infty}^{\infty} F(a-y)\, h(y)\, dy \tag{3.1}$$

The cumulative distribution function F is called the convolution of the distribution g and h. By differentiating equation (3.1) we obtain the probability density function f of $X + Y$ and is given by

$$f(a) = \frac{d}{da}\left[\int_{-\infty}^{\infty} F(a-y)\, h(y)\, dy \right]$$

$$= \int_{-\infty}^{\infty} \left(\frac{d}{da} F(a-y) \right) h(y)\, dy = \int_{-\infty}^{\infty} g(x-a)\, h(y)\, dy \tag{3.2}$$

Recall that a gamma random variable has a density of the form

$$f(x) = \frac{e^{-\lambda x}\, \lambda (\lambda x)^{t-1}}{\Gamma_{(t)}}, \; 0 < x < \infty$$

An important property of this family of distributions is that for a fixed value of λ, it is closed under convolution.

3.1.15 Example

If X and Y are independent gamma random variables with respective parameters (n, λ) and (t, λ), then $X + Y$ is a gamma random variable with parameters $(n + t, \lambda)$.

Solution. Using (3.2) we obtain

$$f(a) = \frac{1}{\Gamma_{(n)}\,\Gamma_{(t)}} \int_0^a \lambda e^{-\lambda(a-y)} \,[\,\lambda (a-y)]^{n-1}\, \lambda e^{-\lambda y} (\lambda y)^{t-1}\, dy$$

$$= Ke^{-\lambda a} \int_0^a (a-y)^{n-1}\, y^{t-1}\, dy,\ K = \frac{\lambda^{n+t}}{\Gamma_{(n)}\,\Gamma_{(t)}}$$

$$= Ke^{-\lambda a}\, a^{n+t-1} \int_0^1 (1-x)^{n-1}\, x^{t-1}\, dx,\ x = \frac{y}{a}$$

$$= Ce^{-\lambda a}\, a^{n+t-1}$$

where C is a constant not depending on a. But $f(a)$ is pdf and thus must integrate to 1, the value of C is determined and we have

$$f(a) = \frac{\lambda e^{-\lambda a}\,(\lambda a)^{n+t-1}}{\Gamma_{(n+t)}}$$

3.1.16 Example

If X and Y are independent Poisson random variables with respective parameters λ_1, and λ_2, calculate the conditional distribution of X, given that $X + Y = n$.

Solution. We calculate conditional probability mass function of X given $X + Y = n$ as

$$P(X = k/X + Y = n) = \frac{P(X = k, X + Y = n)}{P(X + Y = n)}$$

$$= \frac{P(X = k, Y = n - k)}{P(X + Y = n)} = \frac{P(X = k)\, P(Y = n - k)}{P(X + Y = n)}$$

where

$$P\,(X + Y = n) = \sum_{k=0}^{n} P(X = k, Y = n - k)$$

$$= \sum_{k=0}^{n} P(X = k)\, P(Y = n - k)$$

$$= \sum_{k=0}^{n} \frac{e^{-\lambda_1}\,\lambda_1^k}{k!} \frac{e^{-\lambda_2}\,\lambda_2^{n-k}}{(n-k)!} = e^{-(\lambda_1+\lambda_2)} \sum_{k=0}^{n} \frac{\lambda_1^k\,\lambda_2^{n-k}}{k!\,(n-k)!}$$

$$= \frac{e^{-(\lambda_1+\lambda_2)}}{n!} \sum_{k=0}^{n} \frac{n!}{k!\,(n-k)!}\, \lambda_1^k \cdot \lambda_2^{n-k} = \frac{e^{-\lambda_1+\lambda_2}}{n!}\,(\lambda_1 + \lambda_2)^n$$

Now

$$P(X = k/X + Y = n) = \frac{e^{-\lambda_1}\lambda_1^k}{k!}\frac{e^{-\lambda_2}\lambda_2^{(n-k)}}{(n-k)!}\left[\frac{e^{-(\lambda_1+\lambda_2)}(\lambda_1+\lambda_2)^{(n-1)}}{n!}\right]$$

$$= \frac{n!}{(n-k)!k!}\frac{\lambda_1^k\lambda_2^k}{(\lambda_1+\lambda_2)^n} = nC_k\left(\frac{\lambda_1}{\lambda_1+\lambda_2}\right)^k\left(\frac{\lambda_2}{\lambda_1+\lambda_2}\right)^{n-k}$$

In other words, the conditional distribution of X, given $X + Y = n$ is the binomial distribution with parameters n and $\frac{\lambda_1}{\lambda_1+\lambda_2}$.

3.1.17 Example

The joint pdf of a two-dimensional random variable (X, Y) is given by

$$f(x, y) = \begin{cases} 2; & 0 < x < 1, 0 < y < x \\ 0; & \text{elsewhere} \end{cases}$$

Find the

(i) Marginal density function of X and Y.
(ii) Find the conditional density function of Y given $X = x$ and the cdf of X given $Y = y$.
(iii) Check the independence of X and Y.

Solution. Evidently $f(x, y) \geq 0$

$$\int_0^1\int_0^x f(x, y)\, dy\, dx = \int_0^1\int_0^x 2dydx = 2\int_0^1 xdx = 1$$

(i) The marginal pdf of X is

$$g(x) = \int_{-\infty}^{\infty} f(x, y)\, dy = \int_0^x 2dy = 2x$$

Hence

$$g(x) = 2x; \quad 0 < x < 1$$
$$= 0; \quad \text{otherwise}$$

The marginal pdf of Y is

$$h(y) = \int_{-\infty}^{\infty} f(x, y)\, dx = \int_y^1 2dx = 2(1 - y)$$

$$h(y) = \begin{cases} 2(1-y) & 0 < y < x \\ 0 & \text{otherwise} \end{cases}$$

(ii) The conditional density function of Y given X is

$$f(y/x) = \frac{f(x,y)}{g(x)} = \frac{2}{2x} = \frac{1}{x}, \; 0 < x < 1.$$

The conditional density function of X given Y is

$$f(x/y) = \frac{f(x,y)}{h(y)} = \frac{2}{2(1-y)} = \frac{1}{1-y}, \; 0 < y < x.$$

(iii) Since $g(x)\, h(y) = 2x(1 - y) \neq f(x, y)$, X and Y are not independent.

3.1.18 Example

If (X, Y) are two-dimensional random variables having joint density function

$$f(x, y) = \begin{cases} \frac{1}{8}(6 - x - y) & 0 < x < 2 \\ & 2 < y < 4 \\ 0 & \text{otherwise} \end{cases}$$

Find (i) $P(X < 1, Y < 3)$, (ii) $P(X + Y < 3)$, (iii) $P(X < 1/Y < 3)$

Solution.

(i) $$P(X < 1, Y < 3) = \int_{-\infty}^{1}\int_{-\infty}^{3} f(x, y)\, dy\, dx$$

$$= \int_0^1\int_2^3 \frac{1}{8}(6 - x - y)\, dy\, dx = \frac{3}{8}$$

(ii) $$P(X + Y < 3) = \int_0^1\int_2^{3-x} \frac{1}{8}(6 - x - y)\, dy\, dx = \frac{5}{24}$$

(iii) The probability that $X < 1$ when it is known that $Y < 3$ is

$$P(X < 1/Y < 3) = \frac{P(X < 1, Y < 3)}{P(Y < 3)} = \frac{3/8}{5/8} = \frac{3}{5},$$

where $$P(Y < 3) = \int_0^2\int_2^3 \frac{1}{8}(6 - x - y)\, dy\, dx = \frac{5}{8}.$$

3.1.19 Example

Joint pdf of two-dimensional random variable (X, Y) is given by

$$f(x, y) = 4xy\, e^{-(x^2 + y^2)} \quad x \ge 0, y \ge 0$$

Test whether X and Y are independent. For the above pdf, find the conditional density of X given $Y = y$.

Solution. The marginal density function of X is

$$g(x) = \int_0^{\infty} f(x, y)\, dy = \int_0^{\infty} 4xy\, e^{-(x^2 + y^2)}\, dy$$

$$= 4xe^{-x^2}\int_0^{\infty} ye^{-y^2}\, dy = 4xe^{-x^2}\int_0^{\infty} e^{-t}\cdot\frac{dt}{2}, \text{ where } \quad y^2 = t$$

$$= 2xe^{-x^2}\left[-e^{-t}\right]_0^{\infty} = 2xe^{-x^2}$$

Similarly the marginal pdf of Y is given by

$$h(y) = \int_0^{\infty} f(x, y)\, dx = 2ye^{-y^2}, \; y \ge 0$$

Since $f(x, y) = g(x)\, h(y)$, X and Y are independent

Now

$$f(X/Y=y)=\frac{f(x,y)}{h(y)}=\frac{4xy\,e^{-(x^2+y^2)}}{2y\,e^{-y^2}}=2x\,e^{-x^2}$$

3.1.20 Example

Is the function defined as follows a density function

$$f(x)=\begin{cases}0 & \text{for } x<2\\ \frac{1}{18}(3+2x) & 2\le x\le 4\\ 0 & \text{for } x>4\end{cases}$$

Solution. Since $\int_{-\infty}^{\infty} f(x)\,dx=\int_2^4 \frac{1}{18}(3+2x)\,dx=\left[\frac{x}{6}+\frac{x^2}{18}\right]_2^4=1$, the function is a density function.

3.1.21 Example

$$\text{If } f(x,y)=\begin{cases}xe^{-x(y+1)} & x>0,\ y>0\\ 0 & \text{otherwise}\end{cases}$$

Find the moment-generating function of *XY*.

Solution. $M_{XY}(t)=\int_0^{\infty}\int_0^{\infty} e^{xyt}\,xe^{-x(y+1)}\,dy\,dx$

$$=\int_0^{\infty}\left[xe^{-x}\left(\int_0^{\infty} e^{xyt}\,e^{-xy}\,dy\right)dx\right]=\int_0^{\infty} xe^{-x}\left(\int_0^{\infty} e^{-(1-t)xy}\,dy\right)dx$$

$$=\int_0^{\infty} xe^{-x}\left[\frac{e^{-(1-t)xy}}{-(1-t)x}\right]_0^{\infty}dx=\int_0^{\infty} xe^{-x}\left[\frac{1}{(1-t)x}\right]dx$$

$$=\frac{1}{1-t}\int_0^{\infty} e^{-x}\,dx=\frac{1}{1-t}\left[\frac{e^{-x}}{-1}\right]_0^{\infty}=\frac{1}{1-t}$$

3.2. COVARIANCE

Suppose that a function $g(X, Y)$ is associated with a two-dimensional discrete random variable (X, Y) having a joint probability distribution $p(x_i, y_j)$. Then

$$E(g(X,Y))=\sum_i\sum_j p(x_i,y_j)\,g(x_i,y_j)$$

Similarly for continuous case we define

$$E(g(X,Y))=\int_{-\infty}^{\infty}\int_{-\infty}^{\infty} g(x,y)\,f(x,y)\,dx\,dy,$$

where $f(x, y)$ is the joint pdf of (X, Y)

Let $$g(X, Y) = X + Y. \text{ Then}$$

$$E(X+Y) = \sum_i \sum_j p(x_i, y_j)(x_i + y_j)$$

$$= \sum_i \sum_j p(x_i, y_j)\, x_i + \sum_i \sum_j p(x_i, y_j)\, y_j$$

$$= \sum_i x_i \sum_j p(x_i, y_j) + \sum_j y_j \sum_i p(x_i, y_j)$$

$$= \sum_i x_i\, p(x_i) + \sum_j y_j\, q(y_j) = E(X) + E(Y)$$

Similarly

$$E(X_1 + X_2 + \dots X_n) = E(X_1) + E(X_2) + \dots E(x_n)$$

Let $g(X, Y) = XY$. For independent discrete random variables we have $p(x_i, y_j) = p(x_i)\, q(y_j)$ and

$$E(XY) = \sum_i \sum_j x_i y_j\, p(x_i, y_j) = \sum_i \sum_j (x_i p_1(x_i))\,(y_j p_2(y_j))$$

$$= \sum_i x_i p_1(x_i) \sum_j (y_j p_2(y_j)) = E(X)\, E(Y)$$

If X and Y are independent random variables with joint pdf $f(x, y)$, then

$$E(XY) = \int_{-\infty}^{\infty} \int_{-\infty}^{\infty} xy\, f(x, y)\, dx\, dy = \int_{-\infty}^{\infty} \int_{-\infty}^{\infty} xy\, g(x)\, h(y)\, dx\, dy$$

$$= \int_{-\infty}^{\infty} \left(\int_{-\infty}^{\infty} x\, g(x)\, dx \right) (y\, h(y)\, dy) = E(X)\, E(Y)$$

When one of the variables has a constant value k,

$$E(kX) = E(k)\, E(X) = kE(X)$$

Moments of a two-dimensional random variable:

$$\text{Let } g(X, Y) = X^i Y^j \qquad i, j = 1, 2, \dots$$

The moment of order i, j of a two-dimensional random variable is

$$\alpha_{i,j} = E(X^i Y^j) = \sum_i \sum_j x^i y^j\, p(x_i, y_j)$$

if (X, Y) is dicrete, and

$$\alpha_{i,j} = E(X^i Y^j) = \int_{-\infty}^{\infty} \int_{-\infty}^{\infty} x^i y^j\, f(x, y)\, dx\, dy,$$

if (X, Y) is continuous.

The two-dimensional first moment about the point $(E(X), E(Y))$ is

$$\mu_{i,j} = E((X - E(X))^i\, (Y - E(Y))^j$$

Now

$$\alpha_{1,0} = \bar{X} = E(X)$$

$$\alpha_{0,1} = \bar{Y} = E(Y)$$

$$\alpha_{1,1} = \bar{X}\,\bar{Y} = E(XY)$$
$$\mu_{2,0} = E(X - E(X))^2$$
$$\mu_{1,1} = E((X - E(X))(Y - E(Y))$$
$$= E(XY) - E(X)\,E(Y) = \alpha_{1,1} - \alpha_{1,0}\,\alpha_{0,1}$$
$$\mu_{0,2} = E((Y - E(Y))^2$$

The ratio $\dfrac{\mu_{11}}{\sqrt{\mu_{2,0}\,\mu_{0,2}}}$ gives an indication of the degree of linear dependence between the two variables, which is the correlation coeffficient.

3.2.1 Definition

Let X and Y be random variables with joint probability distribution $f(x, y)$. The covariance of X and Y, denoted by cov (X, Y), is

$$E((X - E(X))(Y - E(Y)) = \sum_x \sum_y (x - E(X))(y - E(Y))\,p(x, y)$$

if X and Y are discrete, and

$$E((X - E(X))(Y - E(Y)) = \int_{-\infty}^{\infty}\int_{-\infty}^{\infty} (x - E(X))(y - E(Y))\,f(x, y)\,dx\,dy$$

if X and Y are continuous.

3.2.2 Theorem

The covariance of two random variables X and Y with means $E(X)$, $E(Y)$ respectively, is given by $E(XY) - E(X)\,E(Y)$.

Proof. We prove the theorem for discrete case.

$$\text{cov}(X, Y) = \sum_x \sum_y (x - E(X))(y - E(Y))\,p(x, y)$$
$$= \sum_x \sum_y [xy - E(X)\,y - x\,E(Y) + E(X)\,E(Y)]\,p(x, y)$$
$$= \sum_x \sum_y xy\,p(x, y) - E(X)\sum_x \sum_y y\,p(x, y) - E(Y)\sum_x \sum_y x\,p(x, y) + E(X)\,E(Y)\sum_x \sum_y p(x, y)$$

Since $E(X) = \sum_x \sum_y x\,p(x, y)$ and $E(Y) = \sum_x \sum_y y\,p(x, y)$ by definition, and in addition $\sum_x \sum_y p(x, y) = 1$ for any joint discrete probability distribution, hence

$$\text{cov}(X, Y) = E(XY) - E(X)\,E(Y) - E(Y)\,E(X) + E(X)\,E(Y) = E(XY) - E(X)\,E(Y)$$

For the continuous case, the proof is identical with summations replaced by integrals.

3.2.3 Example

Prove that cov $(aX, bY) = ab$ cov (X, Y)

Solution. $\text{cov}(aX, bY) = E((aX - E(aX))(bY - E(bY))$

$$= E(a(X - E(X))\,b(Y - E(Y))$$
$$= ab\,E((X - E(X))(Y - E(Y)) = ab\,\text{cov}(X, Y)$$

3.2.4. Example

The fraction X of male runners and the fraction Y of female runners who complete marathon races can be described by the joint density function

$$f(x, y) = \begin{cases} 8xy, & 0 \le x \le 1, \ 0 \le y \le x \\ 0, & \text{otherwise} \end{cases}$$

Find the covariance of X and Y.

Solution. The marginal distributions of X and Y are as follows:

$$g(x) = \int_{-\infty}^{\infty} f(x, y)\, dy = \int_{0}^{x} 8xy\, dy = 4x^3, \ 0 < x < 1$$

$$h(y) = \int_{-\infty}^{\infty} f(x, y)\, dx = \int_{y}^{1} 8xy\, dx = 4y(1 - y^2), \ 0 < y < 1$$

$$E(X) = \int_{0}^{1} 4x^4 dx = \frac{4}{5}$$

$$E(Y) = \int_{0}^{1} 4y^2 (1 - y^2)\, dy = \frac{8}{15}$$

$$E(XY) = \int_{0}^{1}\int_{y}^{1} 8x^2y^2 dx\, dy = \frac{4}{9}$$

Then

$$\text{Cov}(X, Y) = E(XY) - E(X)\, E(Y)$$

$$= \frac{4}{9} - \frac{4}{5} \times \frac{8}{15} = \frac{4}{225}$$

Note that if X and Y are independent, then $E(XY) = E(X)\, E(Y)$. Hence $\text{cov}(X, Y) = 0$. But the converse is not true.

Let X be a random variable such that

$$P\{X = 0\} = P\{X = 1\} = P\{X = -1\} = \frac{1}{3}$$

and define

$$Y = \begin{cases} 0 & \text{if } X \ne 0 \\ 1 & \text{if } X = 0 \end{cases}$$

Now, $XY = 0$ and so $E(XY) = 0$. Also, $E(X) = 0$ and thus $\text{Cov}(X, Y) = E(XY) - E(X)\, E(Y) = 0$. However, X and Y are clearly not independent.

$$\begin{aligned} \text{Var}(X + Y) &= E((X + Y - E(X + Y))^2 \\ &= E[(X - E(X)) + (Y - E(Y)]^2 \\ &= E[(X - E(X))^2 + (Y - E(Y))^2 + 2(X - E(X))\,(Y - E(Y)] \\ &= E(X - E(X))^2 + E(Y - E(Y))^2 + 2E((X - E(X))\,(Y - E(Y)) \\ &= \text{Var}(X) + \text{Var}(Y) + 2\text{cov}(X, Y) \end{aligned}$$

Note: $\text{cov}\left(\sum_{i=1}^{n} X_i, \sum_{j=1}^{m} Y_j\right) = \sum_{j=1}^{m}\sum_{i=1}^{n} \text{cov}(X_i, Y_j)$

3.3. CORRELATION AND REGRESSION

In this section we see the "degree of association" between X and Y.

3.3.1 Definition

Let (X, Y) be a two-dimensional random variable. We define ρ_{XY}, the correlation coefficient between X and Y as follows

$$\rho_{XY} = \frac{E((X - E(X))(Y - E(Y))}{\sqrt{\text{var}(X)\,\text{var}(Y)}} = \frac{\text{cov}(X, Y)}{\sigma_x\, \sigma_y} \tag{3.3}$$

The basic properties of ρ_{XY} are given by the following theorems:

3.3.2 Theorem

$$-1 \le \rho_{XY} \le 1. \tag{3.4}$$

Proof. For arbitrary real numbers t and u consider the non-negative expression

$$\begin{aligned} &E[(t(X - E(X)) + u(Y - E(Y))]^2 \\ &= t^2 E(X - E(X))^2 + u^2 E(Y - E(Y))^2 + 2tu\, E((X - E(X))\,(Y - E(Y)) \\ &= t^2 \text{Var}(X) + 2tu \text{ cov}(X, Y) + u^2 \text{Var}(Y) \end{aligned} \tag{3.5}$$

Since the expression (3.5) is always non-negative, we must have

$$\text{cov}(X, Y) \le \text{var}(X).\ \text{var}(Y) \tag{3.6}$$

$$\Leftrightarrow \quad -\sqrt{\text{var}(X)\,\text{var}(Y)} \le \text{cov}(X, Y) \le \sqrt{\text{var}(X)\,\text{var}(Y)}$$

or $\quad -1 \le \dfrac{\text{cov}(X, Y)}{\sqrt{\text{var}(X)\,\text{var}(Y)}} \le 1$

or $\quad -1 \le \rho_{XY} \le 1$ by (3.3)

Note: If the random variables X and Y are independent, then cov $(X, Y) = 0$ and hence $\rho_{XY} = 0$. The converse is not true. Thus, if $\rho_{XY} = 0$, we say X and Y are uncorrelated. The equality (3.6) is called the **Cauchy Schwartz inequality**.

3.3.3 Example

Let X and Y be any random variables having the same distribution. Let $U = X - Y, V = X + Y$. Hence $E(U) = 0$, $E(V) = 0$ and cov $(U, V) = E[(X - Y)\,(X + Y)] = E(X^2 - Y^2) = 0$. Thus U and V are uncorrelated. Even if X and Y are independent, U and V may be dependent. For e.g, let X and Y be the numbers appearing on the first and second fair dice respectively, while tossing. We now find, for e.g, that $P(V = 4/U = 3) = 0$ (since if $X - Y = 3$, $X + Y$ cannot be 4), while $P(V = 4) = \dfrac{3}{36}$. Thus U and V are dependent. Hence being uncorrelated and being independent, are in general, not equivalent.

3.3.4 Theorem

$\rho_{XY}^2 = 1$ if and only if $Y = AX + B$, where A and B are constants. That is, if the correlation coefficient ρ is ± 1. Then Y is a linear function of X and vice versa (with probabilty 1).

Proof. Let $Y = AX + B$. Then $E(Y) = A\,E(X) + B$ and $\text{Var}(Y) = A^2\,\text{Var}(X)$. Also

$$E(XY) = E(X(AX + B)) = AE(X^2) + BE(X)$$

Hence
$$\rho_{XY}^2 = \left[\frac{\text{cov}(X, Y)}{\sqrt{\text{Var}(X)\,\text{Var}(Y)}}\right]^2 = \frac{(E(XY) - E(X)E(Y))^2}{\text{Var}(X)\,\text{Var}(Y)}$$

$$= \frac{[A\,E(X^2) + B\,E(X) - E(X)[A\,E(X) + B]]^2}{\text{Var}(X)\cdot A^2\,\text{Var}(X)}$$

$$= \frac{[A\,E(X^2) + B\,E(X) - A(E(X))^2 - B\,E(X)]^2}{\text{Var}(X)\,A^2\,\text{Var}(X)}$$

$$= \frac{A^2\,[E(X^2) - (E(X))^2]^2}{A^2\,(\text{Var}(X))^2} = 1.$$

Conversely, let $\rho_{XY}^2 = 1$. Consider the function $E(t(X - E(X)) + u(Y - E(Y)))^2$ as in previous theorem. Note that if $E[(t(X - E(X)) + u(Y - E(Y))]^2 > 0$, for all t, u, then $\rho_{XY}^2 < 1$. But, given $\rho_{XY}^2 = 1$. That means there exist atleast one value of t, u say t_0, u_0 such that $E[t_0(X - E(X) + u_0(Y - E(Y))]^2 = 0$. Hence

$$E(t_0(X - E(X)) + u_0(Y - E(Y))) = 0, \quad \text{and}$$

so $\quad \text{Var}\,[t_0(X - E(X)) + u_0(Y - E(Y)] = E((t_0(X - E(X)) + u_0(Y - E(Y))]^2 = 0.$

We know if $\text{Var}(X) = 0$, then using Chebyshevs inequality it is clear that $P(X = E(X)) = 1$. That is if $\text{Var}(X)$ is small, most of the probability distribution of X is 'concentrated' near $E(X)$. Hence using this concept we conclude that the random variable $t_0(X - E(X)) + u_0(Y - E(Y)) = 0$. Rewriting this, we find that $Y = AX + B$ (with probability 1).

Note. The correlation coefficient is a measure of the degree of linearity between X and Y. Values of ρ_{XY} near + 1 or –1 indicate a high degree of linearity, while $\rho_{XY} = 0$ indicates a lack of such linearity. Positive values of ρ_{XY} show that Y tends to increase with increasing X, while negative values of ρ_{XY} show that Y tends to decrease with increasing values of X. A value of ρ_{XY} close to zero indicates the absence of linear relationship between X and Y.

3.3.5 Example

Let (X, Y) be two-dimensional random variables with joint pdf

$$f(x, y) = \begin{cases} 2; & 0 < x < y < 1 \\ 0; & \text{otherwise} \end{cases}$$

The marginal pdf of X and Y are

$$g(x) = \int_x^1 2dy = 2(1 - x) \qquad 0 \le x \le 1$$

$$h(y) = \int_0^y 2dx = 2y \qquad 0 \le y \le 1.$$

$$E(X) = \int_0^1 2x\,(1-x)\,dx = \frac{1}{3}$$

$$E(X^2) = \int_0^1 x^2\, 2(1-x)\,dx = \frac{1}{6}$$

$$E(Y) = \int_0^1 y\, 2y\, dy = \frac{2}{3}$$

$$E(Y^2) = \int_0^1 y^2\, 2y\, dy = \frac{1}{2}$$

$$\text{Var}\,(X) = E(X^2) - (E(X))^2 = \frac{1}{18}$$

$$\text{Var}\,(Y) = E(Y^2) - (E(Y))^2 = \frac{1}{18}$$

$$E(XY) = \int_0^1\int_0^y xy\, 2dx\, dy = \frac{1}{4}$$

Hence

$$\rho_{XY} = \frac{E(XY) - E(X)E(Y)}{\sqrt{\text{Var}\,(X)\ \text{Var}\,(Y)}} = \frac{1}{2}$$

Conditional Expectation. Let (X, Y) be a two-dimensional continuous random variable. The conditional expectation of X for given $Y = y$ is

$$E(X/y) = \int_{-\infty}^{\infty} x\, g(x/y)\, dx$$

and the conditional expectation of X for given $Y = y_j$, if (X, Y) is a discrete random variable is

$$E(X/y_j) = \sum_{i=1}^{\infty} x_i\, p(x_i/y_j)$$

Similarly, we define the conditional expectation of Y for given $X = x$.
The conditional variance of X given $Y = y$ is

$$\text{Var}\,(X/y) = E\{[X - E(X/Y)]^2/y\} = \sum_x [x - E(X/y)]^2\, g(x/y)$$

and can be computed using

$$\text{Var}\,(X/y) = E\,(X^2/y) - [E(X/y)]^2$$

Note that $E(X/y)$ is a function of y alone and hence a random variable. Similarly $E(Y/x)$ is a function of x alone and hence a random variable. Strictly speaking $E(X/y)$ is the value of the random variable $E(X/Y)$. Since $E(X/Y)$ and $E(Y/X)$ are random variables, we talk of their expectations.

3.3.6 Theorem

$$E(E(X/Y)) = E(X)$$
$$E(E(Y/X)) = E(Y)$$

Proof: (Continuous case only): By definition

$$E(X/y) = \int_{-\infty}^{\infty} x\, g(x/y)\, dx = \int_{-\infty}^{\infty} x\,\frac{f(x, y)}{h(y)}\, dx$$

where f is the joint pdf of (X, Y) and h is the marginal pdf of Y. Hence

$$E(E(X/Y)) = \int_{-\infty}^{\infty} E(X/y)\, h(y)\, dy = \int_{-\infty}^{\infty} \left[\int_{-\infty}^{\infty} x \frac{f(x, y)}{h(y)}\, dx \right] h(y)\, dy$$

$$= \int_{-\infty}^{\infty} x \int_{-\infty}^{\infty} f(x, y)\, dy\, dx = \int_{-\infty}^{\infty} x\, g(x)\, dx = E(X)$$

Similarly we prove $E(E(Y/X)) = E(Y)$.

3.3.7 Example

Let X and Y have joint pdf

$$f(x, y) = \frac{x+y}{21}, \; x = 1, 2, 3, \; y = 1, 2.$$

Then the marginal pdf of X and Y is

$$g(x) = \sum_y f(x, y) = \sum_{y=1}^{2} \frac{x+y}{21} = \frac{x+1}{21} + \frac{x+2}{21} = \frac{2x+3}{21}, \; x = 1, 2, 3.$$

$$h(y) = \sum_x f(x, y) = \sum_{x=1}^{3} \frac{x+y}{21}$$

$$= \frac{1+y}{21} + \frac{2+y}{21} + \frac{3+y}{21} = \frac{6+3y}{21}, \; y = 1, 2$$

Thus the conditional pdf of X given $Y = y$ is

$$g(x/y) = \frac{f(x, y)}{h(y)} = \frac{\frac{x+y}{21}}{\frac{6+3y}{21}} = \frac{x+y}{6+3y}, \; x = 1, 2, 3 \text{ when } y = 1 \text{ or } 2$$

Similarly

$$h(y/x) = \frac{x+y}{2x+3}, \; y = 1, 2 \text{ when } x = 1, 2 \text{ or } 3$$

$$E(X/y = 2) = \sum_{x=1}^{3} x\, g(x/2) = \sum_{x=1}^{3} x \left(\frac{x+2}{12} \right) = \frac{13}{6}$$

and

$$\text{Var}(X/2) = E\left(\left(X - \frac{13}{6} \right)^2 \Big/ y = 2 \right) = \sum_{x=1}^{3} \left(x - \frac{13}{6} \right)^2 g(x/2)$$

$$= \sum_{x=1}^{3} \left(x - \frac{13}{6} \right)^2 \left(\frac{x+2}{12} \right) = \frac{23}{36}$$

3.3.8 Theorem

Suppose X and Y are independent random variables. Then

$$E(X/Y) = E(X) \quad \text{and} \quad E(Y/X) = E(Y)$$

Proof:

$$E(X/Y = y) = \int_{-\infty}^{\infty} xg(x/y)\, dx = \int_{-\infty}^{\infty} x \frac{f(x, y)}{h(y)}\, dx$$

$$= \int_{-\infty}^{\infty} x \frac{g(x)\, h(y)}{h(y)}\, dx = \int_{-\infty}^{\infty} xg(x)dx = E(X)$$

Similarly we prove

$$E(Y/X) = E(Y).$$

Regression of the Mean

$E(X/y)$ is the value of the random variable $E(X/Y)$ and is a function of y alone. That is, $E(X/Y) = a + by$. Let us find the constants a and b in terms of the characteristics $E(Y)$, $E(X)$, Var (X), Var (Y) and ρ_{XY}. Assume $\sigma_X = \sqrt{\text{Var}(X)}$, $\sigma_Y = \sqrt{\text{Var}(y)}$ are both positive so that the correlation coefficient will exist. Given

$$E(X/y) = \sum_x x\, g(x/y) = \sum_x x \frac{f(x, y)}{h(y)} = a + by, \quad y \in R_1$$

R_1 is the space of X.

Hence

$$\sum_x x\, f(x, y) = (a + by)\, h(y),\; y \in R,$$

and

$$\sum_{y \in R_1} \sum_x x\, f(x, y) = \sum_{y \in R_1} (a + by)\, h(y)$$

Multiply on both sides by y and sum

$$\sum_{y \in R_1} \sum_x xy\, f(x, y) = \sum_{y \in R_1} (ay + by^2)\, h(y)$$

That is

$$E(XY) = aE(Y) + bE(Y^2)$$

or equivalently

$$\mu_X \mu_Y + \rho_{XY}\, \sigma_X\, \sigma_Y = a\mu_Y + b\left[\mu_Y^2 + \sigma_X^2\right]$$

where $\mu_X = E(X)$, $\mu_Y = E(Y)$, $\sigma_X = \sqrt{\text{Var}(X)}$, $\sigma_Y = \sqrt{\text{Var}(Y)}$ (3.7)

we know

$$E(X/y) = a + by$$

That is,

$$E(E(X/y)) = a + bE(Y)$$

$$\Rightarrow \quad E(X) = a + bE(Y)$$

or

$$\mu_X = a + b\mu_Y \tag{3.8}$$

Solving (3.7) and (3.8), we get

$$a = \mu_X - \rho_{XY} \frac{\sigma_X}{\sigma_Y} \mu_Y$$

$$b = \rho_{XY}\frac{\sigma_X}{\sigma_Y}$$

which implies that if $E(X/y)$ is linear, then

$$E(X/y) = \mu_X + \rho_{XY}\frac{\sigma_X}{\sigma_Y}(y - \mu_Y)$$

which is the 'best' fit line. Similarly if $E(Y/x)$ is linear, then

$$E(Y/x) = \mu_Y + \rho_{XY}\frac{\sigma_Y}{\sigma_X}(x - \mu_X)$$

is the 'best' fit line. The point (μ_X, μ_Y) pass through both of these lines. The slopes of these lines are $\rho_{XY}\frac{\sigma_X}{\sigma_Y}$ and $\rho_{XY}\frac{\sigma_Y}{\sigma_X}$. Note that the product of coefficient of x is $E(Y/x)$ and the coefficient of y is $E(X/y)$ equals ρ_{XY}^2 and the ratio of the coefficients is $\frac{\sigma_{Y^2}}{\sigma_{X^2}}$.

Formally we define.

3.3.9 Definition

Let (X, Y) be a two-dimensional random variable and suppose that

$$E(X) = \mu_X,\ E(Y) = \mu_Y,\ \text{Var}(X) = \sigma_X^2,\ \text{Var}(Y) = \sigma_Y^2$$

Let ρ_{XY} be the correction coefficient of X and Y. If the regression of Y on X is linear, then

$$E(Y/x) = \mu_Y + \rho_{XY}\frac{\sigma_Y}{\sigma_X}(x - \mu_X)$$

If the regression of X on Y is linear, then

$$E(X/y) = \mu_X + \rho_{XY}\frac{\sigma_X}{\sigma_Y}(y - \mu_Y)$$

3.3.10 Example

Suppose that (X, Y) is uniformly distributed over the triangle given in Fig 3.2.

Then $f(x, y) = \frac{1}{2-1} = 1$, and the marginal and conditional *pdf*'s are:

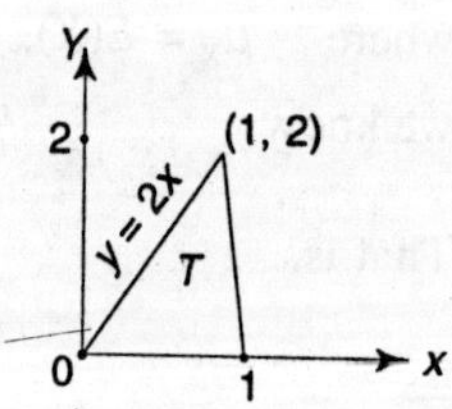

Fig. 3.2

$$g(x) = \int_0^{2x} f(x, y)\,dy = \int_0^{2x} dy = 2x,\ 0 \le x \le 1$$

$$h(y) = \int_{y/2}^{1} f(x, y)\,dx = [x]_{y/2}^{1} = 1 - y/2,\ 0 \le y \le 2$$

$$h(y/x) = \frac{f(x, y)}{g(x)} = \frac{1}{2x},\ 0 \le y \le 2x$$

$$g(x/y) = \frac{f(x, y)}{h(y)} = \frac{2}{2-y}, \; y/2 \le x \le 1$$

$$E(X/y) = \int_{y/2}^{1} x\, g(x/y)\, dx = \int_{y/2}^{1} x\left(\frac{2}{2-y}\right) dx = \frac{2}{2-y}\left[\frac{x^2}{2}\right]_{y/2}^{1} = \frac{2}{2-y}\left[\frac{1}{2} - \frac{y^2}{8}\right]$$

$$= \frac{\frac{1}{2} \times (1 - y/2)(1 + y/2)}{1 - y/2} = \frac{y}{4} + \frac{1}{2}$$

$$E(Y/x) = \int_{0}^{2x} y\, h(y/x)\, dy = \int_{0}^{2x} y\left(\frac{1}{2x}\right) dy = \frac{1}{2x}\left[\frac{y^2}{2}\right]_{1}^{2x} = \frac{1}{2x}\left[\frac{4x^2}{2}\right] = x.$$

Hence both the regression of Y on X and of X on Y are linear (See Fig 3.3)

3.3.11 Example

If the joint *pdf* of X and Y is

$$f(x, y) = \begin{cases} \frac{1}{y}, & 0 < x < y, 0 < y < 1 \\ 0 & \text{elsewhere} \end{cases}$$

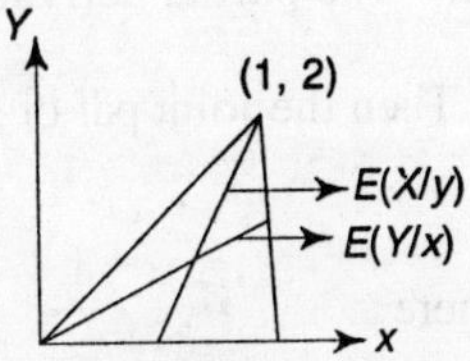

Fig. 3.3

Find $E(X/y)$

Solution. The conditional expectation of X on $Y = y$ is

$$E(X/y) = \int_{-\infty}^{\infty} x\, g(x/y)\, dx$$

$$h(y) = \int_{0}^{y} f(x, y)\, dx = \frac{1}{y}(x)_0^y = 1,$$

and $$g(x/y) = \frac{f(x, y)}{h(y)} = \frac{\frac{1}{y}}{1} = \frac{1}{y}, \; 0 < y < 1$$

Hence $$E(X/y) = \int_{0}^{y} x \cdot \frac{1}{y}\, dx = \frac{1}{y}\left[\frac{x^2}{2}\right]_0^y = \frac{y}{2}$$

3.4 TRANSFORMATION OF RANDOM VARIABLE

Given the probability distribution of one or more random variables we may find distributions of other random variables which are their function. Some such considerations we had in section 2.10. Here we study on two-dimensional random variables.

Let us consider $Z = H_1(X, Y)$, a function of the two random variables X and Y. Clearly $Z = Z(s)$ is again a random variable. Some of the random variables we shall be interested are $X + Y$, XY, X/Y, min (X, Y), max (X, Y) etc.

Now we try to find the answer for the problem: Given the joint pdf of (X, Y), what is the pdf of $Z = H_1(X, Y)$.

3.4.1 Theorem

Suppose that (X, Y) is a two-dimensional discrete random variable with joint pdf P. Let $U = \phi_1(X, Y)$ and $V = \phi_2(X, Y)$ be two discrete random variables such that to each pair of values of X and Y correspond on unique pair of values of U and V and vice versa so that $X = \psi_1(U, V)$ and $Y = \psi_2(U, V)$. Then the joint probability function of U and V is given by

$$g(u, v) = P[\psi_1(u, v), \psi_2(u, v)]$$

Proof.

$$g(u, v) = P(U = u, V = v) = P(\phi_1(X, Y)) = u, \phi_2(X, Y) = v)$$
$$= P(\psi_1(u, v)) = x, \psi_2(u, v) = y) = P(\psi_1(u, v), \psi_2(u, v))$$

3.4.2 Theorem

Suppose that (X, Y) is a two-dimensional continuous random variable with joint pdf f. Let $Z = H_1(X, Y)$ and $W = H_2(X, Y)$, and further assume

(i) The equations $z = H_1(x, y)$ and $w = H_2(x, y)$ may be uniquely solved for x and y in terms of z and w.

(ii) The partial derivatives $\frac{\partial x}{\partial z}, \frac{\partial x}{\partial w}, \frac{\partial y}{\partial z}, \frac{\partial y}{\partial w}$ exist and are continuous.

Then the joint pdf of z and w is given by $g(z, w) = f[\psi_1(z, w), \psi_2(z, w)]\,|J|$

where

$$J = \frac{\partial(x, y)}{\partial(z, w)} = \begin{vmatrix} \frac{\partial x}{\partial z} & \frac{\partial x}{\partial w} \\ \frac{\partial y}{\partial z} & \frac{\partial y}{\partial w} \end{vmatrix}$$

is called the Jacobian of transformation.

Proof. Consider the joint cdf of the two-dimensional random variable (Z, W) say

$$G(z, w) = P(Z \le z, W \le w) = \int_{-\infty}^{w}\int_{-\infty}^{z} g(s, t)\, ds\, dt$$

where g is our required pdf. By assumption (i) we may find the event R equivalent to $\{Z \le z, W \le w\}$ is terms of X and Y. That is $\{(X, Y) \in R\} \Leftrightarrow \{Z \le z, W \le w\}$. Hence

$$\int_{-\infty}^{w}\int_{-\infty}^{z} g(s, t)\, ds\, dt = \iint_C f(x, y)\, dx\, dy$$

Since f is assumed to be known, the integral on the RHS can be evaluated. Differentiating it with respect to z and w will yield the required pdf of Z and W.

Distribution of Sum of Random Variables

Let (X, Y) be two-dimensional random variables. Now we find the cdf F

$$F(Z) = P(X + Y \le Z)$$

If for a given value of z, we draw on the XY- plane the line $x + y = z$, then $F(z)$ will be the probability that the point with the cordinates (x, y) lies below this line

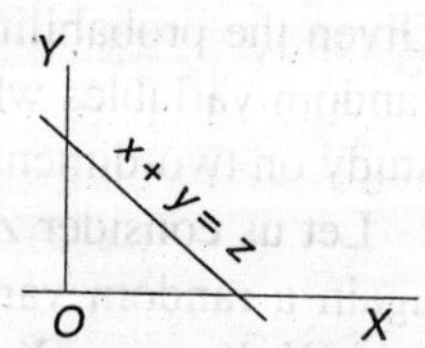

(i) If (X, Y) is a two-dimensional discreate random variable and takes on values (x_i, y_j), then

$$F(z) = \sum_{x_i + y_j \le z} P(X = x_i, Y = y_j)$$

(ii) If (X, Y) is a two-dimensional continuous random variable and let $f(x, y)$, $g(x)$, $h(y)$ denote respectively the densities of the random variables (X, Y), X and Y. In order to apply the previous theorem, let us consider

$$W = X,\ V = Z = X + Y$$

so

$$W = X \text{ and } Y = Z - X = Z - W$$

Now

$$J = \frac{\partial(x, y)}{\partial(z, W)} = \begin{vmatrix} 0 & 1 \\ 1 & -1 \end{vmatrix} = -1$$

Hence

$$g(x, z) = f(x, z - x) \tag{3.9}$$

The pdf of Z is obtained by intergrating (3.9) with respect to x from $-\infty$ to ∞.
That is

$$\psi(z) = \int_{-\infty}^{\infty} f(x, z - x)\, dx$$

Hence $$F(z) = \int_{-\infty}^{z} \psi(z)\, dz = \int_{-\infty}^{z} \left(\int_{-\infty}^{\infty} f(x, z - x)\, dx \right) dz$$

If X and Y are independent, then $f(x, y) = g(x)\, h(y)$. Hence

$$F(z) = \int_{-\infty}^{z} \left(\int_{-\infty}^{\infty} g(x) h(z - x)\, dx \right) dz$$

Distribution of Product of Random Variables

Consider the product of two random variables X and Y with the density function $f(x, y)$ and the marginal density $g(x)$ and $h(y)$ respectively. Let $Z = XY$ and $U = X$. Thus $X = U$ and $Y = \dfrac{Z}{X}$.

The Jacobian is $$J = \begin{vmatrix} 1 & 0 \\ -\dfrac{z}{x^2} & \dfrac{1}{u} \end{vmatrix} = \frac{1}{u} = \frac{1}{x}$$

Hence the Joint pdf of $Z = XY$ and $U = X$ is

$$g(z, u) = f\left(x, \frac{z}{x} \right) \frac{1}{|x|} \tag{3.10}$$

The density $\psi(z)$ of Z can be obtained by integrating (3.10) from $-\infty$ to ∞.

$$\psi(z) = \int_{-\infty}^{\infty} f\left(x, \frac{z}{x} \right) \frac{1}{|x|}\, dx$$

The cdf of Z as $$F(z) = \int_{-\infty}^{z} \left(\int_{-\infty}^{\infty} f\left(x, \frac{z}{x} \right) \frac{1}{|x|}\, dx \right) dz$$

If the random variables are independent, then

$$\psi(z) = \int_{-\infty}^{\infty} g(x) \cdot h\left(\frac{z}{x} \right) \frac{1}{|x|}\, dx$$

and $$F(z) = \int_{-\infty}^{z}\left(\int_{-\infty}^{\infty} g(x)\, h\left(\frac{z}{x}\right)\frac{1}{|x|}\, dx\right) dz.$$

Distribution of Quotient of random variables

Let (X, Y) be a continuous random variable.

Let $Z = \dfrac{X}{Y}$. Then the pdf of Z is

$$\psi(z) = \int_{-\infty}^{\infty} f(yz, y)\, |y|\, dy.$$

Proof. Let $Z = \dfrac{x}{y}$, $W = y$. The Jacobian is

$$J = \begin{vmatrix} y & 0 \\ -\dfrac{x}{z^2} & 1 \end{vmatrix} = y$$

Hence the joint pdf of $Z = \dfrac{X}{Y}$ and $W = Y$ is

$$g(z, w) = f(zy, y)\, |y| \tag{3.11}$$

The density $\psi(z)$ of Z can be obtained by integrating (3.11) from $-\infty$ to ∞. Hence

$$\psi(z) = \int_{-\infty}^{\infty} f(zy, y)\, |y|\, dy$$

the cdf of Z is

$$F(z) = \int_{-\infty}^{z} \psi(z)\, dz = \int_{-\infty}^{z}\left(\int_{-\infty}^{\infty} f(zy, y)\, |y|\, dy\right) dz$$

3.4.3 Example

If the joint pdf of X and Y is

$$f(x, y) = \begin{cases} e^{-(x+y)} & \text{for } x > 0,\ y > 0 \\ 0 & \text{elsewhere} \end{cases}$$

Find the pdf of $Z = \dfrac{X}{X+Y}$.

Solution. Since Z decreases, when Y increases and X is held constant, we can use Theorem 3.4.2 to find the joint density of X and Z. Since $z = \dfrac{x}{x+y}$, yields $y = x\left(\dfrac{1-z}{z}\right)$ and hence $\dfrac{\partial y}{\partial z} = -\dfrac{x}{z^2}$, it follows that

$$g(x, z) = e^{-x/z}\left|-\frac{x}{z^2}\right| = \frac{x}{z^2}\, e^{-x/z} \quad x > 0,\ 0 < z < 1$$

Finally, integrating with respect to x and changing the variable of integration to $u = x/z$, we get

$$h(z) = \int_0^\infty \frac{x}{z^2} e^{-x/z}\, dx = \int_0^\infty u\, e^{-u}\, du = \Gamma_{(2)} = 1,\ 0 < z < 1$$

Hence
$$h(z) = \begin{cases} 1 & 0 < z < 1 \\ 0 & \text{elsewhere} \end{cases}$$

3.4.4 Example

With reference to the random variables X and Y of the above example, find

(i) the joint pdf of $Z_1 = X + Y$ and $Z_2 = \dfrac{X}{X+Y}$

(ii) the marginal density of Z_2.

Solution.

(i) Solving $z_1 = x + y$ and $z_2 = \dfrac{x}{x+y}$, for x and y we get $x = z_1 z_2$ and $y = z_1(1 - z_2)$ and

$$J = \begin{vmatrix} z_2 & z_1 \\ 1 - z_2 & -z_1 \end{vmatrix} = -z_1$$

Then the joint pdf of Z_1 and Z_2 is

$$g(z_1, z_2) = f(x, y)\,|J| = f(z_1 z_2, z_1(1 - z_2))\,|z_1|$$

$$= e^{-(z_1 z_2 + z_1 - z_1 z_2)}|z_1| = e^{-z_1}|z_1| \quad \text{for } z_1 > 0, z_2 > 0$$

$$= e^{-z_1} z_1 \text{ for } z_1 > 0, 0 < z_2 < \infty$$

$$= 0 \text{ otherwise}$$

(ii) Integrating the joint pdf in part (i) with respect to z, we get

$$h(z_2) = \int_0^\infty g(z_1, z_2)\, dz_1 = \int_0^\infty z_1\, e^{-z_1}\, dz_1 = \Gamma_{(2)} = 1$$

3.4.5 Example

If the joint pdf of X and Y is

$$f(x, y) = \begin{cases} 1\ ; & \text{for } 0 < x < 1,\ 0 < y < 1 \\ 0\ ; & \text{elsewhere} \end{cases}$$

find

(i) the joint pdf of $Z = X + Y$ and $W = Y$

(ii) the marginal density of Z.

Solution. Solving for $z = x + y$ and $w = y$ for x and y we get $x = z - w$ and $y = w$,

$$J = \begin{vmatrix} 1 & -1 \\ 0 & 1 \end{vmatrix} = 1$$

and hence the joint pdf of Z and W is

$$g(z, w) = f(x, y)\,|J| = f(z - w, w)\,|J|$$
$$= 1 \text{ for } 0 < w < 1,\ w < z < 1 + w$$
$$= 0 \text{ elsewhere}$$

(ii) The marginal density of Z is

$$h(z) = \begin{cases} 0 & z \le 0 \\ \int_0^z 1\, dw = z & 0 < z < 1 \\ \int_{z-1}^{1} 1\, dw = 2 - z & 1 < z < 2 \\ 0 & z \ge 2 \end{cases}$$

Note 1: If we want to find the joint probability density of the random variables $Z_1 = H_1(X_1, X_2, X_3)$, $Z_2 = H_2(X_1, X_2, X_3)$, $Z_3 = H_3(X_1, X_2, X_3)$, the procedure is same, but the Jacobian is now the 3×3 determinant

$$J = \begin{vmatrix} \frac{\partial x_1}{\partial z_1} & \frac{\partial x_1}{\partial z_2} & \frac{\partial x_1}{\partial z_3} \\ \frac{\partial x_2}{\partial z_1} & \frac{\partial x_2}{\partial z_2} & \frac{\partial x_2}{\partial z_3} \\ \frac{\partial x_3}{\partial z_1} & \frac{\partial x_3}{\partial z_2} & \frac{\partial x_3}{\partial z_3} \end{vmatrix}$$

Note 2. If $Z = aX + bY$ and $W = cX + dY$ and if $ad + bc \neq 0$, then $ax + by = z$, $cx + dy = w$ has one and only one solution.

$$x = Az + Bw,\quad y = Cz + Dw$$

Since $$|J| = \frac{1}{|ad - bc|}$$

$$g(z, w) = f(x, y)\,|J| = \frac{1}{|ad - bc|} f(Az + Bw,\ Cz + Dw)$$

In particular if $Z = X\cos\phi + Y\sin\phi$, $W = -X\sin\phi + Y\cos\phi$, then $a = d = \cos\phi$, $b = -c = \sin\phi$ and $ad - bc = 1$. Hence

$$X = Z\cos\phi - W\sin\varphi$$
$$Y = Z\sin\phi + W\cos\phi$$

and hence

$$g(z, w) = f(z\cos\phi - w\sin\phi,\ z\sin\varphi + w\cos\varphi).$$

Thus if two random variables are rotated by an angle ϕ, the pdf are rotated in the opposite direction by the same angle.

3.4.6 Theorem

Let (X, Y) be two dimensional random variables with a joint pdf f. Let $Z = H_1(X, Y)$, $W = H_2(X, Y)$, then $E(Z + W) = E(Z) + E(W)$.

Proof.

$$E(Z+W) = \int_{-\infty}^{\infty}\int_{-\infty}^{\infty} [H_1(X,Y) + H_2(X,W)]\, f(x,y)\, dx\, dy$$

$$= \int_{-\infty}^{\infty}\int_{-\infty}^{\infty} H_1(X,Y) f(x,y)\, dx\, dy + \int_{-\infty}^{\infty}\int_{-\infty}^{\infty} H_2(X,Y) f(x,y)\, dx\, dy$$

$$= E(Z) + E(W)$$

Corollory: If $H_1(X,Y) = X$, $H_2(X,Y) = Y$, then
$E(X+Y) = E(X) + E(Y)$

3.5 CENTRAL LIMIT THEOREM

Central limit theorems are concerned with determining conditions under which the sum of a large number of random variables has a probability distribution which is approximately a normal distribution.

3.5.1 Theorem

Let $X_1, X_2, \ldots X_n$ be n independent random variables with respective means $\mu_1, \mu_2, \ldots \mu_n$ and variances $\sigma_1^2, \sigma_2^2 \ldots \sigma_n^2$. Then the mean and variance of $Y = \sum_{i=1}^{n} a_i X_i$, where $a_1, \ldots a_n$ are real constants are

$$\mu_Y = \sum_{i=1}^{n} a_i\, \mu_i \text{ and } \sigma_Y^2 = \sum_{i=1}^{n} a_i^2 \sigma_i^2$$

Proof:

$$\mu_Y = E(Y) = E\left(\sum_{i=1}^{n} a_i X_i\right) = \sum_{i=1}^{n} a_i E(X_i) = \sum_{i=1}^{n} a_i \mu_i$$

$$\sigma_Y^2 = E[(Y-\mu_Y)]^2 = E\left[\left(\sum_{i=1}^{n} a_i X_i - \sum_{i=1}^{n} a_i \mu_i\right)\right]^2$$

$$= E\left[\sum_{i=1}^{n} a_i (X_i - \mu_i)\right]^2 = E \sum_{i=1}^{n}\sum_{j=1}^{n} a_i a_j (X_i - \mu_i)(X_j - \mu_j)$$

$$= \sum_{i=1}^{n}\sum_{j=1}^{n} a_i a_j E[(X_i - \mu_i)(X_j - \mu_j)]$$

since E is linear

$$= \sum_{i=1}^{n}\sum_{j=1}^{n} a_i a_j E(X_i - \mu_i) E(X_j - \mu_j)$$

since X_i, X_j are independent

For $i \neq j$,

$$E(X_i - \mu_i) E(X_j - \mu_j) = (\mu_i - \mu_i)(\mu_j - \mu_j) = 0$$

Hence

$$\sigma_Y^2 = \sum_{i=1}^{n} a_i^2 E[(X_i - \mu_i)]^2 = \sum_{i=1}^{n} a_i^2 \sigma_i^2$$

3.5.2 Theorem

Let $X_1, X_2, \ldots X_n$ be independent random variables with respective moment- generating functions $M_{x_i}(t)$, $i = 1,2,3,\ldots n$. Then the moment-generating function of $Y = \sum_{i=1}^{n} a_i X_i$ is

$$M_Y(t) = \prod_{i=1}^{n} M_{Xi}(a_i t)$$

Proof: The moment generating function of Y is given by

$$M_Y = E(e^{tY}) = E\left(e^{t(a_1X_1 + a_2X_2 + \ldots a_nX_n)}\right)$$

$$= E\left(e^{a_1tX_1 + a_2tX_2 + \ldots a_ntX_n}\right) = E\left(e^{a_1tX_1}\right)E\left(e^{a_2tX_2}\right)\ldots E\left(e^{a_ntX_n}\right)$$

$$= \left(M_{X_1}(a_1t)\right)\left(M_{X_2}(a_2t)\right)\ldots\left(M_{Xn}(a_nt)\right) = \prod_{i=1}^{n} M_{X_i}(a_it)$$

If $\quad Y = \sum_{i=1}^{n} \frac{1}{n} X_i$, then

$$M_Y(t) = \prod_{i=1}^{n} M_{X_i}\left(\frac{t}{n}\right) = \left[M\left(\frac{t}{n}\right)\right]^n$$

Central limit theorem

Let $X_1, X_2, \ldots$ be a sequence of independent and identicallly distributed random variables each having mean μ and variance σ^2. Then the distribution of

$$W_n = \frac{(X_1 + X_2 + \ldots X_n) - n\mu}{\sigma\sqrt{n}}$$

is $N(0, 1)$ in the limit as $n \to \infty$.

In order to prove the central limit theorem we define the following lemma.

Lemma. Let $X_1, X_2, \ldots$ be sequence of random variables having distribution function F_{X_i} and moment-generating function $M_{X_i}(t)$, $i \geq 1$; and let X be a random variable having distribution function F_X and moment-generating function $M_X(t)$. If $M_{X_i}(t) \to M_X(t)$ for all t, then $F_{X_i}(t) \to F_X(t)$ for all t at which $F_X(t)$ is continuous.

Proof of central limit Theorem

$$M_{W_n}(t) = E\left(e^{tW_n}\right) = E\left(e^{t\left(\sum_{i=1}^{n} X_i - n\mu\right)\Big/\sigma\sqrt{n}}\right)$$

$$= E\left[e^{\frac{t}{\sigma\sqrt{n}}(X_1 - n\mu)}\right] \cdot E\left(e^{\frac{t}{\sigma\sqrt{n}}(X_2 - n\mu)}\right) \ldots E\left(e^{\frac{t}{\sigma\sqrt{n}}(X_n - n\mu)}\right)$$

since $X_1, X_2, \ldots X_n$ are independent

$$= \left[M\left(\frac{t}{\sqrt{n}}\right)\right]^n, \qquad -h < \frac{t}{\sqrt{n}} < h \tag{3.12}$$

where $M(t) = E\left(e^{t\left(\frac{X_i - \mu}{\sigma}\right)}\right)$, $-h < t < h$, is the common moment-generating function of each $Y_i = \frac{X_i - \mu}{\sigma}$, $i = 1, 2, \ldots n$. Since $E(Y_i) = 0$ and $E\left(Y_i^2\right) = 1$, it must be that

$$M(0) = 1,\ M'(0) = E\left(\frac{X_i - \mu}{\sigma}\right) = 0,\ M''(0) = E\left(\frac{X_i - \mu}{\sigma}\right)^2 = 1.$$

Hence, using Tayler's formula with a remainder, we can find a number t_1 between 0 and t such that

$$M(t) = M(0) + M'(0)\, t + M''(t_1)\frac{t^2}{2}$$

$$= 1 + \frac{t^2}{2}\, M''(t_1) = 1 + \frac{t^2}{2} + \frac{(M''(t_1) - 1)\, t^2}{2}$$

Using this expression of $M(t)$ in (3.12), we get

$$E(e^{tW_n}) = \left\{1 + \frac{1}{2}\left(\frac{t}{\sqrt{n}}\right)^2 + \frac{1}{2}[M''(t_1) - 1]\left(\frac{t}{\sqrt{n}}\right)^2\right\}^n$$

$$= \left\{1 + \frac{t^2}{2n} + \frac{[M''(t_1) - 1]\, t^2}{2n}\right\}^n,\ -\sqrt{nh} < t < \sqrt{nh}$$

$t_1 \in (0, t/\sqrt{n}\,)$. Since $M''(t)$ is continuous at $t = 0$ and $t_1 \to 0$ as $n \to \infty$. Hence

$$\operatorname*{Lt}_{n\to\infty}\ [M''(t_1) - 1] = 1 - 1 = 0.$$

So $$\operatorname*{Lt}_{n\to\infty}\ E(e^{tW_n}) = \operatorname*{Lt}_{n\to\infty}\ \left\{1 + \frac{t^2}{2n} + \frac{(M''(t_1) - 1)}{2n}\, t^2\right\}^n$$

$$= \operatorname*{Lt}_{n\to\infty}\ \left\{1 + \frac{\frac{t^2}{2}}{n}\right\}^n = e^{t^2/2}$$

which completes the proof.

When n is sufficienty large, the distribution of W_n, namely

$$P(W_n \le w) \simeq \int_{-\infty}^{w} \frac{1}{\sqrt{2\pi}}\, e^{-z^2/2}\, dz$$

Note that $E(W_n) = 0$ and Var $(W_n) = 1$.

3.5.3 Example

Let $\bar{X} = \bar{X}_{25}$ denote the mean of a random sample of size 25 from the distribution whose pdf is $f(x) = \frac{x^3}{4}$, $0 < x < 2$. It is easy to show that

$$\mu = \frac{8}{5}, \sigma^2 = \frac{8}{75}. \text{ Thus}$$

$$P(1.5 \le \bar{X} \le 1.65) = P\left(\frac{1.5-1.6}{\sqrt{\frac{8}{75}}\sqrt{25}} \le \frac{\bar{X}-1.6}{\sqrt{\frac{8}{75}}\sqrt{25}} \le \frac{1.65-1.6}{\sqrt{\frac{8}{75}}\sqrt{25}}\right)$$

$$\approx \phi(0.77) - \phi(-1.53) \approx 0.7164 \text{ by normal table}$$

The above example show how the central limit theorem can be used for approximating certain probabilities concerning the mean $\bar{X}_n$ of a random sample. That is, $\bar{X}_n \approx N(\mu, \sigma^2/n)$, when n is sufficiently large.

Central limit theorem for discrete type distributions

Let $X_1, X_2, \ldots X_n$ be a random sample from a Bernoulli distribution with a mean $\mu = p$ and a variance $\sigma^2 = p(1-p)$, where $0 < p < 1$. Then the central limit theorem states that the distribution of

$$W_n = \frac{(X_1 + X_2 + \ldots + X_n) - np}{\sqrt{np(1-p)}}$$

is $N(0, 1)$ in the limit as $n \to \infty$. Thus, if n is "sufficiently large", probabilities Bernoulli distribution can be approximated using $N(np, np(1-p))$. A rule often stated is that n is "suficiently large" if $np \ge 5$ and $n(1-p) \ge 5$.

3.6. SOLVED PROBLEMS

1. The joint density function of a bivariate distribution is given as

$$f(x, y) = \begin{cases} x + y & 0 \le x \le 1,\ 0 \le y \le 1 \\ 0 & \text{otherwise} \end{cases}$$

Determine the marginal distributions of x and y.

Solution. Marginal density function of x is given by

$$g(x) = \int_{-\infty}^{\infty} f(x, y)\, dy = \int_0^1 (x + y)\, dy = \left(x + \frac{1}{2}\right), 0 \le x \le 1$$

Marginal density function of y is given by

$$h(y) = \int_{-\infty}^{\infty} f(x, y)\, dx = \int_0^1 (x + y)\, dx = \left(y + \frac{1}{2}\right); 0 \le y \le 1.$$

2. Let X has density function defined by

$$f(x) = \begin{cases} e^{-x} & x > 0 \\ 0 & x \le 0 \end{cases}$$

Find the density function of $Y = X^2$.

Solution.

$$P(Y \le y) = P(X^2 \le y) = P\left(-\sqrt{y} \le X \le \sqrt{y}\right)$$

$$= \int_{\sqrt{y}}^{0} f(u)\, du + \int_{0}^{\sqrt{y}} f(u)\, du = \int_{0}^{\sqrt{y}} f(-u)\, du + \int_{0}^{\sqrt{y}} f(u)\, du$$

$$= \int_{0}^{\sqrt{y}} [f(u) + f(-u)]\, du = \int_{0}^{\sqrt{y}} h(u)\, du, \text{ where } h(u) = f(u) + f(-u)$$

$$f(y) = F'(y) = \frac{d}{dy}(P(Y \le y)) = \frac{d}{dy}\left\{\int_{0}^{\sqrt{y}} h(u)\, du\right\}$$

$$= \frac{d}{d(\sqrt{y})}\left\{\int_{0}^{\sqrt{y}} h(u)\, du\right\}\frac{d(\sqrt{y})}{dy}$$

$$= \frac{h(\sqrt{y})}{2(\sqrt{y})} \text{ using the fundamental theorem}$$

$$= \frac{1}{2\sqrt{y}}[f(\sqrt{y}) + f(-\sqrt{y})]$$

Hence

$$f(y) = \frac{1}{2\sqrt{y}}[f(\sqrt{y}) + f(-\sqrt{y})] = \frac{e^{-\sqrt{y}}}{2\sqrt{y}} \quad \text{for } y > 0$$

since by hypothesis $f(-\sqrt{y}) = 0$. Therefore

$$f(y) = \begin{cases} \dfrac{e^{-\sqrt{y}}}{2\sqrt{y}} & \text{for } y > 0 \\ 0 & \text{otherwise} \end{cases}$$

3. A coin is tossed 200 times. Let p and q be the probabilities of getting heads and tails respectively. What is the probability that heads will appear more than 140 times and less than 150 times?

Solution. Since $p = q = \frac{1}{2}$, this is a binomial distribution, with mean $= np = 200 \times \frac{1}{2} = 100$ and variance $\sigma^2 = npq = 200 \times \frac{1}{2} \times \frac{1}{2} = 50$. Using De-Moivre-Laplace central limit theorem, binomial distribution tends to normal distribution as $n \to \infty$. Hence

$$P\{140 < X < 150\} = P\left\{\frac{140-100}{\sqrt{50}} < Z < \frac{150-100}{\sqrt{50}}\right\}$$

$$= P\left\{\frac{40}{7.07} < Z < \frac{50}{7.07}\right\} = P\{5.658 < Z < 7.07\}$$

using central limit theorem for large n,

$$P(X_1 < X < X_2) = \frac{1}{2\pi} \int_{y_1}^{y_2} e^{-y^2/2}\, dy$$

we get

$$P(5.658 < Z < 7.07) = \frac{1}{2\pi} \int_{5.658}^{7.07} e^{-y^2/2}\, dy$$

4. A fair coin is tossed 10 times. Find the probability of getting 3 or 4 or 5 heads using central limit theorem.

Solution. Mean = $np = 10 \times \frac{1}{2} = 5$, variance $\sigma^2 = npq = 2.5$

It is to be noted that when approximating a discrete probability distribution to continuous probability distribution, we have to add 0.5 to the upper bound and subtract 0.5 from the lower bound. Hence

$$P(2.5 < X < 5.5) = P\left[\frac{2.5-5}{1.581} < Z < \frac{5.5-5}{1.581}\right]$$

$$= P[-1.58 < Z < 0.32] = \phi(0.32) - \phi(-1.58)$$

$$= \phi(0.32) - [1 - \phi(1.58)] = .6255 - .0571 = .5684$$

5. State and prove the Markov Inequality.
Let Y be a positive-valued random variable with finite expectation $E(Y)$. Then for every $k > 1$,

$$P(Y \geq kE(Y)) \leq \frac{1}{k}$$

Proof. $$\mu = E(Y) = \int_0^\infty yf(y)dy \geq \int_{k\mu}^\infty yf(y)dy \geq k\mu \int_{k\mu}^\infty f(y)dy$$

$$= k\mu\, P(Y \geq k\mu)$$

$$\Rightarrow \quad P\,(Y \geq k\,E(Y)) \leq \frac{1}{k}$$

6. Let X and Y be independent random variables each having density function

$$f(u) = \begin{cases} 2e^{-2u} & u \geq 0 \\ 0 & \text{otherwise} \end{cases}$$

Find (i) $E(X + Y)$, (ii) $E(XY)$

Solution.

(i) $$E(X) = 2\int_0^\infty ue^{-2u}du = \left[-ue^{-2u}\right]_0^\infty + \left[\frac{e^{-2u}}{2}\right]_0^\infty = \frac{1}{2}$$

$$E(X + Y) = E(X) + E(Y) = \frac{1}{2} + \frac{1}{2} = 1$$

(ii) $$E(XY) = E(X)\,E(Y) = \frac{1}{2} \times \frac{1}{2} = \frac{1}{4}$$

7. Let X and Y be random variables with joint density function

$$f(x, y) = \begin{cases} 4xy, & 0 \le x \le 1, \ 0 \le y \le 1 \\ 0 & \text{otherwise} \end{cases}$$

Find (i) $E(X)$, (ii) $E(Y)$, (iii) $E(X + Y)$, (iv) $E(XY)$

Solution.

(i) $E(X) = \int_{-\infty}^{\infty}\int_{-\infty}^{\infty} xf(x, y)\, dx\, dy$

$$= \int_0^1\int_0^1 x^2 4y\, dxdy = 4\int_0^1 x^2 dx \int_0^1 ydy = \frac{2}{3}$$

(ii) $E(Y) = \int_{-\infty}^{\infty}\int_{-\infty}^{\infty} yf(x, y)\, dy\, dx.$

$$= 4\int_0^1 xdx \int_0^1 y^2 dy = \frac{2}{3}$$

(iii) $E(X + Y) = E(X) + E(Y) = \frac{2}{3} + \frac{2}{3} = \frac{4}{3}$

(iv) $E(XY) = \int_{-\infty}^{\infty}\int_{-\infty}^{\infty} xy\, f(x, y)\, dy\, dx = 4\int_0^1 x^2 dx \int_0^1 y^2 dy$

$$= 4\left[\frac{x^3}{3}\right]_0^1 \left[\frac{y^3}{3}\right]_0^1 = \frac{4}{9}$$

Since $E(XY) = \frac{4}{9}$, $E(X)\, E(Y) = \frac{2}{3} \times \frac{2}{3} = \frac{4}{9}$, X and Y are independent.

8. Suppose that 21 slips numbered from 1 to 21 are in an urn from where we draw a slip at random. Let

$$X = \begin{cases} 1 & \text{if the slip bears a number divisible by 2} \\ 0 & \text{otherwise} \end{cases}$$

$$Y = \begin{cases} 1 & \text{if the slip bears a number divisible by 3} \\ 0 & \text{otherwise} \end{cases}$$

Find the joint distribution and the marginal distributions.

Solution. Among 21 numbers we have, the sets as

$A = \{x{:}x \text{ is divisible by both 2 and 3}\} = \{6, 12, 18\}$

$B = \{x{:}x \text{ is divisible by 2 and not by 3}\} = \{2, 4, 8, 10, 14, 16, 20\}$

$C = \{x{:}x \text{ is divisible by 3 and not by 2}\} = \{3, 9, 15, 21\}$

$D = \{x{:}x \text{ is neither divisible by 2 nor divisible by 3}\} = \{1, 5, 7, 11, 13, 17, 19\}$

If the probability of drawing each slip is same then

$$P(X = 1, Y = 1) = \frac{3}{21} \quad ; \quad P(X = 1, Y = 0) = \frac{7}{21},$$

$$P(X = 0,\ Y = 1) = \frac{4}{21}\quad,\quad P(X = 0,\ Y = 0) = \frac{7}{21}.$$

Hence we have the table showing the joint probability distribution of numbers divisible by 2 and 3.

Y \ X	0↓	1	Total
→ 0	$\frac{7}{21}$	$\frac{4}{21}$	$\frac{11}{21}$
1	$\frac{7}{21}$	$\frac{3}{21}$	$\frac{10}{21}$
Total	$\frac{14}{21}$	$\frac{7}{21}$	1

Hence the marginal distribution of divisibility of 2 is

$$P(X = 1,\ Y = 1) + P(X = 1,\ Y = 0) = \frac{3}{21} + \frac{7}{21} = \frac{10}{21}$$

$$P(X = 0,\ Y = 1) + P(X = 0,\ Y = 0) = \frac{4}{21} + \frac{7}{21} = \frac{11}{21}$$

Marginal distribution of divisibility by 3 is

$$P(X = 1,\ Y = 1) + P(X = 0,\ Y = 1) = \frac{3}{21} + \frac{4}{21} = \frac{7}{21}$$

$$P(X = 1,\ Y = 0) + P\ (X = 0,\ Y = 0) = \frac{7}{21} + \frac{7}{21} = \frac{14}{21}$$

9. Given the joint probability distribution as

Y \ X	1	2	3	Total
1	$\frac{1}{12}$	$\frac{1}{6}$	0	$\frac{3}{12}$
2	0	$\frac{1}{9}$	$\frac{1}{5}$	$\frac{14}{45}$
3	$\frac{1}{18}$	$\frac{1}{4}$	$\frac{2}{15}$	$\frac{79}{180}$
Total	$\frac{5}{36}$	$\frac{19}{36}$	$\frac{1}{3}$	1

(i) Find the marginal distribution of X and Y.

(ii) Find the conditional distribution of Y given $X = 3$.

(iii) Find $P(X \leq 2,\ Y = 3)$

Solution. Marginal distribution of X:

$$P(X = 1) = \frac{1}{12} + 0 + \frac{1}{18} = \frac{5}{36}$$

$$P(X = 2) = \frac{1}{6} + \frac{1}{9} + \frac{1}{4} = \frac{19}{36}$$

$$P(X = 3) = 0 + \frac{1}{5} + \frac{2}{15} = \frac{5}{15} = \frac{1}{3}$$

Marginal distribution of Y:

$$P(Y = 1) = \frac{1}{12} + \frac{1}{6} = \frac{1+2}{12} = \frac{3}{12} = \frac{1}{4}$$

$$P(Y = 2) = \frac{1}{9} + \frac{1}{5} = \frac{5+9}{45} = \frac{14}{45}$$

$$P(Y = 3) = \frac{1}{18} + \frac{1}{4} + \frac{2}{15} = \frac{79}{180}$$

(ii) Conditional distribution of Y given $X = 3$

$$P(Y = 1/X = 3) = \frac{P(Y=1, X=3)}{P(X=3)} = 0$$

$$P(Y = 2/X = 3) = \frac{P(Y=2, X=3)}{P(X=3)} = \frac{1/5}{1/3} = \frac{3}{5}$$

$$P(Y = 3/X = 3) = \frac{P(Y=3, X=3)}{P(X=3)} = \frac{2/15}{1/3} = \frac{2}{5}$$

(iii) Find $P(X \le 2, Y = 3) = P(X = 1, Y = 3) + P(X = 2, Y = 3) = \frac{1}{18} + \frac{1}{4} = \frac{2+9}{36} = \frac{11}{36}$

10. Find the acute angle between the two lines of regression.

Solution. Angle between two lines whose slopes are m_1 and m_2 is

$$\tan\theta = \frac{m_1 - m_2}{1 + m_1 m_2}$$

where θ is the acute angle between the lines.

The line of regression of Y on X is

$$Y - \mu_Y = \rho_{XY} \cdot \frac{\sigma_Y}{\sigma_X}(X - \mu_X)$$

The line of regression of X on Y is

$$X - \mu_X = \rho_{XY} \cdot \frac{\sigma_X}{\sigma_Y}(Y - \mu_Y)$$

or

$$Y - \mu_Y = \frac{1}{\rho_{XY}} \frac{\sigma_Y}{\sigma_X}(X - \mu_X)$$

The slopes of these lines are $m_1 = \rho_{XY} \cdot \dfrac{\sigma_Y}{\sigma_X}$ and $m_2 = \dfrac{\sigma_Y}{\rho_{XY} \cdot \sigma_X}$. Therefore

$$\tan\theta = \frac{\rho_{XY}\dfrac{\sigma_Y}{\sigma_X} - \dfrac{\sigma_Y}{\rho_{XY}\sigma_X}}{1 + \left(\rho_{XY}\dfrac{\sigma_Y}{\sigma_X}\right)\left(\dfrac{\sigma_Y}{\rho_{XY}\sigma_X}\right)}$$

$$= \frac{\rho_{XY}^2 - 1}{\rho_{XY}}\left(\frac{\sigma_X\sigma_Y}{\sigma_X^2 + \sigma_Y^2}\right)$$

$$= \frac{1 - \rho_{XY}^2}{\rho_{XY}}\left(\frac{\sigma_X\sigma_Y}{\sigma_X^2 + \sigma_Y^2}\right) \text{ since } \rho_{XY}^2 \le 1 \text{ and } \theta \text{ is accute.}$$

Therefore $$\theta = \tan^{-1}\left[\frac{(1-\rho_{XY}^2)}{\rho_{XY}}\left(\frac{\sigma_X\sigma_Y}{\sigma_X^2+\sigma_Y^2}\right)\right]$$

The obtuse angle between the regression lines is given by

$$\theta = \tan^{-1}\left[\frac{(\rho_{XY}^2-1)}{\rho_{XY}}\left(\frac{\sigma_X\sigma_Y}{\sigma_X^2+\sigma_Y^2}\right)\right]$$

11. Prove that the regression coefficients are independent of the change of origin but dependent on change of scale.

Solution. Let $$u_i = \frac{x_i - A}{h},\; v_i = \frac{y_i - B}{k}$$

Then $x_i = A + hu_i$ and $y_i = B + kv_i$. We know that $\sigma_x = h\sigma_u$; $\sigma_y = k\sigma_v$ and $\rho_{XY} = \rho_{UV}$. Now,

$$b_{YX} = \rho_{XY}\frac{\sigma_Y}{\sigma_X} = \rho_{UV}\left(\frac{k\sigma_v}{h\sigma_u}\right) = \frac{k}{h}b_{VU}$$

Similarly

$$b_{XY} = \frac{h}{k}b_{UV}.$$

Hence we observe that b_{YX} and b_{XY} depend upon the scales h and k but not on the origins A and B.

12. Calculate the correlation coefficient and obtain the lines of regression from the following data. Also obtain the value of Y when $X = 6.2$.

X:	1	2	3	4	5	6	7	8	9
Y:	9	8	10	12	11	13	14	16	15

$$\bar{x} = \frac{45}{9} = 5; \; \bar{y} = \frac{108}{9} = 12$$

X	Y	$X-\bar{X}$	$Y-\bar{Y}$	$(X-\bar{X})^2$	$(Y-\bar{Y})^2$	$(X-\bar{X})(Y-\bar{Y})$
1	9	–4	–3	16	9	12
2	8	–3	–4	9	16	12
3	10	–2	–2	4	4	4
4	12	–1	0	1	0	0
5	11	0	–1	0	1	0
6	13	1	1	1	1	1
7	14	2	2	4	4	4
8	16	3	4	9	16	12
9	15	4	3	16	9	12
				60	60	57

$$\sigma_X = \sqrt{\frac{\sum (x-\bar{x})^2}{n}} = \sqrt{\frac{60}{9}} = 2.58$$

$$\sigma_Y = \sqrt{\frac{\sum (y-\bar{y})^2}{n}} = \sqrt{\frac{60}{9}} = 2.58$$

$$\rho_{XY} = r = \frac{\frac{57}{9}}{(2.58)\,(2.58)} = 0.95$$

Regression line of Y on X is given by

$$(Y-\bar{Y}) = r \cdot \frac{\sigma_Y}{\sigma_X}\,(X-\bar{X})$$

$$\Rightarrow \quad Y - 12 = (0.95)\left(\frac{2.58}{2.58}\right)(X-5)$$

$$\Rightarrow \quad Y = 0.95X - (0.95)\,(5) + 12$$

$$= (0.95)X + 7.25$$

Regression line of X on Y is given by

$$X - \bar{X} = r \cdot \left(\frac{\sigma_X}{\sigma_Y}\right)(Y-\bar{Y})$$

$$\Rightarrow X - 5 = (0.95)\,(Y-12)$$

$$= (0.95)Y - (0.95) \times 12 + 5$$

$$\Rightarrow \quad X = (0.95)Y - 6.4$$

when $\quad X = 6.2$,

$$Y = (0.95)\,(6.2) + 7.25 = 13.14$$

EXERCISES

1. The joint density function of (X, Y) is

$$f(x, y) = Ae^{-x-y}, 0 \le x \le y, 0 \le y < \infty$$
$$= 0 \quad \text{otherwise}$$

   (i) Determine the constant A.
   (ii) Find the marginal density function of Y.

2. Find the joint p.d.f. of (X, Y) if joint distribution is given by

$$F(X, Y) = \begin{cases} \left(1-e^{-x^2}\right)\left(1-e^{-y^2}\right) & \text{for } x>0,\ y>0 \\ 0 & \text{otherwise} \end{cases}$$

3. If $f(x, y) = e^{-(x+y)}$, $x > 0$, $y > 0$, find the density function of $u = \frac{1}{2}(x + y)$.

4. Given the joint pdf of (X, Y) as

$$f(x, y) = \begin{cases} \frac{1}{4}(1+xy), |x|<1, |y|<1 \\ 0 \quad \text{otherwise} \end{cases}$$

   Find the marginal and conditional pdf's of X and Y.

5. Given the Gaussian random variable with pdf $f(x) = \dfrac{e^{-x^2/2\sigma^2}}{\sigma\sqrt{2\pi}}$. Let $Y = X^2$. Find the pdf of Y.

6. Let X_1, X_2 be jointly continuous random variables with pdf $f_{(X_1, X_2)}$. Let $Y_1 = X_1 + X_2$, $Y_2 = X_1 - X_2$. Find the joint pdf of Y_1 and Y_2 in terms of $f_{(X_1, X_2)}$.

7. Using Chebyshev's inequality prove that "If var $(X) = 0$, then $P(X = E(X)) = 1$.

8. Suppose the joint pdf of X and Y is

$$f(x, y) = \frac{e^{-x/y}\, e^{-y}}{y}, 0 < x < \infty, 0 < y < \infty$$

   Find $E(X/Y = y)$

9. Let the independent random variables X_1 and X_2 have respectively the means -4, 3 and variances 4 and 9. Find the mean and variance of $Y = 3X_1 - 4X_2$.

10. A random variable having a Poisson distribution with mean 20 can be thought of as the sum Y_{20} of the items of a random sample of size 20 from a Poisson distribution with mean 1. Then find $\rho(16 \le Y_{20} \le 21)$.

11. Let X and Y have the joint pdf

$$f(x, y) = \frac{x+2y}{18}, x = 1, 2, y = 1, 2$$

   Find the correlation coefficient ρ_{XY}.

12. Let X and Y have the joint pdf

$$f(x, y) = \frac{x+y}{32}, \; x = 1, 2, \; y = 1, 2, 3, 4$$

Find the correlation coefficient ρ_{XY}.

13. Given two discrete random variables X and Y with the joint probability distribution

Y \ X	−1	0	1	
−1	$\frac{1}{6}$	$\frac{1}{3}$	$\frac{1}{6}$	$\frac{2}{3}$
0	0	0	0	0
1	$\frac{1}{6}$	0	$\frac{1}{6}$	$\frac{1}{3}$
	$\frac{1}{3}$	$\frac{1}{3}$	$\frac{1}{3}$	

Show that the covariance is zero even though the two random variables are not independent.

14. If the random variables X, Y, Z have the means $E(X) = 3$, $E(Y) = 5$, $E(Z) = 2$, var $(X) = 8$, var $(Y) = 12$, var $(Z) = 18$, and the covariances cov $(X, Y) = 1$, cov $(X, Z) = -3$, cov $(Y, Z) = 2$, find cov (u, v), where $u = x + 4y + 2z$, $v = 3x - y - z$.

ANSWERS

(1) (i) $A = 2$ (ii) $2e^{-x}$

(2) $f(x, y) = 4xy\, e^{-(x^2 + y^2)}$

(3) $\begin{cases} \frac{u}{2} e^{-2u}, & u \geq 0 \\ 0 & \text{otherwise} \end{cases}$

(4) Marginal pdf's of X and Y are: $\frac{1}{2}, \frac{1}{2}$

Conditional pdf of X given Y is $= \frac{1}{2}(1 + xy)$

(5) $\frac{1}{2\sigma} \frac{e^{-y/2\sigma^2}}{\sqrt{2\pi y}}$

(6) $f(y_1, y_2) = \begin{cases} \frac{1}{2} & 0 \leq y_1 + y_2 \leq 2, \; 0 \leq y_1 - y_2 \leq 2 \\ 0 & \text{otherwise} \end{cases}$

(8) y

(9) mean $= -18$, variance $= 72$

(10) 0.4142

(11) $\rho = -0{,}025$

(12) $\rho = -0.037$

(14) −76

4

Random Processes

4.1. INTRODUCTION

New problems in various branches of technology do not fit into the framework of the classical probability theory. Such problems compel us to study the *Processes,* that is, phenomena that takes place in time. The theory of probability does not have any general procedures or any schemes for solving problems that arise in the study of such phenomena. So it is necessary to develop random processes which is a family of random variables that is indexed by a parameter such as time. Many problems that arise in Physics, Chemistry, and other fields can be solved by using random processes.

An important range of phenomena take place in accordance with the principle of radioactive disintegration. This consists in the atoms of a radioactive substance decaying into the atoms of another element. The decay of every atom occurs instantaneously like an explosion, with release of a certain amount of energy. The decay of different atoms takes place at randomly chosen instants of time according to numerous observations. For studies of the process of radioactive decay it is essential to determine the probability that within a certain time interval a certain quantity of atoms will disintegrate. Some more examples are: number of telephone calls in telephone exchange during a given time interval; changes in the number of particles that are in Brownian motion and that are located in a given region of space at a given instant of time. In this chapter we give a simple solution to the mathematical problems which use random processes technique.

4.1.1 Definition

A **random process** is a family of random variables $\{X(t) : t \in T\}$ defined on a given probability space, indexed by the parameter t, where t varies over an index set T.

The set of all possible values of an individual random variable $X(t)$ of a random process $\{X(t) : t \in T\}$ is called its **state space** and the values of $X(t)$ are called **states**.

Classificiation of Random Process

4.1.2 Definition

The state space is **discrete**, if it contains a finite or countably infinite set of points. In this case, the state space is assumed to be $\{0, 1, 2, \ldots\}$ and we have **discrete state process or a chain**. If the state space is **continuous**, we have a **continuous—state process.**

Note that the parameter t is usually interpreted as 'time', even though it represents, distance, length, thickness, and so on. If the index set T is discrete, then we have a **discrete (time)-parameter process**; otherwise we have a **continuous-parameter process**. For discrete time, the random process is represented by $\{X_n : n = 0, 1, 2, \ldots\}$. For continuous time, the process is $\{X(t) : t \in T, T \text{ is finite or infinite interval}\}$.

A random process can be classified as:

(i) Discrete time, discrete state space,

(ii) Discrete time, continuous state space,

(iii) Continuous time, discrete state space,

(iv) Continuous time, continuous state space.

All the four types may be represented as $\{X(t) : t \in T\}$

4.1.3 Example

Discrete time—discrete state space. Suppose X_n represent the total number of sixes appearing in the first n throws of a dice, then the set of possible values of X_n is the finite set of non-negative integers 0, 1, . . . n. Here the state space is discrete. We assumed that the parameter 'n' of X_n gives the throw number ($n = 0, 1, 2, \ldots$).

4.1.4 Example

Discrete time, continuous state space. Let $X_n = Y_1 + Y_2 + \ldots + Y_n$, where Y_i is a continuous random variable assuming values in $[0, \infty)$. Here the set of possible values of X_n is in the interval $[0, \infty)$ and so the state space is continuous but the parameter n of X_n is restricted to the non-negative integers $n = 0, 1, 2, \ldots$.

4.1.5 Example

Continuous time, discrete state space. Consider a random event occurring in time say the number of telephone calls received at a particular place. Suppose that $X(t)$ is the random variable which represents the number of in- coming calls in an interval $(0, t)$ of duration 't' units. Here the state space is discrete, even though $X(t)$ is defined for a continuous range of time.

4.1.6 Example

Continuous time, continuous state space. Suppose $X(t)$ represents the maximum temperature at a particular place in the time interval $(0, t)$. Then the set of possible values of $X(t)$ is continuous. Then we have a system in continuous time, continuous state space.

4.1.7 Definition

Let E be an experiment and S be the sample space associated with E. A random variable X is a function from S into R. So the random process which is a family of random variables, is nothing but a family of functions $\{X(t, s): s \in S, t \in T\}$.

If $t = t_i$ is fixed, then $X(t_i, s) = X_{t_i}(s)$ is a **random variable** (denoted by $X(t_i)$) as s varies over S. If $s = s_i \in S$ is fixed, then $X(t, s_i) = X_{s_i}(t)$ is a single function of time t, called a **sample function** or **realization of the process**. When both s and t are fixed, then $X(t, s)$ is merely a **number** when both s, t are variables, then we get a family of random variables which constitute a **random process.**

Distribution and Density Functions

For a fixed time $t = t_1$, $X(t_1)$ is a simple random variable that describes the state of the process at time t_1. For a fixed number x_1, the probability of the event $[X(t_1) \leq x_1]$ gives the cumulative distribution of random variable $X(t_1)$ as

$$F(x_1; t_1)) = F_{X(t_1)}(x_1) = P[X(t_1) \le x_1]$$

Here $F(x_1, t_1)$ is known as the **first-order distribution of the process** $\{X(t)\}$. Given two time instants t_1 and t_2, $X(t_1)$ and $X(t_2)$ are the respective random variables on the same probability space. Then joint distribution of the random process $\{X(t): t \in T\}$ is called **second-order distribution** and is given by

$$F(x_1, x_2; t_1, t_2) = P[X(t_1) \le x_1, X(t_2) \le x_2]$$

In general n^{th} **order joint distribution of the process** $\{X(t): t \in T\}$ is given by:

$$F(X; t) = P(X(t_1) \le x_1, \ldots X(t_n) \le x_n),$$

where $X = (x_1, x_2, \ldots x_n) \in R^n$ and $t = (t_1, t_2 \ldots t_n) \in T^n$ such that $t_1 < t_2 < \ldots < t_n$.

The derivative of first order distribution of the process with respect to x_1 gives first-order density function

$$f(x_1, t_1) = \frac{dF(x_1, t_1)}{dx_1}$$

Similarly, we define the second and higher order density functions as:

$$f(x_1, x_2; t_1, t_2) = \frac{\partial^2 F(x_1, x_2; t_1, t_2)}{\partial x_1 \cdot \partial x_2}$$

and $$f(X, t) = \frac{\partial^n F(x, t)}{\partial x_1 \, \partial x_2 \ldots \partial x_n}$$

where $X = (x_1, x_2, \ldots x_n) \in R^n, t = (t_1, t_2, \ldots, t_n) \in T^n.$

4.2 SPECIAL CLASSES OF RANDOM PROCESS

4.2.1 Definition

(Strictly stationary Process). A random (or stochastic) process $\{X(t)\}$ is said to be **stationary** in the **strict sense** if for $n \ge 1$, its nth order joint distribution satisfies the condition:

$$F(X; t) = F(X; t + \tau)$$

for all vectors $X = (x_1, x_2, \ldots, x_n) \in R^n$, $t = (t_1, t_2, \ldots, t_n) \in T^n$ and all scalars τ such that $t_i + \tau \in T$. Here $t + \tau$ implies that the scalar τ is added to all components of vector t.

The time-dependent mean of the stochastic process is said to be **ensemble average** of the stochastic process.

If the mean of the process $\{X(t)\}$ exists, then $E\{X(t)\} = E\{X(t + \tau)\}$, for any scalar τ. That is the mean $E\{X(t)\}$ of the strict-sense stationary process is depending only on time t.

4.2.2 Definition

A random process $\{X(t)\}$ with finite second order moments is called **weakly stationary** or **wide-sense stationary** if its mean $E\{X(t)\} = m$ is independent of t, and its covariance function $E(X(t) X(t + \tau))$ depends only on τ, not on t.

Note that a strict-stationary process is also a weakly-stationary process but the converse does not hold.

A process which is not stationary (in any sense) is said to be **evolutionary**.

4.2.3 Definition

A stochastic process $\{X(t): t \in T\}$ is said to be an **independent process** provided its nth order-joint distribution satisfies:

$$F(X; t) = \prod_{i=1}^{n} F(x_i; t_i) = \prod_{i=1}^{n} P(X(t_i) \le x_i)$$

4.2.4 Example

Consider an experiment of throwing a dice. Let X_n denotes the outcome of nth throw, $n \ge 1$. Then $\{X_n: n \ge 1\}$ is a family of random variables such that for distinct value of n (= 1, 2, . . .) we get distinct random variable X_n. Here $\{X_n; n \ge 1\}$ is an **independent process**.

4.2.5 Definition

A **renewal process** is defined to be a discrete-parameter independent process $\{X_n : n = 1, 2, \ldots\}$, where $X_1, X_2, \ldots$ are independent, identically distributed, non-negative random variables.

Processes with independent increments

If for all $t_1, \ldots, t_n$, $t_1 < t_2 < \ldots < t_n$, the random variables

$$X(t_2) - X(t_1),\ X(t_3) - X(t_2),\ \ldots\ X(t_n) - X(t_n - 1)$$

are independent, then $\{X(t): t \in T\}$ is said to be a process with independent increments.

For discrete-parameter process, we say $\{Z_n : n \ge 0\}$ is a process with independent increments if

$$Z_i = X_i - X_{i-1},\ i = 1, 2, \ldots$$
$$Z_0 = X_0,$$

Markov Process: A stochastic process $\{X(t): t \in T\}$ is called **Markov process** if for $t_1 < t_2 < \ldots t_n < t$, the conditional distribution of $X(t)$ for given values of $X(t_1), X(t_2), \ldots X(t_n)$ depends only on $X(t_n)$; that is,

$$P[X(t) \le x / X(t_n) = x_n, X(t_{n-1}) = x_{n-1}, \ldots X(t_1) = x_1]$$
$$= P[X(t) \le x /_{X(t_n)} = x_n]$$

This definition applies to continuous state space.

A discrete-state Markov process, called *Markov chain* is analogously defined.

4.2.6 Definition

The stochastic process $\{X_n, n = 0, 1, 2 \ldots\}$ is called a **Markov chain**, if for $j, k, j_1, \ldots j_{n-1} \in N$,

$$P\{X_n = k / X_{n-1} = j, X_{n-2} = j_1, \ldots X_0 = j_{n-1}\}$$
$$= P\{X_n = k | X_{n-1} = j\}$$

A Markov chain for which

$P\{X_{n+1} = j /_{X_n} = i\}$ is the same for all $n \ge 0$ has the **time-stationarity property**.

We study more about Markov chain in next chapter.

4.2.7 Definition

A random process $\{X(t): t \in T\}$ is said to be **homogeneous Markov process** if

$$P[X(t) \le x \mid X(t_n) = x_n] = P(X(t - t_n) \le x \mid X(0) = x_n]$$

A homogenous Markov process need not be a stationary random process.

Wiener Processes

Another example of random processes with independent stationary increments is a Wiener process.

4.2.8 Definition

A random Process $\{X(t): t \geq 0\}$ is called a **Wiener process** if

1. $X(t)$ has stationary independent increments.
2. The increment $X(t) - X(s)$ $(t > s)$ normally distributes
3. $E(X(t)) = 0$.
4. $X(0) = 0$.

The Wiener process is also known as the **Brownian motion process.**

Note that Var $X(t) = \sigma^2 t$, where σ^2 is a parameter of the Wiener process which must be determined from observations. When $\sigma^2 = 1$, $X(t)$ is called a **standard wiener process.**

The pdf of Wiener process is

$$f_X(x, t) = \frac{1}{\sqrt{2\pi\sigma^2 t}} \cdot \exp\left(\frac{-x^2}{2\sigma^2 t}\right), \ t > 0$$

The pdf of the increment $\Delta \equiv X(t) - X(\tau)$ for all t > τ is given by

$$f_\Delta(\delta, t - \tau) = \frac{1}{\sqrt{2\pi\sigma^2 (t - \tau)}} \exp\left(\frac{-\delta^2}{2\sigma^2 (t - \tau)}\right)$$

Since

$$E(X(t) - X(\tau)) = E(\Delta) = 0$$

and $$E[(X(t) - X(\tau))^2] = \sigma^2 (t - \tau) \text{ for } t > \tau.$$

The Wiener process is an example of a Gaussian random process. The covariance function of the Wiener process $C_{XX}(t_1, t_2) = \sigma^2 \min(t_1, t_2)$.

To show this take $t_1 \geq t_2$ and note that the increment $X(t_1) - X(t_2)$ is independent of $X(t_2)$ and that they both have zero-mean

$$E[(X(t_1) - X(t_2)\, X(t_2)] = E(X(t_1) - X(t_2)]\, E[X(t_2)] = 0$$

or

$$E[X(t_1)\, X(t_2)] = E(X^2(t_2)] = \sigma^2 t_2$$

Similarly, if $t_2 > t_1$, we get $E(X(t_2)\, X(t_1)] = \sigma^2 t_1$.

Hence $$C_{XX}(t_1, t_2) = \sigma^2 \min(t_1, t_2).$$

White noise Let us consider the Wiener process. Consider the derivative of this process. For any $\alpha > 0$, the covariance function of the Wiener process is $C_{XX}(t_1, t_2) = \sigma^2 \min(t_1, t_2)$ and its mean function $\mu_X = 0$. Let $W(t) = \dfrac{dX(t)}{dt}$. Then

$$\mu_W(t) = E\left(\frac{dX(t)}{dt}\right) = \frac{d}{dt} E(X(t)) = \frac{d}{dt} \mu_X(t) = 0.$$

For the covariance,

$$C_{WW}(t_1, t_2) = \frac{\partial}{\partial t_1}\left(\frac{\partial}{\partial t_2} C_{XX}(t_1, t_2)\right)$$

$$= \frac{\partial}{\partial t_1}\left(\frac{\partial}{\partial t_2}\sigma^2 \min(t_1, t_2)\right) = \frac{\partial}{\partial t_1}\left(\frac{\partial}{\partial t_2}\begin{cases}\sigma^2 t_2, & t_2 < t_1 \\ \sigma^2 t_1, & t_2 \geq t_1\end{cases}\right)$$

$$= \frac{\partial}{\partial t_1}\begin{cases}\sigma^2, & t_2 < t_1 \\ 0, & t_2 > t_1\end{cases} = \frac{\partial}{\partial t_1}\begin{cases}0 & t_1 < t_2 \\ \sigma^2 & t_1 > t_2\end{cases}$$

$$= \frac{\partial}{\partial t_1}\sigma^2 u\,(t_1 - t_2) = \sigma^2 \delta(t_1 - t_2)$$

where $u\,(t_1 - t_2)$ is the unit step function and hence the covariance function of white noise is the impulse function. Since white noise always has zero-mean, the correlation function too is an impulse. Here

$$R_{WW}(t_1, t_2) = \sigma^2 \delta(t_1 - t_2) = C_{WW}(t_1, t_2)$$

$$E(|W(t)|^2) = \sigma^2 \delta(0) \text{ is infinite, not } \sigma^2.$$

4.2.9 Definition

Let $Z_1, Z_2, \ldots$ be independent identically distributed random variables with $P(Z_n = +1) = p$ and $P(Z_n = -1) = q = 1 - p$ for all n. Let

$$X_n = \sum_{i=1}^{n} z_i,\ n = 1, 2, \ldots$$

and $X(0) = 0$. The collection $\{X_n, n \geq 0\}$ is a random process, and is called **simple random walk** $X(n)$ in one dimension.

It is a discrete time, discrete – state random process. The state space is $E = \{\ldots -2, -1, 0, 1, 2, \ldots\}$ and the index parameter set is $T = \{0, 1, 2, \ldots\}$.

4.2.10 Example

Show that a simple random walk $X(n)$ is a Markov Chain.

Solution. $X(n) = \{X_n : n \geq 0\}$ bc

$$X_0 = 0,\ X_n = \sum_{i=1}^{n} Z_i,\ n \geq 1$$

where Z_n $(n = 1, 2, \ldots)$ are independent identically distributed random variables with

$$P(Z_n = k) = a_k,\ (k = -1, 1)$$

$$a_1 = p,\ a_{-1} = q = 1 - p$$

Then $X(n) = \{X_n : n \geq 0\}$ is a Markov chain, since

$$P\,(X_{n+1} = i_{n+1} | X_0 = 0, X_1 = i_1, \ldots X_n = i_n)$$

$$= P\{(Z_{n+1} + i_n = i_{n+1} | X_0 = 0, X_1 = i_1, \ldots X_n = i_n)$$

$$= P(Z_{n+1} = i_{n+1} - i_n) = a_{i_{n+1} - i_n} = P(X_{n+1} = i_{n+1} | X_n = i_n)$$

since Z_{n+1} is independent of $X_0, X_1, \ldots X_n$.

4.2.11 Example

Consider the so-called random-telegraph process. This is a discrete state, continuous parameter process $\{X(t): t \in (-\infty, \infty)$ with the state space $\{-1, 1\}$. Assume that these two values are equally likely; that is

$$P[X(t) = -1] = \frac{1}{2} = P[X(t) = 1], \quad -\infty < t < \infty \tag{4.1}$$

Equation (4.1) implies that the first order distribution function is stationary in time, but since higher-order distributions may be non-stationary, the random process $X(t)$ need not be stationary in the strict sense.

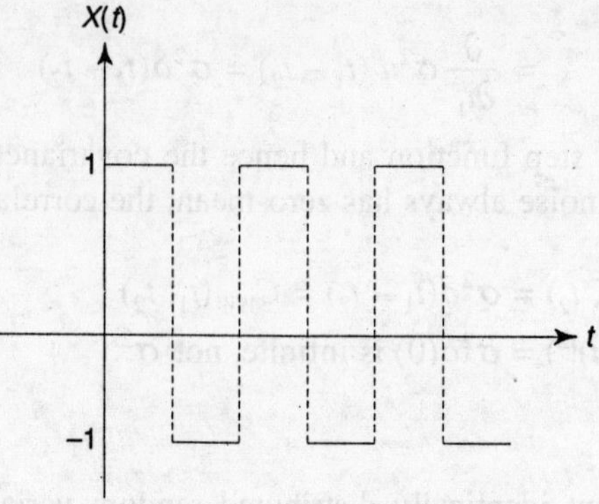

Sample function of the telegraph process.

4.2.12 Example

Prove that the random process $X(t) = A \cos (wt + \theta)$ is wide-sense stationary if it is assumed that A and w are constants and θ is uniformally distributed variable on the interval $(0, 2\pi)$.

Solution. $E(X(t)) = \int_{-\infty}^{\infty} xf(x)\, dx$. For uniform distribution, the distribution function is given by

$$f(x) = \begin{cases} \dfrac{1}{b-a} & a \le x \le b \\ 0 & \text{otherwise} \end{cases}$$

Therefore

$$E(X(t)) = \int_0^{2\pi} \frac{1}{2\pi} A \cos (wt + \theta)\, d\theta$$

$$= \frac{A}{2\pi} \int_0^{2\pi} \cos (wt + \theta)\, d\theta = \frac{A}{2\pi} [\sin (wt + \theta)]_0^{2\pi} = 0$$

Thus $E(X(t)) = 0$, a constant.

$$E(X(t)\, X(t + \tau)) = E\, (A^2 \cos (wt + \theta) \cos (wt + w\tau + \theta))$$

$$= \frac{A^2}{2} E\, (\cos (2wt + w\tau + 2\theta) + \cos (-w\tau)$$

since $2 \cos A \cos B = \cos (A + B) + \cos (A - B)$

$$= \frac{A^2}{2} E\, (\cos (2wt + w\tau + 2\theta)) + \frac{A^2}{2} E\, (\cos (w\tau))$$

$$= \frac{A^2}{2} \int_0^{2\pi} \frac{1}{2\pi} \cos (2wt + w\tau + 2\theta)\, d\theta + \frac{A^2}{2} \cos w\tau$$

The first term integrates to zero and hence

$$E(X(t))\ X(t+\tau) = \frac{A^2}{2}\cos w\tau.$$

Hence the random process is a wide-sense stationary random process.

4.2.13 Example

Let X_n, $n \geq 1$ be uncorrelated random variables with mean 0 and variance 1. Then

$$\begin{aligned} E(X_n, X_m) = E(X_n\ X_m) &= 0 \quad \text{if} \quad m \neq n \\ &= 1 \quad \text{if} \quad m = 1. \end{aligned}$$

Hence $\{X_n, n \geq 1\}$ is wide-sense stationary.

4.2.14 Example

Consider the process $\{X(t) : t \in T\}$ with

$$\begin{aligned} P(X(t) = n) &= \frac{(at)^{n-1}}{(1+at)^{n+1}}, \ n = 1, 2, \ldots \\ &= \frac{at}{1+at} \quad \text{for } n = 0 \end{aligned}$$

Then

$$\begin{aligned} E(X(t)) &= \sum_{n=0}^{\infty} n\ P(X(t) = n) = \sum_{n=1}^{\infty} \frac{n(at)^{n-1}}{(1+at)^{n+1}} \\ &= \frac{1}{(1+at)^2} \sum_{n=1}^{\infty} n\left(\frac{at}{1+at}\right)^{n-1} = 1 \end{aligned}$$

$$\begin{aligned} E(X^2(t)) &= \sum_{n=2}^{\infty} \frac{n^2\ (at)^{n-1}}{(1+at)^{n+1}} \\ &= \frac{at}{(1+at)^3}\left\{\sum_{n=2}^{\infty} n(n-1)\left(\frac{at}{1+at}\right)^{n-2}\right\} + \sum_{n=1}^{\infty} \frac{n(at)^{n-1}}{(1+at)^{n+1}} = 2at + 1 \end{aligned}$$

hence $E(X(t)) = 1$, a constant and the variance increases with t. Hence the process is not stationary.

4.3 BINOMIAL PROCESS

Let the discrete random variable Y_i denote the result of the ith trial, so that the event $Y_i = 1$ when the outcome is a success, denoted by p, and $Y_i = 0$, when the outcome is a failure. Then the family of all independent Bernoulli's trials $\{Y_i : i = 1, 2, \ldots\}$ is a discrete state, discrete parameter, random process which is stationary in the strict sense. Since Y_i's are mutually independent, the above process is an independent process and is called **Bernoulli Process**. Since Y_i is a Bernoulli random variable, the pmf is

$$\left.\begin{aligned} P(X = 0) &= 1 - p \\ P(X = 1) &= p \end{aligned}\right\}$$

And $E(Y_i) = p$, Var $(Y_i) = p(1-p)$, the probability generating function is

$$G_{Y_i}(z) = \sum_{i=0}^{\infty} P(X=i)z^i = \sum_{i=0}^{1} P(X=i)z^i$$

$$= P(X=0)\, z^0 + P(X=1)\, z = (1-p) + pz$$

Based on the Bernoulli process, we define Binomial process $\{S_n : n = 1, 2, \dots\}$, where $S_n = Y_1 + Y_2 + \dots + Y_n$. By rewriting $S_n = S_{n-1} + Y_n$, it is not difficult to see that $\{S_n\}$ is a discrete state, discrete-parameter, Markov process, since

$$P(S_n = k | S_{n-1} = k) = P(Y_n = 0) = 1 - p$$

and

$$P(S_n = k | S_{n-1} = k-1) = P(Y_n = 1) = p$$

Clearly

$$P(S_n = k) = nC_k\, p^k (1-p)^{n-k}$$

$$E(S_n) = np$$

$$\text{Var}(S_n) = np(1-p)$$

And probability generating function is

$$G_{S_n}(Z) = \sum_{i=0}^{n} P(X=i)\, z^i = \sum_{i=1}^{n} nC_i\, p^i (1-p)^{n-i} z^i$$

$$= [(1-p) + pz]^n$$

Some properties of Bernoulli process

1. For a Bernoulli process with parameter p, the probability of success, the interval of time X up to and including the first occurrence of success follows Geometric distribution with mean $\frac{1}{p}$ and variance $\frac{1-p}{p^2}$. For

$$P[X = i] = p(1-p)^{i-1} \qquad i = 1, 2, \dots$$

2. Generalizing the above as: For a Bernoulli Process with parameter p, the interval of time X_r up to and including the rth success follows negative binomial distribution with pmf.

$$P(X_r = i) = (i-1)\, C_{(r-1)}\, p^r (1-p)^{i-r},$$

$$i = r, r+1, \dots$$

$$r = 1, 2, \dots$$

and $\quad E(X_r) = r/p$ and Var $(X_r) = \frac{r(1-p)}{p^2}$

3. Suppose each Bernoulli variable Y_i have a distinct parameter p_i. Assume $Y_1, Y_2, \dots$ are independent random variables. Then the process $\{Y_i : i = 1, 2, \dots\}$ is independent and is called **non-homogeneous Bernoulli process.**
4. Assume that each trial has more than two possible outcomes. Let $\{Y_i : i = 1, 2, \dots n\}$ be a sequence of independent discrete random variables and define the partial sum $S_n = \sum_{i=1}^{n} Y_i$. Then the process $\{S_n : n = 1, 2, \dots\}$ is a Markov chain known as **random walk.**

As the Geometric distribution possesses the memoryless property, the conditional pmf for the remaining number of trials up to and including the next success, given that there were no successes in the first m trials is still geometric with parameter p. Since an arrival gives a change in state of the Binomial process $\{S_n\}$, the occupancy time in state S_n is memoryless.

4.4 POISSON PROCESS

The Poisson process is a continuous parameter, discrete-state process, which apart from their theoretical importants, plays an important role in the study of large number of cases. For example, consider a random event E such that

1. Incoming telephone calls to a switch board
2. Arrival of customers for service at a counter
3. Number of failed components in a large group of initially fault-free components
4. Occurrence of accidents at certain place.

Let us consider the total number $N(t)$ of occurrences of the event E in the time interval $[0, t]$. The stochastic process $\{N(t) : t \geq 0\}$ is said to be a **Poisson process**, if the following conditions are satisfied.

(i) $N(0) = 0$.

(ii) $N(t)$ is independent of the number of occurrences (of the event E) in an interval prior to the interval $(0, t)$ i.e. future changes in $N(t)$ are independent of the past changes (Independence).

(iii) The distribution of the number of events that occur in a given interval depends only on the length of the interval and not on its location (Homogeneity in time).

(iv) In an interval of infinitesimal length h, the probability of exactly one occurrence is $\lambda h + O(h)$ and that of more than one occurrence is of $O(h)$ (where $O(h)$ denotes a quantity having an order of magnitude smaller than h). So

$$\underset{h \to 0}{\text{Lt}} \frac{O(h)}{h} = 0.$$

Let $p_n(t) = P[N(t) = n]$ be the pmf of $N(t)$. Then

$$p_1(h) = p(N(h) = 1) = \lambda h + O(h) \tag{4.2}$$

Since $\sum_{k=2}^{\infty} p_k(h) = O(h)$, $\sum_{n=0}^{\infty} p_n(h) = 1$, we have

$$p_0(h) = (1 - \lambda h) + O(h) \tag{4.3}$$

Also, note that because of condition (i),

$$p_0(0) = 1 \text{ and } p_n(0) = 0 \text{ for } n > 0. \tag{4.4}$$

4.4.1 Theorem

Under the above conditions, $N(t)$ follows Poisson distribution with mean λt. That is, $p_n(t)$ is given by the Poisson law:

$$p_n(t) = \frac{e^{-\lambda t} (\lambda t)^n}{n!} \qquad n = 0, 1, 2, 3, \ldots$$

Proof. Consider $p_n(t + h)$, the probability that n events occur in the interval $(0, t + h]$, and the events happen in the following mutually exclusive ways $A_1, A_2, A_3, \ldots, A_{n+1}$.

For $n \geq 1$, let

A_1: n events by epoch t and no event in the interval $(t, t+h]$

A_2: $(n-1)$ events by epoch t and one event in the interval $(t, t+h]$

A_3: $(n-2)$ events by epoch t and two events in the interval $(t, t+h]$

and so on

Now

$$P(A_1) = P\{N(t) = n\} \;\; P\{ N(h) = 0/N(t) = n\}$$
$$= p_n(t).p_0(h) = p_n(t) \; [(1 - \lambda h) + O(h)]$$
$$P(A_2) = P \; \{N(t) = n - 1\} \;\; P\{N(h) = 1/N(t) = n - 1\}$$
$$= p_{n-1}(t) \; . \; p_1(h) = p_{n-1}(t)[\lambda h + O(h)]$$
$$p(A_3) = p_{n-2}(t) \; \{O(h)\}$$
$$P \; (A_k) = P_{n-k-1}(t) \; O(h) \text{ for } n \geq k \geq 3.$$

Adding the probabilities for these mutually exclusive events, we have

$$p_n(t+h) = p_n(t)\,(1 - \lambda h) + p_{n-1}(t)\,\lambda h + \sum_{i=0}^{n} p_i(t)\; O(h),\; n \geq 1$$
$$= p_n\,(t)\,(1 - \lambda h) + p_{n-1}\,(t)\,\lambda h + O(h) \tag{4.5}$$

Similarly

$$p_0\;(t + h) = (1 - \lambda h)p_0(t) + O(h) \tag{4.6}$$

Now (4.5) implies

$$\frac{p_n\,(t+h) - p_n(t)}{h} = -\lambda p_n(t) + \lambda p_{n-1}\,(t) + \frac{O(h)}{h}$$

Taking limit $h \to 0$ on both sides, we get

$$p_n^1\;(t) = -\lambda[p_n(t) - p_{n-1}(t)],\; n \geq 1 \tag{4.7}$$

(4.6) implies

$$\frac{p_0\;(t+h) - p_0\;(t)}{h} = -\lambda p_0(t) + \frac{O(h)}{h}$$

whence, as $h \to 0$

$$p_0^1(t) = -\;\lambda p_0(t) \tag{4.8}$$

equivalently $\dfrac{p_0^1(t)}{p_0(t)} = -\;\lambda$, which implies by integration, that

$$\log\; p_0(t) = -\;\lambda t + c$$

or $p_0(t) = Ke^{-\lambda t}$

Since $p_0(0) = P\{N(0) = 0\} = 1$, we arrive at

$$p_0(t) = e^{-\lambda t}. \tag{4.9}$$

Consider (4.7) for $n = 1$. Substituting the value of $p_0(t)$ and solving the equation and using (4.4) we get $p_1(t) = \lambda te^{-\lambda t}$. By induction, we prove the theorem for n and hence

$$p_n(t) = \frac{e^{-\lambda t}(\lambda t)^n}{n!}, \ n = 0, 1, 2 \ldots.$$

Therefore the number of events $N(t)$ in the interval $(0, t]$ has a Poisson distribution with parameter λt. We know the mean and variance of the distribution are both equal to λt. That is the mean number of occurrences in an interval of length t is λt, so that the mean number of occurrences per unit time ($t = 1$) is λ. The mean rate λ per unit time is known as the parameter of the Poisson process. Also λ is called the arrival rate of the Poisson process. The Poisson process plays an important role in queueing theory and reliability theory.

The mean and variance of $N(t)$ are dependent on t. The process $\{N(t)\}$ is not stationary – it is evolutionary.

Remark: Condition (ii) says that the Poisson process is Markovian; Condition (iii) says that the Poisson process is time homogeneous. Condition (iv) says that within small length h, the probability of exactly one occurrence is approximately proportional to the length h of that interval and that of the simultaneous occurrence of two or more events is extremely rare.

Hence

$$P\{N(h) = 0|_{N(t) = n}\} = P\{N(h) = 0\} = p_0(h)$$

(here two intervals, one of length h and the other of t, are non-overlapping). In general

$$P\{N(h) = k|_{N(t) = n}\} = P\{N(h) = k\} = p_k(h), \ k \geq 0$$

Hence Poisson process is a Markov process such that the conditional probabilities are constant (independent of t).

Properties of Poisson process

1. **Additive property:** Sum of two independent Poisson processes is a Poisson process.

Proof. The probability generating function of a Poisson random variable is

$$G(z) = \sum_{k=0}^{\infty} \frac{\lambda^k e^{-\lambda} z^k}{k!} = e^{-\lambda} e^{\lambda z}$$

$$= e^{\lambda(z-1)} = e^{-\lambda(1-z)}$$

Let $\{N_1(t)\}$ and $\{N_2(t)\}$ be two Poisson processes with parameters λ_1, λ_2 respectively. Let

$$N(t) = N_1(t) + N_2(t)$$

The probability generating function of $N(t)$ is

$$e^{-\lambda_1(1-z)} \cdot e^{-\lambda_2(1-z)} = e^{-(\lambda_1+\lambda_2)(1-z)}$$

$$P(N(t) = n) = \sum_{j+k=n} P(N_1(t) = j)\, P(N_2(t) = k)$$

$$= \sum_{j+k=n} \frac{e^{-\lambda_1 t} \cdot (\lambda_1 t)^j}{j!} \frac{e^{-\lambda_2 t}(\lambda_2 t)^k}{k!}$$

$$= \frac{e^{-(\lambda_1+\lambda_2)t}}{n!} \sum_{j=0}^{n} nC_j\, (\lambda_1 t)^j (\lambda_2 t)^{n-j}$$

$$= \frac{e^{-(\lambda_1+\lambda_2)t}}{n!} \{(\lambda_1 + \lambda_2)\, t\}^n, \ n \geq 0$$

Hence $N(t)$ is a Poisson process with parameter $(\lambda_1 + \lambda_2)$.

2. Difference of two independent Poisson processes is not a Poisson process.

Let $N(t) = N_1(t) - N_2(t)$. Then probability generating function of $N(t)$ is $e^{-(\lambda_1+\lambda_2)} \cdot e^{\lambda_1 z+\lambda_2/z}$. Now

$$P(N(t) = n) = \sum_{r=0}^{\infty} P(N_1(t) = n + r)\, P(N_2(t) = r)$$

$$= \sum_{r=0}^{\infty} \frac{e^{-\lambda_1 t} \cdot (\lambda_1 t)^{n+r}}{(n+r)!} \frac{e^{-\lambda_2 t} \cdot (\lambda_2 t)^r}{r!}$$

$$= e^{-(\lambda_1+\lambda_2)t} \left(\frac{\lambda_1}{\lambda_2}\right)^{n/2} \sum_{r=0}^{\infty} \frac{(t\sqrt{\lambda_1\lambda_2})^{2r+n}}{r!(r+n)!}$$

$$= e^{-(\lambda_1+\lambda_2)t} \left(\frac{\lambda_1}{\lambda_2}\right)^{n/2} I_{|n|}(2t\sqrt{\lambda_1\lambda_2}) \tag{4.10}$$

where $I_n(x) = \sum_{r=0}^{\infty} \frac{(x/2)^{2r+n}}{\Gamma_{(r+n+1)}}$. Note that $I_{-n}(t) = I_n(t) = I_{|n|}(t)$.

4.4.2 Example

Suppose that customers arrive at a bank according to a Poisson process with a mean rate of 'a' per minute. Then the number of customers $N(t)$ arriving in an interval of duration t minutes follows a Poisson distribution with mean 'at'. It the rate of arrival is 3 per minute, then in an interval of 2 minutes, the probability that the number of customers arriving is

(i) exactly 4 is $\frac{e^{-3\times 2}(6)^4}{4!} = 0.133$

(ii) greater than 4 is $\sum_{k=5}^{\infty} \frac{e^{-6}\, 6^k}{k!} = 0.714$

(iii) less than 4 is $e^{-6} + e^{-6} \cdot 6 + \frac{e^{-6}\, 36}{2} = 0.152.$

4.4.3 Example

If passengers arrive (singly) at a taxi stand in accordance with a Poisson process with parameter λ_1 and taxis arrive in accordance with a Poisson process with parameter λ_2, then $N(t) = N_1(t) - N_2(t)$ gives the excess of passenger over taxis in an interval t. Then

$P(N(t) = n)$, $n = 0; \pm 1, \pm 2, \ldots$ is given by (4.10)

The mean of $N(t)$ is $(\lambda_1 - \lambda_2)t$ and variance $N(t) = (\lambda_1 + \lambda_2)t$.

3. **Decomposition of a Poisson process**
Suppose that $N(t)$, the number of occurrences of an event E in an interval of length t is a Poisson process with parameter λ. Suppose also that each occurrence of E has a constant probability p of being recorded, and that the recording of an occurrence is independent of the occurrence of other and also of $N(t)$.

If $M(t)$ is the number of occurrences recorded in an interval of length t, and the event $\{M(t) = n\}$ can happen in the following mutually exclusive ways.

A_r : E occurs $(n + r)$ times by epoch t and exactly n out of $(n + r)$ occurrences are recorded with probability of each occurrence recorded being $p(r = 0, 1, 2, ..)$

Then

$$P(A_r) = P\{E \text{ occurs } (n + r) \text{ times by epoch } t\}.\ P\{n \text{ occurrences are recorded given that the number of occurrences is } n + r\}$$

$$= \frac{e^{-\lambda t}(\lambda t)^{n+r}}{(n+r)!}(n + r)\, C_n\, p^n q^r$$

Hence

$$P(M(t) = n) = \sum_{r=0}^{\infty} P(A_r) = \sum_{r=0}^{\infty}\frac{e^{-\lambda t}(\lambda t)^{n+r}}{(n+r)!}\, {}^{n+r}C_n\; p^n q^r$$

$$= e^{-\lambda t}\sum_{r=0}^{\infty}\frac{(\lambda pt)^n(\lambda qt)^r}{n!\,r!} = e^{-\lambda t}\cdot\frac{(\lambda pt)^n}{n!}\sum_{r=0}^{\infty}\frac{(\lambda qt)^r}{r!}$$

$$= e^{-\lambda t}\frac{(\lambda pt)^n}{n!}\, e^{\lambda qt} = \frac{e^{-\lambda pt}\cdot(\lambda pt)^n}{n!}$$

Thus $\{M(t)\}$ is a Poisson process with parameter λp. If $M_1(t)$ denote the occurrences not recorded, then $\{M_1(t)\}$ is also a Poisson process with parameter $\lambda q = \lambda\,(1 - p)$ and $M(t)$ and $M_1(t)$ are independent.

Thus by random selection a Poisson process $\{N(t)\}$ of parameter λ is decomposed into two independent Poisson processes $\{M(t)\}$ and $\{M_1(t)\}$ with parameter λp and $\lambda\,(1 - p)$ respectively.

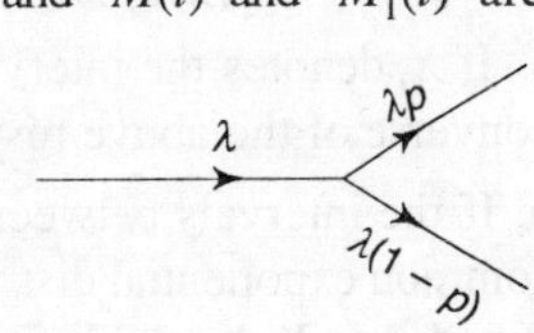

In general if $p_1, p_2, ... p_r$ are the probability of the process decomposing into r independent streams such that $p_1 + p_2 + \ldots + p_r = 1$, then the Poisson process is decomposed into r independent Poisson processes with parameters $\lambda p_1, \lambda p_2, \ldots \lambda p_r$.

4. **Poisson process and bionomial distribution.** If $N(t)$ is a Poisson process, and $s < t$, then

$$P\{N(s) = k|_{N(t) = n}\} = nc_k(s/t)^k\,(1 - s/t)^{n-k}$$

Proof.

$$P\{N(s) = k/N(t) = n\} = \frac{P\{N(s)=k \text{ and } N(t)=n\}}{P(N(t)=n)}$$

$$= \frac{P(N(s)=k \text{ and } N(t-s)=n-k)}{P(N(t)=n)}$$

$$= \frac{P(N(s)=k)\cdot P(N(t-s)=n-k)}{P(N(t)=n)}$$

$$= \frac{\dfrac{e^{-\lambda s}(\lambda s)^k}{k!}\cdot\dfrac{e^{-\lambda(t-s)}(\lambda(t-s))^{n-k}}{(n-k)!}}{\dfrac{e^{-\lambda t}(\lambda t)^n}{n!}}$$

$$= \frac{n!}{k!(n-k)!}\frac{s^k(t-s)^{n-k}}{t^n} = nC_k(s/t)^k(1-s/t)^{n-k}$$

5. The interval between two successive occurrences of a Poisson process $\{N(t)\}$ having parameter λ has a negative exponential distribution with mean $1/\lambda$.

Proof. Let X be the random variable representing the interval between two successive occurrences of $\{N(t)\}$ and let $P(X \le x) = F(x)$ be its distribution function. Let E_i and E_{i+1} be two successive events and let E_i occurred at the instant t_i, then

$$\begin{aligned} P(X > x) &= P[E_{i+1} \text{ didn't occur in } (t_i, t_i + x) \text{ given that } E_i \text{ occurred at the instant } t_i] \\ &= P[E_{i+1} \text{ didn't occur in } (t_i, t_i + x) | N(t_i) = i] \\ &= P\{\text{No occurrence takes place in an interval } (t_i, t_i + x) \text{ of length } x | N(t_i) = i\} \\ &= P(N(x) = 0 | N(t_i) = i) \\ &= p_0(x) \\ &= e^{-\lambda x}, \; x > 0 \end{aligned}$$

Hence

$$F(x) = P(X \le x) = 1 - P\{X > x\} = 1 - e^{-\lambda x}, \; x > 0$$

The density function is

$$f(x) = F^1(x) = \lambda e^{-\lambda x}, \; (x > 0)$$

If X_i denotes the interval between E_i and E_{i+1}, $i = 1, 2, \ldots$, then $X_1, X_2, \ldots$ are also independent. The converse of the above result is as follows.

6. If the intervals between successive occurrences of an event E are independently distributed with a common exponential distribution with mean $1/\lambda$, then the event E form a Poisson process with mean λt.

Proof. Let X_n be the interval between $(n-1)^{\text{th}}$ and n^{th} occurrence of a process $\{N(t)\}$ and let $X_1, X_2, \ldots$ be independent and identically distributed having negative exponential distribution with mean $1/\lambda$. The sum $Z_n = X_1 + X_2 + \ldots + X_n$ is the waiting time upto the nth occurrence. Then Z_n has a gamma distribution with parameteres λ, n. The distribution function is

$$F_{Z_n}(t) = P(Z_n \le t) = \int_0^t f(x)\,dx$$

where
$$f(x) = \frac{\lambda^n x^{n-1} e^{-\lambda x}}{\Gamma_{(n)}} \qquad x > 0$$

The events $\{N(t) < n\}$ and $\{Z_n = X_1 + X_2 + \ldots + X_n > t\}$ are equivalent. Hence

$$F_{Z_n}(t) = P(Z_n < t) = 1 - P(Z_n > t) = 1 - P(N(t) < n)$$
$$= 1 - P(N(t) \le n - 1) = 1 - F_{N(t)}(n-1)$$

Hence $$F_{N(t)}(n-1) = 1 - F_{Z_n}(t) = 1 - \int_0^t \frac{\lambda^n x^{n-1} e^{-\lambda x}}{\Gamma_{(n)}} dx$$

$$= 1 - \frac{1}{\Gamma_{(n)}} \int_0^{\lambda t} y^{n-1} e^{-y} dy, \text{ where } \lambda x = y$$

$$= \frac{1}{\Gamma_{(n)}} \int_{\lambda t}^{\infty} y^{n-1} e^{-y} dy$$

$$= \sum_{j=0}^{n-1} \frac{e^{-\lambda t} \cdot (\lambda t)^j}{j!} \text{ by integration by parts}$$

Now

$$p_n(t) = P[N(t) = n] = F_{N(t)}(n) - F_{N(t)}(n-1)$$

$$= \sum_{j=0}^{n} \frac{e^{-\lambda t} (\lambda t)^j}{j!} - \sum_{j=0}^{n-1} \frac{e^{-\lambda t} \cdot (\lambda t)^j}{j!} = \frac{e^{-\lambda t} \cdot (\lambda t)^n}{n!}, n = 0, 1, 2, \ldots$$

Hence the process $\{N(t)\}$ is a Poisson process with mean λt.

Note that the Poisson process has independent exponentially distributed interarrival times and gamma distributed waiting times.

Note: The Poisson type of occurrences are also called **purely random events** and the Poisson process is called a **process of purely random events**.

4.4.4 Example

Prove that a Poisson process distributes points at random over the infinite interval $[0, \infty]$ in the same way as the uniform distribution distributes points at random over a finite interval $[a, b]$.

Solution.

Poisson process and Geometric distribution

Consider two independent events E and F occurring in accordance with Poisson processes with mean at and bt respectively. The number N of occurrences of E between two successive occurrences of F has a geometric distribution

Proof. The interval between two successive occurrences of F has the density $f(x) = be^{-bx}$ and therefore the probability that k occurrences of E take place during an arbitrary interval (between two successive occurrences of F) is given by

$$P[N = k] = \int_0^{\infty} \frac{e^{-at} (at)^k}{k!} f(t)dt = \int_0^{\infty} \frac{e^{-at} (at)^k}{k!} be^{-bt} dt$$

$$= \frac{ba^k}{k!} \int_0^{\infty} e^{-(a+b)t} t^k dt = \frac{ba^k}{(a+b)^{k+1}}$$

$$= \frac{b}{a+b} \left\{\frac{a}{a+b}\right\}^k, \quad k = 0, 1, 2 \ldots$$

4.4.5 Definition: Poisson Count Process

Let E and F be two random sequences of events occurring at instants say $(t_1, t_2, ...)$ and $(t'_1, t'_2, ..)$ respectively. The number N_n of occurrences of F during the interval (t_{n-1}, t_n) of length $t_n - t_{n-1}$ is called the **count process** of F in E. If E is Poisson, then such a process is called **Poisson count process.** If F is also Poisson, then the count process N_n has a geometrical distribution. N_n $(n = 1, 2, ...)$ are independent and identically distributed with a common geometric distribution.

4.4.6 Definition: Compound Poisson Process (Poisson Cluster Process)

Suppose several events can happen simultaneously at an instant. We assume that

(i) The number $N(t)$ of occurrences at a point in time t constitute a Poisson process with mean rate λ.

(ii) The various numbers of occurrences in different events are mutually independent and follow the same probability distribution.

Then $\{X(t) : t \geq 0\}$, where $X(t)$ = the total number of occurrences in an interval of length t, under the conditions (i) & (ii), is a compound Poisson process.

We have $X(t) = \sum_{i=1}^{N(t)} X_i$, where $N(t)$ is a Poisson process with mean λt. We have $p_k = P\{X_i = k\}$ and

let $P(z) = \sum_{k=1}^{\infty} p_k z^k$, hence the generating function is $e^{\lambda t(P(z)-1)}$. Note that

$$E(X(t)) = \lambda t\, E(X_i)$$
$$\text{Var}\,(X(t)) = \lambda t\, E(X_i^2)$$

4.4.7 EXAMPLE

A person enlists subscriptions to a magazine, the number enlisting being a Poisson process with mean rate 6 per day. Subscribers may subscribe for 1 or 2 years independently of one another with respective probabilities 2/3 and 1/3.

Solution. The number of subscribers $N(t)$ enrolled by the person in time t days is a Poisson process with mean rate $6t$: the number $N_1(t)$ enrolled for 1 year period is a Poisson process with mean $(6 \times 2/3)t = 4t$ and the number $N_2(t)$ enrolled for 2 years is a Poisson process with mean $(6 \times 1/3) = 2t$.

If he received commission x for one year period and y for two years period then the total commission earned in period t is $X(t) = xN_1(t) + yN_2(t) = \sum_{i=1}^{N(t)} X_i$, where X_i is the amount of commission received from a subscription is a random variable such that $P(X_i = x) = \frac{2}{3}$ and $P(X_i = y) = \frac{1}{3}$. Now

$$E(X(t)) = E\{N(t)\}E\{X_i\} = 6t \times \frac{1}{3}(2x + y) = 4xt + 2yt \text{ and}$$

$$\text{Var}\,(X(t)) = 6t\, E(x_i^2) = 6t \times \frac{1}{3}(2x^2 + y^2) = 4x^2t + 2y^2t.$$

Non-homogenous Poisson process

The Poisson process can be generalized to the case where the birth rate λ is varying with time. The probability distribution is given by

$$P_k(t) = \frac{e^{-\Delta(t)}[\Delta(t)]^k}{k!}, \qquad k \geq 0$$

where

$$\Delta(t) = \int_0^t \lambda(x)dx$$

The condition (iii) of the classical Poisson process regarding homogeneity does not hold good. Note that the equations of the process which can be obtained from those of Poisson process can be obtained by replacing λ by $\lambda(t)$. That is,

$$\begin{aligned} P\{N(h) = k|_{N(t) = n}\} &= \lambda(t)\, h + O(h) && k = 1 \\ &= O(h) && k \geq 2 \\ &= 1 - \lambda(t)h + O(h) && k = 0 \\ p'_n(t) &= -\lambda(t)\, p_n(t) + \lambda(t)\, p_{n-1}(t), && n \geq 1 \\ p'_0(t) &= \lambda(t)\, p_0(t). \end{aligned}$$

The probability generating function is

$$e^{m(t)(z-1)}, \; m(t) = \int_0^t \lambda(x)\, dx.$$

The probability of no occurrence in an interval of length t is $e^{-\int_0^t \lambda(x)\,dx}$.

The probability of exactly k occurrences in an interval $(0, t)$ is given by

$$\frac{1}{k!}\left\{\int_0^t \lambda(x)\, dx\right\}^k . e^{-\int_0^t \lambda(x)\,dx}, \qquad k \geq 1.$$

Important points to be noted:

1. A Poisson process has the memoryless property which means that forward recurrence time R_t has the same distribution as the interarrival gaps.
$$P\{R_t \leq \Delta t\} = 1 - e^{-\lambda\Delta t}, \qquad t \geq 0.$$
2. Any arrival counting process in which arrivals occur one at a time and which has independent and stationary increments must be a Poisson process.
3. When the distribution of an interarrival-time gaps changes over time, the process is no longer time stationary and is said to be non-stationary.
4. A Poisson process has the superposition property (Additive property) which means that two independent Poisson processes with arrival rates λ_1, λ_2 that are superposed form a Poisson process with arrival rate $\lambda_1 + \lambda_2$.

4.4.8. Erlang Process

Consider a series of events E' which occur in accordance with a Poisson process $N(t)$ with mean $\lambda' t$.

The random variable X_i which represents the interval between two successive occurrences has density $h(x) = \lambda' e^{-\lambda' x}$ having mean $1/\lambda'$. The interval separating k occurrences is given by the sum $S_k = X_1 + X_2 + \ldots + X_k$ of k independent exponential distributions each with mean $1/\lambda'$. Therefore S_k is a gamma variable with pdf given by

$$f(t) = (\lambda')^{t}\, t^{k-1} . e^{-\lambda' t} \qquad t > 0.$$

and $$E(S_k) = k/\lambda', \text{ Var } (S_k) = k/\lambda'^2$$

Suppose that every k^{th} occurrence of E' is represented by a new series of events E, then E occurs for the second time at the $(2k)^{\text{th}}$ occurrence of E' and E occurs for the third time at the $(3k)^{\text{th}}$ occurrence of E' and so on. If $X(t)$ denote the number of occurrences of E in an interval of duration t, then the probability distribution function is

$$P\,(X(t) = n) = P\{\, nk \le N(t) \le (n+1)\,(k-1)\}, \qquad n \ge 0$$

$$= \sum_{r=nk}^{(n+1)(k-1)} P(N(t) = r)$$

$$= \sum_{r=nk}^{(n+1)(k-1)} \frac{e^{-\lambda^1 t}\,(\lambda^1 t)^r}{r!}, \text{ since } \{N(t)\} \text{ is a Poisson process}$$

For positive k, we have

$$\sum_{n=0}^{\infty} P(X(t) = n) = \sum_{n=0}^{\infty} \left[\sum_{r=nk}^{(n+1)(k-1)} \frac{e^{-\lambda' t}\,(\lambda' t)^r}{r!} \right] = \sum_{s=0}^{\infty} \frac{e^{-\lambda' t}\,(\lambda' t)^s}{s!} = 1$$

For $\lambda' = k\lambda$, the probability distribution function of $X(t)$ is

$$P(X(t) = n) = \sum_{r=nk}^{(n+1)(k-1)} \frac{e^{-\lambda kt} \cdot (k\lambda t)^r}{r!}, \qquad n \ge 0$$

and its density function is

$$f(t) = \frac{(\lambda k)^k\, t^{k-1}\, e^{-k\lambda t}}{\Gamma_{(k)}}, \qquad t > 0$$

$$= 0 \qquad t \le 0$$

$\{X(t) : t \ge 0\}$ is a renewal process which is called **Erlang-k process.**

4.5 ERGODIC RANDOM PROCESS

Let $X(t)$ be a random process . Then we define.

4.5.1 Definition

The time-averaged mean is

$$\mu_T = \underset{T\to\infty}{\text{Lt}} \frac{1}{2T} \int_{-T}^{T} X(t)\, dt, \qquad -T \le t \le T$$

and the time-averaging auto-correlation function is

$$R_{XX}(\tau) = \underset{T \to \infty}{\text{Lt}} \frac{1}{2T} \int_{-T}^{T} X(t)\, X(t+\tau)\, dt, \quad -T \le t \le T$$

4.5.2 Definition

Ensemble Averages or expectations: The mean of $X(t)$ is

$$\mu = E(X(t)) = \int_{-\infty}^{\infty} x\, f(x, t)\, dx$$

and the auto correlation $R_{X_1X_2}(t_1, t_2)$ of the random process $X(t)$ is

$$R_{X_1X_2}(t_1, t_2) = E(X(t_1)\, X(t_2)) = \int_{-\infty}^{\infty}\int_{-\infty}^{\infty} x_1x_2 f(x_1, x_2, t_1, t_2)\, dx_1 dx_2.$$

4.5.3 Definition

A random process is said to be **ergodic** if the time averages of the process is equal to the ensemble averages.

4.5.4 Definition

Mean Ergodic Process.
A random process $X(t)$ is said to be **mean-ergodic** if the time average (mean) is equal to ensemble average (mean) that is, if $\mu_T \to \mu$ as $T \to \infty$.

Variance – Ergodic Process
A stationary random process $X(t)$ is called **variance- ergodic process** if and only if

$$\frac{1}{2T} \int_{-T}^{T} E(X^2(t+\tau)\, X^2(t))\, dt \to C^2_{XX}(0) \text{ as } T \to \infty,$$

where $C_{XX}(t_1, t_2)$ represents the auto-covariance of the random process (see next chapter for definition).

Correlation-Ergodic Process
A stationary random process $X(t)$ is said to be **correlation-ergodic process** if

$$\frac{1}{2T} \int_{-T}^{T} X(t)\, X(t+\tau)\, dt \to R_{XX}(\tau), \text{ as } T \to \infty$$

Regular Random Process
A random process which has the property that any time average of one sample function is equal to the corresponding time average of any other sample function is called **regular random process.**

Deterministic Random Process
If the future values of any sample function can be predicted from a knowledge of the past values, then the random process is called **deterministic random** process. Otherwise it is called **non-deterministic random process.**

For example, consider a random process $\{X(t) = A \cos(wt + \theta): t \ge 0\}$. This consists of family of sine-waves and it is completely specified in terms of the random variables A and θ. Hence it is a deterministic random process.

4.5.5 Example

Consider a random process $\{X(t) = \cos(wt + \theta)\}$, where w is a constant, θ is a random variable with a probability density

$$P(\theta) = \begin{cases} \dfrac{1}{2\pi} & 0 \le \theta \le 2\pi \\ 0 & \text{otherwise} \end{cases}$$

Prove that $X(t)$ is an ergodic random process.

Solution. To prove $X(t)$ is ergodic, we should prove that the time averages and ensemble averages are equal.

Ensemble Averages:

$$E(X(t) = \int_0^{2\pi} \frac{1}{2\pi} \cos(wt + \theta)\, d\theta = \frac{1}{2\pi} [\sin(wt + \theta)]_0^{2\pi}$$

$$= \frac{1}{2\pi} [\sin wt - \sin wt] = 0$$

$$R_{XX}(t, t + \tau) = E(X(t)\, X(t + \tau))$$

$$= E(\cos(wt + \theta) \cos(wt + w\tau + \theta))$$

$$= \frac{1}{2} E\left[\cos(2wt + w\tau + 2\theta) + \cos(-w\tau) \right]$$

$$= \frac{1}{2} \int_0^{2\pi} \frac{1}{2\pi} \cos(2wt + w\tau + 2\theta)\, dt + \frac{1}{2} \cos w\tau$$

$$= \frac{1}{2} \cos w\tau, \text{ as the first term is zero.}$$

Time average:

$$\mu_T = \underset{T \to \infty}{\text{Lt}} \frac{1}{2T} \int_{-T}^{T} X(t)\, dt = \underset{T \to \infty}{\text{Lt}} \frac{1}{2T} \int_{-T}^{T} \cos(wt + \theta) dt$$

$$= \underset{T \to \infty}{\text{Lt}} \frac{1}{2T} \left[\frac{\sin(wt + \theta)}{w} \right]_{-T}^{T}$$

$$= \underset{T \to \infty}{\text{Lt}} \frac{1}{Tw} [\sin(\omega T + \theta) - \sin(-\omega T + \theta)]$$

when $T \to \infty$, $\mu_T \to 0$.

$$R_{XX}(\tau) = E(X(t)\, X(t + \tau))$$

$$= \underset{T \to \infty}{\text{Lt}} \frac{1}{2T} \int_{-T}^{T} X(t)\, X(t + \tau)\, dt$$

$$= \underset{T \to \infty}{\text{Lt}} \frac{1}{2T} \int_{-T}^{T} \cos(wt + \theta) \cos(wt + w\tau + \theta)\, dt$$

$$= \underset{T \to \infty}{\text{Lt}} \frac{1}{2T} \int_{-T}^{T} (\cos(wt + w\tau + 2\theta) + \cos w\tau)\, dt$$

$$= \underset{T \to \infty}{\text{Lt}} \frac{1}{2T} \left\{ \left[\frac{\sin(wt + w\tau + 2\theta)}{w} \right]_{-T}^{T} + [t \cos w\tau]_{-T}^{T} \right\}$$

$$= \underset{T \to \infty}{\text{Lt}} \frac{1}{4T} \left\{ \frac{1}{w} [\sin wT + w\tau + 2\theta) - \right.$$

$$\left. \sin(-wT + w\tau + 2\theta)] + [T \cos w\tau + T \cos w\tau] \right\}$$

$$= \underset{T \to \infty}{\text{Lt}} \left(\frac{0 + \cos w\tau}{2} \right) = \frac{\cos w\tau}{2}.$$

Hence the time averages and the ensembled averages are equal and hence the given process is ergodic.

4.5.6 Example

Prove that the random process $X(t) = A \cos (wt + \theta)$ is not stationary, if it is assumed that A and w are constants and θ is uniformly distributed variable on the interval $(0, \pi)$.

Solution. $$E(X(t)) = \int_0^{\pi} \frac{1}{\pi} A \cos(wt + \theta)\, d\theta = \frac{A}{\pi} \int_0^{\pi} \cos(wt + \theta)\, d\theta$$

$$= \frac{A}{\pi} \int_0^{\pi} [\cos wt \cos \theta - \sin wt \sin \theta]\, d\theta$$

$$= \frac{A}{\pi} [\cos wt \sin \theta + \sin wt \cos \theta]_0^{\pi} = \frac{-2A}{\pi} \sin wt$$

Since $E(X(t))$ is a function of t, the random process $X(t)$ is not a stationary process.

4.5.7 Example

Given the random process $\{X(t) = 10 \cos (100t + \theta)\}$ where θ is uniformly distributed random variable in the interval $(-\pi, \pi)$. Show that the process is correlation ergodic.

Solution. To prove, when $T \to \infty$,

$$\frac{1}{2T} \int_{-T}^{T} X(t)\, X(t + \tau)\, dt \to R_{XX}(\tau)$$

Now

$$R_{XX}(\tau) = E(X(t)\, X(t + \tau))$$
$$= E\,[100 \cos(100\, t + \theta) \cos((t + \tau)\, 100 + \theta)]$$

$$= E\left[100\left[\frac{\cos(200t + 100\tau + 2\theta) + \cos 100\tau}{2}\right]\right]$$

$$= 50\, E(\cos 100\ \tau)$$

since $E(\cos(200t + 100\ \tau + 2\theta)) = 0$

$$= 50 \cos 100\ \tau.$$

$$\frac{1}{2T}\int_{-T}^{T} 10\cos(100\ t + \theta).\ 10\cos(100t + 100\ \tau + \theta)\ dt$$

$$= \frac{1}{2T}\int_{-T}^{T} 50\cos 100\ \tau\, dt$$

$$= \frac{50}{2T}\cos 100\ \tau\,[t]_{-T}^{T} = 50\cos 100\ \tau.$$

Hence $\{X(t)\}$ is correlation Ergodic process.

4.5.8 Example

Consider a random process $\{X(t) = \cos(wt + \theta)\}$ where w is a real constant and θ is a uniform random variable in $(0, \pi/2)$. Then show that $X(t)$ is not a wide-sense stationary.

Solution. $$E(X(t)) = \int_{0}^{\pi/2} \frac{2}{\pi}\cos(wt + \theta)\ d\theta = \frac{2}{\pi}\int_{0}^{\pi/2} (\cos wt \cos\theta - \sin wt \sin\theta)\ d\theta$$

$$= \frac{2}{\pi}[\cos wt \sin\theta + \sin wt \cos\theta]_{0}^{\pi/2} = \frac{2}{\pi}[\cos wt - \sin wt]$$

Since $E(X(t))$ depends on t, $X(t)$ is not a wide-sense stationary process.

4.6 SINE WAVE RANDOM PROCESS

4.6.1. Definition

A sine wave random process $\{X(t)\}$ is defined as $X(t) = A \sin(wt + \theta)$, where the amplitude A or frequency w or phase θ or any of these combination may be random.

4.6.2. Example

For the sine wave process $X(t) = Y \cos w_0 t\ -\infty < t < \infty$, w_0 = constant, the amplitude Y is a random variable with uniform distribution in (0, 1). Check whether the process is stationary.

Solution. $$E(X(t)) = \int_{-\infty}^{\infty} y\cos w_0 t\,.\, f(y)dy$$

$$= \int_{0}^{1} y\cos w_0 t\,.\,1\,.\,dy$$

$$= \cos w_0 t \left[\frac{y^2}{2}\right]_0^1$$

$$= \frac{1}{2} \cos w_0 t$$

Since the mean is time dependent, the process is not stationary.

4.6.3. Example

For a random process $X(t) = Y \sin wt$, Y is a uniformly distributed random variable in $(-1, 1)$. Check whether the process is wide sense stationary or not.

Solution. As Y is uniformly distributed in $(-1, 1)$.

$$E(X(t)) = \int_{-1}^{1} y \sin wt \, . \, f_Y(y)\, dy$$

$$= \int_{-1}^{1} y \sin wt \, . \, \frac{1}{2}\, dy$$

$$= \frac{1}{2} \sin wt \left[\frac{y^2}{2}\right]_{-1}^{1}$$

$$= \frac{1}{2} \sin wt \times 0$$

$$= 0, \text{ a constant.}$$

The auto correlation of the process is :

$$E(X(t_1)\, X(t_2)) = E(y \sin wt_1 \, . \, y \sin wt_2)$$

$$= \sin wt_1 \sin wt_2 \, . \, E(y^2)$$

$$= \sin wt_1 \, . \, \sin wt_2 \int_{-1}^{1} y^2 f_Y(y)\, dy$$

$$= \sin wt_1 \sin wt_2 \, \frac{1}{2}\left[\frac{y^3}{3}\right]_{-1}^{1}$$

$$= \frac{1}{3} \, \frac{\cos w(t_1 - t_2) - \cos(t_1 + t_2)}{2},$$

which is depending on time. Hence the process is not WSS.

4.7 SOLVED PROBLEMS

1. Consider the process $X(t) = A \cos wt + B \sin wt$, where A and B are uncorrelated random variables each with mean 0 and variance 1 and w is a positive constant. Show that the process $X(t)$ is covariance stationary.

Solution. Given $E(A) = E(B) = 0$ and $E(A^2) = E(B^2) = 1$, $E(AB) = 0$.

Now averages are given by:

$$E(X(t)) = E(A \cos wt + B \sin wt)$$

$$= E(A \cos wt) + E(B \sin wt)$$
$$= E(A)\, E(\cos wt) + E(B)\, E(\sin wt) = 0$$
$$E(X(t)\, X(t+\tau)) = E[(A \cos wt + B \sin wt)\,(A \cos (wt + w\tau) + B \sin (wt + w\tau))]$$
$$= E[A^2 \cos wt \cos (wt + w\tau)] + AB \sin wt \cos (wt + w\tau) + BA \sin (wt + w\tau) \cos wt + B^2 \sin wt \sin (wt + w\tau)]$$
$$= E(A^2)\, E(\cos wt \cos (wt + w\tau)) + E(AB)\, E(\sin wt \cos(wt + w\tau)) + E(AB)\, E(\sin (wt + w\tau) \cos wt) + E(B^2)\, E[\sin wt \sin(wt + w\tau)]$$
$$= E(\cos wt \cos (wt + w\tau) + \sin wt \cdot \sin (wt + w\tau)) + 0 = E(\cos(wt - wt - w\tau))$$
$$= E(\cos w\tau) = \cos w\tau$$

and

$$\text{cov}\,(X(t)\, X(t+\tau)) = R_X(t, t+\tau) - E(X(t))\, E(X(t+\tau)) = R_X(\tau) - 0$$

which is depending on τ. Hence $X(t)$ is covariance stationary.

2. Consider the random process

$$Y(t) = (-1)^{N(t)}$$

where $N(t)$ is a Poisson process with rate λ. Thus $Y(t)$ starts at $Y(0) = 1$ and switches back and forth from +1 to –1 at random Poisson time T_i as in figure.

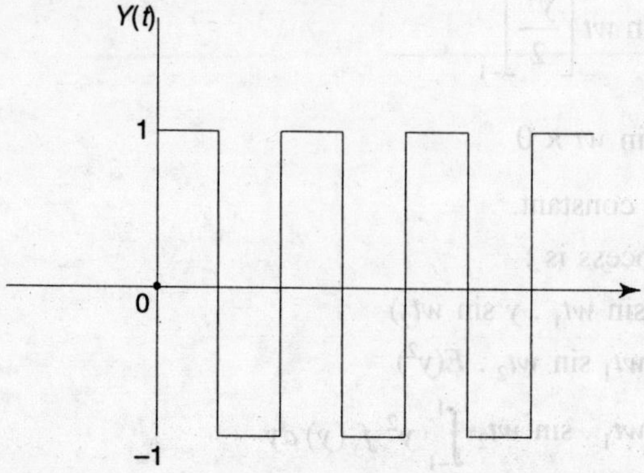

The process $Y(t)$ is known as the **semi-random telegraph signal** because its initial value $Y(0) = 1$ is not random. Find the mean and auto-correlation of $Y(t)$.

Solution. Let

$$Y(t) = \begin{cases} 1 & \text{if } X(t) \text{ is even} \\ -1 & \text{if } X(t) \text{ is odd} \end{cases}$$

Then

$$P(Y(t) = 1) = P\,(N(t) = \text{even integer}) = \frac{e^{-\lambda t}(\lambda t)^n}{n!}, \; n \text{ is even integer}$$

$$= e^{-\lambda t}\left[1 + \frac{(\lambda t)^2}{2!} + \frac{(\lambda t)^4}{4!} + \ldots\right] = e^{-\lambda t} \cos h\, \lambda t$$

$$P(Y(t) = -1) = P\,(N(t) = \text{odd integer}) = \frac{e^{-\lambda t}(\lambda t)^n}{n!}, \; n \text{ is odd}$$

$$= e^{-\lambda t}\left[\lambda t + \frac{(\lambda t)^2}{3!} + \frac{(\lambda t)^5}{5!} + \ldots\right] = e^{-\lambda t} \sin h\ \lambda t.$$

Hence mean = $E(Y(t)) = 1P(Y(t) = 1) + (-1)\ P(Y(t) = -1) = e^{-\lambda t}\ [\cosh \lambda t - \sinh \lambda t] = e^{-2\lambda}$

Similarly, since $Y(t)\ Y(t + \tau) = 1$, if there are an even number of events in $(t, t + \tau)$ for $\tau > 0$ and $Y(t)\ Y(t + \tau) = -1$, if there are an odd number of events, then for $t > 0$, $t + \tau > 0$,

$$R_Y(t, t + \tau) = E(Y(t)\ Y(t + \tau))$$

$$= (1) \sum_{n \text{ even}} e^{-\lambda\tau} \frac{(\lambda\tau)^n}{n!} + (-1) \sum_{n \text{ odd}} e^{-\lambda\tau} \frac{(\lambda\tau)^n}{n!}$$

$$= e^{-\lambda\tau} \sum_{n=0}^{\infty} \frac{(-\lambda\tau)^n}{n!} = e^{-\lambda\tau} e^{-\lambda\tau} = e^{-2\lambda\tau}$$

which gives $R_Y(t, t + \tau) = R_Y(\tau)$ and since $R_Y(-\tau) = R_Y(\tau)$, we have $R_Y(\tau) = e^{-2\lambda|\tau|}$.

Since $E(Y(t))$ is not a constant, $Y(t)$ is not wide-sense stationary.

3. Consider the random process $X(t) = AY(t)$, where $Y(t)$ is the semi-random telegraph signal (of the above problem) and A is a random variable independent of $Y(t)$ and takes on the values ± 1 with equal probability. The process $X(t)$ is known as the random telegraph signal. Show that $X(t)$ is wide-sense stationary.

Solution. Since $E(A) = 0$, and $E(A^2) = 1$,

$$E(X(t)) = E(A\ Y(t)) = E(A)\ E(Y(t)) = 0$$

and

$$E(X(t)\ X(t + \tau)) = E(A^2\ Y(t)\ Y(t + \tau)) = E(A^2)\ E(Y(t)\ Y(t + \tau)) = R_Y(t, t + \tau) = e^{-2\lambda|\tau|}$$

Hence $X(t)$ is wide-sense stationary.

4. Prove that a finite order stationary random process has a constant mean.

Solution. Consider a random process $X(t)$ at two different times t_1 and t_2.

$$E(X(t_1)) = \int_{-\infty}^{\infty} x f(x, t_1)\ dx$$

$$E(X(t_2)) = \int_{-\infty}^{\infty} x f(x, t_2)\ dx$$

Let $t_2 = t_1 + \tau$. Then

$$E(X(t_1 + \tau)) = \int_{-\infty}^{\infty} x f(x, t_1 + \tau)\ dx$$

$$= \int_{-\infty}^{\infty} x f(x, t_1)\ dx \text{ since } X(t) \text{ is first order stationary}$$

$$= E(X(t_1))$$

Hence $E(X(t))$ is constant.

5. Prove that the Random Telegraphic signal process is a covariance stationary process.

Solution. We know from problem 3 that the random telegraphic signal process $\{X(t): t \geq 0\}$ is wide-sense stationary. Now consider

$$\text{Cov}\,(X(t)\,X(t+\tau)) = R_X(t, t+\tau) - E(X(t)\,E\,(X(t+\tau))$$
$$= R_X(\tau) - 0 = R_X(\tau)$$

which indicates that Cov $(X(t), X(t+\tau))$ depends only on τ. Hence $X(t)$ is covariance stationary.

6. If $\{X(t) : t \geq 0\}$ with $X(0) = 0$ is a process having independent increments, then

$$\text{Cov}\,(X(s), X(t)) = \text{Var}\,\{X(s)\} \text{ for } 0 \leq s \leq t$$
$$= \text{Var}\{X\,(\min\,(s, t))\}, \text{ for any } s, t\ (s > 0, t > 0)$$

(such a process is not covariance stationary).

Solution. We first prove the following example.

7. Let $\{X(t)\ ,\ t \geq 0\}$ be a random process with stationary independent increments, and assume that $X(0) = 0$. Show that

(i) $\text{Var}\,(X(t)) = \sigma_1^2\, t$ (4.11)

(ii) $\text{Var}\,(X(t) - X(s)) = \sigma_1^2\,(t - s), \quad t > s$

where $\sigma_1^2 = \text{Var}\,(X(1))$

Solution.

(i) Let $$g(t) = \text{Var}\,(X(t)) = \text{Var}\,(X(t) - X(0))$$

Then for any t, s, and the property of stationary independent increments

$$g(s+t) = \text{Var}\,(X\,(t+s) - X(0))$$
$$= \text{Var}\,(X(t+s) - X(s) + X(s) - X(0))$$
$$= \text{Var}\,(X(t+s) - X(s)) + \text{Var}\,(X(s) - X(0))$$
$$= \text{Var}\,(X(t) - X(0)) + \text{Var}\,(X(s) - X(0)) = g(t) + g(s)$$

which is a functional equation. The only solution to the functional equation is $g(t) = kt$, where k is a constant. Since $k = g(1) = \text{Var}\,(X(1))$, we get

$$\text{Var}\,(X(t) = \sigma_1^2\, t, \sigma_1^2 = \text{var}\,(X(1))$$

(ii) Let $t > s$

$$\text{Var}\,(X(\text{t})) = \text{Var}\,[X(t) - X(s) + X(s) - X(0)]$$
$$= \text{Var}\,(X(t) - X(\text{s})) + \text{Var}\,(X(s) - X(0))$$
$$= \text{Var}\,(X(t) - X(s)) + \text{Var}\,(X(s))$$

This implies

$$\text{Var}\,(X(t) - X(s)) = \text{Var}\,(X(t)) - \text{Var}\,(X(s))$$
$$= \sigma_1^2\, t - \sigma_1^2\, s = \sigma_1^2\,(t - s)$$

Solution to Example 6.

We know $$\text{cov}\,(X(t), X(s)) = \frac{1}{2}\,\{\text{Var}\,(X(t)) + \text{Var}\,(X(s))\} - \text{Var}\,(X(t) - X(s))$$

$$= \begin{cases} \dfrac{1}{2}\sigma_1^2\,[t + s - (t - s)] = \sigma_1^2\, s, & t > s \\ \dfrac{1}{2}\sigma_1^2\,(t + s - (s - t)) = \sigma_1^2\, t, & s > t \end{cases}$$

Hence

$$\text{cov}(X(t), X(s)) = \sigma_1^2 \min(t, s)$$
$$= \text{Var}\{X(\min(t, s))\}, \text{ for any } s, t, \text{ by (4.11)}$$

8. Solve the recurrence-differential equation

$$p'_n(t) = -\lambda[p_n(t) - p_{n-1}(t)], \qquad n \geq 1, \lambda > 0 \tag{4.12}$$

$$p'_0 = -\lambda p_0(t) \tag{4.13}$$

with the initial conditions $p_n(0) = 0$, for $n \neq 0$, $p_0(0) = 1$.

Solution. Let the Laplace transform of $p_n(t)$ be

$$L(p_n(t)) = \int_0^\infty e^{-st} \cdot p_n(t)\, dt = f_n(s) \text{ (say)}$$

Then, we know that the Laplace transform of derivative as

$$L(p'_n(t)) = s\, f_n(s) - p_n(0) = s\, f_n(s)$$

Hence taking Laplace transform on both sides of (4.12) we get

$$sf_n(s) = -\lambda[f_n(s) - f_{n-1}(s)]$$

or

$$f_n(s) = \left(\frac{\lambda}{\lambda + s}\right) f_{n-1}(s), \qquad n \geq 1$$

which is a difference equation in f_n. Its solution is given by

$$f_{n-1}(s) = c\left(\frac{\lambda}{\lambda + s}\right)^{n-1}, \quad n \geq 1.$$

For $n = 1$, $f_0(s) = c = \dfrac{1}{\lambda + s}$, where by taking Laplace transform on both sides of (4.13) gives

$$sf_0(s) - 1 = -\lambda f_0(s)$$

or

$$f_0(s) = \frac{1}{\lambda + s}.$$

Hence

$$f_{n-1}(s) = \frac{\lambda^{n-1}}{(\lambda + s)^n}, \qquad n \geq 1$$

$$f_n(s) = \frac{\lambda^n}{(\lambda + s)^{n+1}}, \qquad n \geq 0$$

Now

$$p_n(t) = L^{-1}(f_n(s)) = L^{-1}\left[\frac{\lambda^n}{(\lambda + s)^{n+1}}\right] = \frac{e^{-\lambda t}(\lambda t)^n}{n!}$$

9. Consider a random process $X(t)$ defined by

$$X(t) = Y \cos wt, \; t \geq 0$$

where w is a constant and Y is uniform random variable over (0, 1). What can you say about $X(t)$.

Solution. The random process $X(t)$ is a continuous-time, continuous state random process.

10. Show that a random process which is stationary to order n is also stationary to all orders lower than n.

Solution.

Let $$P\{X(t_1) \le x_1 \ldots X(t_n) \le x_n\} = P\{X(t_1 + \tau) \le x_1, \ldots X(t_n + \tau) \le x_n\}$$

for any τ. Let $x_n \to \infty$, then

$$P\{X(t_1) \le x_1, \ldots X(t_{n-1}) \le x_{n-1}\} = P\{X(t_1 + \tau) \le x_1, \ldots X(t_{n-1} + \tau) \le x_{n-1}\}$$

and the process is stationary to order $n - 1$. Continuing the same procedure, the process is seem to be stationary to all orders lower than n.

11. Let $\{X_n : n \ge 0\}$ be a sequence of random variables with mean 0 and variance 1. Then show that $\{X_n : n \ge 0)$ is a wide-sense stationary.

Solution. $E(X_n) = 0$ for *all n*

$= a$ constant

$$R_{XX}(\tau) = E(X_n X_{n+\tau}) = \begin{cases} E(X_n^2) = \text{Var}(X_n) = 1, & \tau = 0 \\ E(X_n)\, E(X_{n+\tau}) = 0, & \tau \ne 0 \end{cases}$$

which depends only on τ. Hence $\{X_n\}$ is wide-sense stationary.

12. Show that if a random process $X(t)$ is wide sense stationary, then it is covariance stationary.

Solution. If $X(t)$ is wide-sense stationary, then

$$E(X(t)) = \text{constant} \quad \text{for all } t$$

$$R_X(t, t + \tau) = R_X(\tau) \quad \text{for all } t$$

$$\text{Cov}(X(t)\, X(t + \tau)) = R_X(t, t + \tau) - E(X(t))\, E(X(t + \tau)) = R_X(\tau) - \mu^2$$

which is depending only on τ; thus, $\lambda(t)$ is covariance stationary.

13. Suppose t is not a point at which an event occurs in a Poisson process $X(t)$ with rate λ. Let $W(t)$ be the random variable representing the time until the next occurrence of an event. Show that the distribution of $W(t)$ is independent of t and $W(t)$ is an exponential random variable with parameter λ.

Solution. Let $s (0 \le s < t)$ be the point at which the last event (say $(n - 1)^{st}$ event) occurred. See Fig.

The event $\{W(t) > \tau\}$ is equivalent to the event $\{Z_n > t - s + \tau |_{Z_n > t - s}\}$

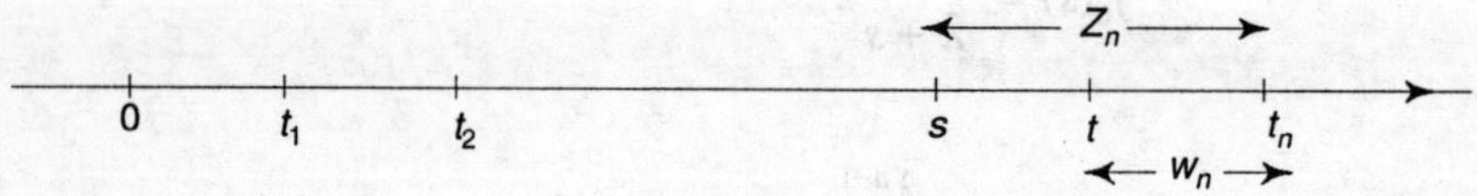

Then

$$P(W(t) > \tau) = P[Z_n > t - s + \tau |_{Z_n > t - s}]$$

$$= \frac{P(Z_n > t - s + \tau)}{P(Z_n > t - s)} = \frac{e^{-\lambda(t - s + \tau)}}{e^{-\lambda(t - s)}} = e^{-\lambda\tau}$$

and

$$P(W(t) \le \tau) = 1 - e^{-\lambda\tau}$$

which says that $W(t)$ is exponential random variable with parameter λ and is independent of t. Note that $W(t)$ is often called *waiting time.*

14. Let T_n denote the time of the nth event of a Poisson process $X(t)$ with rate λ. Suppose that one event has occurred in the interval (0. t). Show that the conditional distribution of arrival time T_1 is uniform over (0, t).

Solution. For $\tau \le t$,

$$P(T_1 \le \tau \mid X(t) = 1) = \frac{P(T_1 \le \tau, X(t)=1)}{P(X(t)=1)}$$

$$= \frac{P(X(\tau)=1, X(t)-X(\tau)=0)}{P(X(t)=1)}$$

$$= \frac{P(X(\tau)=1)\, P(X(t)-X(\tau)=0)}{P(X(t)=1)}$$

$$= \frac{\lambda \tau e^{-\lambda \tau} e^{-\lambda(t-\tau)}}{\lambda t\, e^{-\lambda t}} = \frac{\tau}{t}$$

which says that T_1 is uniform over (0, t).

EXERCISES

1. Let $X(t) = \sum_{r=1}^{k} A_r \cos \theta_r t + B_r \sin \theta_r t$, where A_r, B_r are uncorrelated random variables with mean 0, variance σ^2 and θ_r are constants. Show that $\{X(t)\}$ is covariance stationary.
2. Show that every random process $\{X(t), t = 0, 1, 2...\}$ with independent increments is a Markov process. Is the converse true?
3. Poisson increment process. The process $\{X(t), t \ge 0\}$ defined by

$$X(t) = \frac{N(t+\varepsilon) - N(t)}{\varepsilon}$$

where $\varepsilon > 0$, is a constant, is knowns as Poisson increment process. Show that the process is covariance stationary with

$$E(X(t)) = \lambda$$

and

$$\text{Cov}(s, t) = 0, \quad |t - s| > \varepsilon$$

$$= \lambda/\varepsilon - \frac{\lambda |t-s|}{\varepsilon^2}, \quad |t - s| < \varepsilon.$$

4. If customers arrive at a counter in accordance with a Poisson process with a mean rate of 2 per minute, find the probability that the interval between two consecutive arrivals is
   (i) more than 1 minute
   (ii) between 1 minute and 2 minutes
   (iii) 4 minutes or less.

5. Examine whether the Poisson process $\{X(t)\}$ given by $P\{X(t) = \nu\} = \dfrac{e^{-\lambda t} (\lambda t)^r}{r!}$ $(r = 0, 1, 2, \ldots)$ is of variance stationary.

6. What is the difference between a random variable and random process.

7. Prove that two independent Poisson processes are not a Poisson process.

8. Prove that the random process $X(t) = A \cos (w_0 t + \theta)$ is wide-sense stationary.

9. Determine whether the Stochastic process

$$X(t) = A \cos t + B \sin t$$

is ergodic or not, where A and B are normally distributed independent random variables with means zero and standard deviation equal.

5

Correlation Function and Spectral Densities

5.1 AUTOCORRELATION AND CROSS CORRELATION

Consider two random variables of a random process $X(t)$. The correlation (or inter-relationship) between the two random variables is a measure of the similarity between the variables.

Assume that the two random variables are obtained at two different times say s, t. If the two time instants are very close, then we say X_{t_1} and X_{t_2} are highly correlated. If the two time instants are widley separated, the correlation between the values of the random variables X_{t_1} and X_{t_2} will be less.

5.1.1 Definition Autocorrelation

Consider two random variables X_{t_1} and X_{t_2} obtained by observing a random process $X(t)$ at times t_1 and t_2 respectively. The autocorrelation function R of the random process $X(t)$ is defined by

$$R_{XX}(t_1, t_2) = E(X(t_1), X(t_2))$$

Note that $R_{XX}(t_1, t_1) = E(X^2(t_1))$, is the average power of the random process. The autocorrelation function $R_{XX}(t_1, t_2)$ of a stationary process depends only on the time difference. Thus $R_{XX}(t_1, t_2)$ is a one-dimensional function in this case and is written as $R_X(\tau)$ or $R(\tau)$

5.1.2 Definition Auto covariance

The auto covariance $C_{XX}(t_1, t_2) = C(t_1, t_2)$ of the random process $X(t)$ is defined as the covariance of the random variables $X(t_1)$ and $X(t_2)$

$$C_{XX}(t_1, t_2) = R(t_1, t_2) - E(X(t_1))\, E(X(t_2))$$

The value of auto covariance of two random variables observed at the same time t represents the variance of $X(t)$.

$$\begin{aligned} C_{XX}(t, t) &= R_{XX}(t, t) - E(X(t))\, E(X(t)) \\ &= E(X^2(t)) - E(X(t))^2 \\ &= \text{Var}(X(t)). \end{aligned}$$

5.1.3 Definition Correlation Coefficient

The correlation coefficient of a random process $X(t)$ is defined as

$$\rho_{XX}(t_1, t_2) = \frac{C_{XX}(t_1, t_2)}{\sqrt{C_{XX}(t_1, t_1)\, C_{XX}(t_2, t_2)}}$$

If $X(t)$ is a complex random process, then its auto correlation function $R_X(t, s)$ and auto covariance function $C(t, s)$ are defined, respectively by

$$R_X(t, s) = E(X(t)\, X^*(s))$$
$$C(t, s) = E\{[X(t) - E(X(t))]\,[X(s) - E(X(s))]^*\}$$

where * denotes the complex conjugate.

5.1.4 Definition Cross Correlation

Let $\{X(t): t \geq 0\}$, $\{Y(t), t \geq 0\}$ be two random processes. Then the cross correlation function R i defined by

$$R_{XY}(t_1, t_2) = E(X(t_1)\, Y(t_2))$$

5.1.5 Definition Cross-Covariance

The cross-covariance of the two random processes $X(t)$ and $Y(t)$ is defined as the covariance of twc random variables $X(t_1)$ and $Y(t_2)$

$$C_{XY}(t_1, t_2) = R_{XY}(t_1, t_2) - E(X(t_1))\, E(Y(t_2)).$$

5.1.6 Definition Correlation-coefficient

The cross-correlation coefficient of two random processes $X(t)$ and $Y(t)$ is given by

$$\rho_{XY}(t_1, t_2) = \frac{C_{XY}(t_1, t_2)}{\sqrt{C_{XX}(t_1, t_1)\, C_{YY}(t_2, t_2)}}$$

5.1.7 Definition Orthogonal process

Two random processes $X(t)$ and $Y(t)$ are called **mutually orthogonal process** to each other if thei cross-correlation function is equal to zero, for every t_1, t_2.

$$R_{XY}(t_1, t_2) = 0 \quad \text{for all } t_1, t_2.$$

5.1.8 Definition Uncorrelated Process

Two random processes $X(t)$ and $Y(t)$ are called **uncorrelated** to each other, if their cross-covariance i equal to zero.

$$C_{XY}(t_1, t_2) = 0 \quad \text{for every } t_1, t_2.$$

Important points to be noted

The mean and autocorrelation function only provide a partial description of the distribution of random process $X(t)$.

The conditions $E(X(t))$ = constant and $R_{XX}(t_1, t+\tau) = R_{XX}(\tau)$ involving the mean and autocorrelatio function are not sufficient to guarantee that the random process $X(t)$ is stationary (but only wide sense stationary)

5.1.9 Example

Find the autocorrelation function of the simple random walk $X(n) = \sum_{i=1}^{n} Z_i$, $n = 1, 2, \ldots$ where $P(Z_n = 1) = p$ and $P(Z_n = -1) = q = 1 - p$, for all n with $X(0) = 0$.

Solution.

$$R_X(n, m) = E(X(n)\, X(m)) = E(X_n\, X_m)$$

$$= \sum_{i=0}^{n} \sum_{k=0}^{m} E(Z_i Z_k)$$

$$= \sum_{i=0}^{\min(n,m)} E(Z_i^2) + \sum_{i=0}^{n} \sum_{\substack{k=0 \\ i \neq j}}^{m} E(Z_i)\, E(Z_k)$$

We have
$$E(X_n) = E\left(\sum_{k=1}^{n} Z_k\right) = nE(Z_k) = n(p - q)$$

and
$$\text{Var}(X_n) = \text{Var}\left(\sum_{k=1}^{n} Z_k\right) = n \,\text{Var}(Z_k) = 4\, npq$$

where
$$E(Z_k) = 1 \times p + (-1)\, q = p - q$$

$$E\left(Z_k^2\right) = 1^2 p + (-1)^2 q = p + q = 1$$

$$\text{Var}(Z_k) = E\left(Z_k^2\right) - (E(Z_K))^2 = 1 - (p - q)^2 = 4\, pq$$

Hence
$$R_X(n, m) = \min(n, m) + [nm - \min(n, m)]\, (p - q)^2$$

If
$$p = q = \frac{1}{2}, \text{ then } R_X(n, m) = \min(n, m).$$

5.1.10 Example

Consider the random process $\{X(t) = Y \cos wt, t \geq 0\}$, where w is a constant and Y is a uniform random variable over (0, 1). Find the auto correlation function $R_{XX}(t, s)$ of $X(t)$ and auto covariance $C_{XX}(t, s)$ of $X(t)$.

Solution. We know a random variable Y is uniform over (a, b) if its pdf is

$$f(y) = \begin{cases} \dfrac{1}{b-a} & a < y < b \\ 0 & \text{otherwise} \end{cases}$$

Mean and variance of Y are

$$E(Y) = \int_a^b x \cdot \frac{1}{b-a}\, dx = \frac{1}{b-a}\left[\frac{x^2}{2}\right]_a^b$$

$$= \frac{1}{2(b-a)}\, [b^2 - a^2] = \frac{1}{2}\; b + a$$

$$E(Y^2) = \int_a^b x^2 \cdot \frac{1}{b-a}\, dx = \frac{1}{b-a}\left[\frac{x^3}{3}\right]_a^b$$

$$= \frac{1}{3}\, (b^2 + ab + a^2)$$

$$\text{Var}(Y) = E(Y^2) - E(Y))^2$$
$$= \frac{1}{3}(b^2 + ab + a^2) - \frac{1}{4}(b+a)^2 = \frac{1}{12}(b-a)^2$$

Hence, if Y is a random variable over (0, 1), then

$$E(Y) = \frac{1}{2},\ E(Y^2) = \frac{1}{3}. \text{ Thus}$$

$$E(X(t)) = E(Y \cos wt) = E(Y) \cos wt = \frac{1}{2} \cos wt$$
$$R_{XX}(t, s) = E(X(t)\, X(s))$$
$$= E(Y^2 \cos wt \cos ws)$$
$$= E(Y^2) \cos wt \cos ws$$
$$= \frac{1}{3} \cos wt \cos ws$$
$$C_{XX}(t, s) = R_{XX}(t, s) - E(X(t))\, E(X(s))$$
$$= \frac{1}{3} \cos wt \cos ws - \frac{1}{4} \cos wt \cos ws = \frac{1}{12} \cos wt \cos ws.$$

Properties of Autocorrelation

Let $X(t)$ be a wide-sense stationary random process. Then the autocorrelation function of $X(t)$ has several properties.

1. $R_{XX}(0) = E(X^2(t))$

   Proof: We know $R_{XX}(t_1, t_2) = E(X(t_1)\, X(t_2))$. Let $t_2 = t_1 + \tau$. Then
   $$R_{XX}(\tau) = R_{XX}(t, t+\tau) = E(X(t)\, X(t+\tau)).$$
   For $\tau = 0,\ R_{XX}(0) = R_{XX}(t, t) = E(X^2(t))$

2. The autocorrelation function $R_{XX}(\tau)$ is an even function. That is
   $$R_{XX}(\tau) = R_{XX}(-\tau)$$
   Proof: $R_{XX}(\tau) = E(X(t)\, X(t+\tau))$ and
   $$R_{XX}(-\tau) = E(X(t)\, X(t-\tau))$$
   Let $t - \tau = a$. Then
   $$R_{XX}(-\tau) = E(X(a+\tau)\, X(a)) = R_{XX}(\tau)$$
   hence $R_{XX}(-\tau) = R_{XX}(\tau)$

3. $R_{XX}(0) \geq | R_{XX}(\tau) |$

   Proof: We know $E(X(t_1) \pm X(t_2))^2 \geq 0$. That is
   $$E(X^2(t_1) + X^2(t_2) \pm 2X(t_1)\, X(t_2)) \geq 0$$
   $$E(X^2(t_1)) + E(X^2(t_2)) \pm 2E(X(t_1)\, X(t_2)) \geq 0$$
   Using property 1, we have $R_{XX}(0) = E(X^2(t))$, hence
   $$R_{XX}(0) + R_{XX}(0) \geq \pm 2E(X(t_1)\, X(t_2))$$
   $$\Rightarrow \quad 2R_{XX}(0) \geq 2\, | E(X(t_1)\, X(t_2)) |$$
   $$\geq 2\, | E(X(t_1))\, X(t_1+\tau) | \text{ if } t_2 = t_1 + \tau$$

That is

$$R_{XX}(0) \geq | R_{XX}(\tau) |.$$

4. The autocorrelation function of $X(t)$ is a finite energy function. That is

$$R_{XX}(t_1, t_2) = \int_{-\infty}^{\infty} R_{XX}(\tau)\, d\tau < \infty$$

5. If a random process $X(t)$ has no periodic components and if $E(X(t)) \neq 0$, then

$$\lim_{|\tau| \to \infty} R_{XX}(\tau) = [E(X(t))]^2$$

6. If $X(t)$ has a periodic component, then the autocorrelation function $R_{XX}(\tau)$ will also have a periodic component with the same period.
7. If a random process $X(t)$ is ergodic, zero mean and has no periodic component, then

$$\lim_{|\tau| \to \infty} R_{XX}(\tau) = 0$$

8. If there are two random processes $X(t)$ and $Y(t)$ such that $Z(t) = X(t) + Y(t)$, then

$$R_{ZZ}(\tau) = R_{(X+Y)(X+Y)}(\tau) = R_{XX}(\tau) + R_{YY}(\tau) + R_{XY}(\tau) + R_{YX}(\tau)$$

9. $R_{XX}(\tau)$ cannot have an arbitrary shape.

Properties of cross-correlation

1. $R_{XY}(\tau) = R_{YX}(-\tau)$

Proof: By definition of cross-correlation,

$$R_{XY}(t, t+\tau) = E(X(t)\, Y(t+\tau)) = R_{XY}(\tau)$$

and $$R_{YX}(t, t+\tau) = E_{YX}(Y(t)\, X(t+\tau)) = R_{YX}(\tau)$$

Now $$R_{YX}(-\tau) = E(Y(t)\, X(t-\tau)).$$

Let $t - \tau = a$. Then

$$R_{YX}(-\tau) = E(Y(a+\tau)\, X(a)) = E(X(a)\, Y(a+\tau)) = R_{XY}(\tau)$$

2. The cross-correlation function of two random processes $X(t)$ and $Y(t)$ does not have a maximum value at the origin, i.e. at $\tau = 0$.
3. $| R_{XY}(\tau) | \leq (R_{XX}(0)\, R_{YY}(0))^{1/2}$

The cross-correlation function is bounded by the geometric mean of their values at the origin.

4. If $X(t)$ and $Y(t)$ are two random processes and $R_{XX}(\tau)$ and $R_{YY}(\tau)$ are their respective autocorrelation functions, then

$$| R_{XY}(\tau) | \leq \frac{1}{2}\, [R_{XX}(0) + R_{YY}(0)]$$

Since the geometric mean $\sqrt{R_{XX}(0)\, R_{YY}(0)}$ cannot exceed the arithmetic mean $\left[\frac{R_{XX}(0) + R_{YY}(0)}{2}\right]$, we have

$$\sqrt{R_{XX}(0)\, R_{YY}(0)} \leq \frac{1}{2}\, [R_{XX}(0) + R_{YY}(0)]$$

Hence, by property 3,

$$| R_{XY}(\tau) | \leq \frac{1}{2}\, [R_{XX}(0) + R_{YY}(0)]$$

5. If the random processes $X(t)$ and $Y(t)$ are independent to each other, then their cross-correlations satisfy the relation

$$R_{XY}(\tau) = R_{YX}(\tau) = E(X(t)E(Y(t))$$

6. If the random processes $X(t)$ and $Y(t)$ have periodic components with a period T_0, then their cross-correrealtion functions $R_{XY}(\tau)$ and $R_{YX}(\tau)$ will also contain the same periodic components with the same period T_0.
7. If the mean of two random processes $X(t)$ and $Y(t)$ are zero, then

$$\lim_{|\tau|\to\infty} R_{XY}(\tau) = 0 \quad \text{and} \quad \lim_{|\tau|\to\infty} R_{YX}(\tau) = 0.$$

5.1.11 Example

Given the auto correlation function for a stationary process

$$R_{XX}(\tau) = 25 + \frac{4}{1+6\tau^2},$$

find the mean and variance of the process $X(t)$.

Solution. We know

$$\lim_{|\tau|\to\infty} R_{XX}(\tau) = [E(X(t)]^2$$

That is $$\lim_{|\tau|\to\infty} 25 + \frac{4}{1+6\tau^2} = [E(X(t)]^2$$

$$\Rightarrow \quad 25 = [E(X(t)]^2 \quad \Rightarrow \quad E(X(t)) = \pm 5.$$

$$E(X^2) = R_{XX}(0) = 25 + 4 = 29.$$

$$\text{Var}(X) = E(X^2) - (E(X))^2 = 29 - 25 = 4.$$

5.1.12 Example

Let $X(t) = A \cos wt + B \sin wt$, $Y(t) = B \cos wt - A \sin wt$ where A and B are random variables and w is a constant. Show that $X(t)$ and $Y(t)$ are wide-sense stationary if A and B are uncorrelated; zero mean random variables with the same variance. Prove that $X(t)$ and $Y(t)$ are jointly wide-sense stationary, finding the cross-correlation function.

Solution.

$$\begin{aligned} R_{XY}(t, t+\tau) &= E(X(t)\,Y(t+\tau)) \\ &= E(AB \cos wt.\cos(wt + w\tau) + B^2 \sin wt \cos(wt + w\tau) \\ &\quad - A^2 \cos wt.\sin(wt + w\tau) - AB \sin wt.\sin(wt + w\tau)) \\ &= E(AB)\cos(2\,wt + w\tau) + E(B^2)\sin wt.\cos(wt + w\tau) \\ &\quad - E(A^2)\cos wt.\sin(wt + w\tau) \end{aligned}$$

Since A and B are assumed to be zero-mean, uncorrelated random variables, $E(AB) = 0$, and $E(A) = E(B) = 0$, hence

$$\begin{aligned} \sigma_X^2 &= \text{Var}(X(t)) = E(X^2(t)) - E(X(t))^2 \\ &= E(X^2(t)) = E(A^2)\cos^2 wt + E(B^2)\sin^2 wt \\ &= E(A^2)[\cos^2 wt + \sin^2 wt] \end{aligned}$$

since $E(A^2) = E(B^2)$

$$= E(A^2)$$

$$= E(B^2) \text{ since } A \text{ and } B \text{ have same variance } = \sigma_X^2$$

$$R_{XY}(t, t+\tau) = \sigma_X^2 \ (\sin wt \cos (wt + w\tau) - \cos wt \sin (wt + w\tau))$$

$$= \sigma_X^2 \ (\sin (-w\tau)) = -\sigma_X^2 \ \sin w\tau$$

Therefore $X(t)$ and $Y(t)$ are jointly wide-sense stationary because $R_{XY}(t, t+\tau)$ depends only on τ.

5.1.13 Example

If $\{N(t)\}$ is a Poisson process then the autocorrelation coefficient between $N(t)$ and $N(t+s)$ is

$$\left\{\frac{t}{(t+s)}\right\}^{1/2}$$

Solution. Let λ be the parameter of the process; then

$$E(N(T)) = \lambda T, \qquad \text{Var}\{N(T)\} = \lambda T$$

$$E(N^2(T)) = \lambda T + (\lambda T)^2 \qquad \text{for } T = t \text{ and } t + s$$

Since $N(t)$ and $N(t+s) - N(t)$ are independent, $\{N(t)\}$ being a Poisson process,

$$E\{N(t)\,N(t+s)\} = E[(N(t)\{N(t+s) - N(t) + N(t)\}]$$

$$= E[N(t)\,N(t) + N(t)\{N(t+s) - N(t)\}]$$

$$= E(N^2(t)) + E(N(t))\,E(N(t+s) - N(t))$$

Hence

$$E\{N(t)\,N(t+s)\} = (\lambda t + \lambda^2 t^2) + \lambda t.\lambda s$$

The auto-covariance between $N(t)$ and $N(t+s)$ is given by

$$C_{XX}(t, t+s) = E\{N(t)\,N(t+s)\} - E(N(t))\,E(N(t+s))$$

$$= (\lambda t + \lambda^2 t^2 + \lambda^2 ts) - \lambda t(\lambda t + \lambda s) = \lambda t.$$

Hence the autocorrelation (correlation) coefficient is

$$\rho(t, t+s) = \frac{\text{Auto-covariance}(t, t+s)}{[\text{Var}(N(t)) \cdot \text{Var}(N(t+s))]^{1/2}}$$

$$= \frac{\lambda t}{[\lambda t \cdot \lambda(t+s)]^{1/2}} = \left\{\frac{t}{t+s}\right\}^{1/2}$$

5.1.14 Example

Let $\{X(t): -\infty < t < \infty\}$ be a zero-mean stationary, normal process with the autocorrelation function

$$R_{XX}(\tau) = \begin{cases} 1 - \dfrac{|\tau|}{T} & -T \le \tau \le T \\ 0 & \text{otherwise} \end{cases}$$

Let $\{X(t_i) : i = 1, 2, \ldots, n\}$ be a sequence of n samples of the process taken at the time instants

$$t_i = \frac{iT}{2}, \ i = 1, 2, \ldots n$$

Find the mean and variance of the sample mean.

$$\mu_n = \frac{1}{n}\sum_{i=1}^{n} X(t_i)$$

Solution. Since $X(t)$ is zero-mean, stationary, $E(X(t_i)) = 0$

$$R_{XX}(t_i, t_k) = E(X(t_i)\,X(t_k))$$

$$= R_{XX}(t_k - t_i) = R_{XX}\left((k-i)\frac{T}{2}\right)$$

$$E(\mu_n) = E\left[\frac{1}{n}\sum_{i=1}^{n} X(t_i)\right] = \frac{1}{n}\sum_{i=1}^{n} E(X(t_i)) = 0$$

$$\text{Var}(\mu_n) = E(\mu_n - E(\mu_n))^2$$

$$= E\left(\mu_n^2\right) = E\left[\left\{\frac{1}{n}\sum_{i=1}^{n} X(t_i)\right\}\left\{\frac{1}{n}\sum_{k=1}^{n} X(t_k)\right\}\right]$$

$$= E\left(\frac{1}{n^2}\sum_{i=1}^{n}\sum_{k=1}^{n} X(t_i)\,X(t_k)\right) = \frac{1}{n^2}\sum_{i=1}^{n}\sum_{k=1}^{n} R_{XX}\left[(k-i)\frac{T}{2}\right]$$

By the definition of $R_{XX}(\tau)$, we get

$$R_{XX}\left((k-i)\frac{T}{2}\right) = \begin{cases} 1 & \text{if } k = i \\ \frac{1}{2} & \text{if } |k-i| = 1 \\ 0 & \text{if } |k-i| > 2 \end{cases}$$

Then
$$\text{Var}(\mu_n) = \frac{1}{n^2}\left[n \times 1 + 2(n-1) \times \frac{1}{2} + 0\right] = \frac{1}{n^2}\,(2n-1)$$

5.1.15 Example

Find the autocorrelation function of a random telegraph signal.

Solution. The random telegraph process $\{X(t) : t \geq 0\}$ is defined by $X(t) = AY(t)$, where $Y(t)$ is the semi-random telegraph signal and A is a random variable independent of $Y(t)$ and takes the values -1 and 1 with equal probability. We already proved in previous chapter that $E(X(t)) = 0$. Covariance $(s, t) = e^{-2\lambda|t-s|}$ and $R_{XX}(t, t+\tau) = e^{-2\lambda|\tau|}$.

5.1.16 EXAMPLE

Which of the following are suitable autocorrelation functions?

(a) $A \cos w\tau$, (b) $A \sin w\tau$.

Solution.

(a) $R_{XX}(\tau) = A \cos w\tau$

$$R_{XX}(-\tau) = A \cos w(-\tau) = A \cos w\tau = R_{XX}(\tau)$$

and $$|R_{XX}(\tau)| = A$$

$$R_{XX}(0) = A$$

$\Rightarrow$ $$|R_{XX}(\tau)| = R_{XX}(0)$$

Hence $A \cos w\tau$ is an autocorrelation function.

(b) $R_{XX}(\tau) = A \sin w\tau$

$$R_{XX}(-\tau) = A \sin w(-\tau) = -A \sin w(\tau) = -R_{XX}(\tau)$$

Hence $A \sin w\tau$ is not an autocorrelation function.

5.1.17 Example

The autocorrelation function of a wide-sense stationary random process is given by

$$R_{XX}(\tau) = \frac{4\tau^2 + 100}{\tau^2 + 4}$$

Find the mean and variance of this process.

Solution. $$(\text{Mean})^2 = [E(X(t)]^2 = \underset{|\tau|\to\infty}{\text{Lt}} R_{XX}(\tau)$$

$$= \underset{|\tau|\to\infty}{\text{Lt}} \frac{4\tau^2 + 100}{\tau^2 + 4} = \underset{|\tau|\to\infty}{\text{Lt}} \frac{4 + \frac{100}{\tau^2}}{1 + \frac{4}{\tau^2}} = \frac{4}{1} = 4$$

Hence Mean = 2, variance = $\sigma_X^2 = E(X^2(t)) - E(X(t))^2$

$$E(X^2(t)) = R_{XX}(0) = \frac{100}{4} = 25$$

So $$\sigma_X^2 = 25 - 4 = 21.$$

5.2 POWER SPECTRAL DENSITY

Covariance and correlation functions of stationary processes are functions in the time-domain. An alternative approach for the study of stationary processes is through "frequency domain."

5.2.1 Definition

The power spectral density (or power spectrum) $S_{XX}(w)$ of a continuous time random process $X(t)$ is defined as the Fourier transform of $R_{XX}(\tau)$:

$$S_{XX}(w) = \int_{-\infty}^{\infty} R_{XX}(\tau)\, e^{-iw\tau}\, d\tau \tag{5.1}$$

Thus, taking the inverse Fourier transform of $S_{XX}(w)$, we obtain

$$R_{XX}(\tau) = \frac{1}{2\pi} \int_{-\infty}^{\infty} S_{XX}(w)\, e^{iw\tau}\, dw \tag{5.2}$$

Equations (5.1) and (5.2) are known as the *Wiener-Khinchin relations.*

Properties of $S_{XX}(w)$

1. $S_{XX}(w)$ is real and $S_{XX}(w) \geq 0$
2. $S_{XX}(-w) = S_{XX}(w)$
3. $E(X^2(t)) = R_X(0) = \frac{1}{2\pi} \int_{-\infty}^{\infty} S_{XX}(w)\, dw$

Similarly, the power spectral density $S_{XX}(\Omega)$ of a discrete time random process $X(n)$ is defined as the Fourier transform of $R_X(k)$:

$$S_{XX}(\Omega) = \sum_{k=-\infty}^{\infty} R_{XX}(k)\, e^{-i\Omega k} \tag{5.3}$$

Thus, taking the inverse Fourier transform of $S_{XX}(\Omega)$, we obtain

$$R_{XX}(k) = \frac{1}{2\pi} \int_{-\infty}^{\infty} S_{XX}(\Omega)\, e^{i\Omega k} d\Omega \tag{5.4}$$

Properties of $S_{XX}(\Omega)$:

(1) $S_{XX}(\Omega + 2\pi) = S_{XX}(\Omega)$, since $e^{-i\Omega k}$ is periodic with period 2π.
(2) $S_{XX}(\Omega)$ is real and $S_{XX}(\Omega) \geq 0$
(3) $S_{XX}(\Omega) = S_{XX}(-\Omega)$
(4) $E(X^2(n)) = R_X(0) = \frac{1}{2\pi} \int_{-\pi}^{\pi} S_{XX}(\Omega)\, d\Omega$, is the total power.

That is, the power in a zero mean wide sense stationary random process is proportional to the area under the power spectral density curve.

Cross Power Spectral densities

The cross power spectral density (or cross power spectrum) $S_{XY}(w)$ of two continuous-time random processes $X(t)$ and $Y(t)$ is defined as the Fourier transform of $R_{XY}(\tau)$:

$$S_{XY}(w) = \int_{-\infty}^{\infty} R_{XY}(\tau)\, e^{-iw\tau} d\tau \tag{5.5}$$

Thus taking the inverse Fourier transform of $S_{XY}(w)$, we get

$$R_{XY}(\tau) = \frac{1}{2\pi} \int_{-\infty}^{\infty} S_{XY}(w)\, e^{iw\tau} dw \tag{5.6}$$

Properties of $S_{XY}(w)$

(1) $S_{XY}(w)$ is a complex-valued function
(2) $S_{XY}(w) = S_{YX}(-w)$
(3) $S_{XY}(-w) = S^*_{XY}(w)$

The cross power spectral density $S_{XY}(\Omega)$ of two discrete time random processes $X(n)$ and $Y(n)$ is defined as the Fourier transform of $R_{XY}(k)$:

$$S_{XY}(\Omega) = \sum_{k=-\infty}^{\infty} R_{XY}(k)\, e^{-i\Omega k} \tag{5.7}$$

$$R_{XY}(k) = \frac{1}{2\pi} \int_{-\pi}^{\pi} S_{XY}(\Omega)\, e^{i\Omega k}\, d\Omega \tag{5.8}$$

Properties of $S_{XY}(\Omega)$

(1) $S_{XY}(\Omega)$ is complex valued function

(2) $S_{XY}(\Omega + 2\pi) = S_{XY}(\Omega)$

(3) $S_{XY}(\Omega) = S_{YX}(-\Omega)$

(4) $S_{XY}(-\Omega) = S^{*}_{XY}(\Omega)$

where * denotes the conjugate.

5.2.2 Definition White Noise

A continuous time white noise process, $W(t)$ is a wide-sense stationary, zero mean, continuous-time random process whose autocorrelation function is given by

$$R_{WW}(\tau) = \sigma^2 \delta(\tau) \tag{5.9}$$

where $\delta(\tau)$ is a unit impulse function (or Dirac δ function) defined by

$$\int_{-\infty}^{\infty} \delta(\tau)\, \phi(\tau)\, d\tau = \phi(0)$$

where $\phi(\tau)$ is any function continuous at $\tau = 0$. Now $S_{WW}(w)$ can be obtained by taking the Fourier transform of (5.9),

$$S_{WW}(w) = \sigma^2 \int_{-\infty}^{\infty} \delta(\tau)\, e^{-iw\tau}\, d\tau = \sigma^2 (\text{constant})$$

which shows that $X(t)$ has a constant power spectral density (hence the name white noise).

5.2.3 Example

Two random processes $X(t)$ and $Y(t)$ are given by

$$X(t) = A\cos(wt + \theta), \qquad Y(t) = A\sin(wt + \theta)$$

where A and w are constants and θ is a uniform random variable over $(0, 2\pi)$. Find the cross-correlation function of $X(t)$ and $Y(t)$ and prove $R_{XY}(-\tau) = R_{YX}(\tau)$.

Solution.

$$\begin{aligned} R_{XY}(t, t+\tau) &= E(X(t)\, Y(t+\tau)) \\ &= E(A^2 \cos(wt+\theta)\sin(w(t+\tau)+\theta)) \\ &= \frac{A^2}{2} E[\sin(2wt + w\tau + 2\theta) - \sin(-w\tau)] \\ &= \frac{A^2}{2}\sin w\tau = R_{XY}(\tau) \end{aligned}$$

Similarly we prove

$$R_{YX}(t, t+\tau) = \frac{-A^2}{2}\sin w\tau = R_{YX}(\tau).$$

Hence

$$R_{XY}(-\tau) = \frac{A^2}{2}\sin w(-\tau) = -\frac{A^2}{2}\sin w\tau = R_{YX}(\tau). \text{ That is}$$

$$R_{XY}(-\tau) = R_{YX}(\tau).$$

5.2.4 Example

Show that the power spectrum of a real random process $X(t)$ is real.

Solution.

$$S_{XX}(w) = \int_{-\infty}^{\infty} R_{XX}(\tau)\, e^{-iw\tau}\, d\tau$$

$$= \int_{-\infty}^{\infty} R_{XX}(\tau)\, [\cos w\tau - i \sin w\tau]\, d\tau$$

$$= \int_{-\infty}^{\infty} R_{XX}(\tau) \cos w\tau\, d\tau - i \int_{-\infty}^{\infty} R_{XX}(\tau) \sin w\tau\, d\tau$$

Since $R_{XX}(\tau)$ is even, $R_{XX}(\tau) \cos w\tau$ is even function and $R_{XX}(\tau) \sin w\tau$ is an odd function, hence the imaginary term vanishes. So

$$S_{XX}(w) = \int_{-\infty}^{\infty} R_{XX}(\tau) \cos w\tau\, d\tau$$

which shows that $S_{XX}(w)$ is real.

5.2.5 Definition

A random process $X(t)$ is said to be continuous in mean square or mean square (m.s.) continuous if

$$\lim_{\varepsilon \to 0} E\,(X(t+\varepsilon) - X(t))^2 = 0$$

If $X(t)$ is mean square continuous, then its mean is continuous. That is

$$\lim_{\varepsilon \to 0} E\,(X(t+\varepsilon)) = E\left(\lim_{\varepsilon \to 0} X(t+\varepsilon)\right)$$

5.2.6 Example

Show that the random process $X(t)$ is m.s. continuous if and only if its auto-correlation function $R_{XX}(t, s)$ is continuous.

Solution.

$$E\{[X(t+\varepsilon) - X(t)]^2\} = E\,[X^2(t+\varepsilon) - 2X(t+\varepsilon)\, X(t) + X^2(t)]$$
$$= R_{XX}(t+\varepsilon, t+\varepsilon) - 2R_{XX}(t+\varepsilon, t) + R_{XX}(t, t) \qquad (5.10)$$

If $R_{XX}(t, s)$ is continuous, then

$$\lim_{\varepsilon \to 0} E\{[X(t+\varepsilon) - X(t)]^2\} = \lim_{\varepsilon \to 0} \{R_{XX}(t+\varepsilon, t+\varepsilon) - 2R_{XX}(t+\varepsilon, t) + R_{XX}(t, t)\} = 0$$

and hence $X(t)$ is m.s. continuous.

Conversely, let $X(t)$ is m.s. continuous. Consider

$$R_{XX}(t+\varepsilon_1, t+\varepsilon_2) - R_{XX}(t, t) = E\{(X(t+\varepsilon_1) - X(t))\,(X(t+\varepsilon_2) - X(t))]$$
$$+ E\{[X(t+\varepsilon_1) - X(t)]\, X(t)\} + E\{[X(t+\varepsilon_2) - X(t)]\, X(t)\}$$

Applying Schwartz inequality

$$[E(XY)]^2 \le E(X^2)\, E(Y^2)$$

we get

$$R_{XX}(t+\varepsilon_1, t+\varepsilon_2) - R_{XX}(t, t) \le (E\{[X(t+\varepsilon_1) - X(t)]^2\}\, E\{[X(t+\varepsilon_2) - X(t)]^2\})^{1/2} +$$
$$(E\{[X(t+\varepsilon_1) - X(t)]^2\}\, E(X^2(t)])^{1/2} + (E\{[X(t+\varepsilon_2) - X(t)]^2\}\, E(X^2(t)))^{1/2}$$

Thus if $X(t)$ is m.s. continuous, then for $\varepsilon_1, \varepsilon_2 \to 0$, the R.H.S. of the above inequality tends to zero and hence

$$\underset{\varepsilon_1, \varepsilon_2 \to 0}{\text{Lt}} \; R_{XX}(t+\varepsilon_1, t+\varepsilon_2) - R_{XX}(t, t) = 0$$

That is, $R_{XX}(t, s)$ is continuous.

5.2.7 Example

If $X(t)$ is wide sense stationary (WSS) then it is m.s. continuous if and only if its autocorrelation function $R_{XX}(\tau)$ is continuous at $\tau = 0$.

Solution. If $X(t)$ is WSS, then

$$E\{(X(t+\varepsilon) - X(t)]^2\} = 2\,[R_{XX}(0) - R_{XX}(\varepsilon)] \qquad \text{by (5.10) \& (5.11)}$$

If $R_{XX}(\tau)$ is continuous at $\tau = 0$, that is,

$$\underset{\varepsilon \to 0}{\text{Lt}} \; (R_{XX}(\varepsilon) - R_{XX}(0)) = 0$$

then $\underset{\varepsilon \to 0}{\text{Lt}} \; E(X(t+\varepsilon) - X(t))^2 = 0$.

Hence $X(t)$ is m.s. continuous. Converse part follows from (5.11).

5.2.8 Definition

A random process $X(t)$ is said to have a m.s derivative $X'(t)$ if

$$\underset{\varepsilon \to 0}{\text{l.i.m}} = \frac{X(t+\varepsilon) - X(t)}{\varepsilon} = X'(t)$$

where l.i.m denotes the limit in the mean square; that is

$$\lim_{\varepsilon \to 0} E\left\{\left[\frac{X(t+\varepsilon) - X(t)}{\varepsilon} - X'(t)\right]^2\right\} = 0$$

Note: A random process $X(t)$ has a m.s. derivative $X'(t)$ if $\dfrac{\partial^2 R_{XX}(t, s)}{\partial t \cdot \partial s}$ exists at $s = t$.

5.2.9 Example

Suppose a random process $X(t)$ has a m.s. derivative $X'(t)$.

(i) Find $E(X'(t))$

(ii) Find the cross-correlation of $X(t)$ and $X'(t)$

(iii) Find the autocorrelation function of $X'(t)$

Solution.

$$\text{(i)}\quad E(X'(t)) = E\left(\underset{\varepsilon \to 0}{\text{l.i.m}} \frac{X(t+\varepsilon) - X(t)}{\varepsilon}\right)$$

$$= \underset{\varepsilon \to \infty}{\text{Lt}} \; E\left(\frac{X(t+\varepsilon) - X(t)}{\varepsilon}\right)$$

$$= \underset{\varepsilon\to 0}{\text{Lt}} \frac{E(X(t+\varepsilon)) - E(X(t))}{\varepsilon} = [E(X(t)]'$$

(ii) $R_{XX'}(t, s) = E(X(t) X'(s))$

$$= E\left(X(t) \underset{\varepsilon\to 0}{\text{l.i.m}} \frac{X(s+\varepsilon) - X(s)}{\varepsilon} \right)$$

$$= \underset{\varepsilon\to 0}{\text{Lt}} \frac{E[X(t) X(s+\varepsilon) - E[X(t) X(s)]}{\varepsilon}$$

$$= \underset{\varepsilon\to 0}{\text{Lt}} \frac{R_{XX}(t, s+\varepsilon) - R_{XX}(t, s)}{\varepsilon} = \frac{\partial R_{XX}(t, s)}{\partial s} \tag{5.12}$$

(iii) $R_{X'X'}(t, s) = E(X'(t) X'(s))$

$$= E\left\{ \left(\underset{\varepsilon\to 0}{\text{l.i.m}} \frac{X(t+\varepsilon) - X(t)}{\varepsilon} \right) X'(s) \right\}$$

$$= \underset{\varepsilon\to 0}{\text{Lt}} \frac{E(X(t+\varepsilon) X'(s)) - E(X(t) X'(s))}{\varepsilon}$$

$$= \underset{\varepsilon\to 0}{\text{Lt}} \frac{R_{XX'}(t+\varepsilon, s) - R_{XX'}(t, s)}{\varepsilon}$$

$$= \frac{\partial R_{XX'}(t, s)}{\partial t} = \frac{\partial^2 R_{XX}(t, s)}{\partial t\, \partial s} \tag{5.13}$$

5.2.10 Example

If $X(t)$ is a WSS random process and has a m.s. derivative $X'(t)$, then show that

(1) $R_{XX'}(\tau) = \dfrac{d}{d\tau} R_{XX}(\tau)$

(2) $R_{X'X'}(\tau) = \dfrac{-d^2}{d\tau^2} R_{XX}(\tau)$

Solution.

(1) For a WSS process $X(t)$, $R_{XX}(t, t+\tau) = R_{XX}(\tau)$, hence

$$\frac{\partial R_{XX}(t, s)}{\partial s} = \frac{dR_{XX}(\tau)}{d\tau}$$

by (5.13) of the above example, and hence

$$R_{XX'}(t, t+\tau) = R_{XX'}(\tau) = \frac{d}{d\tau} R_{XX}(\tau)$$

(2) Now $\frac{\partial R_{XX}(s-t)}{\partial t} = -\frac{dR_{XX}(\tau)}{d\tau}$. So $\frac{\partial^2 R_{XX}(s-t)}{\partial t \partial s} = \frac{-d^2 R_{XX}(\tau)}{d\tau^2}$

By (5.13), we get

$$R_{X'X'}(t, t+\tau) = R_{X'X'}(\tau) = \frac{-d^2}{d\tau^2} R_{XX}(\tau).$$

5.2.11 Definition Stochastic Integrals

A m.s. integral of a random process $X(t)$ is defined by

$$Y(t) = \int_{t_0}^{t} X(\alpha)\, d\alpha = \underset{\Delta t_i \to 0}{\text{l.i.m}} \sum_i X(t_i)\, \Delta t_i$$

where $t_0 < t_1 < \ldots t$ and $\Delta t_i = t_{i+1} - t_i$. The m.s. integral of $X(t)$ exists if the following integral exists.

$$\int_{t_0}^{t} \int_{t_0}^{t} R_{XX}(x, y)\, dx\, dy$$

Mean of $Y(t)$ is $E(Y(t)) = E\left(\int_{t_0}^{t} X(\alpha)\, d\alpha\right) = \int_{t_0}^{t} E(X(\alpha))\, d\alpha$

Autocorrelation of $Y(t)$ is

$$R_{YY}(t, s) = \int_{t_0}^{t} \int_{t_0}^{s} R_{XX}(x, y)\, dy\, dx.$$

5.2.12 Example

For a random process $X(t) = A \cos(wt + \theta)$, where A is a real constant, w is a random variable with density $f(w)$ and θ is a random variable uniformly distributed over the interval $(-\pi, \pi)$ and independent of w, calculate the power spectral density.

Solution. By Wiener-Khinchin relation,

$$R_{XX}(\tau) = \frac{1}{2\pi} \int_{-\infty}^{\infty} S_{XX}(w)\, e^{iw\tau}\, dw = \frac{1}{2\pi} \int_{-\infty}^{\infty} S_{XX}(w) \cos w\tau\, dw \tag{5.14}$$

The autocorrelation function

$$R_{XX}(\tau) = E[X(t+\tau)\, X(t)]$$
$$= E(A^2 \cos(wt + w\tau + \theta) \cos(wt + \theta))$$
$$= \frac{A^2}{2} E(\cos w\tau) = \frac{A^2}{2} \int_{-\infty}^{\infty} f(w) \cos w\tau\, dw \tag{5.15}$$

By the uniqueness property of Fourier transform, the RHS of equations (5.14) and (5.15) are equal.

Hence $\frac{S_{XX}(w)}{2\pi} = \frac{A^2}{2} f(w)$

That is $S_{XX}(w) = A^2 \pi f(w)$.

5.2.13 Example

For a random process having $R_{XX}(\tau) = ae^{-b|\tau|}$, find the spectral density function, where a and b are constants.

Solution.

$$S_{XX}(w) = \int_{-\infty}^{\infty} R_{XX}(\tau)\, e^{-iw\tau}\, d\tau = a \int_{-\infty}^{\infty} e^{-b|\tau|}\, e^{-iw\tau}\, d\tau$$

$$= a\int_{-\infty}^{0} e^{b\tau}\, e^{-iw\tau}\, d\tau + a\int_{0}^{\infty} e^{-b\tau}\, e^{-iw\tau}\, d\tau = a\int_{0}^{\infty} e^{-b\tau}\, e^{iw\tau}\, d\tau + a\int_{0}^{\infty} e^{-b\tau}\, e^{-iw\tau}\, d\tau$$

$$= a\left[\int_{0}^{\infty} e^{-(b-iw)\tau}\, d\tau + \int_{0}^{\infty} e^{-(b+iw)\tau}\, d\tau\right] = a\left[\frac{e^{-(b-iw)}}{-(b-iw)}\right]_0^{\infty} + a\left[\frac{e^{-(b+iw)}}{-(b+iw)}\right]_0^{\infty}$$

$$= a\left[\frac{1}{b-iw} + \frac{1}{b+iw}\right] = \frac{2ab}{b^2+w^2}$$

5.2.14 Example

The power spectral density of a random process is given by

$$S_{XX}(w) = \begin{cases} \pi, & |w| < 1 \\ 0, & \text{elsewhere} \end{cases}$$

Find its autocorrelation function.

$$R_{XX}(\tau) = \frac{1}{2\pi}\int_{-\infty}^{\infty} S_{XX}(w)\, e^{iw\tau}\, dw = \frac{1}{2\pi}\int_{-1}^{1} \pi\, e^{iw\tau}\, dw$$

$$= \frac{1}{2}\left[\frac{e^{iw\tau}}{i\tau}\right]_{-1}^{1} = \frac{1}{2}\left[\frac{e^{\varepsilon\tau} - e^{-i\tau}}{i\tau}\right] = \frac{1}{\tau}\sin\tau$$

5.2.15 Example

Find the power spectrum of the random phase sinusoid $x(n) = A\sin(nw_0 + \phi)$, where A and w_0 are constants and ϕ is a random variable that is uniformly distributed between $-\pi$ and π.

Solution. The autocorrelation of $x(n)$ is

$$R_{XX}(k, l) = E(x(k)\, x(l))$$

$$= E(A\sin(kw_0 + \phi)\, A\sin(lw_0 + \phi))$$

$$= \frac{1}{2} E(A^2\, 2\sin(kw_0 + \phi)\sin(lw_0 + \phi))$$

using the formula $2\sin A\sin B = \cos(A - B) - \cos(A + B)$, we get

$$R_{XX}(k, l) = \frac{1}{2} A^2 E\{\cos[(k-l)\, w_0]\} - \frac{1}{2} A^2 E\{\cos[(k+l)\, w_0 + 2\phi]\}$$

Note that the first term is the expected value of a constant and the second term is zero. Hence

$$R_{XX}(k, l) = \frac{1}{2} A^2 \cos[(k-l) w_0].$$

By Wiener-Khinchin relation,

$$S_{XX}(w) = \int_{-\infty}^{\infty} R_{XX}(k, l)\, e^{-iwz}\, d\tau, \; \tau = k - l$$

$$= \frac{A^2}{2} \int_{-\infty}^{\infty} \cos[(k-l) w_0]\, e^{-iw\tau}\, d\tau$$

$$= \frac{A^2}{2} \pi [u_0(w - w_0) + u_0(w + w_0)]$$

since the Fourier transform of $\cos w_0 \tau = \pi [u_0 (w - w_0) + u_0 (w + w_0)]$

5.2.16 Example

Find the autocorrelation and power spectral density of complex harmonic process $x(n) = Ae^{i(nw_0 + \phi)}$, ϕ is a random variable that is uniformly distributed between $-\pi$ and π.

Solution. Mean $= E(x(n)) = E(Ae^{i(nw_0 + \phi)}) = 0$

The autocorrelation is

$$R_{XX}(k, l) = E(x(k)\, x^*(l))$$

$$= E\left(Ae^{i(kw_0 + \phi)}\; A^* e^{-i(lw_0 + \phi)}\right)$$

$$= |A|^2 E\left(e^{i(k-l)w_0}\right) = |A|^2 e^{i(k-l)w_0}$$

The mean is a constant and the autocorrelation is a function of only the difference between k and l. The power spectrum is

$$S_{XX}(w) = |A|^2 \cdot \int_{-\infty}^{\infty} e^{i(k-l)w_0} \cdot e^{-iw\tau}\, d\tau, \qquad \tau = k - l$$

$$= 2\pi |A|^2 u_0(w - w_0), \qquad |w| < \pi$$

where $u_0(w - w_0)$ is used to denote an impulse at frequency $w = w_0$.

5.2.17 Example

If $X(t)$ and $Y(t)$ are uncorrelated random processes, then find the power spectral density of Z if $Z(t) = X(t) + Y(t)$. Also find the cross-spectral density $S_{XZ}(w)$ and $S_{YZ}(w)$.

Solution.

(i) $S_{ZZ}(w) = \int_{-\infty}^{\infty} R_{ZZ}(\tau)\, e^{-iw\tau}\, d\tau$

$$R_{ZZ}(\tau) = E(Z(t)\, Z(t + \tau))$$

$$= E[(X(t) + Y(t))\,(X(t + \tau) + Y(t + \tau))]$$

$$= E[X(t)\,X(t+\tau) + X(t)\,Y(t+\tau) + Y(t)\,X(t+\tau) + Y(t)\,Y(t+\tau)]$$
$$= E[X(t)\,X(t+\tau)] + E(X(t)\,Y(t+\tau)] + E[Y(t)\,X(t+\tau)] + E[Y(t)\,Y(t+\tau)]$$
$$= R_{XX}(\tau) + R_{XY}(\tau) + R_{YX}(\tau) + R_{YY}(\tau)$$

Hence

$$S_{ZZ}(w) = \int_{-\infty}^{\infty} R_{ZZ}(\tau)\, e^{-iw\tau}\, d\tau$$

$$= \int_{-\infty}^{\infty} [R_{XX}(\tau) + R_{XY}(\tau) + R_{YX}(\tau) + R_{YY}(\tau)]\, e^{-iw\tau}\, d\tau$$

$$= \int_{-\infty}^{\infty} R_{XX}(\tau)\, e^{-iw\tau}\, d\tau + \int_{-\infty}^{\infty} R_{XY}(\tau)\, e^{-iw\tau}\, d\tau + \int_{-\infty}^{\infty} R_{YX}(\tau)\, e^{-iw\tau}\, d\tau + \int_{-\infty}^{\infty} R_{YY}(\tau)\, e^{-iw\tau}\, d\tau$$

$$= S_{XX}(w) + S_{XY}(w) + S_{YX}(w) + S_{YY}(w)$$

If $X(t)$ and $Y(t)$ are uncorrelated, then $S_{XY}(w) = S_{YX}(w)$, hence

$$S_{ZZ}(w) = S_{XX}(w) + S_{YY}(w) + 2\, S_{XY}(w)$$

(ii) $R_{XZ}(\tau) = E(X(t)\, Z(t+\tau))$

$$= E[X(t)\,(X(t+\tau) + Y(t+\tau))]$$
$$= E[X(t)\,X(t+\tau) + X(t)\,Y(t+\tau)]$$
$$= E[X(t)\,X(t+\tau)] + E[X(t)\,Y(t+\tau)]$$
$$= R_{XX}(\tau) + R_{XY}(\tau)$$

So $S_{XZ}(w) = \int_{-\infty}^{\infty} R_{XZ}(\tau)\, e^{-iw\tau}\, d\tau$

$$= \int_{-\infty}^{\infty} R_{XX}(\tau)\, e^{-iw\tau}\, d\tau + \int_{-\infty}^{\infty} R_{XY}(\tau)\, e^{-iw\tau}\, d\tau$$

$$= S_{XX}(w) + S_{XY}(w)$$

Similarly we prove,

$$S_{YZ}(w) = S_{YY}(w) + S_{YX}(w).$$

Sum-Decomposition Theorem

Wiener-Khintchine theorem gives us the spectrum of a discrete-time signal. Often power spectral density (PSD) takes the form of a rational function with poles inside and outside of a unit circle in the w-domain. The PSD is generally decomposed to separate out the one half of the poles that lie inside the unit circle. According to the sum-decomposition

$$S_{XX}(w) = \sum_{k=-\infty}^{\infty} R_{XX}(k) e^{-iwk}$$

$$= \sum_{k=0}^{\infty} R_{XX}(k)\, e^{-iwk} + \sum_{k=-\infty}^{0} R_{XX}(k)\, e^{-iwk} - R(0)$$

where $R(0)$ is the correlation with zero delays. Letting $i = -k$ in the second term and since $R_{XX}(k)$ is an even function, we get

$$S_{XX}(w) = \sum_{k=0}^{\infty} R_{XX}(k)e^{-iwk} + \sum_{i=0}^{\infty} R_{XX}(i)\, e^{iwk} - R(0)$$

So

$$S_{XX}(w) = S_{XX}^{+}(w) + S_{XX}^{-}(w) - R(0)$$

The first term of RHS represents a function having poles inside the unit circle and the second term represents a function having poles outside the unit circle.

The spectrum expressed as a rational function of complex frequency plays an important role in signal processing.

5.3 APPLICATIONS—LINEAR SYSTEMS WITH RANDOM INPUTS

5.3.1 Linear Systems

A system is a mathematical model of a physical process that relates the input (signal) X to the output (signal) Y. The system can be viewed as a mapping T of X into Y such that $Y(t) = T(x(t))$.

A discrete-time signal is an indexed sequence of real or complex numbers. That is a discrete-time signal is a function of an integer valued variable, n and is denoted by $x(n)$. Although the independent variable n need not represent time, $x(n)$ is generally referred to as a function of time. The discrete-time signal is undefined for non-integer values of n. The graphical representation of a discrete-time signal $x(n)$ is

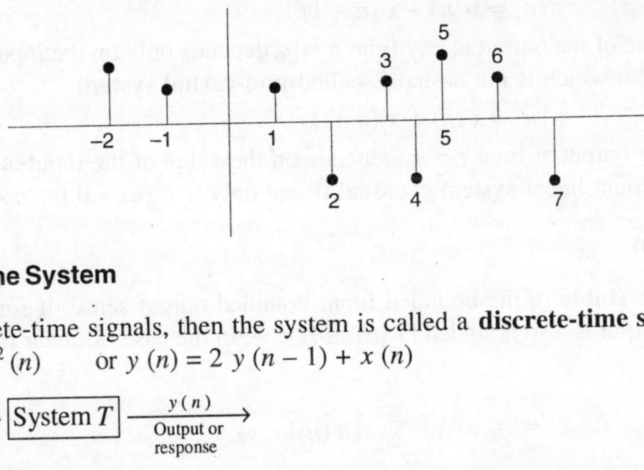

5.3.2 Discrete-time System

If X and Y are discrete-time signals, then the system is called a **discrete-time system.**

e.g $\quad y(n) = x^2(n) \quad$ or $y(n) = 2\,y(n-1) + x(n)$

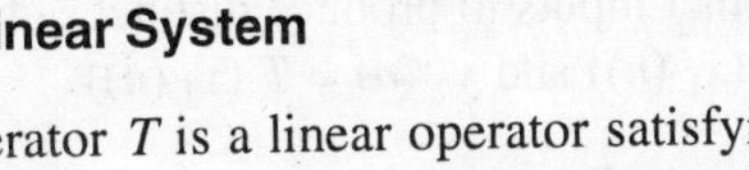

5.3.3 Continuous-time System

If X and Y are continuous time signals, then the system is called a **continuous time system.**

5.3.4 Linear System

If the operator T is a linear operator satisfying

$$T\{x_1 + x_2\} = Tx_1 + Tx_2 = y_1 + y_2 \qquad \text{(Additivity)}$$

$$T(\alpha x) = \alpha Tx \qquad \text{(Homogeneity)}$$

where α is a scalar, then the system T is called a **linear system.**

5.3.5 Time Invariant (Shift Invariant)

A system is called **time-invariant** if a time shift in the input signal causes the same time shift in the output signal. For a continuous-time system

$$T[x(t-t_0)] = y(t-t_0)$$

for any value of t_0, and for a discrete-time system

$$T[x(n-n_0)] = y(n-n_0)$$

for any integer n_0. A continuous-time invariant system

$$y(t) = T(x(t)) = \int_{-\infty}^{\infty} h(\lambda)\, x(t-\lambda)\, d\lambda \tag{5.16}$$

where $h(t) = T(\delta(t))$ is known as the impulse response of a continuous-time invariant system. The R.H.S of (5.16) is the **convolution integral** of $h(t)$ and $x(t)$, denoted by $h(t) * x(t)$. For discrete-time invariant system

$$y(n) = \sum_{i=-\infty}^{\infty} h(i)\, x(n-i) \tag{5.17}$$

where $h(n) = T(\delta(n))$ is the impulse response of a discrete-time-invariant system. The R.H.S of (5.17) is called **convolution sum** of $h(t)$ and $x(t)$.

5.3.6 Casual and Non-casual System

A **casual system** is one whose output at time n_0 depends only upon the values of the input for $n \le n_0$.

e.g. $$y(n) = x(n) + x(n-1)$$

is casual since the value of the output at any time $n = n_0$ depends only on the input $x(n)$ at time n_o and at time $n_0 - 1$. A system which is not casual is called **non-casual system.**

e.g. $$y(n) = x(n) + x(n+1)$$

is non-casual since the output at time $t = n_0$ depends on the value of the input at time $n_0 + 1$.

Note that time-invariant linear system is casual if and only if $h(n) = 0$ for $n < 0$.

5.3.7 Stable system

A system is said to be **stable** in the bounded input bounded output sense, if for any bounded input, $|x(n)| \le A < \infty$, the output is also bounded $|y(n)| \le B < \infty$. In the case of linear time-invariant system, stability will exists if

$$\sum_{n=-\infty}^{\infty} |h(n)| < \infty$$

For example, a linear-time invariant system with $h(n) = a^n u(n)$ is stable whenever $|a| < 1$.

5.3.8 Invertible System

A system is said to be **invertible** if it is necessary for distinct inputs to produce distinct outputs. That is if $x_1(n) \ne x_2(n)$, then $y_1(n) \ne y_2(n)$, where $y_1(n) = T(x_1(n))$ and $y_2(n) = T(x_2(n))$.

5.3.9 Response of a Continuous-Time Linear System to Random Input

When the input of a continuous-time linear system is a random process $\{X(t): t \in T_x\}$, then the ouptut will also be a random process $\{Y(t): t \in T_y\}$. If the system is linear-time invariant system then

$$Y(t) = \int_{-\infty}^{\infty} h(\lambda)\, X(t-\lambda)\, d\lambda \tag{5.18}$$

where (5.18) is stochastic integral. Then

$$E(Y(t)) = \int_{-\infty}^{\infty} h(\lambda)\, E(X(t-\lambda))\, d\lambda$$

when the input random process $X(t)$ is a wide-sense stationary, the mean $E(X(t))$ is a constant and so

$$E(Y(t)) = E(X(t)) \int_{-\infty}^{\infty} h(\lambda)\, d\lambda = E(X(t))\, H(0) \tag{5.19}$$

where $H(0) = [H(w)]_{w=0}$ and $H(w)$ is the frequency response of the system defined by the Fourier transform of $h(t)$. That is

$$H(w) = \int_{-\infty}^{\infty} h(t)\, e^{-iwt}\, dt$$

The auto correlation function of $Y(t)$ is as follows:

$$R_{YY}(t, t+\tau) = E(Y(t)\, Y(t+\tau))$$

$$= E\left(\int_{-\infty}^{\infty} h(\tau_1)\, X(t-\tau_1)\, d\tau_1 \int_{-\infty}^{\infty} h(\tau_2)\, X(t+\tau-\tau_2)\, d\tau_2 \right)$$

since the expected value operation is a linear operation, the expected value and integration operations can be interchanged. So

$$R_{YY}(t, t+\tau) = \int_{-\infty}^{\infty} h(\tau_1)\, d\tau_1 \int_{-\infty}^{\infty} h(\tau_2)\, E[X(t-\tau_1)\, X(t+\tau-\tau_2)]\, d\tau_2$$

$$= \int_{-\infty}^{\infty} h(\tau_1)\, d\tau_1 \int_{-\infty}^{\infty} h(\tau_2)\, R_{XX}(t-\tau_1, t+\tau-\tau_2)\, d\tau_2$$

$$= \int_{-\infty}^{\infty}\int_{-\infty}^{\infty} h(\tau_1)\, h(\tau_2)\, R_{XX}(\tau-\tau_2+\tau_1)\, d\tau_1\, d\tau_2$$

$$= R_{YY}(\tau) \tag{5.20}$$

From (5.19) and (5.20) we see that the output $Y(t)$ is also wide-sense stationary. Taking the Fourier transform of (5.20), the power spectral density of $Y(t)$ is given by

$$S_{YY}(w) = \int_{-\infty}^{\infty} R_{YY}(\tau)\, e^{-iw\tau}\, d\tau$$

$$= \int_{-\infty}^{\infty}\int_{-\infty}^{\infty}\int_{-\infty}^{\infty} h(\tau_1)\, h(\tau_2)\, R_{XX}(\tau-\tau_2+\tau_1)\, d\tau_1\, d\tau_2\, d\tau$$

Let $\tau - \tau_2 + \tau_1 = \lambda$ (say), then

$$S_{YY}(w) = \int_{-\infty}^{\infty}\int_{-\infty}^{\infty}\int_{-\infty}^{\infty} h(\tau_1)\, h(\tau_2)\, R_{XX}(\lambda)\, e^{-iw(\lambda-\tau_1+\tau_2)}\, d\tau_1\, d\tau_2\, d\lambda.$$

$$= \int_{-\infty}^{\infty} h(\tau_1) e^{iw\tau_1}\, d\tau_1 \int_{-\infty}^{\infty} h(\tau_2) e^{-iw\tau_2}\, d\tau_2 \int_{-\infty}^{\infty} R_{XX}(\lambda) e^{-iw\lambda}\, d\lambda$$

$$= H(-w)\, H(w)\, S_{XX}(w) = H^*(w)\, H(w)\, S_{XX}(w) = |H(w)|^2\, S_{XX}(w).$$

Hence we obtain the theorem.

5.3.10 Theorem

The power spectral density of the output is the product of the power spectral density of the input and the magnitude square of the frequency response of the system. We refer $|H(w)|^2$ as power transfer function.

5.3.11 Example

A WSS random process $X(t)$ with auto correlation function $R_{XX}(\tau) = e^{-a|\tau|}$, a is a real positive constant, is applied to the input of an linear time invariant system with impulse response $h(t) = e^{-bt} u(t)$, where b is a real positive constant. Find the auto correlation of the output $Y(t)$ of the system. Note that $u(t)$ is a unit step function.

Solution. The frequency response $H(w) = F(h(t) = \dfrac{1}{iw+b}$ where F is the Fourier transform. The power spectral density of $X(t)$ is

$$S_{XX}(w) = F(R_{XX}(\tau)) = \frac{2a}{w^2+a^2}$$

The power spectral density of $Y(t)$ is

$$S_{YY}(w) = |H(w)|^2 S_{XX}(w) = \frac{1}{w^2+b^2}\,\frac{2a}{w^2+a^2}$$

$$= \frac{a}{(a^2-b^2)b}\left(\frac{2b}{w^2+b^2}\right) - \frac{b}{(a^2-b^2)a}\left(\frac{2a}{w^2+a^2}\right)$$

$$R_{YY}(\tau) = F^{-1}(S_{YY}(w)) = \frac{1}{(a^2-b^2)b}\,[ae^{-b|\tau|} - be^{-a|\tau|}]$$

Note: The average power in the output $Y(t)$ is

$$E(Y^2(t)) = R_{YY}(0) = \frac{1}{2\pi}\int_{-\infty}^{\infty} |H(w)|^2\, S_{XX}(w)\, dw$$

Response of a Discrete-Time Linear System to Random Input:
When the input is a discrete-time random process $X(n)$, then the output $Y(n)$ is

$$Y(n) = \sum_{i=-\infty}^{\infty} h(i)\, X(n-i) \tag{5.21}$$

The auto correlation of $Y(n)$ is

$$R_{YY}(n, m) = \sum_{i=-\infty}^{\infty}\sum_{j=-\infty}^{\infty} h(i)\, h(j)\, R_{XX}(n-i, m-j) \tag{5.22}$$

When $X(n)$ is WSS, then from (5.21) we get

$$E(Y(n)) = E\left(\sum_{i=-\infty}^{\infty} h(i)\, X(n-i)\right)$$

$$= E(X(n)) \sum_{i=-\infty}^{\infty} h(i), \quad \text{since } E(X(n)) = E(X(n-i)) \text{ for any } i$$

$$= E(X(n))\, H(0) \tag{5.23}$$

where $H(0) = (H(\Omega))_{\Omega=0}$, where

$$H(\Omega) = \sum_{n=-\infty}^{\infty} h(n)\, e^{-i\Omega n}$$

The auto correlation function of $Y(n)$ is, from (5.22)

$$R_{YY}(n, m) = \sum_{i=-\infty}^{\infty} \sum_{j=-\infty}^{\infty} h(i)\, h(j)\, R_{XX}(m - n + i - j)$$

$$\Rightarrow \quad R_{YY}(n, n+k) = \sum_{i=-\infty}^{\infty} \sum_{j=-\infty}^{\infty} h(i)\, h(j)\, R_{XX}(k + i - j) = R_{YY}(k) \tag{5.24}$$

From (5.23) and (5.24) we conclude $Y(n)$ is also WSS.

5.4 SOLVED PROBLEMS

1. Define decorrelation time τ_0 of a wide-sense stationary random process.
Solution. The decorrelation time τ_0 of a wide-sense stationary random process $X(t)$ of zero mean as the time taken for the magnitude of the auto correlation function $R_{XX}(\tau)$ to decrease to certain prescribed value of its maximum value $R_{XX}(0)$.
2. Show the auto correlation function of a sinusoidal wave with random phase is another sinusoid at the same frequency in the τ domain.
Solution. Consider a sinusoidal signal with random phase, defined by

$$X(t) = A\cos(2\pi f_c t + \theta)$$

where A and f_c are constants and θ is a random variable that is uniformly distributed over the interval $(-\pi, \pi)$, then

$$f(\theta) = \begin{cases} \dfrac{1}{2\pi}, & -\pi \le \theta \le \pi \\ 0, & \text{otherwise} \end{cases}$$

$$R_{XX}(\tau) = E(X(t+\tau), X(t))$$
$$= E[A^2 \cos(2\pi f_C t + 2\pi f_C \tau + \theta)\cos(2\pi f_C t + \theta)]$$

$$= \frac{A_2}{2}\left\{ E(\cos(4\pi f_C t + 2\pi f_C \tau + 2\theta) + E(\cos(2\pi f_C \tau))\right\}$$

$$= \frac{A^2}{2}\int_{-\pi}^{\pi} \frac{1}{2\pi}\cos(4\pi f_C t + 2\pi f_C \tau + 2\theta)\, d\theta + \frac{A^2}{2}\cos(2\pi f_C \tau)$$

The first term integrates to zero, so

$$R_{XX}(\tau) = \frac{A^2}{2}\cos(2\pi f_c \tau).$$

3. Find the power spectral density of the random telegraph signal.

Solution. We know $X(t) = AY(t)$, where $Y(t)$ is the semirandom telegraph signal and A is a random variable independent of $Y(t)$ and takes the value ±1 with equal probability. Since $E(A) = 0$ and $E(A^2) = 1$, $E(X(t) = 0$ and $R_{XX}(t, t + \tau) = E(A^2Y(t)\,Y(t + \tau)) = E(A^2)\,E(Y(t)\,Y(t + \tau)) = R_{YY}(t, t + \tau) = e^{-2\lambda|\tau|}$ (already proved).

Now Fourier transform of $e^{-2\lambda|\tau|}$ is

$$S_{XX}(w) = \int_{-\infty}^{\infty} e^{-2\lambda|\tau|}\,e^{-iw\tau}\,d\tau = \frac{4\lambda}{w^2 + 4\lambda^2}. \qquad \text{(see Appendix C)}$$

4. Consider a *WSS* process $X(t)$ with auto correlation function $R_{XX}(\tau)$ and power spectral density $S_{XX}(w)$. Let $X'(t) = \dfrac{dX(t)}{dt}$. Show that

(a) $R_{XX'}(\tau) = \dfrac{d}{d\tau} R_{XX}(\tau)$

(b) $S_{X'X'}(w) = w^2 S_{XX}(w)$

Solution.

(a) If $X(t)$ is the input to a differentiator, then its output $Y(t) = X'(t)$. The frequency response of a differentiator is $H(w) = iw$. Then

$$S_{XX'}(w) = H(w)\,S_{XX}(w) = iw\,S_{XX}(w)$$

Taking the inverse Fourier transform on both sides, we obtain

$$R_{XX'}(\tau) = \frac{d}{d\tau} R_{XX}(\tau)$$

(b) $S_{X'X'}(w) = |H(w)|^2\,S_{XX}(w) = |iw|^2\,S_{XX}(w) = w^2 S_{XX}(w)$

5. Let $Y(t)$ be the output of an linear time invariant with impulse response $h(t)$ and when a WSS random process $X(t)$ is applied as an input, show that

$$S_{XY}(w) = H(w)\,S_{XX}(w)$$

Solution. (a) If $X(t)$ is WSS, then

$$R_{XY}(t, s) = E\,(X(t)\,X(s)) = E\left(X(t)\int_{-\infty}^{\infty} h(\alpha)\,X(s - \alpha)\,d\alpha\right)$$

$$= \int_{-\infty}^{\infty} h(\alpha)\,E(X(t)\,X(s - \alpha))\,d\alpha$$

$$= \int_{-\infty}^{\infty} h(\alpha)\,R_{XX}(t, s - \alpha)\,d\alpha$$

$$= \int_{-\infty}^{\infty} h(\alpha)\,R_{XX}(s - t - \alpha)\,d\alpha$$

which indicates that $R_{XY}(t, s)$ is a function of $\tau = s - t$ only. Hence

$$R_{XY}(t, s) = R_{XY}(\tau) = \int_{-\infty}^{\infty} h(\alpha)\,R_{XX}(\tau - \alpha)\,d\alpha \qquad (5.25)$$

Taking Fourier transform on either side of (5.25), we get

$$S_{XY}(w) = \int_{-\infty}^{\infty} R_{XY}(\tau)e^{-iw\tau}\,d\tau$$

$$= \int_{-\infty}^{\infty} \int_{-\infty}^{\infty} h(\alpha)\, R_{XX}(\tau - \alpha)\, e^{-iw\tau}\, d\alpha\, d\tau$$

$$= \int_{-\infty}^{\infty} \int_{-\infty}^{\infty} h(\alpha)\, R_{XX}(\lambda)\, e^{-iw(\lambda + \alpha)}\, d\alpha\, d\lambda$$

put $\tau - \alpha = \lambda$.

$$= \int_{-\infty}^{\infty} h(\alpha)\, e^{-iw\alpha}\, d\alpha \int_{-\infty}^{\infty} R_{XX}(\lambda)\, e^{-iw\lambda}\, d\lambda = H(w)\, S_{XX}(w)$$

6. Consider the figure

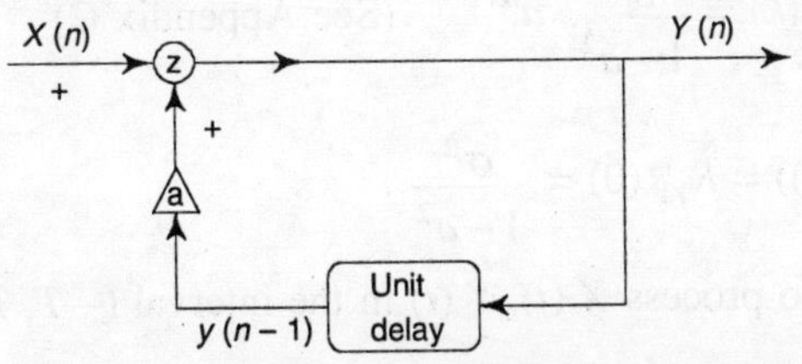

It consists of one unit delay and one scalar multiplier ($a < 1$). The input $X(n)$ is discrete-time white noise with average power σ^2. Find the spectral density and average power of the output $Y(n)$.

Solution. $Y(n)$ and $X(n)$ are related by

$$Y(n) = a\,Y(n-1) + X(n) \tag{5.26}$$

The impulse response $h(n)$ of the system is

$$h(n) = ah(n-1) + \delta(n) \tag{5.27}$$

Solving (5.27) we get

$$h(n) = a^n\, u(n) \tag{5.28}$$

where $u(n)$ is the unit step function given by

$$u(n) = \begin{cases} 1 & n \geq 0 \\ 0 & n < 0 \end{cases}$$

Taking Fourier transform of (5.28) we obtain

$$H(\Omega) = \sum_{n=0}^{\infty} a^n e^{-i\Omega n} = \frac{1}{1 - ae^{-i\Omega}} \quad a < 1,\ |\Omega| < \pi \tag{5.29}$$

We know for a discrete-time white noise $X(t)$, its auto correlation function is given by

$$R_{XX}(k) = \delta^2\, \delta(k)$$

where $\delta(k)$ is a unit impulse function given by

$$\delta(k) = \begin{cases} 1 & k = 0 \\ 0 & k \neq 0 \end{cases}$$

Hence $\quad S_{XX}(\Omega) = F(R_{XX}(k)) = \sigma^2 \sum_{k=-\infty}^{\infty} \delta(k)\, e^{-i\Omega k} = \sigma^2, \quad -\pi < \Omega < \pi \qquad (5.30)$

The power spectral density of $Y(n)$ is

$$S_{YY}(\Omega) = |H(\Omega)|^2 S_{XX}(\Omega) = H(\Omega) H(-\Omega) S_{XX}(\Omega)$$

$$= \frac{\sigma^2}{(1-a e^{-i\Omega})(1-a e^{i\Omega})} \quad \text{by (5.29) and (5.30)}$$

$$= \frac{\sigma^2}{1+a^2-2a\cos\Omega} \qquad -\pi < \Omega < \pi \tag{5.31}$$

Taking the inverse Fourier transform of (5.31) we obtain

$$R_{YY}(k) = \frac{\sigma^2}{1-a^2} a^{|k|} \qquad \text{(See Appendix C)}$$

Average power is $E(Y^2(n)) = R_{YY}(0) = \dfrac{\sigma^2}{1-a^2}$.

7. Define cross power of two process $X(t)$, $Y(t)$ in the interval $(-T, T)$.

Solution.

$$P_{XY}(T) = \frac{1}{2T} \int_{-T}^{T} X(t)\, Y(t)\, dt$$

Using Parseval' theorem

$$P_{XY}(T) = \frac{1}{2\pi} \int_{-\infty}^{\infty} \frac{X_T^*(w)\, Y_T(w)}{2T} dw$$

Average Cross-Power:

$$P_{XY}(T) = \underset{T\to\infty}{\text{Lt}} \frac{1}{2T} \int_{-T}^{T} R_{XY}(t,t)\, dt$$

$$= \frac{1}{2\pi} \int_{-\infty}^{\infty} \underset{T\to\infty}{\text{Lt}} \frac{E(X_T^*(w)\, Y_T(w))}{2T} dw$$

where $\qquad S_{XY}(w) = \underset{T\to\infty}{\text{Lt}} \dfrac{E(X_T^*(w)\, Y_T(w))}{2T}$

8. Consider the random process $X(t) = A\cos(w_0 t + \theta)$ where A and w_0 are real constants and θ is a random variable uniformly distributed on the interval $(0, \pi/2)$. Find the average power in $X(t)$.

Solution. $P_{XX} = \underset{T\to\infty}{\text{Lt}} = \dfrac{1}{2T} \displaystyle\int_{-T}^{T} E(X^2(t))\, dt$

$$E(X^2(t)) = E(A^2(\cos^2(w_0 t + \theta))$$

$$= E\left[A^2\left[\frac{1+\cos(2w_0 t + 2\theta)}{2}\right]\right]$$

$$= \frac{A^2}{2} + \frac{A^2}{2} \int_0^{\pi/2} \frac{\cos(2w_0 t + 2\theta)}{\pi/2} d\theta \qquad \text{since } p(\theta) = \frac{1}{\pi/2 - 0} = \frac{1}{\pi/2}$$

$$= \frac{A^2}{2} + \frac{A^2}{2} \left[\frac{\sin(2w_0 t + 2\theta)}{\pi} \right]_0^{\pi/2}$$

$$= \frac{A^2}{2} + \frac{A^2}{2\pi} [-\sin(2w_0 t) - \sin(2w_0 t)] = \frac{A^2}{2} - \frac{A^2}{\pi} \sin(2w_0 t)$$

Now $$P_{XX} = \underset{T\to\infty}{\text{Lt}} \frac{1}{2T} \int_{-T}^{T} \left[\frac{A^2}{2} - \frac{A^2}{\pi} \sin(2w_0 t) \right] dt$$

$$= \underset{T\to\infty}{\text{Lt}} \frac{1}{2T} \left[\frac{A^2}{2} t + \frac{A^2}{\pi} \frac{\cos(2w_0 t)}{2w_0} \right]_{-T}^{T}$$

$$= \underset{T\to\infty}{\text{Lt}} \left\{ \frac{1}{2T} \left[\frac{A^2}{2} T + \frac{A^2}{\pi} \frac{\cos(2w_0 T)}{2w_0} \right] - \left[-\frac{A^2}{2} T + \frac{A^2}{\pi} \frac{\cos(2w_0 T)}{2w_0} \right] \right\}$$

$$= \underset{T\to\infty}{\text{Lt}} \frac{1}{2T} [A^2 T] = \frac{A^2}{2}.$$

9. If the cross-correlation of two process $X(t)$ and $Y(t)$ is $R_{XY}(t, t+\tau) = \frac{AB}{2}$ $[\sin(w_0 \tau) + \cos w_0 (2t + \tau)]$ where A, B, w_0 are constants. Find the cross power spectrum.

Solution. Time average is $\underset{T\to\infty}{\text{Lt}} \frac{1}{2T} \int_{-T}^{T} R_{XY}(t, t+\tau)\, dt$

$$= \underset{T\to\infty}{\text{Lt}} \frac{1}{2T} \int_{-T}^{T} \left[\frac{AB}{2} \sin(w_0 \tau) + \frac{AB}{2} \cos(w_0 (2t + \tau)) \right] dt$$

$$= \underset{T\to\infty}{\text{Lt}} \left\{ \frac{1}{2T} \left\{ \left[\frac{AB}{2} [\sin(w_0 \tau)]\, t \right]_{-T}^{T} + \frac{AB}{2} \left[\frac{-\sin w_0 (2t + \tau)}{2w_0} \right]_{-T}^{T} \right\} \right\}$$

$$= \underset{T\to\infty}{\text{Lt}} \left\{ \frac{1}{2T} \frac{AB}{2} \sin(w_0 \tau) [T + T] \right.$$

$$\left. + \frac{1}{2T} \cdot \frac{AB}{2} \left[\frac{-\sin w_0 (2T + \tau)}{2w_0} + \frac{\sin w_0 (-2T + \tau)}{2w_0} \right] \right\}$$

$$= \frac{AB}{2} \sin w_0 \tau + \underset{T\to\infty}{\text{Lt}} \frac{AB}{8Tw_0} [-\sin w_0 (2T + \tau) + \sin w_0 (-2T + \tau)]$$

$$= \frac{AB}{2} \sin w_0 \tau.$$

Hence $$S_{XY}(w) = F\left[\frac{AB}{2}\sin w_0\tau\right]$$

$$= -i\,\frac{\pi AB}{2}\,[\delta(w - w_0) - \delta(w + w_0)] \quad \text{by appendix C.}$$

10. Show that the spectra of the time function $e^{-\alpha|t|}[1 + \alpha|t|]$ is equal to $\dfrac{4\alpha^3}{(\alpha^2 + w^2)^2}$.

Solution. $$S_{XX}(w) = \int_{-\infty}^{\infty} e^{-\alpha|t|}[1 + \alpha|t|]\, e^{-iwt}\, dt$$

$$= \int_{-\infty}^{0} e^{\alpha t}(1 + \alpha t)\, e^{-iwt}\, dt + \int_{0}^{\infty} e^{-\alpha t}(1 - \alpha t)\, e^{-iwt}\, dt$$

$$= \int_{-\infty}^{0} e^{\alpha t} e^{-iwt}\, dt + \alpha \int_{-\infty}^{0} te^{\alpha t} e^{-iwt}\, dt + \int_{0}^{\infty} e^{-\alpha t} e^{-iwt}\, dt - \alpha \int_{0}^{\infty} te^{-\alpha t} e^{-iwt}\, dt$$

$$= \left[\frac{e^{(\alpha - iw)t}}{\alpha - iw}\right]_{-\infty}^{0} + \left[\frac{e^{-(\alpha + iw)t}}{-(\alpha + iw)}\right]_{0}^{\infty} - \alpha\left[\int_{-\infty}^{0} t\, e^{(\alpha - iw)t}\, dt - \int_{0}^{\infty} t\, e^{-(\alpha + iw)t}\, dt\right]$$

$$= \left[\frac{1}{\alpha - iw} - 0\right] + \left[\frac{1}{\alpha + iw}\right] - \alpha\left\{\left[\left\{t\cdot\frac{e^{(\alpha - iw)t}}{\alpha - iw}\right\}_{-\infty}^{0} - \int_{-\infty}^{0}\frac{e^{(\alpha - iw)t}}{\alpha - iw}\, dt\right]\right.$$

$$\left. - \left[\left\{t\cdot\frac{e^{-(\alpha + iw)t}}{-(\alpha + iw)}\right\}_{0}^{\infty} - \int_{0}^{\infty}\frac{e^{-(\alpha + iw)t}}{(\alpha + iw)}\, dt\right]\right\}$$

$$= \frac{1}{\alpha - iw} + \frac{1}{\alpha + iw} - \alpha\left\{0 - \left[\frac{e^{(\alpha - iw)t}}{(\alpha - iw)^2}\right]_{-\infty}^{0} - \left[0 - \left[\frac{e^{-(\alpha + iw)t}}{\alpha + iw}\right]_{0}^{\infty}\right]\right\}$$

$$= \frac{1}{\alpha - iw} + \frac{1}{\alpha + iw} + \frac{\alpha}{(\alpha - iw)^2} + \frac{\alpha}{(\alpha + iw)^2}$$

$$= \frac{2\alpha^3 + 2\alpha w^2 + 2\alpha^3 - 2\alpha w^2}{(\alpha + iw)^2\,(\alpha - iw)^2}$$

$$= \frac{4\alpha^3}{(\alpha^2 + w^2)^2}$$

11. Prove that if a Gaussian random process $X(t)$ is applied to a linear-time invariant system, then the output obtained is also Gaussian.

Solution. Assume that the impulse response $h(t)$ is such that the mean square value of the output random process $Y(t)$ is finite for all t, $0 \le t \le \infty$ for which $Y(t)$ is defined. We know

$$Y(t) = \int_0^T h(t-\tau)\, X(\tau)\, d\tau, \qquad 0 \le t \le \infty$$

Let $X(t)$ is Gaussian. To show $Y(t)$ is Gaussian, consider any linear functional $Z = \int_0^\infty g_Y(t)\, Y(t)\, dt$, and we show that Z is a Gaussian random variable. Here $g_Y(t)$ denote the pdf of $Y(t)$.

$$Z = \int_0^\infty g_Y(t)\, Y(t)\, dt = \int_0^\infty g_Y(t) \int_0^T h(t-\tau)\, X(\tau)\, d\tau\, dt$$

$$= \int_0^T \int_0^\infty g_Y(t)\, h(t-\tau)\, X(\tau)\, dt\, d\tau$$

Interchanging the order of integration

$$= \int_0^T g(\tau)\, X(\tau)\, d\tau, \qquad \text{where} \quad g(\tau) = \int_0^\infty g_Y(t)\, h(t-\tau)\, dt.$$

Note that Z is a Gaussian random variable, since $X(t)$ is a Gaussian process. Hence the output $Y(t)$ is a Gaussian process.

12. If $X(t)$ and $Y(t)$ are real input and output random processes of a linear system, show that

(i) $R_{XY}(t_1, t_2) = L_2(R_{XX}(t_1, t_2))$

(ii) $R_{YY}(t_1, t_2) = L_1(R_{XY}(t_1, t_2))$

where L_i represent the time variable of the linear operator L at time t_i, $i = 1, 2$.

Solution.

(i) $$X(t_1)\, Y(t_2) = X(t_1)\, L_2(X(t_2))$$

since when $X(t)$ at time t_2 is applied to linear system, the output random process is $Y(t_2)$. Therefore

$$X(t_1)\, Y(t_2) = X(t_1)\, L_2(X(t_2)) = L_2(X(t_1)\, X(t_2)), \text{ since } L_2\, X(t_1) = X(t_1)$$

Now $$R_{XY}(t_1, t_2) = E(X(t_1)\, Y(t_2))$$

$$= E(L_2(X(t_1)\, X(t_2)) = L_2\, R_{XX}(t_1, t_2)$$

since E and L_2 are linear operators

(ii) $Y(t_1)\, Y(t_2) = L_1(X(t_1)) \cdot Y(t_2) = L_1(X(t_1)\, Y(t_2))$

$$R_{YY}(t_1, t_2) = E(Y(t_1)\, Y(t_2)) = E(L_1(X(t_1)\, Y(t_2)) = L_1(E(X(t_1)\, Y(t_2)) = L_1\, R_{XY}(t_1, t_2)$$

13. If X(t) and $Y(t)$ are the input and output random processes of the linear operator L. Show that

$$R_{XY}{}^*(\tau) = R_{XX}{}^*(\tau) * h^*(-\tau)$$

Solution.

We know $$Y(t) = \int_{-\infty}^{\infty} X(\tau)\, h(t-\tau)\, d\tau$$

or $$Y(t) = \int_{-\infty}^{\infty} h(\alpha)\, X(t-\alpha)\, d\alpha$$

$$Y^*(t) = \int_{-\infty}^{\infty} X^*(t-\alpha)\, h^*(\alpha)\, d\alpha$$

Multiplying both sides by $X(t+\tau)$ we get

$$X(t+\tau)\, Y^*(t) = \int_{-\infty}^{\infty} X(t+\tau)\, X^*(t-\alpha)\, h^*(\alpha)\, d\alpha$$

$$\Rightarrow \quad E(X(t+\tau)\, Y^*(t)) = E\left[\int_{-\infty}^{\infty} X(t+\tau)\, X^*(t-\alpha)\, h^*(\alpha)\, d\alpha\right]$$

$$= \int_{-\infty}^{\infty} E(X(t+\tau)\, X^*(t-\alpha)]\, h^*(\alpha)\, d\alpha$$

Therefore

$$R_{XY}(\tau) = \int_{-\infty}^{\infty} R_{XX}^*(\tau+\alpha)\, h^*(\alpha)\, d\alpha$$

Substituting $\alpha = -\beta$ we get

$$R_{XY}^*(\tau) = \int_{-\infty}^{\infty} R_{XX}^*(\tau-\beta)\, h^*(-\beta)\, d\beta$$

$$= R_{XX}^*(\beta) * h^*(-\beta).$$

Since β is a dummy variable we get

$$R_{XY}(\tau) = R_{XX}(\tau) * h^*(-\tau)$$

14. A stationary random process $X(t)$ having the auto correlation function $R_{XX}(\tau) = A\delta(\tau)$ is applied to a linear system at time $t = 0$ where $\delta(\tau)$ represent the impulse function. The linear system has the impulse response of $h(t) = e^{-bt}\, u(t)$ where $u(t)$ represents unit step function. Show that

$$R_{YY}(t_1, t_2) = \frac{A}{2b}\left(1 - e^{-2bt_1}\right) e^{-b|t_2 - t_1|}, \qquad 0 < t_1 < t_2$$

Solution. Let $t_2 > t_1 > 0$. Then

$$R_{XX}(t_1, t_2) = R_{XX}(t_2 - t_1) = R_{XX}(\tau) = A\delta(\tau) = A\delta(t_2 - t_1).$$ We know that

$$R_{YY}(t_1, t_2) = L_1(R_{XY}(t_1, t_2)$$

and

$$R_{XY}(t_1, t_2) = L_2(R_{XX}(t_1, t_2)) = L_2\,[A\delta(t_2 - t_1)] = A\,[L_2\delta(t_2 - t_1)] = A\,h\,(t_2 - t_1)$$

$$= A\left(e^{-b(t_2 - t_1)}\, u(t_2 - t_1)\right)$$

Hence

$$R_{YY}(t_1, t_2) = \int_{-\infty}^{\infty} R_{XY}(t_1 - \tau, t_2)\, h(\tau)\, d\tau$$

$$= A\int_0^{t_1} e^{-b[t_2 - t_1 + \tau]} \cdot e^{-b\tau}\, d\tau = A\left[\frac{e^{-2b\tau - bt_2 + bt_1}}{-2b}\right]_0^{t_1}$$

$$= -\frac{A}{2b}\left[e^{-2bt_1 - bt_2 + bt_1} - e^{-bt_2 + bt_1}\right] = \frac{A}{2b}\left[e^{-b|t_2 - t_1|}\left[1 - e^{-2bt_1}\right]\right]$$

15. A differentiator is a linear system in which the output is the derivative of the input. That is $Y(t) = L(X(t)) = X'(t)$. Find the mean and auto correlation of the output $Y(t)$.

Solution. Mean $= E(Y(t)) = E(X'(t)) = E\,(L(X(t)) = L\,(E(X(t)) = (E(X(t)]^1$

Autocorrelation function

$$R_{XX'}(t_1, t_2) = L_2(R_{XX}(t_1, t_2)) = \frac{\partial}{\partial t_2}\,[R_{XX}(t_1, t_2)]$$

since L_2 represents the differential operator with respect to time t_2.

Therefore $R_{X'X'}(t_1, t_2) = L_1[R_{XX'}(t_1, t_2)]$

$$= L_1\left[\frac{\partial}{\partial t_2}(R_{XX}(t_1, t_2))\right] = \frac{\partial^2}{\partial t_1 \partial t_2}(R_{XX}(t_1, t_2)).$$

If $X(t)$ is real WSS process, then $E(Y(t)) = 0$. If $X(t)$ is a covariance function of the form $\sigma^2 \cos(w(t_1 - t_2))$, then the covariance of $Y(t) = X'(t)$ is

$$C_{X'X'}(t_1, t_2) = \frac{\partial}{\partial t_1}\left(\frac{\partial}{\partial t_2}(C_{XX}(t_1, t_2))\right)$$

$$= \frac{\partial}{\partial t_1}\left[\frac{\partial}{\partial t_2}\left(\sigma^2 \cos(w(t_1 - t_2))\right)\right]$$

$$= \frac{\partial}{\partial t_1}[w\sigma^2 \sin w(t_1 - t_2)]$$

$$= w^2\sigma^2 \cos w(t_1 - t_2) = w^2 C_{XX}(t_1, t_2)$$

16. Suppose the auto correlation of the input random process $X(t)$ is $R_{XX}(\tau) = 3\delta(\tau)$ and $H(w)$, the impulse response is $\frac{1}{6+iw}$, then find the

(i) Auto correlation of output random process $Y(t)$.

(ii) Power spectral density of $Y(t)$.

Solution.

(i) $S_{XX}(w) = \int_{-\infty}^{\infty} R_{XX}(\tau)\, e^{-iw\tau}\, d\tau$

$$= \int_{-\infty}^{\infty} 3\delta(\tau)\, e^{-iw\tau}\, d\tau$$

$$= 3$$

since $\int_{-\infty}^{\infty} \delta(\tau)\, e^{-iw\tau}\, d\tau = 1$. Also, given $H(w) = \frac{1}{6+iw}$ and so $|H(w)|^2 = H(w)\, H^*(w)$

$= \frac{1}{(6+iw)} \times \frac{1}{(6-iw)} = \frac{1}{36+w^2}$. Therefore $S_{YY}(w) = |H(w)|^2\, S_{XX}(w) = \frac{3}{36+w^2}$.

(ii) $R_{XX}(w) = \frac{1}{2\pi}\int_{-\infty}^{\infty} S_{XX}(w)\, e^{iw\tau}\, dw$

$$= \frac{1}{2\pi}\int_{-\infty}^{\infty} \frac{3}{36+w^2}\, e^{iw\tau}\, dw$$

By known formula $\frac{1}{2\pi}\int_{-\infty}^{\infty} \frac{2\alpha}{\alpha^2+w^2} \cdot e^{iw\tau}\, dw = e^{-\alpha|\tau|}$, we get $R_{XX}(w) = \frac{3}{12}\, e^{-6|\tau|}$.

17. If a white Gaussian noise $X(t)$ with zero-mean and spectral density $\frac{N_0}{2}$ is applied to a low-pass R_C filter shown in the figure, determine the auto correlation of the output $Y(t)$.

Solution. Given $S_{XX}(w) = \frac{N_0}{2}$ and we know R_C filter has filter transfer function as $H(iw) =$

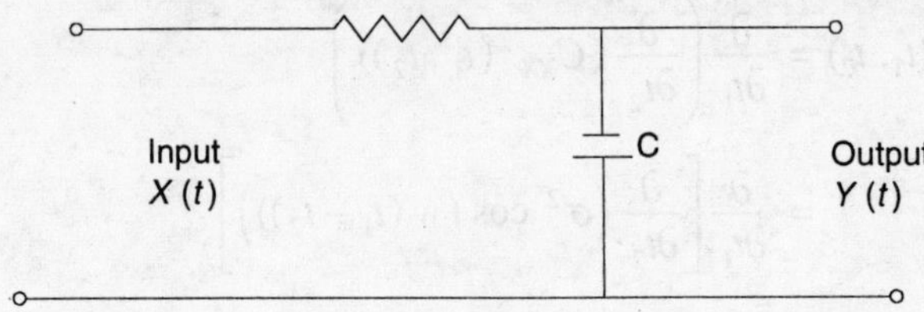

$$\frac{\frac{1}{iwC}}{R+\frac{1}{iwC}} = \frac{1}{1+iw\,RC}.$$

Hence $\quad S_{YY}(w) = |H(w))|^2 S_{XX}(w)$

$$= \left(\frac{1}{1+iw\,RC}\right)\left(\frac{1}{1-iw\,RC}\right)\frac{N_0}{2}$$

$$= \frac{N_0}{2}\left[\frac{1}{2}\left[\frac{1}{1+iw\,RC}+\frac{1}{1-iw\,RC}\right]\right] \text{ by partial fraction}$$

$$= \frac{N_0}{4}\left[\frac{1}{1+iw\,RC}+\frac{1}{1-iw\,RC}\right]$$

To find the auto correlation, take the inverse Fourier transform of both the terms. (see Appendix C).

Hence $\quad R_{YY}(\tau) = F^{-1}\left[\frac{N_0}{4}\left[\frac{1}{1+iw\,RC}+\frac{1}{1-iw\,RC}\right]\right]$

$$= \frac{N_0}{4}\left[F^{-1}\left[\frac{1}{1+iw\,RC}\right]+F^{-1}\left[\frac{1}{1-iw\,RC}\right]\right]$$

$$= \frac{N_0}{4}\left[F^{-1}\left[\frac{1}{RC}\cdot\frac{RC}{1+iw\,RC}\right]+F^{-1}\left[\frac{1}{RC}\cdot\frac{RC}{1-iw\,RC}\right]\right]$$

$$= \frac{N_0}{4}\left[F^{-1}\left[\frac{1}{RC}\cdot\frac{1}{\left(\frac{1}{RC}+iw\right)}\right]+F^{-1}\left[\frac{1}{RC}\cdot\frac{1}{\left(\frac{1}{RC}-iw\right)}\right]\right]$$

$$= \frac{N_0}{4}\left[\frac{1}{RC}\, e^{-\tau/RC} \cdot u(\tau) + \frac{1}{RC}\, e^{\tau/Rc}\, u(\tau)\right]$$

(by Appendix C)

were $u(\tau)$ is the unit step function. Hence

$$R_{YY}(\tau) = \frac{N_0}{4RC}\, e^{-|\tau|/RC}.$$

18. If $X(t)$ is a stationary random process with derivative $X'(t)$, then for a given t, prove that the random processes $X(t)$ and $X'(t)$ are orthogonal and uncorrelated.

Solution.

$$E(X(t)\,X'(t)) = R_{XX'}(0) = \frac{\partial}{\partial t}\,[R_{XX}(0)] = 0$$

since $R_{XX}(0)$ is not a function of t. Therefore $X(t)$ and $X'(t)$ are orthogonal. Moreover, since $X(t)$ is stationary, $E(X'(t) = 0$. Hence, $E(X(t)\,X'(t)) = 0$, as $E(X(t))\,E(X'(t)) = 0$. Hence $X(t)$ and $X'(t)$ are uncorrelated.

19. If a WSS random process $X(t)$ with mean 5, autocorrelation $25 + 4e^{-2|\tau|}$, is applied through a linear system, then find the auto correlation and spectrum of the output $Y(t) = 2X(t) + 3\,X'(t)$.

Solution.

$$E(Y(t)) = E(2X(t) + 3X'(t))$$
$$= 2\,E(X(t)) + 0 = 2 \times 5 = 10$$

since $X(t)$ is WSS process, $E(X(t))$ is constant and hence $E(X'(t)) = 0$. Now

$$R_{YY}(\tau) = E(Y(t)\,Y(t+\tau))$$
$$= E([2X(t) + 3X'(t)]\,Y(t+\tau))$$
$$= E\,\{2X(t)\,Y(t+\tau) + 3X'(t)\,Y(t+\tau)\}$$
$$= 2E(X(t)\,Y(t+\tau)) + 3E(X'(t)\,Y(t+\tau))$$
$$= 2R_{XY}(\tau) + 3R_{X'Y}(\tau)$$
$$R_{XY}(\tau) = E(X(t+\tau)\,Y\,t)]$$
$$= E(X(t+\tau)\,(2X(t) + 3X'(t))$$
$$= E(2X(t)\,X(t+\tau) + 3\,X'(t)\,X(t+\tau))$$
$$= 2E(X(t)\,X(t+\tau)) + 3E(X'(t)\,X(t+\tau))$$
$$= 2R_{XX}(\tau) + 3R_{XX'}(\tau)$$
$$= 2\left[25 + 4e^{-2|\tau|} + 3\left[\frac{\partial}{\partial\tau}\left(25 + 4e^{-2|\tau|}\right)\right]\right]$$
$$= 50 + 8e^{-2|\tau|} - 24e^{-2|\tau|} = 50 - 16\,e^{-2|\tau|}$$
$$R_{X'Y}(\tau) = \frac{\partial}{\partial\tau}\left(50 - 16e^{-2|\tau|}\right) = 32e^{-2|\tau|}$$

Hence

$$R_{YY}(\tau) = 2[50 - 16e^{-2|\tau|}] + 3[32\,e^{-2|\tau|}] = 100 + 64e^{-2|\tau|}$$

Using Wiener-Khinchin relation

$$S_{YY}(w) = \int_{-\infty}^{\infty} R_{YY}(\tau)\,e^{-iw\tau}\,d\tau$$
$$= \int_{-\infty}^{\infty} (100 + 64\,e^{-2|\tau|})\,e^{-iw\tau}\,d\tau$$

$$= \int_{-\infty}^{\infty} 100\, e^{-iw\tau}\, d\tau + 64 \int_{-\infty}^{\infty} e^{-2|\tau|}\, e^{-iw\tau}\, d\tau$$

$$= 100 \times 2\pi\,\delta(w) + 64\left(\frac{2\times 2}{2^2 + w^2}\right)$$

$$= 200\,\pi\,\delta(w) + \frac{256}{w^2+4}.$$

20. A circuit has an impulse response given by

$$h(t) = \frac{1}{T}, \qquad 0 \le t \le T$$

$$= 0 \qquad \text{elsewhere}$$

Find $S_{YY}(w)$ in terms of $S_{XX}(w)$.

Solution.

$$H(w) = \int_0^T h(t)\, e^{-iw\tau}\, d\tau = \int_0^T \frac{1}{T}\, e^{-iw\tau}\, d\tau$$

$$= \frac{1}{T}\left[\frac{e^{-iw\tau}}{-iw}\right]_0^T = \frac{1}{T}\left[\frac{e^{-iwT}-1}{-iw}\right]$$

$$= \frac{2}{wT}\left[\frac{1-e^{-iwT}}{2i}\right] = \frac{2e^{-wT/2}}{wT}\left[\frac{e^{iwT/2}-e^{-iwT/2}}{2i}\right]$$

$$= \frac{2e^{-iwT/2}}{wT}\,[\sin(wT/2)]$$

Therefore $\quad S_{YY}(w) = |H(w)|^2\, S_{XX}(w)$

$$= \left|\frac{2e^{-iwT/2}}{wT}\cdot\sin(wT/2)\right|^2 S_{XX}(w)$$

$$= \left(\frac{2}{wT}\right)^2\cdot\sin^2(wT/2)\cdot S_{XX}(w)$$

21. The Harmonic Process

An important random process that is found in applications like radar and sonar signal processing is the harmonic process. An example of a real-valued harmonic process is the random phase sinusoid which is defined by

$$x(n) = A\sin(nw_0 + \alpha)$$

where A and w_0 are fixed constants and α is a random variable that is uniformly distributed over the interval $-\pi$ to π. That is the *pdf* of α is

$$f(\alpha) = \begin{cases} \dfrac{1}{2\pi}, & -\pi \le \alpha \le \pi \\ 0, & \text{elsewhere} \end{cases}$$

Mean of this process is

$$E(x(n)) = E(A \sin (nw_0 + \alpha))$$

$$= \int_{-\pi}^{\pi} \frac{1}{2\pi} A \sin (nw_0 + \alpha)\, d\alpha = 0$$

Auto correlation function of this process is

$$R_{XX}(t, s) = \frac{1}{2} A^2 \cos [(t - s)\, w_0]$$

Property: If two random process $x(n)$ and $y(n)$ are uncorrelated, then the auto correlation of the sum $z(n) = x(n) + y(n)$ is equal to the sum of the auto correlation of $x(n)$ and $y(n)$.

22. Gaussian Process

A real valued random process $X(t)$ is a Gaussian process if the joint *pdf* of $X(t_1), X(t_2), \ldots X(t_n)$ is

$$f(x_1, x_2, \ldots x_n) = \frac{1}{|C|^{1/2} (2\pi)^{n/2}} e^{\left[-\frac{1}{2}\sum_{i=1}^{n}\sum_{j=1}^{n}(X_i - m_i)\, d_{ij}\, (X_j - m_j)\right]}$$

where $X_i = X(t_i)$, $m_i = E(X(t_i))$, $i = 1$ to n, and

$$C = \begin{bmatrix} C_{XX}(t_1, t_1) & C_{XX}(t_1, t_2) & \ldots & C_{XX}(t_1, t_n) \\ C_{XX}(t_2, t_1) & C_{XX}(t_2, t_2) & \ldots & C_{XX}(t_2, t_n) \\ \vdots & \vdots & & \vdots \\ C_{XX}(t_n, t_1) & C_{XX}(t_n, t_2) & & C_{XX}(t_n, t_n) \end{bmatrix}$$

and $d_{ij} = C^{-1}$.

A Gaussian random process is determined when m_i and d_{ij} are known for all $t_i, t_j \in (-\infty, \infty)$.

23. Power spectral density of triangular auto correlation

Consider an autocorrelation function that is triangular in shape such that the correlation goes to zero at shift $T > 0$,

$$R_{XX}(\tau) = \max\left[1 - \frac{|\tau|}{T}, 0\right].$$

Solution. The triangular correlation function can arise in the running integral average operating on white noise. Consider

$$X(t) = \frac{1}{\sqrt{T}} \int_{t-T}^{t} W(\tau)\, d\tau$$

with $W(\tau)$ a white noise with zero-mean and correlation function $R_{WW}(\tau) = \delta(\tau)$

Then $\mu_X(t) = 0$ and $E(X(t_1)\, X(t_2))$ can be computed as

$$R_{XX}(t_1, t_2) = \frac{1}{T} \int_{t_1-T}^{t_1} \int_{t_2-T}^{t_2} R_{WW}(s_1 - s_2) ds_1 ds_2$$

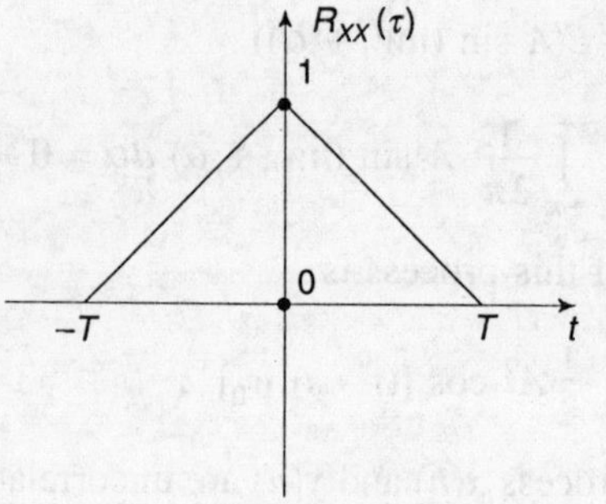

$$= \frac{1}{T} \int_{t_1-T}^{t_1} \left[\int_{t_2-T}^{t_2} \delta(s_2 - s_1)\, ds_2 \right] ds_1$$

where the inner integral is

$$g_{t_2}(s_1) \equiv \int_{t_2-T}^{t_2} \delta(s_2 - s_1) ds_2 = \begin{cases} 1, & t_2 - T \le s_1 \le t_2 \\ 0, & \text{else,} \end{cases}$$

which looks as shown in figure, so

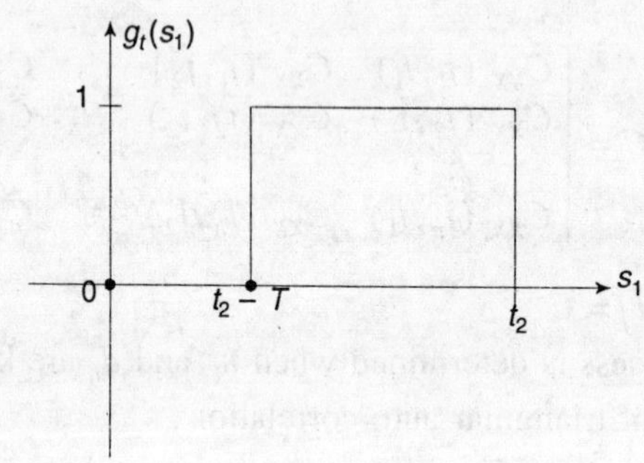

$$R_{XX}(t_1, t_2) = \frac{1}{T} \int_{t_1-T}^{t_1} g_{t_2}(s_1) ds_1 = \max\left[1 - \frac{|t_1 - t_2|}{T}, 0\right].$$

24. Consider the random process

$$X(t) = A \sin(w_0 t + \theta)$$

where A and θ are independent, real valued random variables and θ is uniformly distributed over $[-\pi, \pi]$. The mean function

$$\mu_X(t) = E(A \sin(w_0 t + \theta))$$
$$= E(A)\, E(\sin(w_0 t + \theta))$$

$$= \mu_A \cdot \frac{1}{2\pi} \int_{-\pi}^{\pi} \sin(w_0 t + \theta) d\theta = \mu_A \cdot 0 = 0$$

Then for the correlation

$$R_{XX}(t_1, t_2) = E(X(t_1)\, X(t_2))$$
$$= E(A^2 \sin(w_0 t_1 + \theta) \sin(w_0 t_2 + \theta))$$
$$= E(A^2)\, E(\sin(w_0 t_1 + \theta) \sin(w_0 t_2 + \theta))$$

Now

$E(\sin(w_0 t_1 + \theta) \cdot \sin(w_0 t_2 + \theta))$

$$= \frac{1}{2} \{E[\cos(w_0(t_1 - t_2)] - E[\cos(w_0(t_1 + t_2) + 2\theta)]\}$$

by applying trigonometric identity

$$\sin(B) \sin(C) = \frac{1}{2} \{\cos(B - C) - \cos(B + C)\}$$

and bringing the expectation operator inside.

Then, sine θ is uniformly distributed over $[-\pi, \pi]$, the integral arising from $E(\cos(w_0(t_1 + t_2) + 2\theta))$ is zero and hence we get

$$R_{XX}(t_1, t_2) = \frac{1}{2} E(A^2) \cos w_0(t_1 - t_2)$$

Hence the process is WSS.

25. Consider the above WSS process, which is a random amplitude cosine at frequency f_0 with random phase

$$X(t) = A \cos(2\pi f_0 t + \theta), -\infty < t < \infty$$

where A is $N(0, 1)$, θ is uniformly distributed over $[-\pi, \pi]$ and both A and θ are independent. Then

$$E(X(t)) = E(A)\, E(\cos 2\pi f_0 t + \theta) = 0$$

and

$$E[X(t + \tau)\, X(t)] = \frac{1}{2\pi} \int_{-\pi}^{\pi} \cos(2\pi f_0 t + 2\pi f_0 \tau + \theta) \cos(2\pi f_0 t + \theta) d\theta$$

$$= \frac{1}{2} \cos(2\pi f_0 \tau), \quad -\infty < \tau < \infty$$

Hence $X(t)$ is WSS.

Next we show $X(t)$ is ergodic in the mean.

We know that a WSS random process $X(t)$ is ergodic in the mean if its covariance function $C_{XX}(\tau)$ satisfies

$$\lim_{T \to \infty} \left[\frac{1}{2T} \int_{-2T}^{2T} \left(1 - \frac{|\tau|}{2T}\right) C_{XX}(\tau)\, d\tau \right] = 0$$

Hence we compute

$$\frac{1}{2T} \int_{-2T}^{2T} \left(1 - \frac{|\tau|}{2T}\right) \frac{1}{2} \cos(2\pi f_0 \tau) d\tau$$

If we realize the triangular term in the square brackets can be expressed as the convolution of two rectangular pulses as shown in Figure

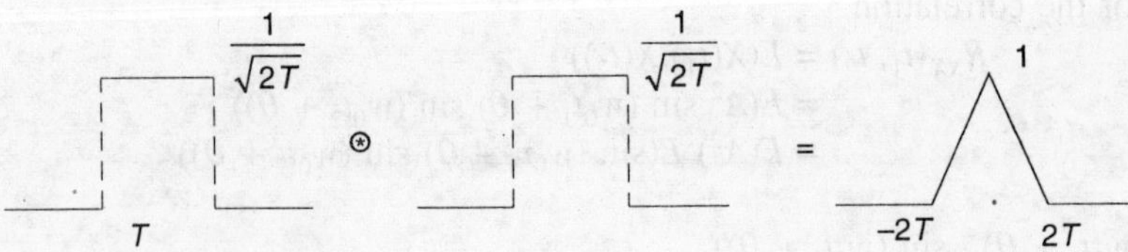

Then we write the Fourier transform of the triangular pulse as the square of the transform of one rectangular pulse of half the width

$$\left(\sqrt{2T}\,\frac{\sin 2\pi fT}{2\pi ft}\right)^2 = 2T\left(\frac{\sin 2\pi fT}{2\pi ft}\right)^2$$

Using Passeval's theorem (see Appendix C) we get

$$\frac{1}{2T}\int_{-2T}^{2T}\left[1-\frac{|\tau|}{2T}\right]\frac{1}{2}\cos 2\pi f_0\tau d\tau = \frac{1}{2T}\int_{-\infty}^{\infty} 2T\left(\frac{\sin 2\pi fT}{2\pi fT}\right)^2 \pi(\delta(f+f_0)+\delta(f-f_0))df$$

$$= \frac{1}{2}\left(\frac{\sin 2\pi f_0 T}{2\pi f_0 T}\right) \to 0$$

as $T \to \infty$, $f_0 \neq 0$. Hence the random process is ergodic in the mean for any $f_0 \neq 0$.
Next we check whether it is ergodic in power. Now

$$\frac{1}{2T}\int_{-T}^{T} X^2(t)dt = A^2\frac{1}{2T}\int_{-T}^{T}\cos^2(2\pi f_0 t+\theta)dt$$

It will converge to $A^2/2$ not to $\frac{E(A^2)}{2}$ since for any θ, the time average of the cosine squared will converge to $\frac{1}{2}$ for $f_0 \neq 0$. Hence it is not ergodic in power and hence not ergodic in correlation.

26. Let $R_{XX}(\tau) = 100e^{-10|\tau|} + 33\cos(5\tau) + 36$. Find the mean, mean square value or power and variance of the random process.

Solution. The mean can be given by the property

$$\bar{X}^2 = \underset{|\tau|\to\infty}{\text{Lt}} R_{XX}(\tau)$$

Hence $$\bar{X}^2 = \underset{|\tau|\to\infty}{\text{Lt}} R_{XX}(\tau) = \underset{|\tau|\to\infty}{\text{Lt}}(100e^{-10|\tau|} + 33\cos 5\tau + 36) = 36 \Rightarrow \bar{X} = \pm 6.$$

Now, the mean square value of the process is:

$$E(X^2(t)) = R_{XX}(0) = \int_{-\infty}^{\infty} x^2 f_{X(t)}(x)dx = \bar{X}^2$$

$$\Rightarrow \quad E(X^2(t)) = 100e^0 + 33\cos 0 + 36 = 100 + 33 + 36 = 169$$

Hence variance of the process is,

$$\text{var } (X(t) = E(X^2(t)) - (E(X(t)))^2 = 169 - 36 = 133.$$

27. Let $X(t) = A \cos (wt + \theta)$ be the random process, θ is uniformly distributed in $(0, 2\pi)$. Find the average power of the process.

Solution. Let $X(t) = A \cos (wt + \theta)$. The autocorrelation of the process is $R_{XX}(\tau) = \frac{A^2}{2} \cos w\tau$.

When $\tau = 0$, $R_{XX}(0) = \frac{A^2}{2}$.

28. Consider the random process $X(t) = A \cos (100t + \theta)$ where A and θ are independent variables and θ is uniformly distributed in $[-\pi, \pi]$. Find mean and autocorrelation.

Solution.

(i) Mean $= E(X(t)) = \int_{-\infty}^{\infty} X(t)\, f(\phi) d\phi$ where $f(\phi) = \frac{1}{2\pi}$, for $\theta \in [-\pi, \pi]$ and have $E(X(t)) =$

$$\int_{-\pi}^{\pi} A \cos (100t + \theta) \frac{1}{2\pi} \cdot d\theta = \frac{A}{2\pi}[\sin (100t + \theta)]_{-\pi}^{\pi} = 0$$

Hence mean is a constant

(ii) Auto correlation $R_{XX}(t, t + \tau) = R_{XX}(\tau) = E(X(t)\, X(t + \tau))$

$$= E(A \cos(100t + \theta)\, A \cos(100(t + \tau) + \theta))$$

$$= E\left(\frac{A^2}{2}[\cos 100(\tau) + \cos (100(2t + \tau) + 2\theta)]\right)$$

$$= E\left(\frac{A^2}{2}\right)\cos 100\tau + E\left(\frac{A^2}{2}\right)\int_{-\pi}^{\pi} \cos (2t + \tau + 2\theta)\frac{1}{2\pi}\, d\theta$$

$$= E\left(\frac{A^2}{2}\right)\cos 100\tau + 0$$

$$= E\left(\frac{A^2}{2}\right)\cos 100\tau, \text{ not depending on time } t.$$

EXERCISES

1. If $X(t)$ and $X'(t)$ are jointly stationary, prove that

$$R_{X'X'}(u) = \frac{-d^2}{du^2} R_{XX}(u).$$

2. Find the power spectrum of the process whose auto correlation is given by $R(u) = e^{-\alpha u^2}$.

3. Find the unit impulse response $h(t)$ of the Wiener filter system with $S_{NN}(w) = N_0$ and $S_{FF}(w) = \frac{3\alpha^2 N_0}{\alpha^2 + w^2}$ and if the signal and noise are uncorrelated.

4. Prove that the random process $X(t) = A_0 \cos(w_0 t + \theta)$ where A_0 and w_0 are constant and θ is uniformly distributed in the interval on $[-\pi, \pi]$ is wide-sense stationary.
5. The auto correlation function of a random process $X(t)$ is given by $R(\tau) = ae^{-b^2\tau^2}$ where a and b are constants. Evaluate the spectral density function.
6. Find the spectral density function $S(w)$ of the auto correlation function of a random process given by

$$R_{XX}(\tau) = \begin{cases} 1 - \dfrac{|\tau|}{T} & \text{for } |\tau| < T \\ 0 & \text{elsewhere} \end{cases}$$

7. The power spectral density of a random process is given by

$$S_{XX}(w) = \begin{cases} \pi & |w| < 1 \\ 0 & \text{elsewhere} \end{cases}$$

Find its auto correlation function.

8. The auto correlation function of a wide sense stationary random process is given by $R_{XX}(\tau) = \dfrac{4\tau^2 + 100}{\tau^2 + 4}$. Find the mean and variance of this process.
9. If $X(t)$ is a WSS random process with auto correlation $R(\tau) = Ae^{-\alpha(\tau)}$, determine the second order moment of the random variable $X(8) - X(5)$.
10. Prove that the random process $X(t) = A\cos(wt + \theta)$ is not stationary if it is assumed that A and w are constants and θ is uniformly distributed variable on the interval $(0, \pi)$.
11. The occurrence of a pulse follows Poisson process. Calculate the auto correlation function of such a random telegraph signal.
12. Calculate the power spectral density of a stationary random process for which the auto correlation is $R_{XX}(\tau) = \sigma^2 e^{-\alpha|\tau|}$.
13. An ergodic process is known to have an auto correlation function of the form

$$R(\tau) = 1 - |\tau|, \quad |\tau| \le 1$$
$$= 0 \quad |\tau| > 1$$

Show that the spectral density is given by

$$S(w) = \left[\frac{\sin w/2}{w/2}\right]^2$$

14. For a process having $R(\tau) = ae^{-b|\tau|}$, find the spectral density function.
15. Determine $R_{XX}(\tau)$ if $S_{XX}(\omega) = \dfrac{1}{(4 + w^2)^2}$.

ANSWERS

(7) $R_{XX}(\tau) = \dfrac{\sin\tau}{\tau}$.

(8) $E[X(t)] = \left[\underset{|\tau| \to \infty}{\text{Lt}} R_{XX}(\tau) \right]^{1/2} = \left[\underset{|\tau| \to \infty}{\text{Lt}} \dfrac{4\tau^2 + 100}{\tau^2 + 4} \right]^{1/2} = (4)^{1/2} = 2.$

$$\begin{aligned} \text{Variance} &= E(X^2(t)) - (E(X(t))^2 \\ &= R_{XX}(0) - (E(X(t))^2 \\ &= \frac{100}{4} - 4 = 21 \end{aligned}$$

(9) $E\{[X(8) - X(5)]^2\} = E\{X^2(8) + X^2(5) - 2X(8)\,X(5)\}$

$$\begin{aligned} &= E(X^2(8)) + E(X^2(5)) - 2E(X(8)\,X(5)) \\ &= A + A - 2R_{XX}(3) \\ &= 2A - 2Ae^{-3}\alpha \end{aligned}$$

since $R_{XX}(0) = E(X^2(t))$

$$= 2A(1 - e^{-3\alpha})$$

(10) $E(X(t)) = \dfrac{-2A \sin wt}{\pi}$, which is a function of t, hence $X(t)$ is not stationary.

(12) $S_{XX}(w) = \int\limits_{-\infty}^{\infty} \sigma^2 e^{-\alpha|\tau|} e^{-iw\tau}\, d\tau$

$$= \sigma^2 \left[\int\limits_{-\infty}^{0} e^{(\alpha - iw)\tau}\, d\tau + \int\limits_{0}^{\infty} e^{-(\alpha + iw)\tau}\, d\tau \right]$$

$$= \sigma^2 \left\{ \left[\frac{e^{(\alpha - iw)\tau}}{\alpha - iw} \right]_{-\infty}^{0} + \left[\frac{e^{-(\alpha + iw)\tau}}{-(\alpha + iw)} \right]_{0}^{\infty} \right\}$$

$$= \sigma^2 \left\{ \frac{1}{\alpha - iw} + \frac{1}{\alpha + iw} \right\} = \frac{2\alpha\sigma^2}{w^2 + \sigma^2}.$$

(13) $S_{XX}(w) = \int\limits_{-\infty}^{\infty} R_{XX}(\tau)\, e^{-iw\tau}\, d\tau$

$$= \int\limits_{-1}^{1} (1 - |\tau|)\, e^{-iw\tau}\, d\tau = \int\limits_{-1}^{0} (1 + \tau)\, e^{-iw\tau}\, d\tau + \int\limits_{0}^{1} (1 - \tau)\, e^{-iw\tau}\, d\tau$$

$$= \int\limits_{-1}^{0} e^{-iw\tau}\, d\tau + \int\limits_{-1}^{0} \tau e^{-iw\tau}\, d\tau + \int\limits_{0}^{1} e^{-iw\tau}\, d\tau - \int\limits_{0}^{1} \tau e^{-iw\tau}\, d\tau$$

$$= \left[\frac{e^{-iw\tau}}{-iw} \right]_{-1}^{0} + \left\{ \left[\tau \left(\frac{e^{-iw\tau}}{-iw} \right) \right]_{-1}^{0} - \int\limits_{-1}^{0} \left(\frac{e^{-iw\tau}}{-iw} \right) d\tau \right\}$$

$$+ \left[\frac{e^{-iw\tau}}{-iw} \right]_{0}^{1} + \left\{ \left[\tau \left(\frac{e^{-iw\tau}}{-iw} \right) \right]_{0}^{1} - \int\limits_{0}^{1} \left(\frac{e^{-iw\tau}}{-iw} \right) d\tau \right\}$$

$$= \frac{-1}{iw} + \frac{1}{w^2}(1 - e^{iw}) + \frac{1}{iw} + \frac{1}{w^2}(1 - e^{-iw})$$

$$= \frac{1}{w^2}[2 - e^{iw} - e^{-iw}] = \frac{2}{w^2}[1 - \cos w]$$

$$= \frac{4}{w^2}(\sin^2 w/2) = \left[\frac{\sin w/2}{w/2}\right]^2$$

(14) $$S_{XX}(w) = a\int_{-\infty}^{\infty} e^{-b|\tau|} e^{-iw\tau} d\tau$$

$$= a\left[\int_{-\infty}^{0} e^{+b\tau} e^{-iw\tau} d\tau + \int_{0}^{\infty} e^{-b\tau} e^{-iw\tau} d\tau\right] = \frac{2ab}{b^2 + w^2}.$$

(15) $$S_{XX}(w) = \frac{1}{(4 + w^2)^2}$$

$$= \left[\frac{1}{(2 + iw)(2 - iw)}\right]^2 = \left[\frac{(2 + iw) + (2 - iw)}{4(2 + iw)(2 - iw)}\right]^2$$

$$= \frac{1}{16}\left\{\frac{1}{2 - iw} + \frac{1}{2 + iw}\right\}^2 = \frac{1}{16}\left\{\frac{1}{(2 - iw)^2} + \frac{1}{(2 + iw)^2} + \frac{2}{4 + w^2}\right\}$$

$$R_{XX}(\tau) = \frac{1}{16}\{(u(\tau)\ \tau(e^{2\tau} + e^{-2\tau})\} + \frac{1}{8}e^{-2|\tau|} \text{ using Appendix C.}$$

6

Markov Process

6.1 MARKOV CHAIN AND TRANSITION PROBABILITIES

6.1.1 Definition

The Stochastic process $\{X_n, n = 0, 1, 2 \ldots\}$ is called a **Markov Chain** if, for $i, k, i_1, i_2, \ldots i_{n-1} \in N$,

$$P\{X_n = k \mid X_{n-1} = i, X_{n-2} = i_1 \ldots X_0 = i_{n-1}\} = P\{X_n = k|_{X_{n-1}} = i\} = p_{ik} \tag{6.1}$$

Equation (6.1) implies that given the 'present' state of the system, the future is independent of its past. If the state space is discrete in a Markov process, then it is called **Markov Chain**. Further, if the parameter space is also discrete, then the Markov Chain is called **discrete-parameter Markov Chain**. Markov Chain can be visualized as the generalization of the notion of independent trials.

Note that the values $i, i_1, i_2, \ldots i_{n-1}$ are called the **states** of the Markov chain. If $X_n = j$, then the process is said to be at state j at n^{th} trial. To state j, there is no longer fixed probability $P(X_n = j)$ but for pair of states (j, k) at the two successive trials say (n and $n + 1$), there is a conditional probability p_{jk}. It is the probability of transition from the state j at n^{th} trial to the state k at the $(n + 1)^{\text{th}}$ trial. We call p_{jk} $(n, n + 1)$ as the **transition probability function** of the Markov chain.

6.1.2 Definition

If the transition probability p_{jk} is independent of n, the Markov chain is said to be **homogeneous** or to have stationary transition probabilities. If it is dependent on n, the chain is said to be **non-homogeneous**.

The transition probability p_{jk} refers to the states (j, k) at two successive trials (say n^{th} and $(n + 1)^{\text{st}}$ trial); the transition is one step and p_{jk} is called **one-step transition probability**. In general, if we are concerned with the pair of states (j, k) at two successive trials say state j at the nth trial and the state k at the $(n + m)^{\text{th}}$ trial, then the corresponding transition probability is called **m-step transition probability** and is denoted by p_{jk} (m), given by

$$P_{jk}(m) = P\{X_{n+m} = k \mid X_n = j\}$$

In other words, $p_{jk}(m)$ is the probability that a Markov chain will move from state j to state k in exactly m steps.

0-Step transition probabilities is given by

$$p_{jk}(0) = \begin{cases} 1, & \text{if } j = k \\ 0, & \text{otherwise} \end{cases}$$

6.1.3 Definition

Transition Matrix. The transition probabilities p_{jk} satisfy

$$p_{jk} \geq 0, \ \sum_k p_{jk} = 1 \text{ for all } j \tag{6.2}$$

The one-step transition probabilities are compactly specified in the matrix form

$$P = [p_{ij}] = \begin{pmatrix} p_{00} & p_{01} & p_{02} & \cdots \\ p_{10} & p_{11} & p_{12} & \cdots \\ p_{20} & p_{21} & p_{22} & \cdots \\ \vdots & \vdots & \vdots & \vdots \end{pmatrix}$$

This is called the **transition probability matrix** of the Markov chain. Any such square matrix that has non-negative entries and unit row sums is called a **Stochastic matrix.**

An equivalent description of one step transition probabilities is given by a directed graph and is called the **state-transition diagram.** *A* vertex i of the state-transition diagram is the state i of the Markov chain and an edge from i to j is p_{ij}, the one-step transition probabilities.

Probability Distribution of $\{X_n : n \geq 0\}$

Equation (6.1) holds true for all n, in Markov chain. By Section 1.4 of Chapter I, we write the joint probability as:

$$\begin{aligned} P(X_0 = i_0, X_1 = i_1, \ldots X_{n-1} = i_{n-1}, X_n = i_n) \\ = P(X_0 = i_0, X_1 = i_1, \ldots X_{n-1} = i_{n-1})\, P(X_n = i_n | X_0 = i_0, \ldots X_{n-1} = i_{n-1}) \\ = P(X_0 = i_0, X_1 = i_1, \ldots X_{n-1} = i_{n-1})\, P(X_n = i_n | X_{n-1} = i_{n-1}) \\ = P(X_0 = i_0, X_1 = i_1, \ldots X_{n-1} = i_{n-1})\, p_{i_{n-1}, i_n} \\ \vdots \\ = p_{i_0}(0)\, p_{i_0, i_1} \cdots p_{i_{n-1}, i_n} \end{aligned}$$

This implies that all joint probabilities of interest are determined from the initial pmf $p_{i_0}(0) = P(X_0 = i_0)$ and the one-step transition probabilities p_{ij}. The pmf of the random variable X_0 is called the **initial distribution** and is specified by the probability vector

$$P(0) = [p_0(0), p_1(0), \ldots]$$

and $P(n)$ is called the **state probability vector after n transitons** or **the probability distributions of** X_n e.g. $\begin{pmatrix} 0.1 & 0 \\ 0.5 & 0 \end{pmatrix}$ is not probability transition matrix since the rows does not sum to 1.

6.1.4 Definition Order of a Markov Chain

A Markov chain $\{X_n\}$ is said to be of order s ($s = 1, 2, \ldots$), if for all n,

$$\begin{aligned} P[X_n = kX_n = k \mid X_{n-1} = i_{n-1}, X_{n-2} = i_{n-2}, \ldots X_{n-s} = i_{n-s}, \ldots] \\ = P[X_n = k \mid X_{n-1} = i_{n-1}, X_{n-2} = i_{n-2}, \ldots X_{n-s} = i_{n-s}] \end{aligned}$$

A Markov chain $\{X_n\}$ is said to be of order one if

$$\begin{aligned} P[X_n = k \mid X_{n-1} = j, X_{n-2} = j_1, \ldots] \\ = p[X_n = k|_{X_{n-1}} = j] \\ = p_{jk} \end{aligned}$$

6.1.5 Example

Let $\{X_n, n \geq 0\}$ be a Markov chain with three states 0, 1, 2 and with the transition matrix

$$\begin{pmatrix} 3/4 & 1/4 & 0 \\ 1/4 & 1/2 & 1/4 \\ 0 & 3/4 & 1/4 \end{pmatrix}$$

and the initial distribution $P(X_0 = i) = \dfrac{1}{3}$, $i = 0, 1, 2$ Then note that

$$\begin{pmatrix} p_{00} & p_{01} & p_{02} \\ p_{10} & p_{11} & p_{12} \\ p_{20} & p_{21} & p_{22} \end{pmatrix} = \begin{pmatrix} 3/4 & 1/4 & 0 \\ 1/4 & 1/2 & 1/4 \\ 0 & 3/4 & 1/4 \end{pmatrix}$$

So we get

$$P(X_1 = 1 | X_2 = 2) = p_{21} = \frac{3}{4}$$

$$P(X_2 = 2 \mid X_1 = 1) = p_{12} = \frac{1}{4}$$

$$P(X_2 = 2, X_1 = 1 | X_0 = 2) = P(X_2 = 2 | X = 1)\ P(X_1 = 1 | X_0 = 2)$$

$$= p_{12} \cdot p_{21} = \frac{1}{4} \times \frac{3}{4} = \frac{3}{16}$$

$$P(X_2 = 2, X_1 = 1, X_0 = 2)$$
$$= P(X_2 = 2, X_1 = 1 | X_0 = 2)\ P(X_0 = 2)$$
$$= \frac{3}{16} \times \frac{1}{3} = \frac{1}{16}$$

$$P(X_3 = 1, X_2 = 2, X_1 = 1, X_0 = 2)$$
$$= P(X_3 = 1 | X_2 = 2, X_1 = 1, X_0 = 2)\ P(X_2 = 2, X_1 = 1, X_0 = 2)$$

$$P(X_3 = 1 | X_2 = 2) \cdot \frac{1}{16} = p_{21} \frac{1}{16} = \frac{3}{4} \times \frac{1}{16} = \frac{3}{64}$$

6.1.6 Example

A particle performs a random walk with absorbing barriers say, at 0 and 4. Whenever it is at any position r $(0 < r < 4)$, it moves to $r + 1$ with probability p or to $(r - 1)$ with probability q, $p + q = 1$. But as soon as it reaches 0 or 4 it remains there itself. Let X_n be the position of the particle after n moves. The different states of X_n are the different positions of the particle. $\{X_n\}$ is a Markov chain whose one-step transition probabilities are given by

$$P\{X_n = r + 1 | X_{n-1} = r\} = p$$
$$P\{X_n = r - 1 | X_{n-1} = r\} = q \qquad 0 < r < 4$$

and

$$P\{X_n = 0 | X_{n-1} = 0\} = 1$$
$$P\{X_n = 4 | X_{n-1} = 4\} = 1$$

The transition matrix is given by

$$\text{States of } X_{n-1} \begin{matrix} \\ 0 \\ 1 \\ 2 \\ 3 \\ 4 \end{matrix} \overset{\text{States of } X_n}{\begin{pmatrix} 0 & 1 & 2 & 3 & 4 \\ 1 & p & 0 & 0 & 0 \\ q & 0 & p & 0 & 0 \\ 0 & q & 0 & p & 0 \\ 0 & 0 & q & 0 & p \\ 0 & 0 & 0 & 0 & 1 \end{pmatrix}}$$

6.1.7 Example

Consider a simple random walk $X(n)$ with absorbing barriers at state 0 and state $N = 3$. Absorbing barrier we mean it as $p_{00} = 1$ and $p_{33} = 1$. Find the transition probability matrix P.

Solution. The collection of random variables $\{X_n : n \geq 0\}$ where $X_n = \sum_{i=1}^{n} Z_i$, $n = 1, 2, \ldots$, and Z_1, $Z_2, \ldots$ are independent, identically distributed random variables with $P(Z_n = 1) = p$ and $P(Z_n = -1) = q$, for all n, is called a **random walk $X(n)$ in one dimension.**

$$X_n = X_{n-1} + Z_n \qquad n = 1, 2, \ldots$$

$$p_{i,\, i+1} = P(X_{n+1} = i + 1 \mid X_n = i) = p$$

$$p_{i,\, i-1} = P(X_{n+1} = i - 1 \mid X_n = i) = q$$

$$p_{i,i} = P(X_{n+1} = i \mid X_n = i) = 0 \qquad i \neq 0, 3$$

$$p_{0,\,0} = P(X_{n+1} = 0 \mid X_n = 0) = 1$$

$$p_{3,\,3} = P(X_{n+1} = 3 \mid X_n = 3) = 1$$

$$P = \begin{matrix} \\ 0 \\ 1 \\ 2 \\ 3 \end{matrix} \overset{\begin{matrix} 0 & 1 & 2 & 3 \end{matrix}}{\begin{pmatrix} 1 & 0 & 0 & 0 \\ q & 0 & p & 0 \\ 0 & q & 0 & p \\ 0 & 0 & 0 & 1 \end{pmatrix}}$$

6.1.8 Example

Suppose that a coin with probability p for a head is tossed indefinitely. Let X_n be the outcome of the nth trial. Let it be k, where k (= 0, 1, 2,...n) denotes there is a run of k successes. $\{X_n\}$ constitutes a Markov chain, with unit-step transition probabilities

$$\begin{aligned} P_{jk} = P(X_n = k \mid X_{n-1} = j) &= p, \quad k = j + 1 \\ &= q, \quad k = 0 \\ &= 0 \quad \text{otherwise} \end{aligned}$$

The transition matrix is given by

$$\begin{array}{r} \\ \\ \text{States of } X_{n-1} \end{array}
\begin{array}{c} \text{States of } X_n \\
\begin{array}{c} \\ 0 \\ 1 \\ 2 \\ 3 \\ k-1 \\ k \\ k+1 \end{array}
\begin{array}{c} 0 \quad 1 \quad 2 \ldots k \; k+1 \ldots \\
\begin{pmatrix} q & p & 0 & 0 & 0 \\ q & 0 & p & 0 & 0 \\ q & 0 & 0 & . & . \\ \vdots & \vdots & \vdots & p & 0 \\ q & . & 0 & 0 & p \\ & & & 0 & 0 \\ . & . & . & & \end{pmatrix} \end{array} \end{array}$$

6.1.9 Example

Consider a two-state Markov chain with the transition probability matrix

$$P = \begin{bmatrix} 1-a & a \\ b & 1-b \end{bmatrix}, \qquad 0 < a < 1, 0 < b < 1$$

Show that the n-step transition probability matrix is P^n is given by

$$P^n = \frac{1}{a+b}\left\{\begin{bmatrix} b & a \\ b & a \end{bmatrix} + (1-a-b)^n \begin{bmatrix} a & -a \\ -b & b \end{bmatrix}\right\}$$

Solution. From matrix theory, the characteristic equation of P is

$$|P - \lambda I| = \begin{vmatrix} (1-a)-\lambda & a \\ b & (1-b)-\lambda \end{vmatrix} = 0, \; I = \begin{bmatrix} 1 & 0 \\ 0 & 1 \end{bmatrix}$$

unit matrix

$\Rightarrow \quad ((1 - a) - \lambda)\,((1 - b) - \lambda) - ab = 0$

$\Rightarrow \quad \lambda^2 - \lambda\,[(1 - a) + (1 - b)] + (1 - a)\,(1 - b) - ab = 0$

$\Rightarrow \quad (\lambda - 1)\,(\lambda - 1 + a + b) = 0$

$\Rightarrow \quad \lambda = 1 \text{ or } \lambda = 1 - a - b$

Hence the eigen values $\lambda_1 = 1$ and $\lambda_2 = 1 - a - b$. Then, using the spectral decomposition method, P^n can be expressed as

$$P^n = \lambda_1^n\, E_1 + \lambda_2^n\, E_2$$

where

$$E_1 = \frac{1}{\lambda_1 - \lambda_2}\,(P - \lambda_1 I)$$

$$E_2 = \frac{1}{\lambda_2 - \lambda_1}\,(P - \lambda_1 I)$$

Substitating $\lambda_1 = 1$ and $\lambda_2 = 1 - a - b$ in the above expressions, we obtain

$$E_1 = \frac{1}{a+b}\begin{bmatrix} b & a \\ b & a \end{bmatrix}$$

$$E_2 = \frac{1}{a+b}\begin{bmatrix} a & -a \\ -b & b \end{bmatrix}$$

Hence

$$P^n = E_1 + (1-a-b)^n E_2$$

$$= \frac{1}{a+b}\left\{\begin{bmatrix} b & a \\ b & a \end{bmatrix} + (1-a-b)^n \begin{bmatrix} a & -a \\ -b & b \end{bmatrix}\right\}$$

when $n \to \infty$, $(1-a-b)^n = 0$, Hence in this case

$$P^n = \frac{1}{a+b}\begin{bmatrix} b & a \\ b & a \end{bmatrix}$$ The state diagram is

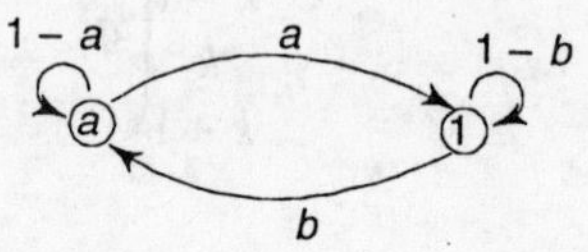

6.1.10 Example

Show that if P is a Markov matrix, then P^n is also a Markov matrix for any positive integer n.

Let
$$P = [p_{ij}] = \begin{bmatrix} p_{11} & p_{12} & \cdots & p_{1m} \\ p_{12} & p_{22} & \cdots & p_{2m} \\ \vdots & \vdots & & \vdots \\ p_{m1} & p_{m2} & \cdots & pmm \end{bmatrix}$$

Since $\sum_{j=1}^{m} p_{ij} = 1,\ i = 1, 2, \ldots m$, we can write

$$\begin{bmatrix} p_{11} & p_{12} & \cdots & p_{1m} \\ p_{12} & p_{22} & \cdots & p_{2m} \\ \vdots & \vdots & & \vdots \\ p_{m1} & p_{m2} & \cdots & pmm \end{bmatrix}\begin{bmatrix} 1 \\ 1 \\ \vdots \\ 1 \end{bmatrix} = \begin{bmatrix} 1 \\ 1 \\ \vdots \\ 1 \end{bmatrix}$$

or $$PA = A \tag{6.3}$$
where $$A^T = [1, 1 \ldots. 1]$$

Pre-multiplying (6.3) by P, we obtain

$$P^2A = PA = A$$

which shows that P^2 is a Markov matrix. Repeated pre-multiplication by P gives

$$P^nA = A$$

which shows that P^n is a Markov matrix.

6.1.11 Show that

$$P(n) = P(0)\ P^n \tag{6.4}$$

where
$$P(n) = [p_0(n), p_1(n), p_2(n) \ldots.] \text{ and}$$
$$P(0) = [p_0(0), p_1(0), p_2(0), \ldots.]$$

with $\sum_k p_k(n) = 1$, and $P(n)$ is the state probability vector after n transitions.

We prove by induction. The state of X_0 is i and the state of X_1 is j if there is a transition from i to j. The events $\{X_0 = i, i = 1, 2\ldots\}$ are mutually exclusive and one of them must occur. Hence by the law of total probability

$$P(X_1 = j) = \sum_i P(X_0 = i)\, P(X_1 = j | X_0 = i)$$

$$\text{or } p_j(1) = \sum_i p_i(0)\, p_{ij} \qquad j = 1, 2, \ldots$$

In matrix notation, we write the above as

$$P(1) = P(0)\, P$$

Thus (6.4) is true for $n = 1$. Assume the result is true for $n = k$. Now we prove that the result is true for $n = k + 1$. By law of total probability

$$P(X_{k+1} = j) = \sum_i P(X_k = i)\, P(X_{k+1} = j | X_k = i)$$

or

$$p_j(k + 1) = \sum_i p_i(k)\, p_{ij} \qquad j = 1, 2, \ldots$$

In matrix notation we write it as

$$P(k + 1) = P(k)\, P$$
$$= P(0)\, P^K P = P(0)\, P^{k+1}$$

Hence the result is true for $k + 1$ and hence by induction it is true for all $n \geq 1$.

6.1.12 Example

Verify the transitivity property of the Markov chain; that is, if $i \to j$, $j \to k$, then $i \to k$.

Solution. By definition, the relations $i \to j$, and $j \to k$ imply that there exist integers n and m such that $p_{ij}(n) > 0$ and $p_{jk}(m) > 0$. Then by the Chapman- Kolmogorov equation, We have

$$p_{ik}(n + m) = \sum_r p_{ir}(n)\, p_{rk}(m) \geq p_{ij}(n)\, p_{jk}(m) > 0$$

hence $i \to k$.

Higher Transition Probabilities

Chapman-Kolmogorov Equation

We have so far discussed one-step transition probabilities. That is, the probability of the outcome at the nth step or trial given the outcome at the previous step. p_{jk} denotes the probability of unit-step transition from the state j at a trial to the state k at the next following trial. The n-step transition probability is given by

$$P_{jk}(n) = P(X_{n+m} = k | X_n = j)$$

p_{jk} gives the probability that from the state j at m trial, the state k is reached at $(n + m)^{th}$ trial through n steps. The one-step transition probabilities $p_{jk}(1)$ are denoted by p_{jk}. Consider P

$$P_{jk}(2) = P(X_{n+2} = k | X_n = j)$$
$$= \sum_r P(X_{n+2} = k,\, X_{n+1} = r | X_n = j)$$
$$= \sum_r P(X_{n+2} = k | X_{n+1} = r)\, P(X_{n+1} = r | X_n = j) = \sum_r p_{jr}\, p_{rk} \qquad (6.5)$$

Now

$$P_{jk}(m + 1) = \sum_r P\{ X_{n+m+1} = k | X_n = j\}$$
$$= \sum_r P\{X_{n+m+1} = k | X_{n+m} = r\}\, P\{X_{n+m} = r | X_n = j\} = \sum_r p_{jr}(m)\, p_{rk}(j)$$

Similarly we have

$$p_{jk}(m+2) = \sum_r p_{rk}(2)\, p_{jr}(m)$$

In general, we have

$$p_{jk}(m+n) = \sum_r p_{jr}(m)\, p_{rk}(n) \tag{6.6}$$

This equation is a special case of Chapman-Kolmogorov equation which is used to calculate the n-step transition probabilities. In matrix notation, we have

$$P^{(m+n)} = P^m \cdot P^n = P^n \cdot P^m$$

6.1.13 Example

Consider the Markov chain of Example 6.5. The two-step transition matrix is given by

$$\begin{pmatrix} \frac{3}{4} & \frac{1}{4} & 0 \\ \frac{1}{4} & \frac{1}{2} & \frac{1}{4} \\ 0 & \frac{3}{4} & \frac{1}{4} \end{pmatrix}\begin{pmatrix} \frac{3}{4} & \frac{1}{4} & 0 \\ \frac{1}{4} & \frac{1}{2} & \frac{1}{4} \\ 0 & \frac{3}{4} & \frac{1}{4} \end{pmatrix} = \begin{pmatrix} \frac{5}{8} & \frac{5}{16} & \frac{1}{16} \\ \frac{5}{16} & \frac{1}{2} & \frac{3}{16} \\ \frac{3}{16} & \frac{9}{16} & \frac{1}{4} \end{pmatrix}$$

Hence

$$P_{01}(2) = P\{X_{n+2} = 1 | X_n = 0\} = \frac{5}{16} \text{ for } n \geq 0.$$

Thus $$P(X_2 = 1 | X_0 = 0) = \sum_{r=0}^{2} P(X_2 = 1 | X_1 = r)\, P(X_1 = r | X_0 = 0)$$

$$= \sum_{r=0}^{2} p_{r1}\, p_{0r} = [p_{01}\, p_{00} + p_{11}\, p_{01} + p_{21}\, p_{02}]$$

$$= \left[\frac{1}{4} \times \frac{3}{4} + \frac{1}{2} \times \frac{1}{4} + \frac{3}{4} \times 0\right] = \frac{5}{16} = p_{01}(2).$$

By multiplication law of probability, we have

$$P\{X_2 = 1, X_0 = 0\} = P\{X_2 = 1|_{X_0=0}\}\, P\{X_0 = 0\}$$

$$= \frac{5}{6} \times \frac{1}{3} = \frac{5}{48}$$

6.1.14 Example

Two-state Markov Chain. Suppose that the probability of a dry day (state 0) following a rainy day (state 1) is $\frac{1}{3}$ and that the probability of a rainy day following a dry day is $\frac{1}{2}$. We have a two state Markov chain such that $p_{10} = \frac{1}{3}$, $p_{01} = \frac{1}{2}$ and transition matrix is

$$P = \begin{matrix} & \begin{matrix} 0 & 1 \end{matrix} \\ \begin{matrix} 0 \\ 1 \end{matrix} & \begin{pmatrix} \frac{1}{2} & \frac{1}{2} \\ \frac{1}{3} & \frac{2}{3} \end{pmatrix} \end{matrix}$$

Then

$$P^2 = \begin{pmatrix} \frac{5}{12} & \frac{7}{12} \\ \frac{7}{18} & \frac{11}{18} \end{pmatrix}$$

6.2 STATE CLASSIFICATION AND LIMITING DISTRIBUTION

The states j, $j = 0, 1, 2\ldots$of a Markov chain $\{X_n, n \geq 0\}$ can be classified as follows.

6.2.1 Definition

If the probability $p_{jk}(n)$ is non-zero for some $n \geq 1$, then we say that the state k can be reached from the state j. If every state can be reached from every other state (in any number of transitions) then the chain is said to be **irreducible**. The transition matrix is irreducible. State j is said to be **accessible** from state i, if for some $n \geq 0$, $p_{ij}(n) > 0$ and we write $i \to j$.

6.2.2 Definition

If C is a set of states such that no states outside of C can be reached from any state of C, then the set of states C is said to be **closed**. If C is a closed set and suppose $j \in C$ and $k \notin C$, then $p_{jk} = 0$ and hence $p_{jk}(n) = 0$ for $n \geq 1$.

6.2.3 Definition

If a closed set contains only one state i, then the state is called an **absorbing state**. That is, a state i is said to be absorbing if $p_{ii} = 1$ and $p_{ik} = 0$ for $i \neq k$.

In Example 6.1.6, the states 0 and 4 are absorbing states.

Suppose that a system starts with the state j. Let $f_{jk}(n)$ be the probability (first passage probability) that it reaches the state k, starting from state j for the *first* time at the nth step and let $p_{jk}(n)$ be the probability that it reaches the state k (not necessarily for the first time) after n steps. The probability that starting with j, the state k is reached for the first time at the rth step and again after that at the $(n - r)^{\text{th}}$ step is given by $f_{jk}(r)\, p_{kk}(n - r)$, $n \geq 1$

with

$$P_{kk}(0) = 1,\ f_{jk}(0) = 0,\ f_{jk}(1) = p_{jk}$$

Let F_{jk} denote the probability of ever visiting state k, starting from state j. Then:

$$F_{jk} = \sum_{n=1}^{\infty} f_{jk}(n), \text{ where } f_{jk}(n) = \sum_{i \neq k} p_{ji}\, f_{ik}(n-1) \qquad n = 2, 3, \ldots$$

The mean (first passage) time from the state j to the state k is given by

$$\mu_{jk} = \sum_{n=1}^{\infty} n\, f_{jk}(n)$$

$$\mu_{jj} = \sum_{n=1}^{\infty} n\, f_{jj}(n) \text{ is known as the mean recurrence time for } j.$$

6.2.4 Definition

A state i is said to be **recurrent** if and only if starting from state i, the process eventually returns to state i with probability 1. That is, if $F_{ii} = 1$ (i.e. return to the state i is certain). A state i is said to be **transient** (or non-recurrent) if and only if there is a positive probability that the process will not return to this state. That is the state i is transient if $F_{ii} < 1$.

6.2.5 Definition

A recurrent state is said to be **null**, if $\mu_{ii} = \infty$, that is, if the mean recurrence time is infinite. A recurrent state is said to be **non-null** if $\mu_{ii} < \infty$.

6.2.6 Definition

For a recurrent state i, $p_{ii}(n) > 0$ for some $n \geq 1$. Define the **period** of state i, by $t > 1$ which is the greatest common divisor of the set of positive integers n such that $p_{ii}(n) > 0$.

6.2.7 Definition

A *recurrent state i* is said to be **aperiodic** if no such $t(>1)$ exists. e.g $t = 1$ corresponds to the aperiodic case.

A recurrent state i is said to be **periodic** if $t > 1$.

6.2.8 Example

Consider a cascade of binary channels that are so noisy that the digit transmitted is always complemented. In other words, take $a = b = 1$ in

$$P = \begin{bmatrix} 1-a & a \\ b & 1-b \end{bmatrix}, 0 \leq a, b \leq 1.$$

The matrix becomes

$$P = \begin{bmatrix} 0 & 1 \\ 1 & 0 \end{bmatrix}$$

Example 6.1.9 is not applicable here. The state diagram is given below. Starting in state 0 (or 1), we return to state 0(or 1) after an even number of steps. Therefore the time between visits to a given state exhibits a periodic behaviour. Hence the given chain is a periodic Markov chain with period 2. But the Markov chain is irreducible.

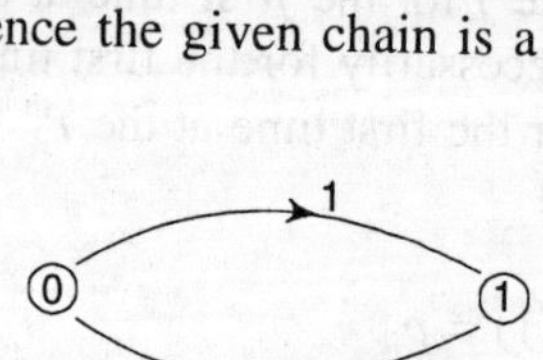

Note: All states of an irreducible Markov chain are of the same type. Thus if one state of an irreducible chain is aperiodic, then all the states are aperiodic. Such a Markov chain having all state aperiodic is said to be aperiodic.

6.2.9 Definition

A recurrent non-null and aperiodic state of a Markov chain is said to be **ergodic**. A Markov chain all of whose states are ergodic is said to be an **ergodic chain**. Ergodic chains are nothing but the irreducible chains.

If one state of an irreducible chain is periodic, then all states are *periodic* and have the same period. If one state is transient, then all states are transient. If one state is recurrent, then all are recurrent.

Let $p_{ij}(n)$ be the n-step transition probabilities which become independent of both n and i, if $n \to \infty$. That is, all rows of matrix P^n converge toward a common limit. Now by total probability theorem, we have

$$P_j(n) = P(X_n = j) = \sum_i p_i(0)\, p_{ij}(n)$$

Since $p_{ij}(n)$ is not depending on n and i, $p_j(n)$ approaches a constant as $n \to \infty$. This constant is independent of the initial distribution. Denote the limiting state probabilities by

$$v_j = \lim_{n\to\infty} p_j(n) = \lim_{n\to\infty} p_{ij}(n) \qquad j = 0, 1, \ldots$$

Suppose $P = \begin{array}{c} \\ 1 \\ 2 \end{array}\!\!\begin{array}{c} \begin{array}{cc} 1 & 2 \end{array} \\ \begin{pmatrix} 0 & 1 \\ 1 & 0 \end{pmatrix} \end{array}$. Since the process alternates between states 1 and 2, $p_{11}(n) = 0$ for $n = 1, 3, 5 \ldots$ and $p_{11}(n) = 1$ for $n = 2, 4, 6 \ldots$. Therefore $\lim_{n\to\infty} p_{11}(n)$ does not exist.

6.2.10 Definition

Stationary Distribution. Consider a Markov chain with transition probability p_{jk}. A probability distribution $\{v_j\}$ is called **stationary** (or invariant) for the given chain if

$$v_k = \sum_j v_j\, p_{jk} \tag{6.7}$$

such that $v_j \geq 0$, $\sum_j v_j = 1$.

If the numbers v_j, $j \in I$, are such that $\sum_{j\in I} v_j = 1$, then v_j's are said to form a **steady-state distribution.**

Again
$$v_k = \sum_j v_j\, p_{jk} = \sum_j \left(\sum_i v_i\, p_{ij} \right) p_{jk} = \sum_i v_i\, p_{ik}(2)$$

In genereal

$$v_k = \sum_i v_i\, p_{ik}(n),\ n \geq 1.$$

Note that the probability v_j is sometimes interpreted as the *long-run proportion* of time, the Markov chain spends in state j.

6.2.11 Definition

A Markov chain is called **regular** if there is a finite positive integer m such that after m time-steps, every state has a non-zero chance of being occupied, no matter what the initial state is. Let $A > 0$ denotes that every element a_{ij} of A satisfies the condition $a_{ij} > 0$. Then, for a regular Markov chain with transition probability P, there exists $m > 0$ such that $P^m > 0$.

For a regular homogeneous Markov chain $\lim_{n \to \infty} P^n = \hat{P}$, where $\hat{P}$ is a matrix whose rows are identical and equal to the stationary distribution $\{v_j\}$

Determination of Steady-State Conditions

Every regular ergodic chain will have steady-state condition. The existence of these conditions can be demonstrated by computing P^n for various values of n. For example, consider the following transition matrix:

$$P = \begin{bmatrix} 0.4 & 0.4 & 0.2 \\ 0.3 & 0.5 & 0.2 \\ 0.3 & 0.3 & 0.4 \end{bmatrix}$$

We compute P^n, for $n = 1, 2, \ldots 8$. We see that

$$P^3 = \begin{bmatrix} 0.334 & 0.418 & 0.248 \\ 0.333 & 0.419 & 0.248 \\ 0.333 & 0.411 & 0.256 \end{bmatrix}$$

$$P^4 = \begin{bmatrix} 0.3334 & 0.4170 & 0.2502 \\ 0.3333 & 0.4155 & 0.2496 \\ 0.3333 & 0.4155 & 0.2552 \end{bmatrix}$$

$$P^8 = \begin{bmatrix} 0.33334 & 0.41610 & 0.25016 \\ 0.33333 & 0.41595 & 0.25102 \\ 0.33333 & 0.41595 & 0.25202 \end{bmatrix}$$

As n becomes larger, the transition matrix tends to a fixed limit and each probability vector v_i^n tends to become equal for all the states. Hence the conditions can be stated as:

(1) For sufficiently large value of n, the probability vector v_i^n becomes equal for all i and does not change for large value of n.

(2) Since $v_i^{n+1} = v_i^n\, P$ and $v_i^{n+1} = v_i$, there exists a vector V^* such that $V^* = V^*\, P$.

The vector V^* contains the probabilities which exist at steady-state conditions. As V^* is a probability vector, we have $\sum_{j=1}^{m} v_j = 1$. From conditon 2, we get $(V_1, V_2, \ldots V_m)\, P = (V_1, V_2, \ldots V_m)$. From this equation we can solve for the m unknowns.

6.2.12 Example

Let $$P = \begin{bmatrix} 0.4 & 0.4 & 0.2 \\ 0.3 & 0.5 & 0.2 \\ 0.3 & 0.3 & 0.4 \end{bmatrix}$$

Let $V^* = (V_1, V_2, V_3)$. We know that $(V_1, V_2, V_3)\, P = (V_1, V_2, V_3)$. This leads to four equations

$$V_1 + V_2 + V_3 = 1$$
$$0.4\, V_1 + 0.3\, V_2 + 0.3\, V_3 = V_1$$
$$0.4\, V_1 + 0.5\, V_2 + 0.3\, V_3 = V_2$$
$$0.2\, V_1 + 0.2\, V_2 + 0.4\, V_3 = V_3$$

consider any three equations and solving, we get

$$V_1 = 0.333,\ V_2 = 0.417,\ V_3 = 0.250$$

Absorption Probabilities

Consider a Markov chain $\{X(n);\ n = 0, 1, 2...\}$ with finite state space $S = \{1, 2,...N\}$ and transition probability matrix P. Let $A = \{1, 2,...m\}$ be the set of absorbing states and $B = \{m + 1,...N\}$ be a set of non-absorbing states. Then P can be expressed as

$$P = \begin{bmatrix} 1 & 0 & \cdots & 0 & \cdots & 0 & \cdots & 0 \\ 0 & 1 & \cdots & 0 & \cdots & 0 & \cdots & 0 \\ \cdots & \cdots & \cdots & \cdots & \cdots & \cdots & \cdots & \cdots \\ 0 & 0 & \cdots & 1 & \cdots & 0 & \cdots & 0 \\ p_{m+1,1} & \cdots & p_{m+1,m} & \cdots & \cdots & \cdots & \cdots & p_{m+1,N} \\ \vdots & \cdots & \cdots & \cdots & \cdots & \cdots & \cdots & \cdots \\ p_{N,1} & \cdots & p_{N,m} & \cdots & \cdots & & & p_{N,N} \end{bmatrix} = \begin{bmatrix} I & O \\ R & Q \end{bmatrix} \quad (6.8)$$

where I is an $m \times m$ identily matrix, 0 is $m \times (N - m)$ zero matrix

$$R = \begin{bmatrix} p_{m+1,1} & p_{m+1,2} & \cdots & p_{m+1,m} \\ p_{m+2,1} & p_{m+2,2} & \cdots & p_{m+2,m} \\ \cdots & \cdots & \cdots & \cdots \\ p_{N,1} & p_{N,2} & \cdots & p_{N,N} \end{bmatrix} \quad (6.9)$$

$$Q = \begin{bmatrix} p_{m+1,m+1} & \cdots & \cdots & p_{m+1,N} \\ p_{m+2,m+1} & \cdots & \cdots & p_{m+2,N} \\ p_{N,m+1} & \cdots & \cdots & p_{N,N} \end{bmatrix} \quad (6.10)$$

Note that the elements of R are the one-step transition probabilities from non absorbing to absorbing states, and the elements of Q are the one-step transition probabilities among the non-absorbing states.

The matrix $(I - Q)^{-1}$ is known as the **fundamental matrix** of the Markov chain $X(n)$. Also note that $U = (I - Q)^{-1} R$ is an $(N - m) \times m$ matrix and its elements are the absorption probabilities for the various absorbing states, where $U = (u_{kj})$, $u_{kj} = P\{X_n = j\ (\in A)/X_0 = k\ (\in B)\}$.

Absorbing Markov Chains

A Markov chain is absorbing if

(i) it has atleast one absorbing state

(ii) it is possible to go from every non-absorbing state to atleast one absorbing state (not necessarily in one step).

6.2.13 Example

Check whether the given matrix is from absorbing Markov chain.

$$\begin{array}{c} \\ S_1 \\ S_2 \\ S_3 \\ S_4 \end{array} \begin{array}{c} \begin{array}{cccc} S_1 & S_2 & S_3 & S_4 \end{array} \\ \begin{bmatrix} 1 & 0 & 0 & 0 \\ \frac{1}{2} & 0 & 0 & \frac{1}{2} \\ \frac{1}{4} & \frac{1}{4} & \frac{1}{4} & \frac{1}{4} \\ 0 & 0 & 0 & 1 \end{bmatrix} \end{array}$$

Solution. First we draw the state transition diagram (given below).

Here S_1 and S_4 are absorbing states and these can be reached from S_2 and S_3. Hence this chain is from an absorbing Markov chain.

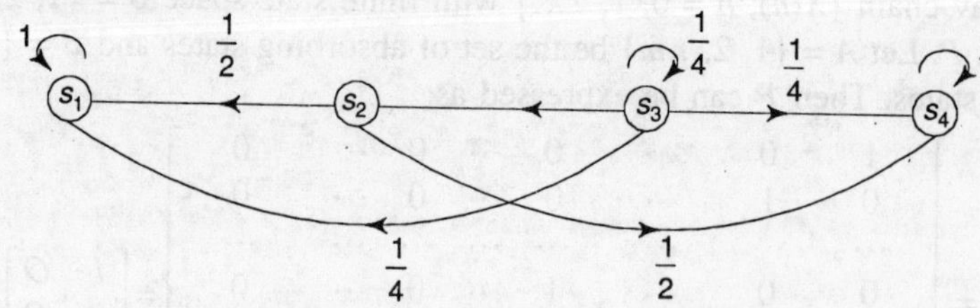

6.2.14 Example

Consider Example 6.1.7. Find the probabilities of absorption into states 0 and 3.

Solution. The transition probability matrix is

$$P = \begin{array}{c} \\ 0 \\ 1 \\ 2 \\ 3 \end{array}\begin{array}{c} \begin{array}{cccc} 0 & 1 & 2 & 3 \end{array} \\ \begin{bmatrix} 1 & 0 & 0 & 0 \\ q & 0 & p & 0 \\ 0 & q & 0 & p \\ 0 & 0 & 0 & 1 \end{bmatrix} \end{array}$$

Rearrange the transition probability matrix P as (6.8)

$$P = \begin{array}{c} \\ 0 \\ 3 \\ 1 \\ 2 \end{array}\begin{array}{c} \begin{array}{cccc} 0 & 3 & 1 & 2 \end{array} \\ \begin{bmatrix} 1 & 0 & 0 & 0 \\ 0 & 1 & 0 & 0 \\ q & 0 & 0 & p \\ 0 & p & q & 0 \end{bmatrix} \end{array} = \begin{bmatrix} I & 0 \\ R & Q \end{bmatrix}$$

where
$$I = \begin{bmatrix} 1 & 0 \\ 0 & 1 \end{bmatrix}, 0 = \begin{bmatrix} 0 & 0 \\ 0 & 0 \end{bmatrix}$$

$$R = \begin{bmatrix} q & 0 \\ 0 & p \end{bmatrix} = \begin{bmatrix} p_{10} & p_{13} \\ p_{20} & p_{23} \end{bmatrix}$$

$$Q = \begin{bmatrix} 0 & p \\ q & 0 \end{bmatrix} = \begin{bmatrix} p_{11} & p_{12} \\ p_{21} & p_{22} \end{bmatrix}$$

Then
$$I - Q = \begin{bmatrix} 1 & 0 \\ 0 & 1 \end{bmatrix} - \begin{bmatrix} 0 & p \\ q & 0 \end{bmatrix} = \begin{bmatrix} 1 & -p \\ -q & 1 \end{bmatrix}$$

and
$$(I - Q)^{-1} = \frac{1}{1 - pq}\begin{bmatrix} 1 & p \\ q & 1 \end{bmatrix}$$

$$\left[\text{since inverse of } A = \begin{pmatrix} a & b \\ c & d \end{pmatrix} \text{ is } A^{-1} = \frac{1}{ad - bc}\begin{pmatrix} d & -b \\ -c & a \end{pmatrix}\right]$$

We know

$$U = (I - Q)^{-1} R \text{ is the probability of absorption}$$

$$U = \begin{bmatrix} u_{10} & u_{13} \\ u_{20} & u_{23} \end{bmatrix} = \frac{1}{(1-pq)} \begin{bmatrix} 1 & p \\ q & 1 \end{bmatrix} \begin{bmatrix} q & 0 \\ p & q \end{bmatrix} = \frac{1}{1-pq} \begin{bmatrix} q & p^2 \\ q^2 & p \end{bmatrix}$$

Thus, the probabilities of absorption from state 1 into state 0 is $u_{10} = \dfrac{q}{1-pq}$; and the probability of absorption from state 2 into 0 is $u_{20} = \dfrac{q^2}{1-pq}$; the probability of absorption from state 1 into state 3 is $\dfrac{p^2}{1-pq}$; the probability of absorption from state 2 into state 3 is $u_{23} = \dfrac{p}{1-pq}$.

Note that $u_{10} + u_{13} = 1$ and $u_{20} + u_{23} = 1$.

Note: Let T_R denote the total time units (or steps) to absorption from state k. Let $T = [T_{m+1}, T_{m+2}, \ldots T_N]$. Then the expected time (or steps) to absorption from state k is

$$E(T_k) = \sum_{i=m+1}^{N} \phi_{ki},\ k = m+1, \ldots, N \tag{6.11}$$

where ϕ_{ki} is the $(k, i)^{\text{th}}$ element of the fundamental matrix $(1 - Q)^{-1}$.

6.2.15 Example

Consider Example 6.1.7. Find the expected time to absorption when $X_0 = 1$ and when $X_0 = 2$.

Solution. $$(1 - Q)^{-1} = \begin{bmatrix} \phi_{11} & \phi_{12} \\ \phi_{21} & \phi_{22} \end{bmatrix} = \frac{1}{1-pq} \begin{bmatrix} 1 & p \\ q & 1 \end{bmatrix}$$

Let T_i be the time to absorption when $X_0 = i$. Then by (6.2.5), we have

$$E(T_1) = \sum_{i=1}^{2} \phi_{1i} = \phi_{11} + \phi_{12} = \frac{1}{1-pq}[1 + p]$$

$$E(T_2) = \sum_{i=1}^{2} \phi_{2i} = \phi_{21} + \phi_{22} = \frac{1}{1-pq}[q + 1]$$

6.2.16 Example

Consider a Markov chain with two states and transition probability matrix

$$P = \begin{bmatrix} \frac{3}{4} & \frac{1}{4} \\ \frac{1}{2} & \frac{1}{2} \end{bmatrix}$$

(a) Find the stationary distribution of the chain.

(b) Find $\lim_{n \to \infty} P^n$.

Solution.

(a) By (6.7), we have, $v_1 + v_2 = 1$ and

$$v_1 = v_1 p_{11} + v_2 p_{21} = v_1\left(\frac{3}{4}\right) + v_2\left(\frac{1}{2}\right)$$

and

$$v_2 = v_1 p_{12} + v_2 p_{22} = v_1\left(\frac{1}{4}\right) + v_2\left(\frac{1}{2}\right)$$

Solving for v_1 and v_2, we get

$$v_1 = \left(\frac{2}{3}\right) \text{ and } v_2 = \left(\frac{1}{3}\right)$$

(b) Since the Markov chain is regular, $\underset{n\to\infty}{\text{Lt}}\ P^n = \hat{P}$, where $\hat{P}$ is a matrix whose rows are identical and equal to the stationary distribution $\{v_j\}$. Hence

$$\underset{n\to\infty}{\text{Lt}}\ P^n = \underset{n\to\infty}{\text{Lt}}\begin{bmatrix} \frac{3}{4} & \frac{1}{4} \\ \frac{1}{2} & \frac{1}{2} \end{bmatrix}^n = \begin{bmatrix} v_1 & v_2 \\ v_1 & v_2 \end{bmatrix} = \begin{bmatrix} \frac{2}{3} & \frac{1}{3} \\ \frac{2}{3} & \frac{1}{3} \end{bmatrix}$$

6.2.17 Example

Consider a Markov chain with two states and transition probability matrix $P = \begin{bmatrix} 0 & 1 \\ 1 & 0 \end{bmatrix}$. Find the stationary distribution of the chain.

Solution. By (6.2.1), we have $v_1 + v_2 = 1$, $v_1 = v_1 p_{11} + v_2 p_{21} = v_2$ and $v_2 = v_1 p_{12} + v_2 p_{22} = v_1$; hence $v_1 = \frac{1}{2}$ and $v_2 = \frac{1}{2}$.

Next, we state a lemma without proof which are used in a sequel.

6.2.18 Lemma

Let $\{f_n\}$ be a sequence such that $f_o = 0, f_n \geq 0$ and $\sum_{n=1}^{\infty} f_n = 1$. Let t be the g.c.d of those n for which $f_n > 0$. Let $\{u_n\}$ be another sequence such that $u_0 = 1$, $u_n = \sum_{r=1}^{n} f_r u_{n-r}$ $(n \geq 1)$. Then $\underset{n\to\infty}{\text{Lt}}\ u_n = \frac{t}{\mu}$, *where* $\mu = \sum_{n=1}^{\infty} n f_r$. Whenever N is not divisible by t, $\underset{n\to\infty}{\text{Lt}}\ u_N = 0$.

6.2.19 Theorem

If a state j is recurrent, then as $n \to \infty$,

$$p_{jj}(nt) \to \frac{t}{\mu_{jj}} \tag{6.12}$$

when j is periodic with period t.

Proof: Let the state j be recurrent. Then $\mu_{jj} = \sum_n n\, f_{jj}(n)$ is defined. Since $p_{jj}(n) = \sum_{r=1}^{n} f_{jj}(r)\, p_{jj}(n-r)$, we replace f_{jj} by f_n, $p_{jj}(n)$ by u_n and μ_{jj} by μ in Lemma 6.2.18. Then we get

$$\operatorname{Lt}_{n\to\infty} p_{jj}(nt) = \frac{t}{\mu_{jj}}$$

6.2.20 Theorem

If a state j is recurrent and aperiodic, then

$$p_{jj}(n) \to \frac{1}{\mu_{jj}} \text{ as } n \to \infty \tag{6.13}$$

Proof: When state is aperiodic, $t = 1$, then (6.12) becomes $p_{jj}(n) \to \dfrac{1}{\mu_{jj}}$ as $n \to \infty$

6.2.21 Theorem

If a state j is recurrent null (whether periodic or aperiodic), then

$$p_{jj}(n) \to 0 \text{ as } n \to \infty \tag{6.14}$$

Proof. In the case of j as recurrent null, $\mu_{jj} = \infty$. So (6.12) becomes

$$p_{jj}(n) \to 0 \text{ as } n \to \infty$$

6.2.22 Theorem

It state k is either transient or recurrent null, then for every j,

$$p_{jk}(n) \to 0 \qquad \text{as } n \to \infty$$

Proof: We know $p_{jk}(n) = \sum_{r=1}^{n} f_{jk}(r)\, p_{kk}(n-r)$

Let $n > m$ then

$$\begin{aligned} p_{jk}(n) &= \sum_{r=1}^{m} f_{jk}(r)\, p_{kk}(n-r) + \sum_{r=m+1}^{n} f_{jk}(r)\, p_{kk}(n-r) \\ &\le \sum_{r=1}^{m} f_{jk}(r)\, p_{kk}(n-r) + \sum_{r=m+1}^{n} f_{jk}(r) \end{aligned} \tag{6.15}$$

We know $\sum_{m=1}^{\infty} f_{jk}(m) < \infty$ and $\sum_{r=m+1}^{n} f_{jk}(r) \to 0$ and $p_{kk}(n-r) \to 0$ as $n \to \infty$, we get

$$P_{jk}(n) \to 0 \text{ as } n \to \infty.$$

6.2.23 Theorem

If state k is aperiodic, recurrent non-null, then

$$p_{jk}(n) \to \frac{F_{jk}}{\mu_{kk}} \text{ as } n \to \infty$$

Proof. From (6.15), we have

$$p_{jk}(n) = \sum_{r=1}^{m} f_{jk}(r)\, p_{kk}(n-r) \le \sum_{r=m+1}^{n} f_{jk}(r) \tag{6.16}$$

Since j is aperiodic recurrent and non-null, then by Theorem 6.2.20,

$$P_{kk}(n-r) \to \frac{1}{\mu_{kk}} \text{ as } n \to \infty.$$

From (6.16), we get as $n, m \to \infty$,

$$P_{jk}(n) - \frac{\sum_{r=1}^{m} f_{jk}(r)}{\mu_{kk}} \to 0$$

$$\Rightarrow \qquad p_{jk}(n) \to \frac{F_{jk}}{\mu_{kk}} \text{ when } n \to \infty$$

6.2.24 Example

Consider the Markov chain with transition probability matrix

$$P = \begin{matrix} & \begin{matrix} 0 & 1 & 2 \end{matrix} \\ \begin{matrix} 0 \\ 1 \\ 2 \end{matrix} & \begin{pmatrix} 0 & 1 & 0 \\ \frac{1}{2} & 0 & \frac{1}{2} \\ 0 & 1 & 0 \end{pmatrix} \end{matrix}$$

The chain is irreducible as the matrix is so.

$$P^2 = \begin{pmatrix} \frac{1}{2} & 0 & \frac{1}{2} \\ 0 & 1 & 0 \\ \frac{1}{2} & 0 & \frac{1}{2} \end{pmatrix}$$

$$P^3 = P$$

In general

$P^{2n} = P^2$, $P^{2n+1} = P$, so that $p_{ii}(2n) > 0$, $p_{ii}(2n+1) = 0$ for each i.

The states are periodic with period 2.

Also we find that

$$f_{11} = 0, f_{11}(2) = 1, \text{ so that } F_{11} = \sum_n f_{11}(n) = 1.$$

That is, state 1 is recurrent, and hence other states are also recurrent. Now

$$\mu_{11} = \Sigma n f_{11}(n) = 1 \times 0 + 2 \times 1 = 2$$

That is state 1 is non-null. Thus the states of the chain are periodic, recurrent non-null. Further

$$p_{11}(2n) \to \frac{t}{\mu_{11}} = \frac{2}{2} = 1 \text{ for all } n.$$

Note: Another way of determining $p_{jj}(n)$ or P^n is by means of spectral theory of matrices. Find the eigen values t_i, $i = 1, 2, \ldots m$ of the matrix P. Assume all are distinct and are different from zero. Let $X = (X_1, X_2, \ldots X_m)$ be the associated matrix of eigen vector. Let $D = (d_{ij})$ be the diagonal matrix such that $d_{ii} = t_i$, $d_{ij} = 0$, $j \neq i$. Then the spectral theorem states that $P = XDX^{-1}$ and $P^n = XD^nX^{-1}$.

Next we state a theorem which will be needed to prove Theorem 6.2.26.

6.2.25 Theorem

If the state j is recurrent, then for every state k that can be reached from state j, $F_{jk} = 1$.

6.2.26 Theorem

For an irreducible ergodic chain, the limits $v_k = \lim_{n\to\infty} p_{jk}(n)$ exist and are independent of the initial state j. The limits v_k are such that $v_k \geq 0$, $\Sigma v_k = 1$. Furthermore, the limiting probability distribution $\{v_k\}$ is identical with the stationary distribution for the given chain, so that

$$v_k = \sum_j v_j p_{jk}, \qquad \Sigma v_k = 1 \tag{6.17}$$

Proof. Since the states are ergodic, i.e. aperiodic, recurrent non-null, $\lim_{n\to\infty} p_{jk}(n)$ exists and is equal to $\frac{F_{jk}}{\mu_{kk}}$. By Theorem 6.2.25, we find $F_{jk} = 1$, so that

$$v_k = \lim_{n\to\infty} p_{jk}(n) = \frac{1}{\mu_{kk}} > 0$$

and is independent of j, we have

$$p_{jk}(n+m) = \sum_i p_{ji}(n)\, p_{ik}(m)$$

$$v_k = \lim_{n\to\infty} p_{jk}(n+m)$$

$$= \lim_{n\to\infty} p_{ji}(n)\, p_{ik}(m)$$

$$\geq \sum_i \left\{\lim_{n\to\infty} p_{ji}(n)\right\} p_{ik}(m) \text{ by Fatou's lemma.}$$

$$= \sum_i v_i\, p_{ik}(m) \tag{6.18}$$

Suppose, if possible that $v_k > \sum_i v_i p_{ik}(m)$; then

$$\sum_k v_k > \sum_k \left(\sum_i v_i\, p_{ik}(m)\right) = \sum_i \left(\sum_k v_i\, p_{ik}(m)\right) = \sum_i v_i$$

which is not possible. Hence

$$v_k = \sum_i v_i\, p_{ik}(m), \text{ for all } m \geq 1.$$

In particular, when $m = 1$

$$v_k = \sum_i v_i\, p_{ik}$$

For large m, we have

$$v_k = \sum_i v_i \cdot v_k = \left(\sum_i v_i\right) v_k$$

implying $\quad \Sigma v_i = 1$

Note: Converse of the above theorem is also true.

6.2.27 Example

Consider the two-state Markov chain

$$P = \begin{pmatrix} 1-a & a \\ b & 1-b \end{pmatrix} \qquad 0 < a, b < 1.$$

The chain is irreducible and consists of two recurrent non-null states. Equation (6.17) gives

$$v_1 + v_2 = 1$$
$$v_1 = v_1 p_{11} + v_2 p_{21} = v_1 (1 - a) + v_2 b$$
$$v_2 = v_1 p_{12} + v_2 p_{22} = v_1 a + v_2 (1 - b)$$

solving the above equations, we get

$$v_1 a - v_2 b = 0$$

and $$v_1 + v_2 = 1$$

which implies $$v_1 = \frac{b}{a+b}, \ v_2 = \frac{a}{a+b}$$

6.2.28 Distribution of Time between State Changes

Assume that the state at the nth step is $X_n = i$. The probability that the next state $X_{n+1} = j$ should depend only upon the current state i and not upon the time the chain has spent in the current state i. Let the random variable T_i denote the time the Markov chain spends in state i during a single visit to state i. It follows that the distribution of T_i should be memoryless for $\{X_n, n = 0, 1, 2,\ldots\}$ to form a homogeneous Markov chain.

Suppose the chain has just entered the state i at the nth step and it will remain at the step with probability p_{ii} and it will leave the state at the next step with probability $\sum_{j \neq i} p_{ij} = 1 - p_{ii}$. Now, if the next state is also i that is $X_{n+1} = i$, then the same two choices are available at the next step. Since $\{X_n\}$ is Markov chain, the probability of events at $(n + 1)^{\text{st}}$ step is independent of probability of events at n^{th} step. This reduces to a sequences of Bernoulli trials with probability of success $1 - p_{ii}$, where success is defined to be event that the chain leaves state i.

The event $T_i = n$ corresponds to n trials upto and including the first success and is given by geometric distribution as

$$P(T_i = n) = (1 - p_{ii})\, p_{ii}^{-1}$$

and $$E(T_i) = \frac{1}{1 - p_{ii}} \quad i \in I, \ \text{Var}\,(T_i) = \frac{p_{ii}}{1 - p_{ii}^2} \quad i \in I$$

$E(T_i)$ denote the expected number of steps the chain spends in state i, per visit to state i.

Markov chains with continuous state space.

6.2.29 Definition

If for all m, and for all possible values of $\{X_m\}$ in $(-\infty, \infty)$

$$P\{X_{m+1} \le x / X_m = y, X_{m-1} = y_1, \ldots X_0 = y_m\} = P\{X_{m+1} \le x |_{X_m} = y\} \tag{6.19}$$

Then $\{X_m\}$ is said to be a Markov chain with continuous state space. If (6.19) is independent of m, then the chain is homogeneous. The n step transition probability distribution function is defined by

$$P\{X_{m+n} \le x | X_m = y, \ldots X_0 = y_m\}$$
$$= P\{X_{m+n} \le x |_{X_m} = y\}$$
$$= P_n(y; x)$$

Denote $P_1(y, x) = P\{X_{m+1} \le x |_{X_m} = y\} = P\{X_{n+1} \le x |_{X_m} = y\}$ by simply $P(y, x)$. Let $P_n(x) = P\{X_n \le x\}$. The transition distribution function $P_n(y; x)$ and the initial distribution $P\{X_0 \le x\} = P_0(x)$ can uniquely determine $P_n(x)$. Then Chapman-Kolmogorov equation (6.6) takes the form

$$P_{n+m}(y; x) = \int_{-\infty}^{\infty} P_n(y, z)\, P_m(z, x) dz,\; m, n \ge 0 \tag{6.20}$$

For $m = 1$, we have

$$P_{n+1}(y; x) = \int P_n(y; z)\, P_1(z; x) dz$$

when $n \to \infty$, $P_n(y; x) \to P(x)$ independent of initial value. The limiting distribution $P(x)$ satisfies the integral equation

$$P(x) = \int P(z)\, P(z, x) dz$$
$$= \int P(z; x)\, d(p(z))$$

6.2.30 Example

Random Walk. Suppose that the steps of the unrestricted random walk are independent identically distributed continuous random variables $Z_1, Z_2, \ldots$ with a pdf $f(Z)$, $-\infty < Z < \infty$. We have $X_n = Z_1 + Z_2 + \ldots + Z_n$, so that $X_{n+1} = X_n + Z_{n+1}$. Clearly $\{X_n\}$ is a Markov chain with a continuous state space. One-step transition function is

$$P(y; x) = P\{X_{n+1} \le x |_{X_n} = y\}$$
$$= P\{X_n + Z_{n+1} \le x |_{X_n} \le y\}$$

hence we get

$$f(x) = \int f_n(Z)\, f(x - Z) dZ$$

where $f_n(Z)$ is the n-fold convolution of $f(Z)$. Here $f(x - Z)$ is the pdf of $X_{n+1} = X_n + Z_{n+1}$ given that $X_n = Z$.

Non-stationary or Non-homogeneous chains

Suppose the transition probabilities p_{ij} are functions of n i..e. $P\{X_n = j | X_{n-1} = I\} = p_{ij}(n)$. Then the probabilistic situation is described by a sequence of transition probability matrices $P_1, P_2, \ldots P_n \ldots$ Now introduce a sequence of matrices called causative matrices $C_1, C_2, \ldots C_n \ldots$ to describe the change in the transition probability matrices, given by

$$P_1C_1 = P_2,\; P_2C_2 = P_3, \ldots\; P_nC_n = P_{n+1} \tag{6.21}$$

From (6.21), we get $C_n = P_n P_{n+1}$ provided the matrix P_n is non-singular.
Assume all the causative matrices C_n are equal.
That is, $C_n = C$ for all n. We then have

$$P_{n+1} = P_1C^n \tag{6.22}$$

In general,

$$P_{n+m} = P_mC^n,\; n = 0, 1, \ldots$$

Further, if the matrices are non-singular, then

$$P_{n+1} = P_n P_{n-1}^{-1} P_n$$

and

$$P_n = \left(P_2 P_1^{-1}\right)^{n-2} P_2$$

$$= P_2 \left(P_2^{-1} P_1^{-1}\right)^{n-2} \qquad n = 2, 3, \ldots$$

If the transition probability matrix from period x to y is denoted by $P(x, y)$ and $P(0, 1)$ by Q, then

$$P(n, n+1) = P_{n+1}$$
$$= QC^n \qquad n = 0, 1, 2\ldots$$

and

$$P(0, n+1) = P(0, 1)\, P(1, 2)\ldots P(n, n+1)$$
$$= (Q)\,(QC)\ldots(QC^n)$$
$$= T_n \text{ (say)}$$

The limiting properties of a constant chain with $C_1 = C_2 = \ldots = C$ depends on the convergence of C^n and T_n as $n \to \infty$.

6.3 CONTINUOUS PARAMETER MARKOV CHAINS

So far we have dealt with discrete parameter Markov chains whose state space is also discrete. Consider $X(t)$ a continuous parameter markov chain with the state space being discrete. Let the states be tabled by the set of positive integers 0, 1, 2,...Further, let $X(t)$ be time-homogeneous. Then the probability of a transition from state i to state j during the time interval $(T, T + t)$ does not depend on the initial time T but depends only on t and on the initial and terminal states i and j. We thus write

$$P\{X(T+t) = j \mid X(T) = i\} = p_{ij}(t)\ ,\ i, j = 0, 1, 2, \ldots,\ t \geq 0$$

with $0 \leq p_{ij}(t) \leq 1$ for each i, j and t,

and

$$\sum_j p_{ij}(t) = 1$$

Let $p_j(t) = P\{X(t) = j\}$ be the state probability at t. Then

$$p_j(t) = P\{X(t) = j\}$$
$$= \sum_i P(X(t) = j \text{ and } X(0) = i)$$
$$= \sum_i P(X(0) = i)\, P(X(t) = j \mid X(0) = i)$$
$$= \sum_i p_i(0) \cdot p_{ij}(t)$$

Denote the transition probability matrix by

$$P(t) = (p_{ij}(t))$$

Waiting time for a change of states

Suppose $X(t)$ is a homogeneous Markov process and that at time $t_0 = 0$ the state of the process $X(t_0) = X(0) = i$ is known. The time taken for a change of state from state i is a random variable, say τ, which is called the waiting time for a change of state from state i, then

$$P\{\tau > s + t \mid X(0) = i\}$$
$$= P\{\tau > s + t \mid X(0) = i,\ \tau > s\}\ P\{\tau > s \mid X(0) = i\}$$
$$= P\{\tau > s + t \mid X(s) = i\}\ P\{\tau > s \mid X(0) = i\}$$

The above is satisfied if and only if

$$P(\tau > t | X(0) = i) = e^{-\lambda t},\ t > 0,\ \ i > 0$$

Thus the waiting time τ has exponential distribution with parameter λ, which is called the *transition density from state i.*

6.3.1 Chapman-Kolmogorov Equations

The transition probability $p_{ij}(t + T)$ is the probability that given at time 0, the state is i, and at time $t + T$, the state is j. But in passing from state i to state j in time $(t + T)$ the process passes through some state k in time t. Thus

$$p_{ij}(t + T) = \sum_k P(X(t + T) = j,\ X(t) = k | X(0) = i)$$

$$= \frac{\sum_k P(X(0)=i,\ X(t)=k,\ X(t+T)=j)}{P(X(0)=i)}$$

$$= \sum_k \frac{(X(0)=i,\ X(t)=k)}{P(X(0)=i)} \cdot \frac{P(X(0)=i,\ X(t)=k,\ X(t+T)=j)}{P(X(0)=i,\ X(t)=k)}$$

$$= \sum_k P(X(t) = k | X(0) = i)\ P(X(t + T) = j) |\ X(0) = i,\ X(t) = k)$$

$$= \sum_k p_{ik}(t) \cdot p_{kj}(T), \tag{6.23}$$

and for all $i, j, t \geq 0$.

As $X(t)$ is a Markov process, we have since

$$P(X(t + T) = j | X(0) = i,\ X(t) = k)$$

$$= P(X(t + T = j | X(t) = k) = p_{kj}(T).$$

Equation (6.23) is called the *Chapman-Kolmogorov Equation*

In matrix notation

$$P(t + T) = P(t)\ P(T)$$

Transition density matrix or *probability rates*

Let $\quad a_{ij} = \left[\frac{d}{dt}(p_{ij}(t))\right]_{t=0},\ i \neq j$

and

$$a_{ii} = \left[\frac{d}{dt}(p_{ii}(t))\right]_{t=0}$$

Then

$$= \lim_{\Delta t \to 0} \frac{p_{ij}(\Delta t) - p_{ij}(0)}{\Delta t}$$

$$= \lim_{\Delta t \to 0} \frac{p_{ij}(\Delta t)}{\Delta t} - \operatorname{lt}_{\Delta t \to 0} \frac{p_{ij}(0)}{\Delta(t)}$$

$$\Rightarrow p_{ij}(\Delta t) = a_{ij}\Delta(t) + O(\Delta(t)),\ i \neq j \tag{6.24}$$

Similarly

$$a_{ii} = \lim_{\Delta t \to 0} \frac{p_{ii}(\Delta t) - p_{ij}(0)}{\Delta t}$$

$$= \lim_{\Delta t \to 0} \frac{p_{ii}(\Delta t) - 1}{\Delta t}$$

$\Rightarrow \qquad p_{ii}(\Delta(t)) = 1 + a_{ii}(\Delta(t)) + O(\Delta(t)) \qquad (6.25)$

From (6.25) and (6.26) it is seen that

$$a_{ij} \geq 0, \text{ for } i \neq j$$
$$a_{ii} < 0, \text{ otherwise}$$

As $\qquad \sum_j p_{ij}(t) = 1$, we get $\sum_j a_{ij} = 0$

The quantities a_{ij} are called **transition densities** and the matrix $A = (a_{ij})$ is called the *transition density matrix* or *rate matrix of the process.*

Properties of rate matrix

- its off-diagonal elements are non negative
- its diagonal elements are negative
- sum of elements of each row is zero
- sum of the off-diagonal elements are equal in magnitude but opposite in sign to the diagonal elements.

6.3.2 Forward and Backward Kolmogorov Equations

Consider the Chapman-Kolmogorov equation

$$P_{ij}(t + T) = \sum_k p_{ik}(t)\, p_{kj}(T)$$

Differentiate with respect to T. Then

$$p'_{ij}(t + T) = \frac{\partial}{\partial T}(p_{ij}(t + T))$$

$$= \sum_k p_{ik}(t).\ \frac{d}{dt} p_{kj}(T)$$

At $T = 0$,

$$p'_{ij}(t) = \sum_k p_{ik}(t) \left[\frac{d}{dt}(p_{kj}(T))\right]_{T=0}$$

$$p'_{ij}(t) = \sum_k p_{ik}(t)\, a_{kj} \qquad (6.26)$$

In matrix notation

$$P'(t) = P(t)\, A \qquad (6.27)$$

Similarly we get

$$\frac{d}{dT}(p_{ij}(T)) = \sum_k a_{kj}\, p_{kj}(T)$$

If we replace T by t, then

$$p'_{ij}(t) = \sum_k a_{ik}\, p_{kj}(t) \tag{6.28}$$

or

$$P'(t) = AP(t) \tag{6.29}$$

Equation (6.26) is called *Forward Kolmogorov equation* and (6.28) is called *Backward Kolmogorov equation*

Next we see how to obtain the solution to $P(t)$. From (6.27), it is easy to see that $P(t) = e^{At}$

where
$$e^{At} = \sum_{n=0}^{\infty} \frac{(At)^2}{n!} = I + \sum_{n=0}^{\infty} \frac{A^n t^n}{n!}$$

Assume the eigen values of A, $d_1, d_2, \ldots d_m$ are all different from zero and distinct. Let 0 be an eigen value of A. By spectral theorem of matrices we have

$$A = H\, D\, H^{-1}$$

where D is the diagonal matrix and N is a non-singular matrix. As $A^n = H\, D^n\, H^{-1}$, we have

$$P(t) = I + \sum_{n=0}^{\infty} \frac{(H\,D\,H^{-1})^n\, t^n}{n!}$$

$$= I + \sum_{n=1}^{\infty} \frac{(H\, D^n\, H^{-1})\, t^n}{n!}$$

$$= H \left\{ I + \sum_{n=1}^{\infty} \frac{D^n\, t^n}{n!} \right\} H^{-1} = H\, e^{Dt}\, H^{-1}$$

with

$$e^{Dt} = \begin{pmatrix} 1 & 0 & \cdots & 0 \\ 0 & e^{d_1 t} & \cdots & 0 \\ \cdots & \cdots & \cdots & \cdots \\ 0 & 0 & \cdots & e^{d_n t} \end{pmatrix}, \text{ where } D = \begin{pmatrix} 0 & 0 & 0 & 0 & \cdots \\ 0 & d_1 & \cdots & 0 & \cdots \\ 0 & 0 & d_2 & \cdots & 0 \\ \cdots & \cdots & \cdots & \cdots & \cdots \\ 0 & 0 & 0 & \cdots & d_m \end{pmatrix}$$

6.3.3 Example

Poisson process. If events occur in accordance with Poisson process $N(t)$ with mean λt, then

$$\begin{aligned} p_{i,\,i+1}(\Delta t) &= P\,\{\text{the process goes to state } i+1 \text{ from state } i \text{ in time } \Delta t\} \\ &= P\{\text{one event occurs in time } \Delta t\} \\ &= P\{N(\Delta t) = 1\} = \lambda \Delta t + O(\Delta t) \end{aligned}$$

$$p_{ii}(\Delta t) = 1 - \lambda\,\Delta t + O(\Delta t)$$

$$p_{ij}(\Delta t) = O(\Delta(t)) \quad j \neq i,\ i+1$$

By comparing with (6.24) and (6.25) we get

$$a_{i,\,i+1} = \lambda,\ a_{i,\,i} = -\lambda,\ a_{i,\,j} = 0 \quad \text{for } j \neq i,\ i+1$$

Hence Kolmogorov forward equation becomes

$$p'_{ii}(t) = -\lambda p_{ii}(t)$$

$$p_{ij}(t) = -\lambda\, p_{ij}(t) + \lambda\, p_{i,\,j-1}(t), \quad j = i+1,\ i+2, \ldots$$

6.3.4 Example

Two-state process. Consider two state Markov processes $\{X(t) : t \geq 0\}$ where the state space consists of two points 0 and 1. An example of such a process is:

(i) Telephone line with state of the line is 0 if no calls are occurring and the state is 1 if the line is busy.

(ii) Random telegraph signal where the signal is positive and negative.

Let the probability of transition from 0 to 1 in a small time interval of time h is $\lambda t + O(h)$ and it remains in the state is $1 - \lambda h + O(h)$, and the probability of transition from 1 to 0 in a small interval of time is $\mu h + O(h)$, and the probability that it remains in the state is $1 - \mu h + O(h)$. Then the Chapman-Kolmogorov relation for time homogeneous process is given by

$$P_{ij}(t + h) = \sum_{k=0}^{1} p_{ik}(t)\, p_{kj}(h), \qquad k = 0, 1$$

For $i, j = 0, 1$, we get four equations and

$$p_{00}(t + h) = p_{00}(t)\, p_{00}(h) + p_{01}(t)\, p_{10}(h)$$

We know

$$p_{ij}(t) = P[X(t) = j | X(0) = i], \qquad i, j = 0, 1$$

So

$$p_{00}(t + h) = (1 - \lambda h + O(h))\, p_{00}(t) + p_{01}(t)\mu h + O(h)$$

Limit $h \to 0$, we get

$$\frac{dp_{00}(t+h)}{dt} = -\lambda p_{00}(t) + \mu p_{01}(t)$$

But $p_{00}(t) + p_{01}(t) = 1$, so

$$\frac{d}{dt} p_{00}(t + h) = \mu(1 - p_{00}(t)) - \lambda p_{00}(t) = \mu - (\lambda + \mu)\, p_{00}(t) \tag{6.30}$$

The solution of (6.30) is

$$p_{00}(t) = \frac{\mu}{\lambda + \mu} + \frac{\lambda}{\lambda + \mu}\, e^{-(\lambda + \mu)t}$$

with $p_{00}(0) = 1$. Similarly we can find the values for $p_{10}(t)$ and $p_{11}(t)$ Hence the transition matrix is given by

$$\begin{bmatrix} p_{00}(t) & p_{01}(t) \\ p_{10}(t) & P_{11}(t) \end{bmatrix} = \begin{bmatrix} \dfrac{\mu}{\lambda+\mu} & \dfrac{\lambda}{\lambda+\mu} \\ \dfrac{\mu}{\lambda+\mu} & \dfrac{\lambda}{\lambda+\mu} \end{bmatrix} + e^{-(\lambda+\mu)t} \begin{bmatrix} \dfrac{\lambda}{\lambda+\mu} & \dfrac{-\lambda}{\lambda+\mu} \\ \dfrac{-\mu}{\lambda+\mu} & \dfrac{\mu}{\lambda+\mu} \end{bmatrix}$$

The expectied length of time $\mu_{ij}(t)$ spent in a particular state between $(0, t)$ is given by

$$\mu_{ij}(t) = E\left(\int_0^t Y_{ij}(\tau)\, d\tau\right), \text{ where } Y_{ij}(t) = 1 \text{ if } X(t) = j, \text{ given } X(0) = i$$

$$= 0 \text{ if } X(t) \neq j, \text{ given } X(0) = i$$

$$= \int_0^t E(Y_{ij}(\tau)d\tau = \int_0^t p_{ij}(\tau)d\tau,$$

for all $i, j = 0, 1$.

Limiting Distribution: Let $P(t) = (p_{ij}(t))$, $i, j = 0, 1, 2,\ldots$ and the matrix of transition density is $A = (a_{ij})$, where $a_{ij} = (p^1_{ij}(t))_{t=0}$. Then given any state j

$$\lim_{t \to \infty} p_{ij}(t) = v_j$$

exists and is the same for all initial states $i = 0, 1, 2,\ldots$, v_j represents a probability distribution with

$$0 \le v_j \le 1, \sum_j v_j = 1$$

The values v_j can be determined as solution of the system of linear equations

$$\sum_j v_i\, a_{ij} = 0, j = 0, 1, 2,\ldots$$

6.3.5 Example

Consider Example 6.3.4

$$v_0 = \lim_{t \to \infty} p_{00}(t) = \lim_{t \to \infty} p_{10}(t) = \frac{\mu}{\lambda + \mu}$$

$$v_1 = \lim_{t \to \infty} p_{01}(t) = \lim_{t \to \infty} p_{11}(t) = \frac{\lambda}{\lambda + \mu}$$

6.4 PURE BIRTH, PURE DEATH PROCESS

6.4.1 Pure Birth Process

The simplest generalization of the Poisson process is obtained by permitting the probabilities of jumps to depend on the actual state of the system.

Let the probability that r events occur between t and $t + h$ given that n events occurred by epoch t is given by

$$\begin{aligned} p_r(h) &= P\{N(h) = r \mid N(t) = n\} \\ &= \lambda h + O(h) && r = 1 \\ &= O(h) && r \ge 2 \\ &= 1 - \lambda h + O(h), && r = 0 \end{aligned}$$

$p_r(h)$ is independent of n as well as t. We generalize the process by considering that λ is not a constant but is a function of n or t or both. The resulting process is still Markovian. Let λ be a function of n. Assume

$$\begin{aligned} p_r(h) = P\{N(h) = k \mid N(t) = n\} &= \lambda_n h + O(h) && k = 1 \\ &= O(h) && k \ge 2 \\ &= 1 - \lambda_n h + O(h) && k = 0 \end{aligned}$$

Then proceeding as in Theorem 4.4.1, we get

$$p_n(t + h) = p_n(t)\,(1 - \lambda_n h) + p_{n-1}(t)\,\lambda_{n-1}\,h + O(h), \qquad n \ge 1$$

$$\left.\begin{aligned} p_n'(t) &= -\lambda_n\, p_n(t) + \lambda_{n-1}\, p_{n-1}(t) \\ p_0'(t) &= -\lambda_0\, p_0\,(t) \end{aligned}\right\} \qquad (6.31)$$

In the Poisson process it is natural to assume that the system starts from the intial state E_0 at time 0. Assume more generally that the system starts from an arbitrary initial state E_i. Then

$$p_i(0) = 1,\ p_n(0) = 0, \qquad \text{for } n \neq i \tag{6.32}$$

The initial conditions determine uniquely the solution $\{p_n(t)\}$ of (6.31).

The above process is called the **pure birth process.**

6.4.2 Yule Process

Consider a population of members who can (by splitting or otherwise) give birth to new members but cannot die. Assume that during any short time interval of length say h each member has probability $\lambda h + O(h)$ to create a new one. Then, if n individuals are present at time t, the probability that there will be one birth between t and $t + h$ is $n\lambda h + O(h)$.

If $N(t)$ denotes the total number of members by epoch t and $p_n(t) = P\{N(t) = n\}$, then by putting $\lambda_n = n\lambda$ in (6.31), we obtain

$$\left.\begin{aligned} p_n'(t) &= -n\lambda\, p_n(t) + (n-1)\lambda\, p_{n-1}(t) \\ \text{and} \quad p_0'(t) &= 0 \end{aligned}\right\} \quad n \geq 1 \tag{6.33}$$

we find the solution $\{p_n(t)\}$ by induction. Assume $p_1(0) = 1$ and $p_i(0) = 0$ for $i \neq 1$,

For $n = 1$, $p_1'(t) = -\lambda\, p_1(t)$

That is $p_1'(t) + \lambda\, p_1(t) = 0$. The auxilary equation is $m + \lambda = 0$, which implies $m = -\lambda$. Hence the solution is $p_1(t) = C_1 e^{-\lambda t}$. When $t = 0$, $p_1(0) = 1 = C_1$. Hence $p_1(t) = e^{-\lambda t}$.

For $n = 2$

$$p_2'(t) = -2\lambda p_2(t) + \lambda p_1(t)$$

or

$$p_2'(t) + 2\lambda p_2(t) = \lambda p_1(t) = \lambda e^{-\lambda t}$$

This linear equation has integrating factor $e^{2\lambda t}$ and hence the solution is

$$e^{2\lambda t}\, p_2(t) = \int \lambda\, e^{2\lambda t}\, e^{-\lambda t}\, dt = \int \lambda\, e^{\lambda t}\, dt$$

$$= \lambda\left[\frac{e^{\lambda t}}{\lambda}\right] + C_2 = e^{\lambda t} + C_2$$

when $t = 0$, $p_2(0) = 0$, hence

$$0 = p_2(0) = e^0 + C_2 = C_2 + 1 \Rightarrow C_2 = -1$$

So $e^{2\lambda t}p_2(t) = e^{\lambda t} - 1$ or $p_2(t) = e^{-2\lambda t}(e^{\lambda t} - 1) = e^{-\lambda t}\,(1 - e^{-\lambda t})$

For $n = 3$,

$$p_3'(t) = -\,3\lambda p_3(t) + 2\lambda\, p_2(t)$$

or

$$p_3'(t) + 3\lambda p_3(t) = 2\lambda[e^{-\lambda t}(1 - e^{-\lambda t})]$$

This linear equation has integrating factor $e^{3\lambda t}$ and hence the solution is

$$e^{3\lambda t}\, p_3(t) = \int 2\lambda(e^{-\lambda t}\,(1 - e^{-\lambda t}))\, e^{3\lambda t}\, dt$$

$$= 2\lambda\left\{\int e^{2\lambda t}\, dt - \int e^{\lambda t}\, dt\right\}$$

$$= 2\lambda\left\{\frac{e^{2\lambda t}}{2\lambda} - \frac{e^{\lambda t}}{\lambda}\right\} + C_3$$

$$= e^{2\lambda t} - 2e^{\lambda t} + C_3$$

As $p_3(0) = 0$, we have $C_3 = 1$. So

$$e^{3\lambda t} p_3(t) = e^{2\lambda t} - 2e^{\lambda t} + 1$$

$$\Rightarrow \quad p_3(t) = e^{-\lambda t} - 2e^{-2\lambda t} + e^{-3\lambda t} = e^{-\lambda t}[1 - e^{-\lambda t}]^2$$

In general

$$p_n(t) = e^{-\lambda t}(1 - e^{-\lambda t})^{n-1}, \qquad n \geq 1 \tag{6.34}$$

Suppose the initial population size is i, then

$$p_i(0) = 1 \text{ and } p_k(0) = 0 \text{ for } k \neq i$$

In this case, proceeding as above, it is easily verified that for $n \geq i > 0$

$$p_n(t) = (n-1)C_{(n-i)}\, e^{-i\lambda t}(1 - e^{-\lambda t})^{n-i} \tag{6.35}$$

and $\quad p_n(t) = 0 \quad$ for all $n < i$ and for all t.

From (6.12), we write the probability generating function as

$$\sum_{n=1}^{\infty} p_n(t) z^n = \sum_{n=1}^{\infty} \{e^{-\lambda t}(1 - e^{-\lambda t})^{n-1}\} z^n$$

$$= ze^{-\lambda t}\{1 + (1 - e^{-\lambda t})\, z + (1 - e^{-\lambda t})^2 z^2 + \ldots\}$$

$$= \frac{ze^{-\lambda t}}{1 - z(1 - e^{-\lambda t})}$$

The mean of the process is $e^{\lambda t}$ and variance is $e^{\lambda t}(e^{\lambda t} - 1)$

This type of process was first studied by Yule in connection with the mathematical theory of evolution. The population consists of the species within a genus and the creation of a new element is due to mutations. The assumption that each species has the same probability of throwing out a new species neglects the difference in species sizes. Since we are assuming that die may not be possible, (6.13) gives only an approximate answer.

6.4.3 Example

For the pure birth process show that the interval T_K between the Kth and $(K+1)^{\text{st}}$ birth has an exponential distribution with parameter $\lambda_n = \lambda$.

Solution: Let the birth rates be constant. That is,

$\lambda_n = \lambda$(for $n = 0, 1, 2, \ldots$). By (6.9), we have

$$p_n'(t) = -\lambda p_n(t) + \lambda\, p_{n-1}(t) \tag{6.36}$$

Assume, further that

$$p_0(0) = 1,\ p_n(0) = 0, \qquad \text{for } n \geq 1$$

Using Laplace transform, we solve the differential equation as follows. The Laplace transform of any function $f(t)$ is

$$L(f(t))) = F(s) = \int_0^{\infty} e^{-st} f(t)\, dt$$

Hence, the Laplace transform of $p_n(t)$ is

$$L(p_n(t)) = \int_0^{\infty} e^{-st} p_n(t)\, dt$$

Take Laplace transform on both sides of (6.36)

$$L(p_n'(t)) = -\lambda L(p_n(t)) + \lambda L(p_{n-1}(t))$$

We know $L(f'(t)) = sL(f(t)) - f(0)$. Hence the above equation becomes

$$sL(p_n(t)) - p_n(0) = -\lambda L(p_n(t)) + \lambda L(p_{n-1}(t))$$

$$\Rightarrow \quad (s+\lambda)\, L(p_n(t)) = \lambda L(p_{n-1}(t)), \quad \text{since } p_n(0) = 0$$

$$\Rightarrow \quad L(p_n(t)) = \frac{\lambda}{s+\lambda} L(p_{n-1}(t)) = \frac{\lambda^2}{(s+\lambda)^2} L(p_{n-2}(t))$$

$$\dots \quad \dots \quad \dots$$

$$= \frac{\lambda^n}{(s+\lambda)^n} p_0(0) = \frac{\lambda^n}{(\lambda+s)^n}, \text{ since } p_0(0) = 1$$

Hence

$$p_n(t) = L^{-1}\left[\frac{\lambda^n}{(\lambda+s)^n}\right] = \lambda^n L^{-1}\left[\frac{1}{(\lambda+s)^n}\right]$$

$$= \lambda^n \cdot e^{-\lambda t} \cdot \frac{t^n}{n!} \text{ by shifting property}$$

$$= \frac{(\lambda t)^n \cdot e^{-\lambda t}}{n!}$$

Therefore

$$p_n(t) = P(X(t) = n) = \frac{(\lambda t)^n \cdot e^{-\lambda t}}{n!}, \quad n \geq 0,\ t \geq 0$$

Thus, $X(t)$ is a Poisson process with parameter λt.

6.4.4 Pure Death Process

An individual cannot give birth to a new individual and the probability of a death of an individual in $(t, t+h)$ is $\mu h + O(h)$. Then, if n individuals are present at time t, the probability of one death in $(t, t+h)$ is $n\mu h + O(h)$.

6.4.5 Polya Process

The Polya process is another example of a non-homogeneous birth process where the occurrence of an event is allowed to depend both on time and on the number. The assumptions for this process are same as for the birth process except that $\lambda_n(t) = \dfrac{\lambda + \lambda an}{1 + \lambda at}$, where n and a are non-negative constants.

Proceeding as before we see that the Polya process is described by the following differential equation

$$p_n'(t) = -\frac{\lambda + \lambda an}{1 + \lambda at} p_n(t)$$

and

$$p_m'(0) = -\frac{\lambda + \lambda am}{1 + \lambda at} p_n(t) + \frac{\lambda + \lambda a(m-1)}{1 + \lambda at} p_{n-1}(t) \quad \text{provided } m > n$$

6.5 BIRTH AND DEATH PROCESS

In the last section, we considered pure birth process that cannot serve as a realistic model for changes in the size of populations whose members can die (or drop out). Here we consider events that are characterised by birth (increase in population) and death (decrease in population)

Consider pure birth process, where

P\{Number of births between t and $t + h$ is r, given that the number of individuals at epoch t is n\}

$$= \begin{cases} \lambda_n h + O(h), & r = 1 \\ O(h), & r \geq 2 \\ 1 - \lambda_n h + O(h), & r = 0, \text{for all } n \geq 0 \end{cases}$$

λ_0 may or may not be equal to zero.

Note that r is a non-negative integer which implies that there can only be an increase by r. That is only births are considered possible.

Consider death's are also possible. That is, there could be a decrease by r. Assume \{Number of deaths between t and $t + h$ is r, given that the number of individual at epoch t is n\}

$$= \begin{cases} \mu_n h + O(h), & r = 1 \\ O(h), & r \geq 2 \\ (1 - \mu_n h) + O(h), & r = 0 \quad \text{for } n \geq r \end{cases}$$

Assume that the probability of more than one birth or death in an interval $(t, t + h)$ is $O(h)$. Further, let $\mu_0 = 0$. Through a birth, there is an increase of one individual and through a death there is a decrease of one individual in the total number of individuals.

Now we derive differential equations for the probabilities $p_n(t) = P\{N(t) = n\}$, where $N(t)$ denotes the total number of individuals at epoch t starting from $t = 0$. In order to calculate $p_n(t + h)$ split the interval $(0, t + h)$ as $(0, t)$ and $(t, t + h)$. The event $\{N(t + h) = n, n \geq 1\}$ can occur in a number of mutually exclusive ways. Let $A_{ij} = (n - i + j)$ individuals by epoch t, i birth and j death between t and $t + h$, $i, j = 0, 1$.

Then consider the following possible cases:

(i) A_{00}: n individuals at epoch t, with no births and no deaths between t and $t + h$.
(ii) A_{10}: $n - 1$ individuals at epoch t, with one birth and 0 death between t and $t + h$.
(iii) A_{01}: $n + 1$ individuals at epoch t with no birth but one death between t and $t + h$.
(iv) A_{11}: n individuals at epoch t, with one birth and one death between t and $t + h$.

Then

$$\begin{aligned} p(A_{00}) &= p_n(t)\{1 - \lambda_n h + O(h)\}\ \{1 - \mu_n h + O(h)\} \\ &= p_n(t)\ \{1 - (\lambda_n + \mu_n)\ h + O(h)\} \\ p(A_{10}) &= p_{n-1}(t)\{\lambda_{n-1}\ h + O\ (h)\}\ \{1 - \mu_n h + O(h)\} \\ &= p_{n-1}(t)\ \lambda_{n-1}\ h + O(h) \\ p(A_{01}) &= p_{n-1}(t)\ \{1 - \lambda_{n+1}\ h + O(h)\}\ \{\mu_{n+1}\ (h) + O(h)\} \\ &= p_{n+1}(t)\ \mu_{n+1}\ h + O(h) \\ p(A_{11}) &= p_n(t)\ (\lambda_n h + O(h)\ (\mu_n h + O(h)) \\ &= O(h) \text{ by our assumption} \end{aligned}$$

Hence

$$\begin{aligned} p_n(t + h) = p_n(t)\ \{1 - (\lambda_n + \mu_n)h\} + \{&p_{n-1}\ (t)\ \lambda_{n-1} h \\ &+ p_{n+1}(t)\ \mu_{n+1} h + O(h)\} \end{aligned}$$

or

$$\frac{p_n(t+h)-p_n(t)}{h} = -(\lambda_n + \mu_n)\, p_n(t) + p_{n-1}(t)\, \lambda_{n-1} + \mu_{n+1}\, p_{n+1}(t) + \frac{O(h)}{h}$$

Taking limit $h \to 0$, we have

$$p_n'(t) = -(\lambda_n + \mu_n)\, p_n(t) + \lambda_{n-1}\, p_{n-1}(t) + \mu_{n+1}\, p_{n+1}(t), \qquad n \geq 1 \tag{6.37}$$

For $n = 0$,

$$p_0(t+h) = p_0(t)\{1 - \lambda_0 h + O(h)\} + p_1(t)\,\{1 - \lambda_0 h + O(h)\}\,\{\mu_1 h + O(h)\}$$
$$= p_0(t) - \lambda_0 h\, p_0(t) + \mu_1 p_1(t) h$$

or $$\frac{p_0(t+h)-p_0(t)}{h} = \lambda_0 p_0(t) + \mu_1 p_1(t)$$

or $$p_0'(t) = -\lambda_0 p_0(t) + \mu_1\, p_1(t) \tag{6.38}$$

If at time $t = 0$ there were $i(\geq 0)$ individuals, then the initial condition is

$$p_i(0) = 1 \text{ and } p_n(0) = 0 \quad \text{for } n \neq i$$

Equation (6.37) and (6.38) are the equations of birth and death process. It play an important role in queueing theory and reliability theory.

Steady-state Birth-death processes

When $t \to \infty$, $\frac{dp_j(t)}{dt} = 0$ and $p_j(t) = p_j$,

then

$$0 = -(\lambda_j + \mu_j)\, p_j + \lambda_{j-1}\, p_{j-1} + \mu_{j+1} p_{j+1},\ j \geq 1$$
$$0 = -\lambda_0 p_0 + \mu_1 p_1$$

Rewriting, we get

$$p_{j+1} = \frac{\lambda_j + \mu_j}{\mu_{j+1}}\, p_j - \frac{\lambda_{j-1}}{\mu_{j+1}}\, p_{j-1},\ j \geq 1$$

$$p_1 = \frac{\lambda_0}{\mu_1}\, p_0$$

It follows that

$$p_2 = \frac{\lambda_1 + \mu_1}{\mu_2}\, p_1 - \frac{\lambda_0}{\mu_2}\, p_0$$
$$= \frac{\lambda_1 + \mu_1}{\mu_2} \cdot \frac{\lambda_0}{\mu_1}\, p_0 - \frac{\lambda_0}{\mu_2}\, p_0$$
$$= \frac{\lambda_1 \lambda_0}{\mu_2 \mu_1}\, p_0$$

$$p_3 = \frac{\lambda_2 + \mu_2}{\mu_3}\, p_2 - \frac{\lambda_1}{\mu_3}\, p_1$$
$$= \frac{\lambda_2 + \mu_2}{\mu_3} \cdot \frac{\lambda_1 \lambda_0}{\mu_2 \mu_1}\, p_0 - \frac{\lambda_1 \lambda_0}{\mu_3 \mu_1}\, p_0$$
$$= \frac{\lambda_2 \lambda_1 \lambda_0}{\mu_3 \mu_2 \mu_1}\, p_0$$

Hence $$p_n = \frac{\lambda_{n-1}\, \lambda_{n-2} \ldots \lambda_0}{\mu_n\, \mu_{n-1} \ldots \mu_1}\, p_0, \quad (n \geq 1)$$

$$= p_0 \prod_{i=1}^{n} \frac{\lambda_{i-1}}{\mu_i}$$

By mathematical induction also, we can prove the result.

Since the probabilities must sum to 1, it follows that

$$p_0 = \left(1 + \sum_{n=1}^{\infty} \prod_{i=1}^{n} \frac{\lambda_{i-1}}{\mu_i}\right)^{-1} \tag{6.39}$$

From equation (6.39), we see that a necessary and sufficient condition for the existence of a steady state solution is the convergence of the infinite series

$$1 + \sum_{n=1}^{\infty} \prod_{i=1}^{n} \frac{\lambda_{i-1}}{\mu_i} \tag{6.40}$$

Equations (6.39) and (6.49) are extremely useful in generating a variety of queueing models. For example, if $\lambda_n = \lambda$, $\mu_n = \mu$, for all n, then we obtain $M \mid M \mid 1$ queue.

Note: The existence and uniqueness of solutions in this cases are not trivial. In a pure birth process, solving the diffential equation (6.31) is easy. So we state properties of the solution without proof.

For arbitrarily prescribed coefficients $\lambda_n \geq 0$, $\mu_n \geq 0$, there always exists a positive solution $\{p_n(t)\}$ such that $\Sigma p_n(t) \leq 1$. If λ_n, μ_n are bounded, the solution is unique and satisfies $\Sigma p_n(t) = 1$.

Particular cases:

(i) When $\lambda_n = \lambda$, that is, λ_n is independent of the population size n, then the increase may be thought of as due to a external source such as **immigration**.

(ii) When $\mu_n = \mu$, the decrease may be thought of as due to a factor such as **emigration**.

(iii) When $\lambda_n = n\lambda$, we have the case of linear birth. In this case the rate of birth in unit interval is λ per individual. Here $\lambda_0 = 0$.

(iv) When $\mu_n = n\mu$, we have the case of linear death, the rate of death in unit interval being μ per individual.

(v) For $\lambda_n = \lambda$, $\mu_n = \mu$, we have **immigration-emigration process**. The process associated with the simple queueing model $M/M/1$ (which we discuss in Sec. 7.)

(vi) **Linear growth:** Suppose that a population consists of individuals which can split or die. During any short interval $(t, t + h)$ the probability of individuals which live and split into two is $\lambda h + O(h)$, whereas the corresponding probability of dying is $\mu h + O(h)$. Here λ and μ are two constants characteristic of population. Thus, if n members are present at the instant t, the probability of one birth between t and $t + h$ is $n\lambda h + O(h)$ and that of one death is $n\mu h + O(h)$, $n \geq 1$. We thus have a birth and death process with

$$\lambda_n = n\lambda,\ \mu_n = n\mu\ (n \geq 1),\ \lambda_0 = \mu_0 = 0$$

If $X(t)$ denotes the total number of members at time t, then from (6.37) and (6.38) we have

$$p_0'(t) = \mu p_1(t)$$

$$p_n'(t) = -(\lambda + \mu)n\, p_n(t) + \lambda(n - 1)\, p_{n-1}(t) + \mu(n + 1)\, p_{n+1}(t)$$

If the intial population size is i, i.e., $N(0) = i$, then the initial condition is $p_i(0) = 1$ and $p_n(0) = 0$ for $n \neq i$.

Linear Growth with Immigration.

(vii) In the above case $\lambda_0 = 0$ and, as a result, if the population size reaches zero at any time, it remains at zero thereafter. That is, 0 is the absorbing state. If $\lambda_n = n\lambda + a$ $(a > 0)$, $\mu_n = n\mu$ $(n \geq 0)$, we get a process called **linear growth process with immigration**, where 0 is not an absorbing state.

(viii) **Immigration Death Process**

If $\lambda_n = \lambda$ and $\mu_n = n\mu$, we get a process called an **immigration death process**. This corresponds to the Markovian queue with infinite number of channels i.e. the queue $M/M/\infty$.

(ix) **Pure Death Process**

Assume the birth rates are all assumed to be zero. That is $\lambda_n = 0$ for all n. Assume at time $t = 0$ there are n individuals and eventually decays to $n = 0$. Thus, 0 is the absorbing state. We consider two cases.

Death process with a constant rate. Let $\lambda_i = 0$ for all i and $\mu_i = \mu$ for all i. Then (6.37) and (6.38) reduce to

$$p_k'(t) = -\mu p_k(t) + \mu p_{K+1}(t),\ 1 \le k \le n-1 \tag{6.41}$$

For $k = 0$,

$$p_0'(t) = \mu p_1(t), \tag{6.42}$$

For $k = n$

$$p_n'(t) = -\mu p_n(t) \tag{6.43}$$

where the initial number of individuals at time $t = 0$ is n, so that $p_n(0) = 1$ and $p_k(0) = 0$ for $0 \le k \le n-1$.

Taking Laplace transform on both sides of (6.41), (6.42) and (6.43), we get

$$sL(p_K(t) - p_K(0)) = \mu L(p_K(t)) + \mu L(p_{K+1}(t))$$

$$\Rightarrow \quad (s+\mu)\, L(p_K(t)) = \mu\, L(p_{K+1}(t))$$

$$\Rightarrow \quad L(p_K(t)) = \begin{cases} \dfrac{\mu}{s+\mu} L(p_{K+1}(t)), & 1 \le k \le n-1 \\ \dfrac{1}{s+\mu}, & k = n \\ \dfrac{\mu}{s} L(p_1(t)), & k = 0 \end{cases}$$

so $$L(p_k(t)) = \frac{1}{\mu}\left(\frac{\mu}{s+\mu}\right)^{n-k+1},\ 1 \le k \le n.$$

Taking inverse Laplace transform, we get

$$p_k(t) = L^{-1}\left[\frac{1}{\mu}\left(\frac{\mu}{s+\mu}\right)^{n-k+1}\right]$$

$$= e^{-\mu t}\,\frac{(\mu t)^{n-k}}{(n-k)!},\ 1 \le k \le n$$

We know $\sum_{k=0}^{n} p_k(t) = 1$, hence

$$p_0(t) = 1 - \sum_{k=1}^{n} p_k(t)$$

$$= 1 - \sum_{k=1}^{n} e^{-\mu t} \frac{(\mu t)^{n-k}}{(n-k)!} = 1 - \sum_{k=0}^{n-1} e^{-\mu t} \frac{(\mu t)^{k}}{k!}$$

Death process with a linear rate. Assume $\lambda_i = 0$ for all i and $\mu_i = i\mu$, $i = 1, 2,...n$. The basic differential equation in this case is:

$$p_k'(t) = -k\ \mu p_k(t) + (k+1)\ \mu p_{k+1}(t), \qquad \leq k \leq n-1$$

$$p_0'(t) = \mu p_1(t) \qquad k = 0$$

$$p_n'(t) = -n\mu p_n(t) \qquad k = n$$

where the number of individuals at $t = 0$ is $N(0) = 0$ so that $p_n(0) = 1$, $p_k(0) = 0$, $0 \leq k \leq n-1$. Using method of Laplace transforms we obtain the solution to the above system of equations as:

$$p_k(t) = nC_k(e^{-\mu t})^k (1 - e^{-\mu t})^{n-k}, 0 \leq k \leq n, t \geq 0 \tag{6.44}$$

The state diagram of a pure death process with a linear death rate is:

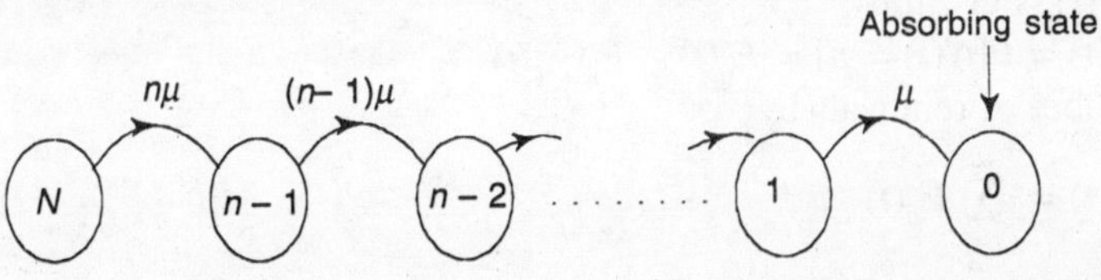

Note: Not all Markov chains satisfy the restriction of nearest-neighbour-only transitions. That is, there are process, which are not Birth-death process.

6.6 RENEWAL PROCESS

In the classical Poisson process, the intervals between successive occurrences are independently and identically distributed with a negative exponential distribution. Suppose that there is a sequence of events E such that the intervals between successive occurrences of E are distributed independently and identically but have a distributive not necessarily negative exponential. This general independent process $\{X_i : i = 1, 2,...\}$ is a **renewal process** or a **recurrent process.** Here X_i' are independent, identically distributed, non-negative random variables. The random variable X_i interpreted as the time between the occurrence of ith event and $(i-1)^{st}$ event. Renewal process is a discrete parameter independent process.

Examples:

1. Consider a system in which the repair (or replacement) after a failure is performed, requiring negligible time. Now the times between successive failures be independent, identically distributed (i.i.d) random variables $\{X_n, n = 1, 2...\}$ is a renewal process.
2. Consider an electric bulb which either works or fails completely. Suppose that the detection of the failure of a bulb and its replacement by a new bulb takes negligible time and suppose that the life times of bulbs are i.i.d random variables, then we have a renewal process.
3. Job arrival in a computer system or a telephone-call arrival to a switch board.

We confine here to the more important case of continuous random variable X_i.

Let X_i be a continuous random variable with the distribution function $F(x)$, called the underlying distribution of the renewal process.

Let $S_k = X_1 + X_2 + ... + X_k$, $k \geq 1$ and $S_0 = 0$. Let $F_k(x) = P\{S_k \leq x\}$ be the distribution function of S_k. Define the random variable

$$N(t) = \sup\{n : S_n \leq t\}$$

That is, $N(t)$ denotes the number of renewals in the interval $(0, t)$. The discrete state, continuous-parameter process $\{N(t) : t \geq 0\}$ is called a **renewal counting process.** $N(t)$ is called the **renewal random variable**. Note that $N(t) \geq n$ if and only if $S_n \leq t$.

If, for some n, $S_n = t$ then a renewal is said to occur at t; S_n gives the time (epoch) of the nth renewal, and is called the nth **renewal**. The random variable X_n gives the inter-arrival time (or waiting time) between $(n-1)^{th}$ and n^{th} renewals.

6.6.1 Definition

The average number of renewals in the interval, denoted by $M(t)$, is called the **renewal function.**

6.6.2 Theorem

The distribution of $N(t)$ is given by

$$P_n(t) = P[N(t) = n] = F_n(t) - F_{n+1}(t) \tag{6.45}$$

and the expected number of renewals by

$$M(t) = \sum_{n=1}^{\infty} F_n(t)$$

Proof: We have

$$\begin{aligned} P[N(t) = n] &= P[N(t) \geq n] - P[N(t) \geq n+1] \\ &= P[S_n \leq t] - P[S_{n+1} \leq t] = F_n(t) - F_{n+1}(t) \end{aligned}$$

Now

$$\begin{aligned} M(t) = E(N(t)) &= \sum_{n=0}^{\infty} np_n(t) \\ &= \sum_{n=0}^{\infty} n[F_n(t) - F_{n+1}(t)] \\ &= \sum_{n=1}^{\infty} F_n(t) = F(t) + \sum_{n=1}^{\infty} F_{n+1}(t) \end{aligned} \tag{6.46}$$

From the above, note that F_{n+1} is a convolution of F_n and F. Let $f(x)$ be the density function (*pdf*) of X_n.

Renewal Equation

For the renewal function $M(t) = E(N(t))$ we obtain on integral equation

$$M(t) = F(t) + \int_0^t M(t-x)\, f(x) dx \tag{6.47}$$

called **fundamental renewal equation**, which gives the expected number of renewals in $[0, t]$.

6.6.3 Theorem

The renewal function M satisfies the renewal equation (6.47).

Proof: $M(t) = F(t) + \sum_{n=1}^{\infty} F_{n+1}(t)$

$$= F(t) + \sum_{n=1}^{\infty} \left(\int_0^t F_n(t-x)\, f(x) dx \right)$$

$$= F(t) + \int_0^t \left[\sum_{n=1}^{\infty} F_n(t-x) \right] f(x)\,dx$$

$$= F(t) + \int_0^t M(t-x)\, f(x)dx,$$

by assuming the validity of the change of order of integration and summation. Define the **renewal density** ***m*(*t*)** to be the derivative of the renewal function $M(t)$; that is

$$m(t) = \frac{d}{dt}(M(t))$$

For small h, $m(t)h$ is interpreted as the probability of occurrence of a renewal in the interval $(t, t + h]$. Thus in the case of Poisson process, renewal density $m(t)$ equals the Poisson rate λ. Now

$$m(t) = \sum_{n=1}^{\infty} f_n(t) \tag{6.48}$$

and using (6.47), we have

$$m(t) = f(t) + \int_0^t m(t-x)\, f(x)dx \tag{6.49}$$

called **renewal equation**. To solve this, take Laplace transform on both sides of (6.48). Then we get

$$L(m(t)) = L(f(t)) + L(m(t))\, L(f(t))$$

So that

$$L(m(t)) = \frac{L(f(t))}{1 - L(f(t))} \tag{6.50}$$

and

$$L(f(t)) = \frac{L(m(t))}{1 - L(m(t))} \tag{6.51}$$

Thus $m(t)$ may be determined from $f(t)$ and conversely, $f(t)$ can be determined from $m(t)$.

6.6.4 Example

Consider $f(x) = \lambda e^{-\lambda x}$, then $L(f(x)) = \lambda L(e^{-\lambda x}) = \lambda\left(\frac{1}{s+\lambda}\right)$ and $L(m(x)) = \frac{L(f(x))}{1-L(f(x))} = \frac{\frac{\lambda}{s+\lambda}}{1-\frac{\lambda}{s+\lambda}} = \frac{\lambda}{s}$

Since the Laplace transform of a constant K is $\frac{K}{s}$, we have

$$m(t) = L^{-1}\left(\frac{\lambda}{s}\right) = \lambda, \qquad t \geq 0$$

$$M(t) = \lambda t \qquad t \geq 0$$

In this case, $F_n(t)$ is a convolution of n identical exponential distributions and hence

$$F_n(t) = 1 - \sum_{k=0}^{n-1} \frac{(\lambda t)^k}{k!} e^{-\lambda k}$$

Then

$$P(N(t) = n) = F_n(t) - F_{n+1}(t) = \frac{(\lambda t)^n}{n!} e^{-\lambda t}$$

Thus $N(t)$ has a Poisson distribution with parameter λt.

6.6.5 Theorem

With probability 1,

$$\frac{N(t)}{t} \to \frac{1}{\mu} \text{ as } t \to \infty$$

where $\mu = E(X_n) \le \infty$

Proof: Consider an interval (0, t]. We have

$$S_{N(t)} \le t < S_{N(t)+1} \tag{6.52}$$

By strong law of large numbers for the sequence $\{S_n\}$, we have

$$\frac{S_n}{n} = \frac{X_1 + X_2 + ... + X_n}{n} \to E(X_n) = \mu$$

with probability 1. As $t \to \infty$, $N(t) \to \infty$ and,

$$\frac{S_{N(t)}}{N(t)} \to \mu \text{ as } t \to \infty \tag{6.53}$$

with probability 1. Similarly,

$$\frac{S_{N(t)+1}}{N(t)} = \frac{S_{N(t)+1}}{N(t)+1} \cdot \frac{N(t)+1}{N(t)} \to \mu \text{ as } t \to \infty \tag{6.54}$$

with probability 1. From (6.52), (6.53), (6.54), we get $\frac{S_{N(t)+1}}{N(t)} \ge \frac{t}{N(t)}$, which tends to μ as $t \to \infty$.

Hence $\frac{N(t)}{t} \to \frac{1}{\mu}$ as $t \to \infty$.

An integer valued random variable N is said to be a **stopping time** for the sequence $\{X_i\}$ if the event $\{N = n\}$ is independent of $X_{n+1}, X_{n+2}, \ldots$ for all $n = 1, 2, \ldots$

6.6.6 Example

Consider an experiment of tossing a coin and X_i denotes the event of ith toss with $X_i = 1$ or 0 depending on the result being head or tail respectively. Let $S_n = X_1 + X_2 + \ldots + X_n$ denotes the number of heads in the first n toss. Suppose that m is a given integer, then $N = \min\{n \text{ such that } S_n = m\}$ is a stopping time.

If $N(t) = n$ whenever $S_n \le t$ and $S_{n+1} > t$, then $N(t) = n$ depends not only on $X_1, X_2, \ldots X_n$ but also on X_{n+1}. The event $N(t) + 1 = n$ implies $S_{n-1} \le t$ and $S_n > t$, so that the event $\{N(t) + 1 = n\}$ is independent of $X_{n+1}, X_{n+2}, \ldots$. Thus $N(t) + 1$ is a stopping time for the sequence $\{X_i\}$ while $N(t)$ is not.

If $S_N = \sum_{i=1}^{N} X_i$, where X_i are *i.i.d* random variables and N is independent of $X_i's$, then

$$E(S_n) = E\left(\sum_{i=1}^{N} X_i\right) = E(X_i)\ E(N)$$

The result holds when N is a stopping time for the sequence $\{X_i\}$. This equation is called **Wald's equation.**

6.6.7 Definition

Let $\{X_i : i = 1,2,\ldots\}$ be a sequence of independent non-negative random variables such that the first inter-arrival time X_1 has the distribution G and the subsequent inter-arrival times X_n, $n = 2, 3,\ldots$ have the identical distribution F. Let $S_0 = 0$, $S_n = \sum_{i=1}^{n} X_i$ and $N_D(t) = \sup\{n : S_n \le t\}$.

The process $\{N_D(t);\ t \ge 0\}$ is called a **modified delayed** or **a general renewal process.**

6.6.8 Definition

Let $\{X(t)\}$ be a random process having a countably infinite number of states $\{0, 1, 2,\ldots k\ldots\}$. Assume with probability one there exists an epoch t, such that the process beyond t_1 is a probabilistic replica of the whole process starting at $t = 0$. The same property holds for $t_2, t_3,\ldots$Such a process is known as **regenerative process**. Clearly $\{t_n\}$ forms a renewal process.

Next, we state a theorem without proof.

6.6.9 Theorem

Let $\{X(t)\}$ be a regenerative process and let $p_k(t) = P[X(t) = k]$. Then $\lim_{t\to\infty} p_k(t) = p_k$ exists and $p_k \ge 0$ with $\sum_{k=0}^{\infty} p_k = 1$.

6.6.10 Definition

Let $X(t)$ be a Markov process with discrete state space $E = \{0, 1, 2,..\}$ and let the transitions of the process occur at epochs $t_0 = 0, t_1, t_2,\ldots(t_n < t_n + 1)$. Let X_n denote the transition occurring at epoch t_n. If

$$P\{X_{n+1} = k,\ t_{n+1} - t_n \le t \mid X_0 = i_1, X_1 = i_2,\ldots X_n = i_n;\ t_0, t_1,\ldots t_n\}$$
$$= P\{X_{n+1} = k,\ t_{n+1} - t_n \le t \mid X_n = i_n\}$$

then $\{X_n, t_n\}$ $n = 0, 1, 2,\ldots$is said to constitute a **Markov renewal process** with state space E.

6.6.11 Definition

Let $\{X_n\}$ be a Markov chain with state space $E = \{0, 1, 2,\ldots\}$ and transition probability matrix $P = (p_{jk})$. The continuous parameter process $Y(t)$ with state space E defined by

$$Y(t) = X_n \text{ on } t_n \le t < t_{n+1}$$

is called a **semi-Markov process**. If $N_j(t)$ denotes the number of transitions into state j in $(0, t]$, then

$$N(t) = \{N_1(t), N_2(t),\ldots N_m(t)\}$$

is called a **Markov renewal process.**

6.6.12 Definition

Suppose that the initial state in a Markov renewal process $\{X_n, t_n\}$ is i. The state i is said to be **recurrent** if and only if

$G_{ii}(\infty) = 1$, where $G_{ij}(t) = P\{N_j(t) \geq 1 | Y(0) = i\} = P\{S_1 \leq t | Y(0) = i\}$. The state i is **transient** if and only if $G_{ii}(\infty) < 1$. The functions

$$M(i, j, t) = \sum_n Q_{ij}^{(n)}(t)$$

where $Q_{ij}^{(n)}(t) = P[X_n = j, t_n \leq t | X_0 = i\}$, is called the **Markov renewal functions.** Note that

$M(i, j, t) = E(N_j(t) | X_0 = i),\ j \in E$

6.7 SOLVED PROBLEMS

1. Consider a communication system that transmits the digits 0 and 1 through several stages. At each stage the probability that the same digit will be received by the next stage, as transmitted, is 0.75. What is the probability that a 0 that is entered at the first stage is received as a 0 by the fifth stage?

Solution. We have to find $p_{00}(5)$. The transition probability matrix P is given by

$$P = \begin{bmatrix} 0.75 & 0.25 \\ 0.25 & 0.75 \end{bmatrix}$$

Hence

$$P^2 = \begin{bmatrix} 0.625 & 0.375 \\ 0.375 & 0.625 \end{bmatrix}$$

$$P^4 = P^{(2)}P^{(2)} = \begin{bmatrix} 0.53125 & 0.46875 \\ 0.46875 & 0.53125 \end{bmatrix}$$

$$P^5 = P^{(1)}\,P^{(4)} = \begin{bmatrix} 0.515625 & 0.484375 \\ 0.484375 & 0.515625 \end{bmatrix}$$

Therefore, the probability that a zero will be transmitted through five stages as a zero is

$$p_{00}(5) = 0.515625.$$

2. Suppose a Markov chain with three states has the probability transition matrix

$$P = \begin{bmatrix} 0 & 1 & 0 \\ 0.5 & 0 & 0.5 \\ 1 & 0 & 0 \end{bmatrix}$$

Determine whether or not this Markov chain is irreducible.

Solution.

$$P^{(2)} = \begin{bmatrix} 0.5 & 0 & 0.5 \\ 0.5 & 0.5 & 0 \\ 0 & 1 & 0 \end{bmatrix},\ P^{(3)} = \begin{bmatrix} 0.5 & 0.5 & 0 \\ 0.25 & 0.5 & 0.25 \\ 0.5 & 0 & 0.5 \end{bmatrix}$$

$$P^{(4)} = \begin{bmatrix} 0.25 & 0.5 & 0.25 \\ 0.5 & 0.25 & 0.25 \\ 0.5 & 0.5 & 0 \end{bmatrix}, P^{(5)} = \begin{bmatrix} 0.5 & 0.25 & 0.25 \\ 0.375 & 0.5 & 0.125 \\ 0.25 & 0.5 & 0.25 \end{bmatrix}$$

Since $p(5) > 0$, a transition is made between any two states in five steps, so the Markov is irreducible.

3. A customer owing a Maruti car right now has got the option to switch over to Maruti, Ambassador or Fiat next time with the probability of (0.2, 0.5, 0.30). Given the transition martix

$$P = \begin{bmatrix} 0.40 & 0.30 & 0.30 \\ 0.20 & 0.50 & 0.30 \\ 0.25 & 0.25 & 0.50 \end{bmatrix}$$

Find the probabilities with his fourth purchase.

Solution.

Given $\quad v_i = (0.20 \quad 0.5 \quad 0.30)$

$$P = \begin{bmatrix} 0.40 & 0.30 & 0.30 \\ 0.20 & 0.50 & 0.30 \\ 0.25 & 0.25 & 0.50 \end{bmatrix}$$

$v_i^4 = v_i P^3$, where P^3 is calculated as

$$P^3 = \begin{bmatrix} 0.277 & 0.351 & 0.372 \\ 0.269 & 0.359 & 0.372 \\ 0.275 & 0.345 & 0.380 \end{bmatrix}$$

$$v_i^4 = (0.2 \quad 0.5 \quad 0.3) \begin{bmatrix} 0.277 & 0.351 & 0.372 \\ 0.269 & 0.359 & 0.372 \\ 0.275 & 0.345 & 0.380 \end{bmatrix}$$

$$= (0.272 \quad 0.353 \quad 0.375).$$

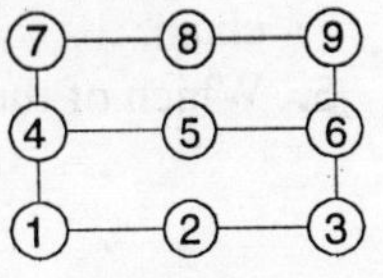

4. Consider a person walking in an area with roads laid out as indicated in Fig. Check whether the travelling from any state to any other is ergodic.

Solution. If a person is at corner 1 (State 1), he can move to right or above with probability $\frac{1}{2}$. If he is in corner 2, he has three options to reach 1, 3 or 5 with a probability of $\frac{1}{3}$. If he is at corner 5, he may go next to 2 or 4 or 6 or 8 with a probability $\frac{1}{4}$ etc. There are totally nine corners (states) and the transition matrix can be given as:

$$
\begin{array}{c} \\ 1 \\ 2 \\ 3 \\ 4 \\ 5 \\ 6 \\ 7 \\ 8 \\ 9 \end{array}
\begin{array}{c}
\begin{array}{ccccccccc} 1 & 2 & 3 & 4 & 5 & 6 & 7 & 8 & 9 \end{array} \\
\begin{bmatrix}
0 & \frac{1}{2} & 0 & \frac{1}{2} & 0 & 0 & 0 & 0 & 0 \\
\frac{1}{3} & 0 & \frac{1}{3} & 0 & \frac{1}{3} & 0 & 0 & 0 & 0 \\
0 & \frac{1}{2} & 0 & 0 & 0 & \frac{1}{2} & 0 & 0 & 0 \\
\frac{1}{3} & 0 & 0 & 0 & \frac{1}{3} & 0 & \frac{1}{3} & 0 & 0 \\
0 & \frac{1}{4} & 0 & \frac{1}{4} & 0 & \frac{1}{4} & 0 & \frac{1}{4} & 0 \\
0 & 0 & \frac{1}{3} & 0 & \frac{1}{3} & 0 & 0 & 0 & \frac{1}{3} \\
0 & 0 & 0 & \frac{1}{2} & 0 & 0 & 0 & \frac{1}{2} & 0 \\
0 & 0 & 0 & 0 & \frac{1}{3} & 0 & \frac{1}{3} & 0 & \frac{1}{3} \\
0 & 0 & 0 & 0 & 0 & \frac{1}{2} & 0 & \frac{1}{2} & 0
\end{bmatrix}
\end{array}
$$

Thus he can travel from any state to any other state in one or many steps. Thus this is an ergodic chain.

5. Which of the following matrices are from Absorbing Markov Chain?

(a) $P = \begin{array}{c} \\ S_1 \\ S_2 \end{array} \begin{array}{c} \begin{array}{cc} S_1 & S_2 \end{array} \\ \begin{bmatrix} 1 & 0 \\ \frac{1}{2} & \frac{1}{2} \end{bmatrix} \end{array}$ (b) $\begin{array}{c} \\ S_1 \\ S_2 \\ S_3 \end{array} \begin{array}{c} \begin{array}{ccc} S_1 & S_2 & S_3 \end{array} \\ \begin{bmatrix} 1 & 0 & 0 \\ 0 & \frac{1}{2} & \frac{1}{2} \\ 0 & \frac{1}{3} & \frac{2}{3} \end{bmatrix} \end{array}$

Solution.

(a) Since $p_{11} = 1$, $p_{12} = 0$, S_1 is the absorbing state and it can be reached from S_2 also. Hence the given matrix is from absorbing Markov chain.

(b) Consider the transition diagram below.
S_1 is the only absorbing state and this absorbing state cannot be reached from S_2 and S_3. Hence it is not from absorbing Markov chain.

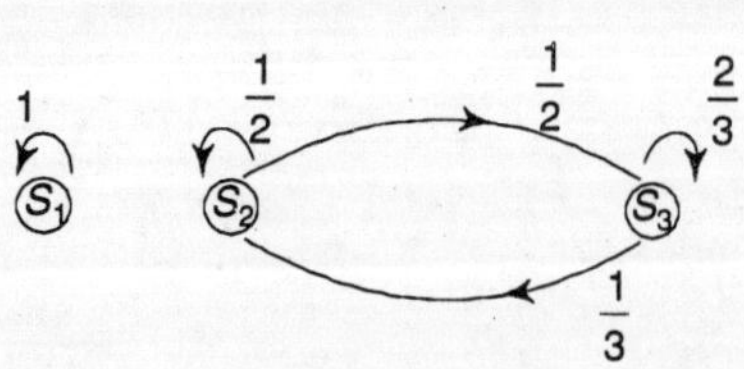

6. *Queuing problem:*

A transport maintenance station services the vehicles as they arrive for repairs. There is only space for two vehicles to be lined up before service. That is, the system capacity is 3. The following data on arrivals are available.

No. of vehicles arriving in one hr.	Probability of occurrence
0	0.5
1	0.4
2	0.1

Regarding service we have the data that the probability of a completion during one hour is 0.6 given that a unit is there to be serviced. Since all service times equal one hour or more the probability of more than one serivce is zero. Formulate a Markov chain model for the given queueing problem.

Solution. Define a state as the number of vehicles in the system. The transition matrix is a 4×4 matrix with p_{ij} as elements of the matrix. A step of the transition is one hour.

Assume the arrivals take place at the beginning of each period and the arrival can be serviced during the period if the queue was previously empty. This means that the service is complete before an arrival appears. Now we find the probabilities of all possible events.

$$p_{00} = P\{\text{No. vehicles waiting, no arrival}\}$$
$$+ P\{\text{No. vehicles waiting, one arrival, one service}\}$$
$$= 0.50 + (0.4)\,(0.6) = 0.75$$

$$p_{01} = P\{\text{No. Vehicles waiting, one arrival, no service}\}$$
$$+ P\{\text{No. vehicles waiting, two arrivals, one service}\}$$
$$= (0.4)\,(0.4) + (0.1)\,(0.6) = 0.22$$

$$p_{02} = P\{\text{No. vehicles waiting, two arrivals, one service}\}$$
$$= (0.1)\,(0.4) = 0.04$$

$$P_{03} = \text{Not possible}$$

Similarly we calculate

$$p_{10} = (0.5)\,(0.6) = 0.30,\ p_{11} = 0.44,\ p_{12} = 0.22,\ p_{13} = 0.04$$
$$p_{20} = 0,\ p_{21} = 0.30,\ p_{22} = 0.20,\ p_{23} = 0.20,\ p_{30} = 0$$
$$p_{31} = 0,\ p_{32} = 0.6,\ p_{33} = 0.4.$$ Hence the transition matrix is

$$\begin{bmatrix} 0.74 & 0.22 & 0.04 & 0 \\ 0.30 & 0.44 & 0.22 & 0.04 \\ 0 & 0.30 & 0.50 & 0.20 \\ 0 & 0 & 0.60 & 0.40 \end{bmatrix}$$

7. The amount of stored water in a reservoir may be idealized into three states, namely full, half-full, and empty. Because of the probabilistic nature of the inflowing water to the reservoir as well as the outflow from the reservoir to meet demands, for water, the amount of water storage may shift from one state to another between seasons. Suppose the transition probabilities from one state to another is given in Fig. Denote empty, half-full and full storage with the states 1, 2, 3 respectively.

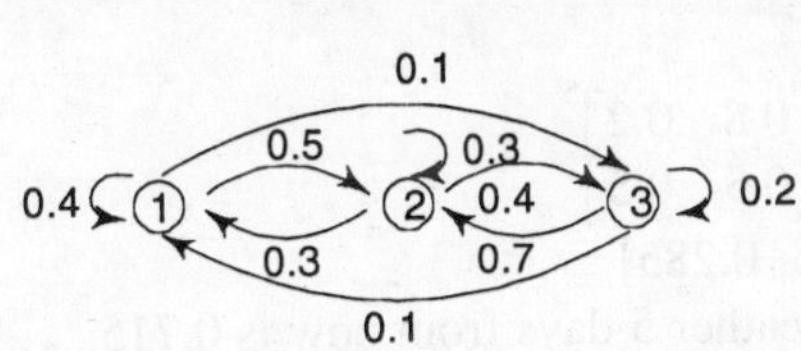

The corresponding transition probability matrix is

$$P = \begin{matrix} & 1 & 2 & 3 \\ 1 & 0.4 & 0.5 & 0.1 \\ 2 & 0.3 & 0.3 & 0.4 \\ 3 & 0.1 & 0.7 & 0.2 \end{matrix}$$

Suppose two weeks before the end of a season, it is predicted that there is 80% probability the reservoir will be full at the beginning of the next season; if it is not full, the reservoir will equally likely be empty or half-full. What is the probability that the reservoir will be full at the end of the next season?

Solution. The initial state probabilities are $P_{(0)} = (0.1, 0.1, 0.8)$ therefore after 1 season the state probabilities will becomes

$$P(1) = [0.1\ 0.1\ 0.8] \begin{bmatrix} 0.4 & 0.5 & 0.1 \\ 0.3 & 0.3 & 0.4 \\ 0.1 & 0.7 & 0.2 \end{bmatrix}$$

$$= [0.15\ 0.64\ 0.21]$$

The probability that the reservoir will be full at the end of the following season is only 21%.

8. The transition of wet and dry days in a town is modeled as a homogeneous Markov chain with the following transition probability matrix;

$$\begin{matrix} & \text{Dry} & \text{Wet} \\ \text{Dry} & 0.8 & 0.2 \\ \text{Wet} & 0.5 & 0.5 \end{matrix}$$

Let the dry and wet days be denoted as states 1 and 2 respectively.

(a) If it is dry today, find the probability that it will be dry after two days and after five days. (i.e. find $p_{1,1}$ (2) and $p_{1,1}$ (5)).

(b) If today is wet, then find the probability that it will be dry after two days and after five days. (i.e. find $p_{2,1}$ (2) and $p_{2,1}$ (5)).

Solution. We have to calculate $P(2)$.

(a) The probabilities of the initial states of a system may be denoted by a row matrix as $P(0) = [p_1(0), p_2(0)] = [1, 0]$. Then,

$$P(2) = P(0)\ P^2$$

$$= [1, 0] \begin{bmatrix} 0.8 & 0.2 \\ 0.5 & 0.5 \end{bmatrix} \begin{bmatrix} 0.8 & 0.2 \\ 0.5 & 0.5 \end{bmatrix}$$

$$= [0.8\ 0.2] \begin{bmatrix} 0.8 & 0.2 \\ 0.5 & 0.5 \end{bmatrix}$$

$$= [0.74, 0.26]$$

Hence it is 74% probable that after two days from today the day will be dry. Similarly

$$P(5) = P(0)\ P^5$$

$$= [1\ 0] \begin{bmatrix} 0.8 & 0.2 \\ 0.5 & 0.5 \end{bmatrix}^5$$

$$= [0.715, 0.285]$$

Hence, the probability of dry weather 5 days from now is 0.715.

(b) $$P(2) = [0\ 1]\begin{bmatrix} 0.8 & 0.2 \\ 0.5 & 0.5 \end{bmatrix}^2$$

$$= [0.65\ 0.35]$$

$$P(5) = (0\ 1)\begin{bmatrix} 0.8 & 0.2 \\ 0.5 & 0.5 \end{bmatrix}^5$$

$$= (0.713, 0.287)$$

The probability that 2 days from now it will be dry is 0.65. However the probability that it will be dry 5 days from now is same as that in part (a).

9. For the probability matrix given above, determine the following.

(a) Suppose it is wet today. What is the probability that it will be wet again for the first time the day after tomorrow?

(b) Suppose it is wet today. What is the probability that there will be a dry day within the next five days?

(c) What is the expected length of dry spell and a wet spell.

(d) If it is wet today, what is the expected length of time (in days) until the next wet day?

Solution.

(a) We know that the first passage probability that it will be wet for the first time two days from now is

$$f_{22}(2) = p_{21}\, p_{12}$$
$$= (0.5)\,(0.2)$$
$$= 0.1$$

(b) $$F_{21}(m) = \sum_{n=1}^{m} f_{21}(n)$$

$$= \sum_{n=1}^{m} p_{22}\,(n-1)p_{21}$$

$$= p_{21}\frac{1 - p_{22}\,(m)}{1 - p_{22}}$$

$$= 1 - p_{22}(m)$$

So $$F_{2,1}(5) = 1 - p_{22}(5)$$

$$= 1 - (0.5)^5 = 0.96875$$

(c) The expected length of occupation in state i is the weighted average period until the system first leaves state i. We know that the expected occupation time is

$$l_i = E(T_i) = \frac{1}{1 - p_{ii}}$$

So the average length of dry spell is

$$l_1 = \frac{1}{1 - p_{11}} = \frac{1}{p_{12}} = \frac{1}{0.2} = 5 \text{ days}$$

The average length of wet spell is

$$l_2 = \frac{1}{1-p_{22}} = \frac{1}{p_{21}} = \frac{1}{0.5} = 2 \text{ days}$$

(d) We know

$$\mu_{jk} = \sum_{n=1}^{\infty} nf_{jk}(n).$$

Then
$$\mu_{11} = \sum_{n=1}^{\infty} nf_{11}(n)$$

$$= p_{11} + \sum_{n=2}^{\infty} n\, p_{12}\, p_{22}\,(n-2)\, p_{21}$$

$$= p_{11} + p_{12}\, p_{21} \sum_{i=1}^{\infty} (i+1)\, p_{22}\,(i-1)$$

$$= p_{11} + p_{12}\, p_{21} \left[\frac{1}{(1-p_{22})^2} + \frac{1}{1-p_{22}}\right]$$

$$= 1 + \frac{p_{12}}{P_{21}}.$$

Similarly, we prove

$$\mu_{22} = 1 + \frac{p_{21}}{P_{12}}.$$

Hence the expected recurrence time of wet days is,

$$\mu_{22} = 1 + \frac{p_{21}}{P_{12}} = 1 + \frac{0.5}{0.2} = 3.5 \text{ days}.$$

10. A quality control program is set up to inspect the quality of aggregates that are produced daily from a rock core. Suppose the daily aggregate quality is divided into two classes, namely, satisfactory (S) and unsatisfactory (U). It is found that the probability of a given quality of aggregates in a given day depends on the quality of the product for the previous day, with probability matrix

$$\begin{array}{c} \\ S \\ U \end{array} \begin{array}{c} \begin{array}{cc} S & U \end{array} \\ \begin{bmatrix} 0.9 & 0.1 \\ 0.4 & 0.6 \end{bmatrix} \end{array}$$

(a) If the aggregate is of satisfactory quality today, what is the probability that it will become unsatisfactory within the next week.

(b) What is the average period that satisfactory quality aggregates will be produced.

Solution. Let states 1 and 2 denote satisfactory and unsatisfactory aggregate quality respectively.

(a) $$F_{12}(m) = \sum_{n=1}^{m} f_{12}(n)$$

$$= \sum_{n=1}^{m} p_{11}\,(n-1)\, p_{12}$$

$= 1 - p_{11}(m)$

$F_{12}(7) = 1 - p_{11}(7)$

$= 1 - (0.9)^7$

$= 0.522$

(b) The expected period of satisfactory quality aggregate is

$$l_1 = \frac{1}{p_{12}} = \frac{1}{0.1} = 10 \text{ days}$$

11. Consider a Markov chain with state space $\{0, 1\}$ and transition probability matrix $p = \begin{array}{c} \\ 0 \\ 1 \end{array}\begin{array}{c} \begin{array}{cc} 0 & 1 \end{array} \\ \begin{bmatrix} 1 & 0 \\ \frac{1}{2} & \frac{1}{2} \end{bmatrix} \end{array}$

(a) Show that the state 0 is recurrent

(b) Show that the state 1 is transient

Solution. The state transition diagram is

(a) 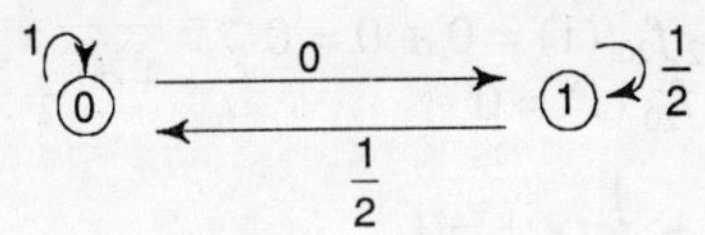

We know

$$f_{ij}(1) = p\{X_1 = j/X_0 = i\} = p_{ij} \text{ and}$$

$$f_{ij}(n) = \sum_{k \neq j} p_{ik} f_{kj}(n-1),\ n = 2, 3, \ldots$$

So

$$f_{00}(1) = p_{00} = 1$$

$$f_{10}(1) = p_{10} = 1$$

$$f_{00}(2) = p_{01} f_{10}(1) = (0)\frac{1}{2} = 0 \text{ and } f_{00}(n) = 0, \text{ for } n \geq 2.$$

Hence

$$F_{00} = \sum_{n=1}^{\infty} f_{00}(n)$$

$$= f_{00}(1) + f_{00}(2) + f_{00}(3) + \ldots$$

$$= f_{00}(1)$$

$$= 1$$

Hence the state 0 is recurrent

(b) $f_{11}(1) = p_{11} = \frac{1}{2}$

$$f_{01}(1) = p_{01} = 0$$

$$f_{11}(2) = p_{10} f_{01}(1)$$

$$= \left(\frac{1}{2}\right)0$$

$$= 0$$

$$f_{11}(n) = 0, \text{ for } n \geq 2,$$

and $$F_{11} = \sum_{n=1}^{\infty} f_{11}(n)$$

$$= f_{11}(1) + f_{11}(2) + f_{11}(3) + \ldots$$

$$= \frac{1}{2} < 1$$

Hence the state 1 is transient.

12. Consider a Markov chain with state space {0, 1, 2} and transition probability martix

$$P = \begin{array}{c} \\ 0 \\ 1 \\ 2 \end{array}\begin{array}{c} \begin{array}{ccc} 0 & 1 & 2 \end{array} \\ \begin{bmatrix} 0 & \frac{1}{2} & \frac{1}{2} \\ 1 & 0 & 0 \\ 1 & 0 & 0 \end{bmatrix} \end{array}$$

. Show that state 0 is periodic with period 2.

Solution.

$f_{00}(1) = p_{00} = 0$

$f_{10}(1) = p_{10} = 1; f_{10}(2) = p_{11}\, f_{10}(1) + p_{12} f_{20}(1) = 0 + 0 = 0$

$f_{20}(1) = p_{20} = 1; f_{20}(2) = p_{21} f_{10}(1) + p_{22} f_{20}(1) = 0$

$$f_{00}(2) = p_{01} f_{10}(1) + p_{02} f_{20}(1) = \frac{1}{2} \times 1 + \frac{1}{2} \times 1 = 1$$

$$f_{00}(3) = p_{01} f_{10}(2) + p_{02} f_{20}(2) = \left(\frac{1}{2}\right) 0 + \left(\frac{p}{2}\right) \times 0 = 0$$

$$F_{00} = f_{00}(1) + f_{00}(2) + f_{00}(3) = 1$$

Hence the state 0 is recurrent. The greatest common divisor of the set of positive integers n such that $p_{ii}(n) > 0$ gives the period of a periodic state i.

We know $p_{ij}(n) = P\{X_n = j\,|_{X_0 = i}\}$ can be computed by matrix powers

$$P^n = \begin{cases} \begin{pmatrix} 1 & 0 & 0 \\ 0 & \frac{1}{2} & \frac{1}{2} \\ 0 & \frac{1}{2} & \frac{1}{2} \end{pmatrix} & \text{if } n \text{ is even} \\ \begin{pmatrix} 0 & \frac{1}{2} & \frac{1}{2} \\ 1 & 0 & 0 \\ 1 & 0 & 0 \end{pmatrix} & \text{if } n \text{ is odd} \end{cases}$$

Then g.c.d. $\{\, n \geq 1 : p_{00}(n) > 0\} =$ g.c.d $\{2, 4, 6, \ldots\} = 2$.

The state diagram is given by

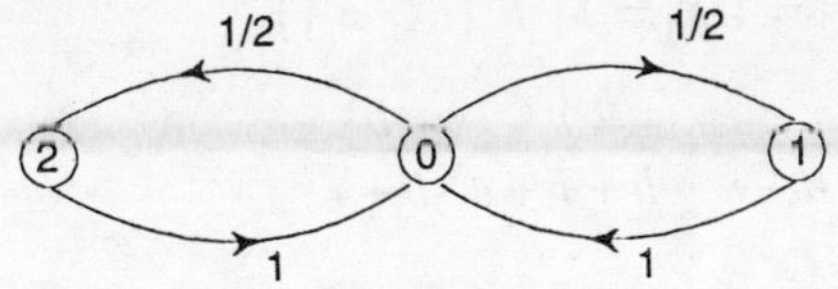

It is clear that 0 is periodic with period 2.

13. Consider a cascade of binary channels that are so noisy that the digit transmitted is always complemented. That is substitute $a = b = 1$ in Example 6.1.9. Then matrix P is given by

$$P = \begin{bmatrix} 0 & 1 \\ 1 & 0 \end{bmatrix}$$

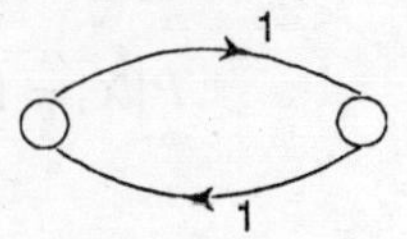

By induction it is verified that

$$P^n = \begin{cases} \begin{pmatrix} 1 & 0 \\ 0 & 1 \end{pmatrix} \text{if } n \text{ is even} \\ \begin{pmatrix} 0 & 1 \\ 1 & 0 \end{pmatrix} \text{if } n \text{ is odd} \end{cases}$$

Starting in state 0 (or 1) we return to state 0 (or 1) after an even number of steps. Now

$$gcd\ \{n \geq 1;\ p_{00}(n)\} = gcd\ \{2, 4, 6, \ldots\} = 2.$$

The state diagram of a two-state periodic Markov chain is:
Hence the given chain is periodic with period 2.

14. Consider a two-state Markov chain with a = 0, b = 1, so that

$$P = \begin{bmatrix} 1 & 0 \\ 1 & 0 \end{bmatrix}$$

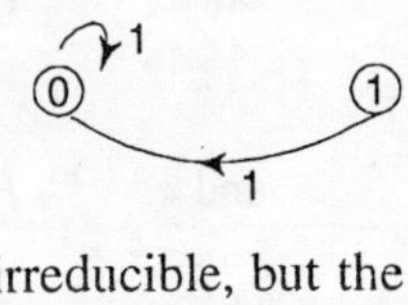

with the state diagram
In this case the state 1 is transient and state 0 is absorbing. The chain is not irreducible, but the limiting state probabilities exist for this aperiodic chain and is given by $V_j = \underset{n \to \infty}{\text{Lt}}\ p_j(n)$. We see that $V_0 = 1$ and $V_1 = 1$.
That is, the chain will remain in state 0 after atmost one transition

15. Polya's Urn Model: Non-Markov Process

An urn contains b block and r red balls. A ball is drawn at random and is replaced after the drawing. The outcome at the nth drawing is either a black ball or a red ball. Let the random variable X_n be defined as

$X_n = 1$ if n^{th} drawing results in a black ball
$X_n = 0$ if it result in a red ball.

The model is such that after each drawing not only that the ball drawn is replaced but c (> 0) balls of the colour drawn (at that drawing) are added to the urn so that the number of balls of the colour drawn increases, while the number of balls of the other colour remains unchanged as at the drawing. That is, if the nth ball drawn is a black, then the number of black balls for $(n + 1)^{st}$ drawing will be c more than the number of black balls at the nth drawing, while the number of red balls will be the same for both the drawing. We have

$$P\{X_2 = 1\} = P\{X_2 = 1, X_1 = 0\} + P\{X_2 = 1, X_1 = 1\}$$

$$= P\{ X_2 = 1/X_1 = 0\}\ P\{ X_1 = 0\} + P\{ X_2 = 1/X_1 = 1\}\ P\{ X_1 = 1\}$$

$$= \frac{b}{b+r+c}\cdot\frac{r}{b+r} + \frac{b+c}{b+r+c}\cdot\frac{b}{b+r}$$

$$= \frac{b}{b+r}$$

We can show that $P\{X_n = 1\} = \dfrac{b}{b+r}$ and $P\{X_n = 0\} = \dfrac{r}{b+r}$ for all n. Again we have

$$P\{X_3 = 1, X_2 = 1\} = P\{ X_3 = 1, X_2 = 1, X_0 = 0\} + P\{ X_3 = 1, X_2 = 1, X_1 = 1\}$$
$$= P\{X_3 = 1/X_2 = 1, X_0 = 0\}\ P\{ X_2 = 1/X_0 = 0\}\ P\{X_0 = 0\}$$
$$+ P\{ X_3 = 1/ X_2 = 1, X_1 = 1\}\ P\{ X_2 = 1/X_1 = 1\}\ P\{X_1 = 1\}$$
$$= \frac{b+c}{b+r+2c}\cdot\frac{b}{b+r+c}\cdot\frac{r}{b+r} + \frac{b+2c}{b+r+2c}\cdot\frac{b+c}{b+r+c}\cdot\frac{b}{b+r}$$
$$= \frac{b(b+c)}{(b+r)(b+r+c)}$$

Hence $\quad P\{X_3 = 1|X_2 = 1\} = \dfrac{P\{X_3 = 1, X_2 = 1\}}{P\{X_2 = 1\}} = \dfrac{b+c}{b+r+c}$

but $\quad P\{X_3 = 1/X_2 = 1, X_1 = 1\} = \dfrac{b+2c}{b+r+2c}$

and $\quad P\{X_3 = 1|X_2 = 1, X_1 = 0\} = \dfrac{b+c}{b+r+2c}$

Hence $\quad P\{X_3 = 1|X_2 = 1\} \neq P\{X_3 = 1/X_2 = 1, X_1 = k\}, k = 0, 1.$

Here the conditional probability of X_3 given X_2 is not equal to that of X_3 given X_2 and X_1. Hence it is not a Markov-process.

16. Consider the above example. If Y_n denotes the number of black balls at the time of nth drawing. Y_1 takes only one value b, Y_2 takes values b and $b + c$, Y_3 takes values b, $b + c$ and $b + 2c$. In general Y_n can assume values $b, b + c, b + 2c, \ldots b + (n - 1)\ c$. Given that Y_{n-1} assume one of the possible values $b, b + c, \ldots b + (n - 2)\ c$, $i = b + kc$, $k = 0, 1, \ldots n - 2$, then Y_n assumes the value $j = i$ with probability a_1 and the value $j = i + c$ with probability b_1, where $a_1 = P\{Y_{n-1} = 0\} = \dfrac{r}{b+r}$ and $b_1 = P\{Y_{n-1} = 1\} = \dfrac{b}{b+r}$. Then $\{Y_n\}$ is an ordinary Markov chain with constant transition probabilities.

17. Moving averages: Let $\{Y_n\}$ be a sequance of mutually independent random variables. Moving average of order r are defined by

$X^{(n)} = \dfrac{Y_n + Y_{n+1} + \ldots + Y_{n+r-1}}{r}$. Here $\{X^{(n)}\}$ is not a Markov process.

18. Random Walk with absorbing barriers: Consider a particle that moves along the x-axis in such a way that its absolute speed remains constant but the direction of the motion can be reversed. The system is said to be in state E_0, if the particle moves in the positive direction, and in state E_1 if it moves in the negative direction. Then p is the probability of a reversal when the particle moves to the right and q is the probability of a reversal when it moves to the left.
Let the possible states be $E_0, E_1, \ldots E_k$. Consider the transition probability matrix P as

$$P = \begin{bmatrix} 1 & 0 & 0 & 0 & \cdots & 0 & 0 & 0 \\ q & 0 & p & 0 & \cdots & 0 & 0 & 0 \\ 0 & q & 0 & p & \cdots & 0 & 0 & 0 \\ & \cdots & & \cdots & & \cdots & & \\ & \cdots & & \cdots & & \cdots & & \\ 0 & 0 & 0 & 0 & \cdots & q & 0 & p \\ 0 & 0 & 0 & 0 & \cdots & 0 & 0 & 1 \end{bmatrix}$$

From each of the "interior" states $E_1, \ldots E_{k-1}$ transitions are possible to the right and the left neighbours (with $p_{i,\,i+1} = p$ and $p_{i,\,i-1} = q$). No transition is possible from either E_0 or E_k to any other state; the system may move from one state to another but once E_0 or E_k is reached, the system stays there fixed forever with probability one. The states E_0, E_k are called the **absorbing barriers**.

The elements of the one-step transition probability matrix P are given by

$$p_{ij} = p\delta_{i,\,j-i} + q\delta_{i,\,j+1}, \qquad i, j = 1, 2, \ldots k-1$$

$$p_{ik} = \delta_{ik} \text{ and } p_{oj} = \delta_{oj}$$

where the Kronecker δ is defined by

$$\delta i, k = \begin{cases} 1 & \text{if } i \neq j \\ 0 & \text{if } i = j, i, j = 1, 2, \ldots k-1 \end{cases}$$

19. Random Walk with Reflecting barriers: In the above example, we can introduce a variant by making the boundary reflect the particle back. In other words, from state E_0 the particle has probability p to pass to state 1 and probability q to stay at 0, Similarly, if the particle is in state E_k, it remains in E_k with probability p and moves to E_{k-1} with probability q. In this case the matrix P is given by

$$P = \begin{bmatrix} q & p & 0 & 0 & 0 & \cdots & 0 & 0 & 0 \\ q & 0 & p & 0 & 0 & \cdots & 0 & 0 & 0 \\ 0 & q & 0 & p & 0 & \cdots & 0 & 0 & 0 \\ & \cdots & & \cdots & & & \cdots & & \\ & \cdots & & \cdots & & & \cdots & & \\ 0 & 0 & 0 & 0 & 0 & \cdots & q & 0 & p \\ 0 & 0 & 0 & 0 & 0 & \cdots & 0 & q & p \end{bmatrix}$$

21. Random walk with no barriers. Consider a process $\{Y_n : n \geq 0\}$, discrete parameter Markov chain with state space $\{0, \pm 1, \pm 2, \ldots\}$. Rearrange the states as $\{0, 1, -1, 2, -2, 3, -3, \ldots\}$. Then the elements of one-step transition probability matrix P are given by

$$p_{ij} = p\delta_{i,j-1} + q\delta_{i;j+1}, \qquad i, j = 0, \pm 1, \pm 2, \ldots$$

Thus P is an infinite matrix with non-zero elements only along the super and sub-diagonols.

22. Cyclical Random Walk. Let the possible states be $E_1, E_2, \ldots E_\rho$. Order the states such that E_ρ has neighbours $E_{\rho-1}$ and E_1. Let $q_0, q_1, \ldots q_{\rho-1}$ be, respectively, the probability of staying fixed or moving 1, 2, ... $\rho - 1$ units to the right (where k units to the right is the same as $\rho - k$ units to the left). Then P is the cyclical matrix.

$$P = \begin{bmatrix} q_0 & q_1 & \cdots & q_{\rho-2} & q_{\rho-1} \\ q_{\rho-1} & q_0 & \cdots & q_{\rho-3} & q_{\rho-2} \\ q_{\rho-2} & q_{\rho-1} & q_0 & q_{\rho-4} & q_{\rho-3} \\ \cdots & & \cdots & & \cdots \\ \cdots & & \cdots & & \cdots \\ q_1 & q_2 & \cdots & q_{\rho-1} & q_0 \end{bmatrix}$$

23. Ehrenfest model of diffusion: Consider a chain with the $\rho + 1$ states $E_0, E_1, \ldots, E_\rho$ and transitions possible only to the right and to the left neighbour. Put $p_{j,j+1} = 1 - j/\rho$ and $p_{j,j-1} = j/\rho$ so that the matrix P is:

$$P = \begin{bmatrix} 0 & 1 & 0 & 0 & \cdots & 0 & 0 & 0 \\ 1/\rho & 0 & 1-1/\rho & 0 & \cdots & 0 & 0 & 0 \\ 0 & 2/\rho & 0 & 1-2/\rho & \cdots & 0 & 0 & 0 \\ & \cdots & & \cdots & & & \cdots & \\ & \cdots & & \cdots & & & \cdots & \\ 0 & 0 & 0 & \cdots & \cdots & 0 & 0 & 1/\rho \\ 0 & 0 & 0 & \cdots & \cdots & 0 & 1 & 0 \end{bmatrix}$$

24. Gambler's Ruin Problem: Let two gamblers, A and B initially have k dollars, and m dollars, respectively. Suppose that at each round of theii game, A wins one dollar from B with probability p and loses one dollar to B with probability $q = 1 - p$. Assume that A and B play until one of them has no money left. Let X_n be A's capital after round n, where $n = 0, 1, 2, \ldots$ and $X_0 = k$. Then $\{X_n : n \geq 0\}$ is a Markov chain with absorbing states. Find the probability transition matrix P.

Solution. Let Z_n be independent random variable with $P(Z_n = 1) = p$ and $P(Z_n = -1) = q = 1 - p$ for all n. Then

$$X_n = X_{n-1} + Z_n \qquad n = 1, 2, \ldots$$

and $X_0 = k$. The game ends when $X_n = 0$ or $X_n = N$.

$$\begin{aligned} P(X_{n+1} = i_{n+1}/X_0 = k, X_1 = i_1, \ldots X_n = i_n) \\ = P(Z_{n+1} + i_n = i_{n+1}/X_0 = k, X_1 = i_1, \ldots X_n = i_n) \\ = P(Z_{n+1} = i_{n+1} - i_n) \text{ since } Z_{n+1} \text{ is independent of } X_0, X_1, \ldots X_n \\ = P(X_{n+1} = i_{n+1}/X_n = i) \end{aligned}$$

Hence $\{X_n\}$ is a Markov chain. The one step transition probabilities are given by

$$p_{jk} = P(X_n = k/X_{n-1} = j) = \begin{cases} p, & k = j+1 \\ q = 1 - p, & k = j - 1 \\ 0, & \text{otherwise} \end{cases}$$

which do not depend on n. Hence $X(n)$ is a homogeneous Markov chain. Now

$$p_{i,\,i+1} = P(X_{n+1} = i+1/X_n = i) = p$$
$$p_{i,\,i-1} = P(X_{n+1} = i-1/X_n = i) = q$$
$$p_{i,\,i} = P(X_{n+1} = i/X_n = i) = 0, \qquad i \neq 0, N$$
$$p_{0,\,0} = P(X_{n+1} = 0/X_n = 0) = 1$$
$$p_{N,\,N} = P(X_{n+1} = N/X_n = N) = 1$$

Hence the transition probability P is

$$P = \begin{bmatrix} 1 & 0 & 0 & & \ldots & & 0 \\ q & 0 & p & & \ldots & & 0 \\ 0 & q & 0 & p & \ldots & & 0 \\ & \ldots & & & \ldots & & \ldots \\ & \ldots & & & \ldots & & \ldots \\ 0 & 0 & 0 & \ldots & q & 0 & p \\ 0 & 0 & 0 & \ldots & 0 & 0 & 1 \end{bmatrix}$$

25. Find periodic (or aperiodic) states of the *tpm* $\begin{bmatrix} 0 & 1 \\ 1 & 0 \end{bmatrix}$.

Solution. Here starting in state 0 (or 1), we return back to state 0 (or 1) after an even number of transitions. So states are with period 2.

26. Check whether the following vectors are probability vectors. If not, explain

1. $u = \left(\frac{1}{4}, 0, \frac{-3}{4}, \frac{1}{2}\right)$

2. $v = \left(\frac{1}{4}, \frac{1}{2}, \frac{1}{3}, 0\right)$

3. $w = \left(\frac{1}{2}, \frac{1}{4}, 0, \frac{1}{4}\right)$

Solution.

(1) u is not a probability vector, since p_3 is negative. (2) v is not a probability since the sum $\frac{1}{4}+\frac{1}{2}+\frac{1}{3} \neq 1$. (3) w is a probability vector since the sum $\frac{1}{2}+\frac{1}{4}+\frac{1}{4} = 1$.

27. A fair dice is tossed repeatedly. If X_n denotes the maximum of the numbers occurring in the first n tosses, find the transition probability matrix P of the Markov Chain $\{X_n\}$. Find also P^2 and $P(X_2 = 6)$.

Solution. Let the state space = [1, 2, 3, 4, 5, 6}.

Let X_n = the maximum of the numbers occurring in the first n trials = 3 say. Then

$$X_{n+1} = 3 \text{ if the } (n+1)^{\text{th}} \text{ trial results in 1, 2 or 3}$$
$$= 4 \text{ if the } (n+1)^{\text{th}} \text{ trial results in 4}$$
$$= 5 \text{ if the } (n+1)^{\text{th}} \text{ trial results in 5}$$
$$= 6 \text{ if the } (n+1)^{\text{th}} \text{ trial results in 6}$$

Therefore $\quad P\{X_{n+1} = 3/X_n = 3\} = \frac{1}{6} + \frac{1}{6} + \frac{1}{6} = \frac{3}{6}$

$$P\{X_{n+1} = i/X_n = 3\} = \frac{1}{6} \text{ when } i = 4, 5, 6$$

Therefore, the transition probability matrix of the chain is

$$P = \begin{bmatrix} \frac{1}{6} & \frac{1}{6} & \frac{1}{6} & \frac{1}{6} & \frac{1}{6} & \frac{1}{6} \\ 0 & 2/6 & 1/6 & 1/6 & 1/6 & 1/6 \\ 0 & 0 & 3/6 & 1/6 & 1/6 & 1/6 \\ 0 & 0 & 0 & 4/6 & 1/6 & 1/6 \\ 0 & 0 & 0 & 0 & 5/6 & 1/6 \\ 0 & 0 & 0 & 0 & 0 & 1 \end{bmatrix}$$

$$P^2 = \frac{1}{36}\begin{bmatrix} 1 & 3 & 5 & 7 & 9 & 11 \\ 0 & 4 & 5 & 7 & 9 & 11 \\ 0 & 0 & 9 & 7 & 9 & 11 \\ 0 & 0 & 0 & 16 & 9 & 11 \\ 0 & 0 & 0 & 0 & 25 & 11 \\ 0 & 0 & 0 & 0 & 0 & 36 \end{bmatrix}$$

Initial state probability distribution is

$$p(0) = \left(\frac{1}{6}, \frac{1}{6}, \frac{1}{6}, \frac{1}{6}, \frac{1}{6}, \frac{1}{6}\right) \text{ since all the values } 1, 2, \ldots, 6 \text{ are equally likely.}$$

$$P\{X_2 = 6\} = \sum_{i=1}^{6} P(X_2 = 6/X_0 = i) \times P(X_0 = i)$$

$$= \frac{1}{6}\sum_{i=1}^{6} p_{i6}^{(2)}$$

$$= \frac{1}{6} \times \frac{1}{36} \times (11 + 11 + 11 + 11 + 11 + 36)$$

$$= \frac{91}{216}$$

28. Three boys A, B and C are throwing a ball to each other. A always throw the ball to B and Balways throw the ball to C, but C is just as likely to throw the ball to B as to A. Show that the process is Markovian. Find the transition matrix and classify the states.

Solution. The transition probability matrix of the process $\{X_n\}$ is given below.

$$\text{state of } X_{n-1} \begin{matrix} \\ A \\ B \\ C \end{matrix}\overset{\text{State of } X_n}{\overset{\begin{matrix} A & B & C \end{matrix}}{\begin{bmatrix} 0 & 1 & 0 \\ 0 & 0 & 1 \\ \frac{1}{2} & \frac{1}{2} & 0 \end{bmatrix}}}$$

states of X_n depends only on states of X_{n-1} but no states of $X_{n-2}, X_{n-3}, \ldots X_1$. Therefore $\{X_n\}$ is a Markov chain. Now

$$P^2 = \begin{bmatrix} 0 & 0 & 1 \\ \frac{1}{2} & \frac{1}{2} & 0 \\ 0 & \frac{1}{2} & \frac{1}{2} \end{bmatrix}; P^3 = \begin{bmatrix} \frac{1}{2} & \frac{1}{2} & 0 \\ 0 & \frac{1}{2} & \frac{1}{2} \\ \frac{1}{4} & \frac{1}{4} & \frac{1}{2} \end{bmatrix}$$

$P_{11}^{(3)} > 0,\ P_{13}^{(2)} > 0, P_{21}^{(2)} > 0, P_{22}^{(2)} > 0, P_{33}^{(2)} > 0$ and all other $P_{ij}^{(1)} > 0$. Therefore, the chain is irreducible.

$$P^4 = \begin{bmatrix} 0 & \frac{1}{2} & \frac{1}{2} \\ \frac{1}{4} & \frac{1}{4} & \frac{1}{2} \\ \frac{1}{4} & \frac{1}{2} & \frac{1}{4} \end{bmatrix}; P^5 = \begin{bmatrix} \frac{1}{4} & \frac{1}{4} & \frac{1}{2} \\ \frac{1}{4} & \frac{1}{2} & \frac{1}{4} \\ \frac{1}{8} & \frac{3}{8} & \frac{1}{2} \end{bmatrix}$$

Note that $p_{ii}^{(2)}, p_{ii}^{(3)}, \ldots > 0$ for $i = 2, 3$ and G.C.D of 2, 3 = 1. Therefore the states 2 and 3 are periodic with period 1. Also $p_{11}^{(3)}, p_{11}^{(5)}, p_{11}^{(6)} \ldots > 0$ and GCD (3, 5) = GCD (5, 6) = 1, the state 1(i.e. state A) is periodic with period 1. That is aperiodic. Since the chain is finite and irreducible, all its states are non null persistent. Moreover all the states are ergodic.

29. A salesman's territory consists of A, B, C. He never sells in the same city on successive days. If he sells in city A then the next day he sells in B. However if he sells either in B or C, then the next day he is twice as likely to sell to city A as in other city. How after does he sell in each of the cities in the steady state.

Solution. The state space of $X_n = \{A, B, C\}$ and the transition probability distribution is

$$P = \begin{matrix} \\ A \\ B \\ C \end{matrix}\overset{\begin{matrix} A & B & C \end{matrix}}{\begin{bmatrix} 0 & 1 & 0 \\ \frac{2}{3} & 0 & \frac{1}{3} \\ \frac{2}{3} & \frac{1}{3} & 0 \end{bmatrix}}$$

Let $\pi = (\pi_1, \pi_2, \pi_3)$ be the limiting form of the stationary probability distribution. By the property of π, $\pi p = \pi$ and $\pi_1 + \pi_2 + \pi_3 = 1$

That is, $$(\pi_1, \pi_2, \pi_3)\begin{pmatrix} 0 & 1 & 0 \\ \frac{2}{3} & 0 & \frac{1}{3} \\ \frac{2}{3} & \frac{1}{3} & 0 \end{pmatrix} = (\pi_1, \pi_2, \pi_3)$$

$$\Rightarrow \quad \frac{2}{3}\pi_2 + \frac{2}{3}\pi_3 = \pi_1$$

$$\pi_1 + \frac{1}{3}\pi_3 = \pi_2$$

$$\frac{1}{3}\pi_2 = \pi_3$$

Solving we get $\pi_1 = 2/5$, $\pi_2 = 9/20$, $\pi_3 = 3/20$. Therefore the probability of selling in the 3 cities is (0.4, .45, .15)

30. Suppose that the probability of a dry day following a rainy day is $\frac{1}{3}$ and that the probability of a rainy day following a dry day is $\frac{1}{2}$. Given that May 1 is a dry day. Find the probability that May 3 is a dry day.

Solution. The state space = $\{D, R\}$ and the transition probability matrix is

$$P = \begin{array}{c} \\ D \\ R \end{array}\begin{array}{c} \begin{array}{cc} D & R \end{array} \\ \begin{pmatrix} \frac{1}{2} & \frac{1}{2} \\ \frac{1}{3} & \frac{2}{3} \end{pmatrix} \end{array}$$

$P(0) = (1, 0)$. Since May 1 is a dry day,

$$P^2 = P(0) \cdot P = (1, 0)\begin{pmatrix} \frac{1}{2} & \frac{1}{2} \\ \frac{1}{3} & \frac{2}{3} \end{pmatrix} = (1/2, 1/2)$$

$$P^3 = P^2 \cdot P = (1/2, 1/2)\begin{pmatrix} \frac{1}{2} & \frac{1}{2} \\ \frac{1}{3} & \frac{2}{3} \end{pmatrix} = (5/12, 7/12)$$

$P(\text{May 3 is a dry day}) = 5/12$.

31. A housewife buys 3 kinds of cereals A, B, C. She never buys the same cereal in successive weeks. If she buys cereal A, the next week she buys cereal B. However if she buys B or C the next week she is 3 times as likely to buy A as the other cereal. How often she buys each of the 3 cereals?
Solution. The state space is {A, B, C} and the transition probability matrix is

$$\begin{array}{c} X_{n-1} \backslash X_n \rightarrow \quad A \quad B \quad C \\ \downarrow \end{array}$$

$$P = \begin{matrix} A \\ B \\ C \end{matrix}\begin{pmatrix} 0 & 1 & 0 \\ \frac{3}{4} & 0 & \frac{1}{4} \\ \frac{3}{4} & \frac{1}{4} & 0 \end{pmatrix}$$

Let $\pi = (\pi_1, \pi_2, \pi_3)$ be the limiting form of the state probability distribution. Hence $\pi P = \pi$ and $\pi_1 + \pi_2 + \pi_3 = 1$. Now

$$\pi P = \pi$$

$$\Rightarrow \quad (\pi_1, \pi_2, \pi_3)\begin{pmatrix} 0 & 1 & 0 \\ \frac{3}{4} & 0 & \frac{1}{4} \\ \frac{3}{4} & \frac{1}{4} & 0 \end{pmatrix} = (\pi_1, \pi_2, \pi_3)$$

$$\Rightarrow \quad \frac{3}{4}\pi_2 + \frac{3}{4}\pi_3 = \pi_1$$

$$\pi_1 + \frac{\pi_3}{4} = \pi_2$$

$$\frac{\pi_2}{4} = \pi_3$$

or $\quad \pi_2 = 4\pi_3.$

Now

$$\pi_1 + \frac{\pi_3}{4} = \pi_2 = 4\pi_3 \Rightarrow 15\pi_3 - 4\pi_1 = 0$$

$$\pi_1 + \pi_2 + \pi_3 = 1 \Rightarrow \pi_1 + \pi_3 + 4\pi_3 = 1 \quad \Rightarrow \quad \pi_1 + 5\pi_3 = 1$$

Solving $\quad 4\pi_1 - 15\pi_3 = 0$

$$\pi_1 + 5\pi_3 = 1$$

We get $\quad \pi_3 = \frac{4}{35}, \; \pi_1 = \frac{3}{7} \text{ and } \pi_2 = \frac{16}{35}.$

Hence the probability of buying the cereals is (3/7, 16/35, 4/35).

32. A gambler has Re 2/-. He bets Re 1 at a time and wins Re 1 with probability $\frac{1}{2}$. He stops playing if he loses Rs 2 or wins Rs 4.

(a) What is the transition probability matrix of the related Markov chain?
(b) What is the probability that he has lost his money at the end of 5 plays?
(c) What is the probability that the game lasts more that 7 plays?

Solution. State space of $\{X_n\} = \{0, 1, 2, 3, 4, 5, 6\}$, as the game ends, if the player loses all the money ($X_n = 0$) or wins Rs 4 i.e. Rs 6 ($X_n = 6$). The transition probability matrix of the Markov chain is

$$P = \begin{array}{c} \\ 0 \\ 1 \\ 2 \\ 3 \\ 4 \\ 5 \\ 6 \end{array} \begin{array}{c} \begin{array}{ccccccc} 0 & 1 & 2 & 3 & 4 & 5 & 6 \end{array} \\ \begin{bmatrix} 1 & 0 & 0 & 0 & 0 & 0 & 0 \\ \frac{1}{2} & 0 & \frac{1}{2} & 0 & 0 & 0 & 0 \\ 0 & \frac{1}{2} & 0 & \frac{1}{2} & 0 & 0 & 0 \\ 0 & 0 & \frac{1}{2} & 0 & \frac{1}{2} & 0 & 0 \\ 0 & 0 & 0 & \frac{1}{2} & 0 & \frac{1}{2} & 0 \\ 0 & 0 & 0 & 0 & \frac{1}{2} & 0 & \frac{1}{2} \\ 0 & 0 & 0 & 0 & 0 & 0 & 1 \end{bmatrix} \end{array}$$

This is called a random walk with absorbing barriers at 0 and 6, since the chain cannot come out of the states 0 and 6 cannot come out of the states 0 and 6 once it entered.
The initial probability distribution of $\{X_n\}$ is $p(0) = (0, 0, 1, 0, 0, 0, 0)$ as the player has got Rs 2/ - to start with.

$$p^{(1)} = p^{(0)}\ P = \left(0, \frac{1}{2}, 0, \frac{1}{2}, 0, 0, 0\right)$$

$$p^{(2)} = p^{(1)}\ P = \left(\frac{1}{4}, 0, \frac{1}{2}, 0, \frac{1}{4}, 0, 0\right)$$

$$p^{(3)} = p^{(2)}\ P = \left(\frac{1}{4}, \frac{1}{4}, 0, \frac{3}{8}, 0, \frac{1}{8}, 0\right)$$

$$p^{(4)} = P^{(3)}\ P = \left(\frac{3}{8}, 0, \frac{5}{16}, 0, \frac{1}{4}, 0, \frac{1}{16}\right)$$

$$p^{(5)} = p^{(4)}\ P = \left(\frac{3}{8}, \frac{5}{32}, 0, \frac{9}{32}, 0, \frac{1}{8}, \frac{1}{16}\right)$$

P\{the man has lost his money at the end of 5 plays\} = $P\{X_5 = 0\}$ = the entry corresponding to state 0 in $p^{(5)} = 3/8$.

Again $$p^{(6)} = p^{(5)}\ P = \left(\frac{29}{64}, 0, \frac{7}{32}, 0, \frac{13}{64}, 0, \frac{1}{8}\right)$$

$$p^{(7)} = p^{(6)}\ P = \left(\frac{29}{64}, \frac{7}{64}, 0, \frac{27}{128}, 0, \frac{13}{128}, \frac{1}{8}\right)$$

P\{the game lasts more than 7 rounds\} =
P\{the system is neither in state 0 nor in 6 at the end of the seventh round\}

$$= P\{X_7 = 1, 2, 3, 4 \text{ or } 5\}$$

$$= \frac{7}{64} + 0 + \frac{27}{128} + 0 + \frac{13}{128} = \frac{27}{64}.$$

33. There are 2 white marbles in urn A and 3 red marbles in urn B. At each step of the process, a marble is selected from each urn and the 2 marbles selected are interchanged. Let the state a_i of the system be the number of red marbles in A after 3 steps? In the long run what is the probability that there are 2 red marbles in urn A?

Solution. State space of the chain $\{X_n\} = \{0, 1, 2\}$, since the number of marbles in the urn A is always 2. Let the transition probability matrix of $\{X_n\}$ is

$$P = \begin{matrix} & 0 & 1 & 2 \\ 0 & p_{00} & p_{01} & p_{02} \\ 1 & p_{10} & p_{11} & p_{12} \\ 2 & p_{20} & p_{21} & p_{22} \end{matrix}$$

$p_{00} = 0$ (since the state cannot remain at 0 after interchange of marbles)
$p_{02} = p_{20} = 0$ (since the number of red marbles in urn cannot increase or decrease by 2 in one interchange.

To start with, A contains 0 red marble. After an interchange, A will contain 1 red marble (and 1 white marble) certainly. Hence $p_{01} = 1$.

Let $X_n = 1$. That is, A contains 1 red marble (and 1 white marble) and B contains 1 white and 2 red marbles.

If 1 red marble is chosen from A and 1 white marble is chosen from B and interchanged, then $X_{n+1} = 0$. Hence

$$P(X_{n+1} = 0/X_n = 1) = p_{10} = 1/2 \times 1/3 = 1/6.$$

Similarly we get $\qquad p_{12} = \frac{1}{2} \times \frac{2}{3} = \frac{1}{3}.$

Since P is a probability matrix $p_{10} + p_{11} + p_{12} = 1$. Hence $p_{11} = \frac{1}{2}$. Similarly $p_{21} = \frac{2}{3}$ and $p_{22} = 1 - (p_{20} + p_{21}) = \frac{1}{3}$. So

$$P = \begin{bmatrix} 0 & 1 & 0 \\ \frac{1}{6} & \frac{1}{2} & \frac{1}{3} \\ 0 & \frac{2}{3} & \frac{1}{3} \end{bmatrix}$$

Now $p^{(0)} = (1, 0, 0)$, as there is no red marble in A in the beginning. Now

$$p^{(1)} = p^{(0)} P = (0, 1, 0)$$
$$p^{(2)} = p^{(1)} P = (1/6, 1/2, 1/3)$$
$$p^{(3)} = p^{(2)} P = \left(\frac{1}{12}, \frac{23}{26}, \frac{5}{18}\right)$$

Hence P(there are 2 red marbles in A after 3 steps) = $P(X_3 = 2) = p_2^{(3)} = \frac{5}{18}$

Let the state probability distribution has the limiting form as $\pi = (\pi_0, \pi_1, \pi_2)$. By the property of π, $\pi P = \pi$, $\pi_0 + \pi_1 + \pi_2 = 1$. That is

$$(\pi_0, \pi_1, \pi_2)\begin{bmatrix} 0 & 1 & 0 \\ \frac{1}{6} & \frac{1}{2} & \frac{1}{3} \\ 0 & \frac{2}{3} & \frac{1}{3} \end{bmatrix} = (\pi_0, \pi_1, \pi_2)$$

$$\Rightarrow \qquad \frac{\pi_1}{6} = \pi_0 \Rightarrow \pi_1 = 6\pi_0$$

and $$\pi_0 + \frac{\pi_1}{2} + \frac{2\pi_2}{3} = \pi_1$$

$$\frac{\pi_1}{3} + \frac{\pi_2}{3} = \pi_2$$

and $$\pi_0 + \pi_1 + \pi_2 = 1$$

Now

$$\frac{\pi_1}{3} + \frac{\pi_2}{3} = \pi_2 \qquad \Rightarrow \pi_1 = 2\pi_2$$

$$\pi_0 + \frac{\pi_1}{2} + \frac{2\pi_2}{3} = \pi_1 \qquad \Rightarrow \pi_0 + \pi_2 + \frac{2\pi_2}{3} = 6\pi_0$$

$$\Rightarrow \frac{5\pi_2}{3} = 5\pi_0$$

$$\Rightarrow \pi_2 = 3\pi_0.$$

Now $$\pi_0 + \pi_1 + \pi_2 = 1$$

$$\Rightarrow \qquad \pi_0 + 6\pi_0 + 3\pi_0 = 1 \qquad \Rightarrow 10\pi_0 = 1 \Rightarrow \pi_0 = \frac{1}{10}.$$

Hence $$\pi_1 = 6\pi_0 = \frac{6}{10} \quad \text{and} \quad \pi_2 = 3\pi_0 = \frac{3}{10}.$$

So P{ there are 2 red marbles in A in the long run} = 0.3.

34. A man drives a car or catches a train to go to office each day he never goes 2 days in a row by a train but if the drives one day, then the next day he is just as likely to drive again as he is to travel by train. Now suppose that on the first day of the week the man tossed a **fair coin** and drive to work if head appeared. Find the probability that he takes a train on the third day.

Solution. The travel pattern is a Markov chain with state space {car, train}. The transition probability matrix is

$$P = \begin{matrix} & T & C \\ T & 0 & 1 \\ C & \frac{1}{2} & \frac{1}{2} \end{matrix}$$

The initial probability is $p^{(1)} = \left(\frac{1}{2}, \frac{1}{2}\right)$, since the probability of going by car is the probability of getting a head is equal to $\frac{1}{2}$. Now

$$p^{(2)} = P^{(1)}\,P = \left(\frac{1}{2}, \frac{1}{2}\right)\begin{bmatrix} 0 & 1 \\ \frac{1}{2} & \frac{1}{2} \end{bmatrix} = \left(\frac{1}{4}\ \frac{3}{4}\right)$$

$$p^{(3)} = \left(\frac{1}{4}\ \frac{3}{4}\right)\begin{bmatrix} 0 & 1 \\ \frac{1}{2} & \frac{1}{2} \end{bmatrix} = \left(\frac{3}{8}, \frac{5}{8}\right)$$

P(travelling by train on the third day) $= \frac{3}{8}$.

35. For a cascade of binary communication channels let $P(X_0 = 1) = a$ and $P(X_0 = 0) = 1 - a$, $a \geq 0$. Compute $P\,(X_0 = 1/X_n = 1)$

Solution. Given that

$$P(X_0 = 1) = P_1(0) = a$$
$$P(X_0 = 0) = P_0(0) = 1 - a$$

The transition probability matrix is

$$P = \begin{bmatrix} 1-a & a \\ a & 1-a \end{bmatrix}$$

$$P(n) = P^n = \begin{bmatrix} 1 & 0 \\ 0 & 1 \end{bmatrix}$$

Now

$$P(X_0 = 1/X_n = 1) = P_{11}{}^{(n)} = p_{11}^{(n)} = 1.$$

36. The transition probability matrix of a Markov Chain $\{X_n\}$, $n = 1, 2, \ldots$ having 3 states 1, 2 and 3 is

$$P = \begin{bmatrix} 0.1 & 0.5 & 0.4 \\ 0.6 & 0.2 & 0.2 \\ 0.3 & 0.4 & 0.3 \end{bmatrix}$$

and the initial distribution is $p^{(0)} = (0.7, 0.2, 0)$. Find

(i) $P\{X_2 = 3\}$,

(ii) $P(X_3 = 2, X_2 = 3, X_1 = 3, X_0 = 2)$.

Solution.
$$P^2 = \begin{bmatrix} 0.1 & 0.5 & 0.4 \\ 0.6 & 0.2 & 0.2 \\ 0.3 & 0.4 & 0.3 \end{bmatrix}\begin{bmatrix} 0.1 & 0.5 & 0.4 \\ 0.6 & 0.2 & 0.2 \\ 0.3 & 0.4 & 0.3 \end{bmatrix}$$

$$= \begin{bmatrix} 0.43 & 0.31 & 0.26 \\ 0.24 & 0.42 & 0.34 \\ 0.36 & 0.35 & 0.29 \end{bmatrix}$$

(i) $P\{X_2 = 3\} = \sum_{i=1}^{3} P\{X_2 = 3/X_0 = i\}\ P(X_0 = i)$

$= p_{13}^{(2)} P(X_0 = 1) + p_{23}^{(2)} P(X_0 = 2) + p_{33}^{(2)} P(X_0 = 3)$

$= (0.26) \times (0.7) + (0.34) \times (0.2) + (0.29) \times (0.1),$ where $P^{(0)} = (0.7, 0.2, 0.1)$

$= 0.279.$

(ii) $P(X_1 = 3/X_0 = 2) = p_{23} = 0.2$

$P(X_1 = 3, X_0 = 2) = P(X_1 = 3/X_0 = 2) \times P(X_0 = 2)$

$= (0.2)\ (0.2) = 0.04$

$P\{X_2 = 3, X_1 = 3, X_0 = 2\}$

$= P\{X_2 = 3/X_1 = 3, X_0 = 2\} \times P\{X_1 = 3, X_0 = 2\}$

$= P\{X_2 = 3/X_1 = 3\} \times P\{X_1 = 3, X_0 = 2\}$ by Markov property

$= p_{33}^{(2)} \times 0.04 = (0.3) \times (0.04) = 0.012$

Now

$P(X_3 = 2, X_2 = 3, X_1 = 3, X_0 = 2)$

$= P\{X_3 = 2/X_2 = 3, X_1 = 3, X_0 = 2\} \times P\{X_2 = 3, X_1 = 3, X_0 = 2\}$

$= P\{X_3 = 2/X_2 = 3\}\ P(X_2 = 3, X_1 = 3, X_0 = 2\}$ by Markov property

$= p_{33}^{(2)} \times 0.012 = (0.4) \times (0.012) = 0.0048.$

37. A man either drives a car or catches a train to go to office each day. He never goes 2 days in a row by train but if he drives one day, then the next day he is just as likely to drive again as he is to travel by train. Now suppose that on the first day of the week, the man tossed a **fair dice** and drove to work if and only if a 6 appeared. Find (i) the probability that he takes a train on the third day and (ii) the probability that he drives to work in the long run.

Solution. The travel pattern is a Markov chain with state space = (train, car) = (T, C). The transition probability matrix of the chain is

$$P = \begin{matrix} & \begin{matrix} T & C \end{matrix} \\ \begin{matrix} T \\ C \end{matrix} & \begin{bmatrix} 0 & 1 \\ \frac{1}{2} & \frac{1}{2} \end{bmatrix} \end{matrix}$$

The initial state probability distribution is

$$p^{(1)} = \left(\frac{5}{6}, \frac{1}{6}\right),$$

since P(travelling by car) $= P$ (getting 6 in the toss of the dice) $= \frac{1}{6}$

and P (travelling by train) $= \frac{5}{6}$.

Now

$$p^{(2)} = p^{(1)}\,P = \left(\frac{5}{6}, \frac{1}{6}\right)\begin{pmatrix} 0 & 1 \\ \frac{1}{2} & \frac{1}{2} \end{pmatrix} = \left(\frac{1}{12}, \frac{11}{12}\right)$$

$$p^{(3)} = p^{(2)}\,P = \left(\frac{1}{12}, \frac{11}{12}\right)\begin{pmatrix} 0 & 1 \\ \frac{1}{2} & \frac{1}{2} \end{pmatrix} = \left(\frac{11}{24}, \frac{13}{24}\right)$$

Hence the probability the man travels by train on the third day is $\frac{11}{24}$.

Let $\pi = (\pi_1, \pi_2)$ be the limiting form of the state probability distribution or stationary state distribution of the Markov chain. So $\pi P = \pi$ and $\pi_1 + \pi_2 = 1$. Now $\pi P = \pi$ implies

$$(\pi_1, \pi_2)\begin{pmatrix} 0 & 1 \\ \frac{1}{2} & \frac{1}{2} \end{pmatrix} = (\pi_1, \pi_2)$$

$$\Rightarrow \quad \frac{1}{2}\pi_2 = \pi_1 \text{ and } \pi_1 + \frac{\pi_2}{2} = \pi_2$$

solving with $\pi_1 + \pi_2 = 1$, we get $\pi_1 = \frac{1}{3}$, $\pi_2 = 2/3$. Hence the probability the man travels by car in the long run is $\frac{2}{3}$.

38. Prove that the time that a continuous parameter Markov chain spends in a given state has an exponential distribution.

Solution. Since a homogeneous Markov chain is memory less,

$$P(Y \le r + t / Y \le t) = P(Y \le r)$$

$$\Rightarrow \quad \frac{P((Y \le r + t) \cap Y \le t)}{P(Y \le t)} = p(Y \le r)$$

$$\Rightarrow \quad \frac{P(t \le Y \le r + t)}{P(Y \le t)} = P(Y \le r)$$

$$\Rightarrow \quad \frac{F_Y(t + r) - F_Y(t)}{1 - F_Y(t)} = F_Y(r)$$

As $r \to 0$, we have

$$\underset{r \to 0}{\text{lt}} \frac{F_Y(t + r) - F_Y(t)}{r} \cdot \frac{1}{1 - F_Y(t)} = \underset{r \to 0}{\text{lt}} \frac{F_Y(r)}{r}$$

$$\Rightarrow \qquad F_Y^1(0) = \frac{F_Y^1(t)}{1 - F_Y(t)}$$

$$\Rightarrow \qquad F_Y^1(0) = d\left(\frac{1 - F_Y(t)}{1 - F_Y^1(t)}\right)$$

Integrating with respect to t, we get

$$F_Y^1(0)t = -\log(1 - F_Y^1(t))$$

$$\Rightarrow \qquad 1 - F_Y(t) = e^{-F_Y(0)t}$$

$$\Rightarrow \qquad F_Y(t) = 1 - e^{-F_Y(0)t}$$

39. Let A = $\begin{pmatrix} 0 & 1 \\ \frac{1}{2} & \frac{1}{2} \end{pmatrix}$ be a probability matrix. Prove that A is regular.

Solution. A is a probability matrix since the sum of the elements of any row of the matrix is 1. Now

$$A^2 = \begin{pmatrix} 0 & 1 \\ \frac{1}{2} & \frac{1}{2} \end{pmatrix}\begin{pmatrix} 0 & 1 \\ \frac{1}{2} & \frac{1}{2} \end{pmatrix} = \begin{pmatrix} \frac{1}{2} & \frac{1}{2} \\ \frac{1}{4} & \frac{3}{4} \end{pmatrix}$$

Since each entry in A^2 is positive and sum of the elements of any row of the matrix is 1, A^2 is regular.

40. Assume that a man is at an integral point of the x-axis between the origin and the point $x = 3$. He takes a unit step either to the right with probability 0.6 or to the left with probability 0.4. When he is at the origin he takes a step to the right to reach x = 1 or he is at the point $x = 3$ when he takes a step to the left to reach $x = 2$.

Solution. The transition matrix is

		states of X_n			
		0	1	2	3
States of X_{n-1}	0	0	1	0	0
	1	0.4	0	0.6	0
	2	0	0.4	0	0.6
	3	0	0	1	0

41. Verify Chapman Kolmogorov Theorem for the probability matrix

$$P = \begin{pmatrix} 0 & 1 & 0 & 0 \\ 0.3 & 0 & 0.7 & 0 \\ 0 & 0.3 & 0 & 0.7 \\ 0 & 0 & 1 & 0 \end{pmatrix}$$

Solution. For a given P, P^2 is

$$P^2 = \begin{pmatrix} 0 & 1 & 0 & 0 \\ 0.3 & 0 & 0.7 & 0 \\ 0 & 0.3 & 0 & 0.7 \\ 0 & 0 & 1 & 0 \end{pmatrix} \begin{pmatrix} 0 & 1 & 0 & 0 \\ 0.3 & 0 & 0.7 & 0 \\ 0 & 0.3 & 0 & 0.7 \\ 0 & 0 & 1 & 0 \end{pmatrix}$$

$$= \begin{pmatrix} 0.3 & 0 & 0.7 & 0 \\ 0 & 0.51 & 0 & 0.49 \\ 0.09 & 1 & 0.91 & 0 \\ 0 & 0.3 & 0 & 0.7 \end{pmatrix}$$

In this matrix consider $p_{22}^{(2)} = 0.91$. Now in order to reach from state 2 to 2, the intermediate states are k = 0, 1, 2, 3. That is $p_{20}\, p_{02}$; $p_{21}\, p_{12}$; $p_{22}\, p_{22}$; $p_{23}\, p_{32}$. Hence $p_{22}^{(2)}$ is obtained by

$$\begin{aligned} p_{22}^{(2)} &= p_{20}\, p_{02} + p_{21}\, p_{12} + p_{22}\, p_{22} + p_{23}\, p_{32} \\ &= (0)(0) + (0.3)(0.7) + (0)(0) + (0.7) \\ &= 0.21 + 0.7 = 0.91 \end{aligned}$$

Similarly we prove $\quad P_{ij}^{(n)} = [P_{ij}]^n$.

Hence the Chapman Kolomogorov Theorem is verified.

42. Explain the Sine Wave random process.

Solution. The random process $X(t) = r\cos(\omega t + \phi)$ where r is the random amplitude, ϕ, phase, w $= 2\pi f$, $f = \dfrac{1}{T}$, r and ϕ are random variables, where $\phi \in [0, 2\pi]$.

43. Consider the Markov chain with transition probability matrix

$$P = \begin{bmatrix} 0 & 0 & 1 & 0 \\ 0 & 0 & 0 & 1 \\ 0 & 1 & 0 & 0 \\ \frac{1}{2} & \frac{1}{8} & \frac{1}{8} & \frac{1}{4} \end{bmatrix}$$

Show that it is ergodic.

Solution. Since $p_{44}(4) = \dfrac{1}{4} > 0$, the state 4 is aperiodic. We prove the recurrent state 4, is non-null. That is, to prove $\mu_{44} < \infty$. Now

$$\begin{aligned} \mu_{44} &= \sum_{n=1}^{\infty} n\, f_{44}(n) \\ &= 1f_{44}(1) + 2f_{44}(2) + 3f_{44}(3) + 4f_{44}(4) \\ &= 1 \times \frac{1}{2} + 2 \times \frac{1}{8} + 3 \times \frac{1}{8} + 4 \times \frac{1}{4} = \frac{17}{8} < \infty \end{aligned}$$

Hence 4th state is ergodic. As all the states are ergodic, the transition probability matrix is ergodic.

44. Consider the Markov chain with transition probability matrix

$$\begin{bmatrix} 0.4 & 0.6 & 0 & 0 \\ 0.3 & 0.7 & 0 & 0 \\ 0.2 & 0.4 & 0.1 & 0.3 \\ 0 & 0 & 0 & 1 \end{bmatrix}$$

is it irreducible ? If not, find the classes. Find the nature of states.

Solution. As all states do not communicate with each other, the chain is not irreducible or ergodic. The classes of the chain are {0, 1}, {2}, {3}. State 3 is an absorbing state, as no other state is accessible from it and $P_{33} = 1$.

EXERCISES

1. Consider the Markov chain having state space {0, 1, 2} and the transition matrix

$$P = \begin{bmatrix} 0 & 1 & 0 \\ 1-p & 0 & p \\ 0 & 1 & 0 \end{bmatrix}$$

(a) Find P^2 and show it is equal to P'.

(b) Find P^n, $n \geq 1$.

2. Assume that a computer system is in one of three states; busy, idle or undergoing repair, respectively denoted by states 0, 1, 2. Observing its state at 2 p.m each day, we believe that the system approximately behaves like a homogeneous Markov chain with the transition probability matrix

$$P = \begin{bmatrix} 0.6 & 0.2 & 0.2 \\ 0.1 & 0.8 & 0.1 \\ 0.6 & 0.0 & 0.4 \end{bmatrix}$$

Prove that the chain is irreducible, and determine the steady-state probabilities.

3. The number of accidents in a town follows a Poisson process with a mean of 2 per day and the number X_i of people involved in the accident has the distribution

$$P\{X_i = k\} = \frac{1}{2^k} \qquad (k \geq 1)$$

Find the mean and the variance of the number of people involved in accidents per week.

4. If $N_1(t)$, $N_2(t)$ are two independent Poisson processes with parameters λ_1, λ_2 respectively, then show that

$$P\{N_1(t) = k/N_1(t) + N_2(t) = n\} = {}^nC_k\, p^k q^{n-k}$$

where $$p = \frac{\lambda_1}{\lambda_1 + \lambda_2};\ q = \frac{\lambda_2}{\lambda_1 + \lambda_2}.$$

5. If $X(t)$ with $X(0)$ and $\mu = 0$ is a Wiener process, show that $Y(t) = \sigma_x(t/\sigma^2)$ is also a Wiener process. Find its covariance function.

6. Show that the renewal function of a renewal process generated by X_n, $n = 1, 2, \ldots$ is given by

$$M(t) = \frac{t}{\mu} + \frac{\sigma^2 - \mu^2}{2\mu^2} + O(1)$$

where $\mu < \infty$ and σ^2 are the mean and variance of X_n.

7. Suppose that the probability of a dry day (state 0) following a rainy day (state 1) is $\frac{1}{3}$ and the probability of a rainy day following a dry day is $\frac{1}{2}$. Given that May 1 is dry day, find the probability that (*i*) May 3 is also a dry day. (*ii*) May 5 is also dry day.

8. The three-state Markov chain is given by the *tpm*

$$P = \begin{bmatrix} 0 & 2/3 & 1/3 \\ 1/2 & 0 & 1/2 \\ 1/2 & 1/2 & 0 \end{bmatrix}$$

Prove that the chain is irreducible and all the states are aperiodic and null persistent. Find also the steady state distribution of the chain.

9. Two boys B_1, B_2, two girls G_1, G_2 are throwing a ball from one to another. Given the probability that a boy throws to another boy as $\frac{1}{2}$ and to each girl as probability $\frac{1}{4}$. On the other hand, each girl throws the ball to each boy with probability $\frac{1}{2}$, and never to the other girl. In the long run, how often does each receive the ball ?

10. Prove that the matrix $\begin{pmatrix} 0 & 1 & 0 \\ 0 & 0 & 1 \\ \frac{1}{2} & \frac{1}{2} & 0 \end{pmatrix}$ is the *tpm* of an irreducible Markov chain.

11. For a cascade of binary communication channels let $P(X_0 = 1) = a$, $P(X_0 = 0) = 1 - a$, $a \geq 0$. Compute the probability that a one was transmitted given that a one was received after the *n*th stage.

12. If the *tpm* of a Markov chain is $\begin{pmatrix} 0 & 1 \\ \frac{1}{2} & \frac{1}{2} \end{pmatrix}$, find the steady state distribution of the chain.

13. Assume that the weather in a certain locality can be modeled as the homogeneous Markov chain whose transition probability matrix is given below :

Today's weather	*Tomorrow's weather*		
	Fair	*Cloudy*	*Rainy*
Fair	0.8	0.15	0.05
Cloudy	0.5	0.3	0.2
Rainy	0.6	0.3	0.1

If the initial state distribution is given by $p^{(0)} = (0.7, 0.2, 0.1)$ find $p^{(2)}$ and $\underset{n \to \infty}{\text{Lt}}\ p^{(n)}$.

Answers

7. (*i*) $tpm = \begin{matrix} & D & R \\ D & \frac{1}{2} & \frac{1}{2} \\ R & \frac{1}{3} & \frac{1}{3} \end{matrix}$, $P(\text{May 3 is a dry day}) = \dfrac{5}{12}$

(*ii*) $P(\text{May 5 is also a dry day}) = \dfrac{173}{432}$.

8. $\left(\dfrac{9}{27}, \dfrac{10}{27}, \dfrac{8}{27}\right)$

9. $\left(\dfrac{1}{3}, \dfrac{1}{3}, \dfrac{1}{6}, \dfrac{1}{6}\right)$

10. $P(X_0 = 1/X_n = 1) = p_{11}^{(n)} = 1$.

11. If (π_1, π_2) is the steady-state distribution of the chain, then $\pi_1 = \dfrac{1}{3}$, $\pi_2 = \dfrac{2}{3}$.

13. $p^{(2)} = (.7245, .1920, .0835)$; $\underset{n \to \infty}{\text{Lt}}\ p^{(n)} = \left(\dfrac{114}{157}, \dfrac{30}{157}, \dfrac{13}{157}\right)$.

7
Queueing Models

Queue is a common word that means either a waiting line or the act of joining a line. It occurs whenever arriving customers wait in a line to get service at one or more service point. It is formed when the number of customers arriving is greater than the number of customers being served during a period of time.

Various Queueing situations are: Vehicles waiting in a petrol bunk, patients waiting for doctor's clinic, planes arriving in an airport for landing, customers waiting at barber shop for hair cut, subscribers waiting for telephone connection, passengers waiting at railway booking counter, customers waiting at ration shop, students waiting at college fee counter.

The above situations have some common features like customers' arrival, formation of a queue, waiting for service, provision of service to the customers in a certain order, departure after service etc.

In this chapter, we learn how to construct and solve equations that describe queueing behaviour for the above discussed situations.

To minimise the total cost of queue and the cost of time, with the help of a suitable mathematical model is the aim of developing queueing model.

7.1 QUEUEING SYSTEM

A queueing system consists of one or more servers, an arrival pattern of customers, service pattern, queue discipline, the order in which the service is provided and customer behaviour. The word 'queue' is sometimes used to describe the whole system, but it is a part of the system that holds the excess customers who cannot be immediately served. Hence, the number of customers in the system at any given time will equal to the number of customers in the queue plus the number of customers in service. Of course, these numbers will vary with time due to customers' arrival and departure, so they are formally random process.

(i) **Arrival Pattern:** The pattern of arrival of customers like patients, telephone calls, cars at service station is known as **arrival pattern of a queue**. The arrival pattern is either regular or irregular (or at random intervals of time). A regular arrival pattern is not common in the case of customers coming for service. Generally it is completely random.

In the case of number of customers is very large, the probability of an arrival in next interval of time does not depend upon the number of customers already in the system. Hence the arrival is completely random and it follows Poisson process with mean equals average number of arrivals per unit of time. Also we can view the times between arrivals as independent negative exponentially distributed random variables. The same process, viewed another way, gives a Poisson distributed random variables for the number of customers arriving in any time interval. We use Poisson arrival process as a convenient one in queueing models because the data requirements are minimal. A single parameter, the arrival rate λ, is enough to describe the entire process. The reciprocal of λ gives the mean time between arrivals.

Another arrival process is renewal process which means that the times between arrivals are independent random variables.

(ii) **Service Pattern:** The service pattern is similar to arrival pattern, but there are some important differences. Strictly speaking, it is not a random process because it is interrupted by the idle periods when no customers arrives.

The servers may work fast when the queue is long. In the this case, the easiest service time distribution to work with is the negative exponential distribution, which is fully characterized by a single parameter, the mean service rate μ or its mean service time $1/\mu$.

(iii) **Service Arrangements:** For providing service to the incoming customers one or more service points (or servers) are established. The number depends on the number of customers, rate of arrivals, time taken for providing service to a single customer, etc. Depending on these variables, a service channel is a single or a multiple.

Single servers: In the case of a single channel service, there is one queue and one service counter, where one customer gets service at one.

Pictorial representation of a queueing system are shown in the Figure.

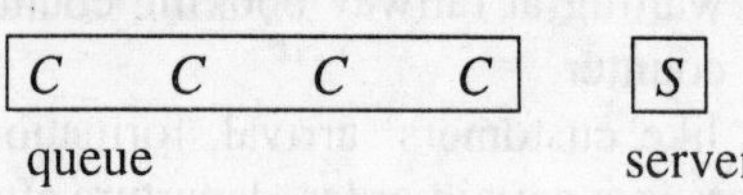

Series channels: A customer must pass successively through all the ordered channels before service is completed. For example, in public offices where part of the service are done at different service counters.

Parallel channels: A number of channels providing identical service facilities so that several customers may be serviced simultaneously.

For example, in a barber shop many chairs are considered as different service channels and the customers may wait in a single queue until one of the service channel is ready to serve.

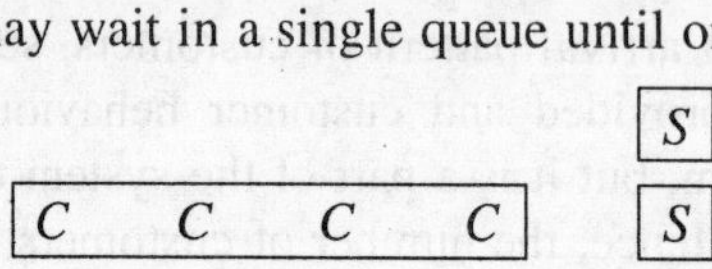

three server, one queue

Multiple-server Model: A queueing system is called a **multiple-server model** when the system has a number of parallel channels with one server.

Example Ticket counters at railway station, airports, etc.

Modeling the multiple-channel server is very difficult. We study more about multiple-channel server in section 3.

(iv) Service time: The time required for servicing a customer is called **service time**. Service time may be either a constant or it vary with the customer. However, for sake of simplicity of queue model, it is assumed that the time required for servicing is constant for all customers.

Moreover, since the arrival pattern is assumed to be random, the service time is also taken as random. Hence the service time follows exponential distribution with mean equal to reciprocal of the mean rate of service. In case where the assumption of service time follows exponential fails, Erlang distribution is applied.

(v) Queue Discipline: The order in which members of the queues are selected for service is known as **queue discipline**. This can be one of the following types:

(a) **First in First served (out) (FIFO):** Here the selection of customers for service is done in the order in which they arrive and form the queue.
Example: Customers served in ration shop.

(b) **Last in First out (LIFO):** Selection of customers for service is done in the reverse order in which they arrive and the person entering last is the first to be selected for service.
Example: In a big godown, the items which come last are taken out first.

(c) **Service in Random Order (SIRO):** The selection of customer for servicing is random at any particular time.

(d) **Service in Priority (SIP):** Certain customers are given priority over others in selection for service. It is of two types: Non-emptive priority, emptive priority.

Non-emptive priority: The customer already getting service is allowed to continue till it is completed even if a priority customer comes mid-way during the service.

Emptive priority: The service to non-priority customer is stópped as soon as a priority customer arrives.

e.g.: Repairing of some critical machine when it breaks in a factory.

(v) **Customer Behaviour**

Customers generally behaves in 4 ways.

(a) **Balking:** Some customers show reluctance for waiting in the queue. They do not join the queue at their correct position and attempt to jump the queue and reach the service centre by passing other ahead of them.

(b) **Reneging:** Some customers after waiting some time in the queue leaves the queue without getting the service due to impatience.

(c) **Collusion:** Some of the customers join together and only one of them, instead of all, stay in the queue. However, when their turn comes for service, the customers who were in collusion demand service.

(d) **Jockeying:** In case there is more than one queue for similar type of service, some customers keep on shifting from one queue to another queue to improve their position and to get immediate service.

Output: The output per unit of time will depend upon the total time spent in the system, both while waiting and for getting service. The output is inversely proportional to the total system time. The output rate or production rate of the queueing system is the mean number of service completions per unit and is denoted by R.

Definition of Transient and Steady State

A system is said to be in **transient state** when its operating characteristics (like input, output, mean queue length etc.) are dependent upon time.

A system is said to be in steady state when the behaviour of the system is independent of time. If $P_n(t)$ denotes the probability that there are n customers in the system at time t, then in the steady state case,

$$\lim_{t \to \infty} P_n(t) = P_n \text{ (independent of } t\text{)}$$

This implies

$$\lim_{t \to \infty} \frac{d}{dt} P_n(t) = 0.$$

The queue models can be of the following

(i) **Probabilistic Models:** When both arrival rate and service rate are unknown and are assumed to be random variables. The probability distribution obtained is based on past experience.
(ii) **Deterministic Models:** When both arrival rate and service rate are exactly known.
(iii) **Mixed Models:** When either the arrival rate or the service rate is exactly known and the other is not known.

Terminology used in the text:

(i) **Customer:** A unit coming for service to the service station is called a **customer**.
Example: persons, machines, telephone calls, demand for some commodity. Sometimes number of customes in the system are mentioned as states of the system.
(ii) **Waiting line:** A line formed by customers waiting to receive service is known as **waiting line.**
(iii) **Arrival rate:** The arrival rate is calculated by dividing the total number of arrivals by the total units of time. It is denoted by λ.
If the rate of arrivals is in Poisson distribution, then the distribution of time between two arrivals will be exponential.
(iv) **Service rate:** Average number of customers being served per unit of time. It is denoted by μ.
(v) **Traffic intensity or Utilisation Rate:**

$$\text{Utilisation rate } (\rho) = \frac{\text{Mean arrival rate}}{\text{Mean service rate}} = \frac{\lambda}{\mu}.$$

This is the probability of having to queue on arriving. If $\rho > 1$, the traffic intensity is very high and consequently the waiting time will be more.

(vi) **Idle Rate:** Idle rate is $p_0 = 1 - \rho = 1 - \frac{\lambda}{\mu}$.
Expected idle time (for service facility)

$$= \left(1 - \frac{\lambda}{\mu}\right) \times \text{Total service time}$$

(vii) **Expected number of customers in the system, both waiting and in service *E* (*n*)**

$$E(n) = \frac{\lambda}{\mu - \lambda} = \frac{\rho}{1 - \rho}$$

(viii) **Expected number of customers in queue or waiting line *E* (*m*)**

$$E(m) = E(n) - \rho = \frac{\lambda}{\mu - \lambda} - \frac{\lambda}{\mu} = \frac{\lambda^2}{\mu(\mu - \lambda)}$$

(ix) **Expected time spent by the customers in system *E* (*T*)**
$E(T)$ = expected time spent while waiting for service + the service time

$$= E(w) + \frac{1}{\mu}$$

Since the rate of arrivals is λ, the expected number of customers in the system $E(n)$ will be equal to $\lambda \times E(T)$. So

$$E(T) = \frac{E(n)}{\lambda} = \frac{\frac{\lambda}{\mu - \lambda}}{\lambda} = \frac{1}{\mu - \lambda}$$

(x) **Expected waiting time in the queue $E(w)$:**

$$E(w) = E(T) - \frac{1}{\mu}$$

$$= \frac{1}{\mu - \lambda} - \frac{1}{\mu} = \frac{\lambda}{\mu(\mu - \lambda)}$$

(xi) (a) Probability that there are n customers in the system at any time t, both waiting and in service is denoted by $P_n(t)$.

(b) Probability of zero customers waiting

$$p_0 = 1 - \rho = 1 - \frac{\lambda}{\mu}.$$

(c) Probability of 1, 2, . . . n persons waiting in the queue for service are

$$p_1 = p_0\left(\frac{\lambda}{\mu}\right);\ p_2 = p_0\left(\frac{\lambda}{\mu}\right)^2,$$

$$p_n = p_0\left(\frac{\lambda}{\mu}\right)^n$$

(d) Probability that the number of customers in the system greater than or equal to k is

$$P(n \geq k) = \sum_{k=n}^{\infty} p_k$$

$$= \sum_{k=n}^{\infty}\left(1 - \frac{\lambda}{\mu}\right)\left(\frac{\lambda}{\mu}\right)^k$$

$$= \left(\frac{\lambda}{\mu}\right)^k$$

(e) Probability of spending more than w units of time is $e^{-(\mu-\lambda)w}$.

Distribution of Arrivals

The model in which only arrivals are counted and no departure takes place are called pure birth model. We proved already that the number of arrivals in time interval $(0, t)$ is a random variable which follows a Poisson distribution with parameter $\lambda(t)$ is $p_n(t) = \frac{(\lambda t)^n}{n!} e^{-\lambda t}$, for $n \geq 0$.

Distribution of Inter-arrival Times (Exponential process)

Inter-arrival times are defined as the time intervals between two successive arrivals. If the arrival process follows the Poisson distribution, an associated random variable defined as the time between successive arrivals follows the exponential distribution $f(t) = \lambda e^{-\lambda t}$ and vice versa. That is;

Let the random variable T be the time between successive arrivals; then

$$\begin{aligned} P(T > t) &= P\,(\text{no arrival in time } t) \\ &= p_0(t) \\ &= e^{-\lambda t} \end{aligned}$$

The cumulative distribution function of T is

$$\begin{aligned} F(t) = P(T \le t) &= 1 - P(T > t) \\ &= 1 - e^{-\lambda t} \qquad 0 \le t \le \infty \end{aligned}$$

The density function $f(t)$ of inter-arrival times is

$$f(t) = \frac{d}{dt}F(t) = \begin{cases} \lambda e^{-\lambda t}, & 0 \le t \le \infty \\ 0, & t < 0 \end{cases}$$

The expected time of inter-arrival is given by

$$\begin{aligned} E(t) &= \int_0^\infty t\, f(t)\, dt \\ &= \int_0^\infty \lambda t e^{-\lambda t}\, dt = \frac{1}{\lambda}. \end{aligned}$$

Thus T has the exponential distribution with mean $\frac{1}{\lambda}$.

Distribution of service times

The time t to complete the service on a customer follows an exponential distribution which is given by

$$s(t) = \begin{cases} \mu e^{-\mu t}, & 0 \le t \le \infty \\ 0, & t < 0 \end{cases}$$

where μ is the mean service rate for a particular service channel.

Markovian Queueing Models

To be a Markovian queueing system, both inter-arrival times and service times must be negative exponentially distributed, so that it is fully characterized by only one parameter, which is related to the mean.

When the queueing model is Markovian, we can solve a set of linear equations whose solution provides the steady-state distribution of the number of customers in the system. All of the other performance measures we have mentioned (*v*) – (*xi*) can be computed. The simplest Markovian model to set up and solve is the single server, infinite capacity model.

Kendal's Notation for Representing Queueing Models

Generally queueing model may be completely specified in the following symbol form:

$$(a/b/c) : (d/e)$$

The first and second symbols denote the type of distributions of inter-arrival times and of inter-service times respectively. Third symbol specifies the number of servers, whereas fourth symbol stands for the capacity of the system and the last symbol denotes the queue discipline.

If we specify the following letters as:

$\mu \equiv$ Poisson arrival or departure distribution
$E_K \equiv$ Erlang or Gamma inter-arrival or service time distribution
$GI \equiv$ General input distribution
$G \equiv$ General service time distribution,

then $(M/E_K/I)$: (∞/FIFO) defines a queueing system in which arrivals follow Poisson distribution, service times are Erlang, single server, infinite capacity and 'first in first out' queue discipline.

M/M/I Queueing System: This queueing system deals with the process in which arrivals and services occur randomly over time. Arrivals can be considered as **births** to the system, since if the system is in state E_n and an arrival occurs, the state is changed to E_{n+1}. If departure occurs the state E_n is changed down to E_{n-1} and can be looked upon as a death. This type process is a birth-death process.

Model I (M/M/I): (∞/FIFO): This model deals with a queueing system having single server, Poisson arrival, Exponential services and infinite capacity and the service is on the basis "First in First out".

State diagram:

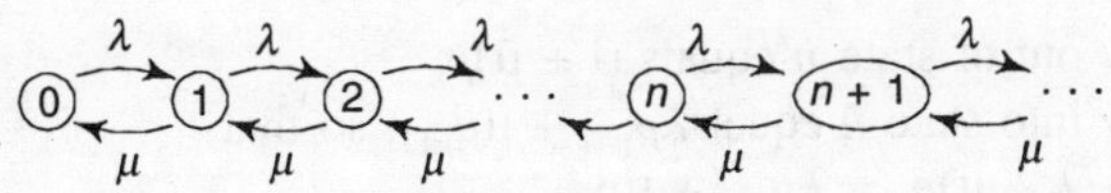

7.1.1 Theorem

With usual notation, show that probability distribution of queue length n is given by $\rho^n (1 - \rho)$.

Proof: Let $p_n(t)$ be the probability that there n customers in the system at time t. To be in state E_n at time $t + \Delta t$,

(i) the system could have been in state E_n at time t and have no arrivals or service completions in the time Δt.

(ii) the system be in E_{n-1} at time t and during Δt, one arrival and no service completions.

(iii) the system be in state E_{n+1} at time t and during Δt time, one service completed and no arrivals.

Hence, as arrivals and service are both independent of each other, for $n \geq 1$, we have

$$\begin{aligned} p_n(t + \Delta t) &= p_n(t) \cdot P(\text{no arrival in } \Delta t)\, P(\text{no service completed in } \Delta t) \\ &\quad + p_n(t)\, P(\text{one arrival in } \Delta t)\, P(\text{one service completed in } \Delta t) \\ &\quad + p_{n+1}(t)\, P(\text{one service completed in } \Delta t).\ P(\text{no arrival in } \Delta t). \\ &\quad + p_{n-1}(t)\, P(\text{one arrival in } \Delta t)\, P(\text{no service in } \Delta t) \\ &= p_n(t)\,[1 - \lambda\Delta t]\,[1 - \mu\Delta t] + p_n(t)\,(\lambda\Delta t)\,(\mu\,\Delta t) \\ &\quad + p_{n+1}(t) \cdot \mu\Delta t\,(1 - \lambda\,\Delta t) + p_{n-1}(t)\,\lambda\,\Delta(t)\,(1 - \mu\Delta t) \\ &= p_n(t)\,[1 - (\lambda + \mu)\,\Delta t] + p_{n-1}(t)\,\lambda\Delta t\,(1 - \mu\,\Delta t) + p_{n+1}(t)\,\mu\,\Delta t\,(1 - \lambda\,\Delta t) \end{aligned}$$

This implies

$$\lim_{\Delta t \to 0} \frac{p_n(t + \Delta t) - p_n(t)}{\Delta t} = -(\lambda + \mu)\, p_n(t) + \lambda\, p_{n-1}(t) + \mu\, p_{n+1}(t) \tag{7.1}$$

Next we find $p_0(t + \Delta t)$ separately as $p_{n-1}(t)$ does not exist for $n = 0$. The system can be in state E_0 at time $t + \Delta t$ if it were in E_0 at t and no arrival during Δt (no service is possible since the system is empty) or the system can be in E_1 at time t and no arrivals but one service completed in Δt. Thus

$$p_0(t + \Delta t) = p_0(t)\,[1 - \lambda\Delta t] + p_1(t)\,\mu\,\Delta t\,(1 - \lambda\,\Delta t)$$

$$\Rightarrow \quad \lim_{\Delta t \to 0} \frac{p_0(t + \Delta t) - p_0(t)}{\Delta t} = -\lambda p_0(t) + p_1(t)\,\mu \tag{7.2}$$

(7.1) and (7.2) are called **differential difference equations**

Steady State Equations

In the steady state, $p_n(t) \to p_n$ and $p'_n(t) \to 0$, as $t \to \infty$. The steady state solution exists only when $\frac{\lambda}{\mu} < 1$ as $t \to \infty$. If $\lambda = \mu$, there is no queue and if $\frac{\lambda}{\mu} > 1$, the state is explosive state. Throughout this chapter we assume the steady state conditions prevails. Then (7.1) and (7.2) becomes

$$\left.\begin{aligned} 0 &= -(\lambda + \mu)\, p_n + \mu\, p_{n+1} + \lambda\, p_{n-1}, \quad n \geq 1 \\ 0 &= -\lambda\, p_0 + \mu p_1 \;\Rightarrow\; p_1 = \frac{\lambda}{\mu}\, p_0. \end{aligned}\right\} \tag{7.3}$$

The above equations are called **balance** (or equilibrium or conservation flow) **equations.**

Rate Equality Principle. In the steady state no build up occurs at any state. As such the two rates, i.e. the rate of flow out of an arbitrary state k and the rate of flow into the same state n, $n = 0, 1, 2, \ldots$ are equal. In equilibrium, $\lambda p_0 = \mu p_1$.

For $n \neq 0$,

rate of flow out of state n equals $(l + \mu)p_n$

rate of flow into state n equal $\lambda p_{n-1} + \mu p_{n+1}$, so that

$$(\lambda + \mu)p_n = \lambda p_{n-1} + \mu p_{n+1}$$

which is equation (7.3). The state-transition diagram is given by

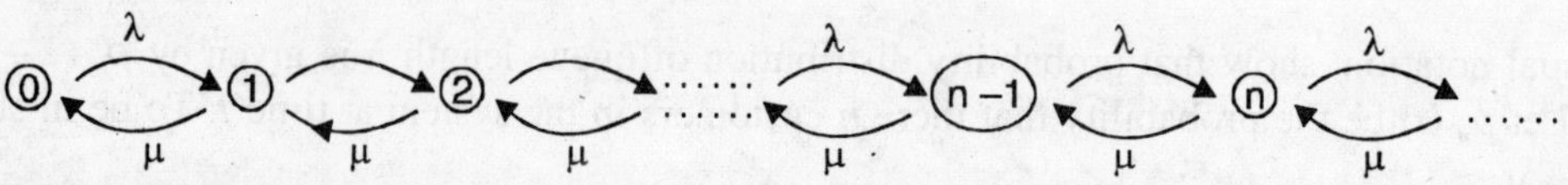

In the state-transition rate diagram of continuous parameters Markov chain, the arcs are labelled with state transition rates, and for the discrete parameter Markov chain, the arcs are labeled with conditional probablities.

For $n = 1$, in (7.3), we get $0 = -(\lambda + \mu)\, p_1 + \mu p_2 + \lambda p_0$

$$\Rightarrow \qquad p_2 = \frac{\lambda + \mu}{\mu}\, p_1 - \frac{\lambda}{\mu}\, p_0 = \left(\frac{\lambda}{\mu}\right)^2 p_0.$$

In general, we get $\quad p_n = \left(\frac{\lambda}{\mu}\right)^n p_0$, for $n \geq 0$.

To obtain the value of p_0, we make use of the boundary condition

$$\sum_{n=0}^{\infty} p_n = 1 \;\Rightarrow\; p_0 + \left(\frac{\lambda}{\mu}\right) p_0 + \left(\frac{\lambda}{\mu}\right)^2 p_0 + \ldots = 1$$

$$\Rightarrow \qquad p_0 \left[1 + \frac{\lambda}{\mu} + \left(\frac{\lambda}{\mu}\right)^2 + \ldots\right] = 1 \;\Rightarrow\; p_0 \left[\frac{1}{1 - \frac{\lambda}{\mu}}\right] = 1, \quad \text{since } \frac{\lambda}{\mu} < 1$$

$$\Rightarrow \qquad p_0 = 1 - \frac{\lambda}{\mu}$$

Hence $$p_n = \left(\frac{\lambda}{\mu}\right)^n p_0 = \left(\frac{\lambda}{\mu}\right)^n \left(1-\frac{\lambda}{\mu}\right) = \rho^n (1-\rho), \text{ where}$$

$$\rho = \frac{\lambda}{\mu} < 1 \text{ and } n \geq 0.$$

Average number of customers in the system is given by

$$E(n) = \sum_{n=0}^{\infty} n\, p_n = (1-\rho) \sum_{n=0}^{\infty} n\rho^n$$

$$= (1-\rho)\,\rho \sum_{n=0}^{\infty} n\,\rho^{n-1} = (1-\rho)\,\rho\,(1-\rho)^{-2} \quad \text{using Binomial series}$$

$$= \frac{\rho}{1-\rho} = \frac{\lambda}{\mu-\lambda}$$

The formulae as listed from (v) to (xi) above are used for obtaining various results in single channel models.

7.1.2 Example

A telephone exchange receives one call for every 4 minutes and connects one call for every 3 minutes. If the rate of arrivals follows Poisson distribution and service time follows exponential distribution find out

(a) Expected waiting time for a call
(b) Expected time in the system
(c) Expected number of customers in the system
(d) Expected number of customers in queue.

Solution: Since the arrival is one per every 4 minutes and the service is one for every 3 minutes,

Arrival rate $\lambda = 1/4$ per minute

Service rate $\mu = 1/3$ per minute

(a) Expected waiting time $$E(w) = \frac{\lambda}{\mu(\mu-\lambda)} = \frac{\frac{1}{4}}{\frac{1}{3}\left(\frac{1}{3}-\frac{1}{4}\right)} = 9 \text{ minutes}$$

(b) Expected time in system $$E(T) = \frac{1}{\mu-\lambda} = \frac{1}{1/3-1/4} = 12 \text{ minutes}$$

(c) Expected number of customers in the system $$= \frac{\lambda}{\mu-\lambda} = 3 \text{ calls}$$

(d) Expected number of customers in queue

$$= \frac{\lambda^2}{\mu(\mu-\lambda)} = \frac{\left(\frac{1}{4}\right)^2}{\frac{1}{3}\left(\frac{1}{3}-\frac{1}{4}\right)} = \frac{9}{4} \text{ calls.}$$

7.1.3 Example

In a super market, the average arrival rate of customer is 10 per every 30 minutes following Poisson process. The average time taken by the cashier to list and calculate the customers purchases is 2.5 minutes following exponential distribution. What is the probability that the queue length exceeds 6. What is the expected time spent by a customer in the system?

Solution. Here the mean arrival rate is $\lambda = \dfrac{10}{30}$ per minute

The mean service rate $= \dfrac{1}{2.5}$ per minute

$$\Rightarrow \quad \rho = \lambda/\mu = \frac{\frac{1}{3}}{\frac{1}{2.5}} = 0.8333$$

(i) The probability of queue size $> n = \rho^n$
when $n = 6$, $\rho^n = (0.8333)^6 = 0.3346$.

(ii) $E(T) = \dfrac{1}{\mu - \lambda} = 14.96$ minutes.

7.1.4 Example

For a single queue, single service point system with Poisson arrivals and exponential service times, show how the average number of people in the system would vary if an average of 12 people needed the service every hour and μ took alternative values of 24, 18 and 15 per hour. What is the probability that a person coming in would have to be in queue?

Solution. Given $\lambda = 12$,

(a) Average number of customers in the system $= \dfrac{\lambda}{\mu - \lambda}$

when $\mu = 24$, the average number of customers is $\dfrac{12}{24-12} = 1$,

when $\mu = 18$, the average number of customers is $\dfrac{12}{18-12} = 2$,

when $\mu = 15$, the average number of customers is $\dfrac{12}{15-12} = 4$

(b) The probability of having to queue on arriving $= \rho = \dfrac{\lambda}{\mu}$.

when $\mu = 24$, $\rho = \dfrac{12}{24} = 0.5$

when $\mu = 18$, $\rho = \dfrac{12}{18} = 0.67$

when $\mu = 15$, $\rho = \dfrac{12}{15} = 0.80$

7.1.5 Example

At a one man barber shop, it takes on the average half an hour for a hair cut, and customers arrive at an average rate of one per every 45 minutes. If the Poisson Exponential model is assumed, at any point in time, on the average,

(a) How many customers will there be in the shop?
(b) What time will a customer spend in the shop?
(c) Of this time, what part be spent waiting to be served?
(d) What portion of the day will barber be idle?
(e) What is the probability that a customer has to wait for 30 minutes?

Solution. λ = mean arrival rate per hour = $\frac{60}{45} = \frac{4}{3}$

μ = mean service rate per hour = $\frac{60}{30} = 2.$

$$\rho = \frac{\lambda}{\mu} = \frac{4}{3 \times 2} = \frac{2}{3}.$$

(a) Expected number of customers in the shop

$$= \frac{\lambda}{\mu - \lambda} = \frac{\frac{4}{3}}{2 - 4/3} = \frac{\frac{4}{3}}{2/3} = 2 \text{ customers.}$$

(b) Average time spent by customer in the shop

$$= \frac{1}{\mu - \lambda} = \frac{1}{2 - \frac{4}{3}} = \frac{1}{2/3} = 3/2 = 1\frac{1}{2} \text{ hours}$$

(c) Average time spent waiting in the shop to be served = $1\frac{1}{2}$ hours – $\frac{1}{2}$ hour = 1 hour

or $$= \frac{\lambda}{\mu(\mu - \lambda)} = \frac{\frac{4}{3}}{2(2/3)} = 1 \text{ hour}$$

(d) Idle time = $1 - \lambda/\mu = 1 - \frac{\frac{4}{3}}{2} = 1 - \frac{4}{6} = \frac{1}{3}$ of the day.

(e) Probability that a customer has to wait for 1/2 hour is $e^{-(\mu - \lambda)w} = e^{-\left(2 - \frac{4}{3}\right)\frac{1}{2}} = e^{-5/6}$.

7.1.6 Example

Patients arrive at a doctor's clinic at mean intervals of 20 minutes, and spend a mean of 15 minutes in consultation, both times exponentially distributed. The doctor wishes to have enough seats in the waiting room so that no more than about 1% of arriving patients will have to stand. How many seats should be provided?

Solution. Here $\lambda = 3$ per hour, $\mu = 4$ per hour

$$\rho = \frac{3}{4} = 0.75$$

Let C be the number of seats required.

Probability of C or fewer patients in the system should be about $1 - .01 = 0.99$.

That is $$p(\le C) = \sum_{n=0}^{C} p_n = 0.99$$

We know $$p_n = (1-\rho)\rho^n, \text{ so that}$$

$$p(\le C) = \sum_{n=0}^{C} p_n = \sum_{n=0}^{C} (1-\rho)\rho^n = (1-\rho)\sum_{n=0}^{C} \rho^n$$

$$= (1-\rho)(1 + \rho + \rho^2 + \ldots + \rho^C) = (1-\rho)\left(\frac{1-\rho^{C+1}}{1-\rho}\right)$$

by Geometric series expansion

$$= 1 - \rho^{C+1}$$

This implies

$$1 - \rho^{C+1} = 0.99 \quad \Rightarrow \quad \rho^{C+1} = 0.01$$

$$\Rightarrow \quad (0.75)^{C+1} = 0.01 \quad \Rightarrow \quad (C+1)\log(0.75) = \log(0.01)$$

$$\Rightarrow \quad C + 1 = \frac{\log(0.01)}{\log(0.75)} = \frac{-2}{-0.1249} \quad \Rightarrow \quad C = \frac{-2}{-0.1249} - 1 \cong 16 \text{ seats.}$$

Waiting Time Distribution for Model I

Discrete distribution: (i) when $w = 0$, i.e. the system is empty and the new arrival is immediately taken into service. Waiting time is zero. Hence

$$P(w = 0) = p_0 = 1 - \rho$$

(ii) when $w > 0$, then $P(w > 0) = 1 - P(w = 0) = 1 - (1 - \rho) = \rho$.

Continuous Distribution

Probability of waiting time in the system $\ge t$ is

$$P(w \ge t) = \int_t^\infty (\mu - \lambda)\, e^{-(\mu-\lambda)w}\, dw$$

Probability of waiting time in the queue $\ge t$ is

$$\int_t^\infty \rho(\mu - \lambda)\, e^{-(\mu-\lambda)w}\, dw.$$

7.1.7 Example

Arrivals at a telephone booth are considered to be Poisson with an average time of 10 minutes between one arrival and the next. The length of a phone call is assumed to be distributed exponential, with 3 minutes.

(a) What is the probability that a person arriving at the booth will have to wait?

(b) The telephone department will install a second booth when convinced that an arrival would expect waiting for at least 3 minutes for phone. By how much should the flow of arrivals increase in order to justify a second booth?

(c) Find the average number of units in the system?

(d) Estimate the fraction of a day that the phone will be in use.

(e) What is the probability that it will take him more than 10 minutes altogether to wait for phone and complete his call?

Solution. Here $\lambda = 1/10 = 0.10$ per minute, $\mu = 1/3 = 0.33$ per minute

(a) $P(w > 0) = 1 - p_0 = 1 - (1 - \lambda/\mu) = \lambda/\mu = \dfrac{0.10}{0.33} = 0.3.$

(b) The installation of second booth will be justified if the arrival rate λ' is greater than waiting time. Now

$$E(w) = \frac{\lambda'}{\mu(\mu - \lambda')}$$

$$\Rightarrow \quad 3 = \frac{\lambda'}{0.33(0.33 - \lambda^1)} = \frac{\lambda'}{0.1089 - 0.33\lambda^1} \quad \Rightarrow \quad \lambda^1 = .164$$

Hence the arrival rate should become 0.16 person per minute to justify the second booth.

(c) Average number of units in the system is given by

$$E(n) = \frac{\rho}{1-\rho} = \frac{0.3}{1-0.3} = 0.43 \text{ customers}$$

(d) The fraction of a day that the phone will be busy $= \rho = \lambda/\mu = 0.3$

(e) $P(w \geq 10) = \int_{10}^{\infty} (\mu - \lambda)e^{-(\mu - \lambda)w}\, dw$

$$= (\mu - \lambda)\left[\frac{e^{-(\mu-\lambda)w}}{-(\mu-\lambda)}\right]_{10}^{\infty} = -\left[e^{-(\mu-\lambda)w}\right]_{10}^{\infty} = e^{-(\mu-\lambda)10} = e^{-2.3} = 0.10.$$

7.1.8 Example

In the production shop of a company the breakdown of the machines is found to be Poisson with an average rate of 3 machines per hour. Break down time at one machine costs Rs. 40 per hour to the company. There are two choices before the company for hiring the repairman. One of the repairman is slow but cheap, the other is fast but expensive. The slow-cheap repairman demands Rs. 20 per hour and will repair the broken down machines exponentially at the rate of 4 per hour. The fast-expensive repairman demands Rs. 30 per hour and will repair machines exponentially at an average rate of 6 per hour. Which repairman should be hired?

Solution. Compare the total expected daily cost for both the repairmen. This equal the total wages paid plus the down time cost.

Case 1: Slow-cheap repairman.

$\lambda = 3$ machines per hour, $\mu = 4$ machines per hour

$$\text{Average down time of a machine} = \frac{1}{\mu\ \lambda} = \frac{1}{4-3} = 1 \text{ hour}$$

The down time of 3 machines that arrive in an hour = 1 × 3 = 3 hours.
Down time cost = Rs. 40 × 3 = Rs. 120. Charges paid to the repairman = Rs. 20 × 3 = Rs. 60
Total cost = Down time cost + charges paid = Rs. 120 + Rs. 60 = Rs. 180.

Case 2: Fast-expensive repairman

λ = 3 machines per hour, μ = 6 machines per hour

$$\text{Average down time of machine} = \frac{1}{\mu - \lambda} = \frac{1}{3} \text{ hours}$$

$$\text{Down time of 3 machines that arrive in an hour} = \frac{1}{3} \times 3 = 1 \text{ hour.}$$

Down time cost = Rs. 40 × 1 = Rs. 40
Charges paid to the repairman = Rs. 30 × 1 = Rs. 30
Total cost = Rs. 40 + Rs. 30 = Rs. 70
Comparing the total cost of both the repairmen's the fast. expensive repairman is to be hired.

7.1.9 Relations between Average Queue Length and Average Waiting Time–Little's formula

Mean number in the system equals the mean number in the queue plus the mean number in service. That is

$$E(n) = E(m) + \rho$$

A similar relation in terms of time is

$$E(T) = E(w) + \frac{1}{\mu}$$

Another relation is

$$U = \frac{\rho}{c}$$

where the average utilization equals the mean number of busy servers divided by number of servers c. If there is only one server $c = 1$, then

$$U = \rho = 1 - p_0.$$

If the arrival rate is constant λ and if every arrival is accepted, then the output rate

$$R = \lambda$$

However, finite capacity will not accept every arrival. Only those customers who arrive when the system is less than full are allowed to the system. In such cases, the pure arrival rate have to be adjusted to get an effective arrival rate.

If arrivals see random states, the pure arrival rate that are successful is $1 - p_N$. Hence, in this case

$$R = \lambda(1 - p_N)$$

Little's formula gives the relationship between $E(n)$, $E(T)$ and R as

$$E(n) = R\,E(T)$$

If any two of these quantities are fixed, the third can be determined. In an infinite capacity system, where

$$R = \lambda,$$

$$E(n) = \lambda\,E(T)$$

Any system that transforms input to output over time and possesses a steady state conditions corresponding to $E(n)$, R, $E(T)$ will obey Little's formula.

Model II (M/M/1): (∞/SIRO). This model is same as Model I, except the service discipline follows the SIRO rule (service in random order) instead of FIFO-rule. For SIRO-case, we must have

$$p_n = (1 - \rho)\rho^n, \quad n \geq 0$$

The average number of customers in the system is $E(n)$ and p_n remains unchanged. So

$$E(T) = \frac{1}{\lambda} E(n)$$

under the SIRO-rule and is given by

$$E(T) = \frac{1}{\mu - \lambda}$$

This result can be extended to any queue discipline as long as p_n remains unchanged. The three queue disciplines FIFO, LIFO, SIRO differ only in the distribution of waiting time where the probability of long and short waiting time change depending upon the discipline used.

7.1.10 Limited Queue Capacity

Model III (M/M/1): (C/FIFO). When waiting space is limited, a number of minor modifications must be made to the previous models. In this model the maximum number of customers in the system is limited to C.

The steady-state equations are identical to those used previously with additional difference equation for $n = C$, then

$$p_C(t + \Delta t) = p_C(t)[1 - \mu\Delta t] + p_{C-1}(t)(\lambda\Delta t)(1 - \mu\Delta t) + O(\Delta t)$$

This gives after simplification, the equation

$$\frac{d}{dt} p_C(t) = -\mu p_C(t) + \lambda p_{C-1}(t)$$

from which the steady-state equation is obtained as

$$0 = -\mu p_C + \lambda p_{C-1}$$

The complete set of steady-state equations for this model, therefore, can be written as

$$\mu p_1 = \lambda p_0$$

$$\mu p_{n+1} = (\lambda + \mu) p_n - \lambda p_{n-1}, \quad 1 \leq n \leq C - 1,$$

$$\mu p_C = \lambda p_{C-1}$$

Using the iterative procedure (as was done in Model I), the first two difference equations give

$$p_n = (\lambda/\mu)^n p_0, \; n \leq C$$

Applying the normalizing equation to determine p_0, we get

$$1 = p_0 + p_1 + p_2 + \ldots + p_C = p_0 + \left(\frac{\lambda}{\mu}\right) p_0 + \left(\frac{\lambda}{\mu}\right)^2 p_0 + \ldots + \left(\frac{\lambda}{\mu}\right)^C p_0$$

$$= p_0 \left[1 + \left(\frac{\lambda}{\mu}\right) + \left(\frac{\lambda}{\mu}\right)^2 + \ldots + \left(\frac{\lambda}{\mu}\right)^C\right] = p_0 \left[\frac{1 - \left(\frac{\lambda}{\mu}\right)^{C+1}}{1 - \frac{\lambda}{\mu}}\right]$$

$$\Rightarrow \quad p_0 = \frac{1-\dfrac{\lambda}{\mu}}{1-\left(\dfrac{\lambda}{\mu}\right)^{C+1}} = \frac{1-\rho}{1-\rho^{C+1}}, \text{ for } \rho \neq 1$$

Thus
$$p_n = \begin{cases} \dfrac{(1-\rho)\rho^n}{1-\rho^{C+1}}, & \rho \neq 1 \\ \dfrac{1}{C+1} & \rho = 1 \end{cases}$$

(i) Average number of customers in the system is given by

$$E(n) = \sum_{n=0}^{C} n\, p_n = p_0 \sum_{n=0}^{C} n\, \rho^n = p_0\, \rho \sum_{n=0}^{C} \frac{d}{d\rho}(\rho^n) = p_0\, \rho\, \frac{d}{d\rho} \sum_{n=0}^{C} \rho^n$$

$$= p_0\, \rho\, \frac{d}{d\rho}\left[\frac{1-\rho^{C+1}}{1-\rho}\right] \qquad \text{by using Geometric series expansion}$$

$$= p_0 \rho \left[\frac{1-(C+1)\rho^C + C\rho^{C+1}}{(1-\rho)^2}\right]$$

$$= p_0 \rho \left[\frac{1-(C+1)\rho^C + C\rho^{C+1}}{(1-\rho)(1-\rho^{C+1})}\right] \text{ for } \rho \neq 1$$

$$= \frac{C}{2} \text{ for } \rho = 1.$$

(ii) Average queue length is given by

$$E(m) = \sum_{n=1}^{C} (n-1)p_n = E(n) - \sum_{n=1}^{C} p_n = E(n) - (1-p_0) = E(n) - \left[1 - \left(\frac{1-\rho}{1-\rho^{C+1}}\right)\right]$$

$$= E(n) - \left[\frac{\rho-\rho^{C+1}}{1-\rho^{C+1}}\right] = E(n) - \left[\frac{\rho(1-\rho^C)}{1-\rho^{C+1}}\right]$$

$$= \rho^2 \left[\frac{1-C\rho^{C-1} + (C-1)\rho^C}{(1-\rho)(1-\rho^{C+1})}\right], \rho \neq 1$$

(iii) The average waiting time in the system can be obtained by using Little's formula

$$E(T) = \frac{1}{\lambda'} E(n)$$

where λ' is the mean rate of customers entering the system and is equal to $\lambda(1-p_N)$.

(iv) The average waiting time in the queue is

$$E(w) = E(T) - \frac{1}{\mu} \text{ or } E(w) = E(m)/\lambda'.$$

7.1.11 Example

If for period of 2 hours in a day (8-10 a.m) trains arrive at the yard every 20 minutes but the service time continues to remain 36 minutes then calculate for this period.

(a) the probability that the yard is empty

(b) average queue length assuming the capacity of the yard is 4 trains only.

Solution. Here $\rho = \lambda/\mu = \dfrac{1/20}{1/36} = \dfrac{36}{20} = 1.8$ per minute

$C = 4$

(a) $p_0 = \dfrac{\rho - 1}{\rho^5 - 1} = .04$

(b) average queue size $= p_0 \sum\limits_{n=0}^{4} n\,\rho^4 = p_0\,[\rho + 2\rho^2 + 3\rho^3 + 4\rho^4]$

$$= .04\,[1.8 + 3.24 \times 2 + 5.832 \times 3 + 10.4976 \times 4] = 2.7106 \cong 3 \text{ trains}$$

7.1.12 Example

In a railway marshalling yard, goods trains arrive at a rate of 30 trains per day. Assume that the inter-arrival-time follows, an exponential distribution and the service time distribution is also exponential with an average of 36 minutes, calculate

(a) the probability that the yard is empty

(b) average queue length assuming that the line capacity of the yard is 9 trains.

Solution. Here $\rho = \dfrac{\lambda}{\mu} = \dfrac{\left(\dfrac{30}{24\times 60}\right)}{\left(\dfrac{1}{36}\right)} = \dfrac{36}{48} = 0.753$ per minute

(a) The probability that the yard is empty is

$$p_0 = \frac{1-\rho}{1-\rho^{C+1}}, \text{ where } C = 9 = \frac{1-.075}{1-(0.75)^{10}} = \frac{0.25}{0.90} = 0.2649$$

(b) Average queue length is

$$E(m) = \frac{\rho^2\,[1 - C\rho^{C-1} + (C-1)\,\rho^C\,]}{(1-\rho)\,(1-\rho^{C+1})}$$

$$= \frac{(0.75)^2\,(1-9(0.75)^8 + 8(0.75)^9\,)}{(0.25)\,(1-(0.75)^{10})} = \frac{(0.75)^2\,(0.6997)}{(0.25)\,(0.9437)} = 1.668$$

7.2 MULTI-SERVER QUEUES

Multiserver system is designated by *M/M/s*, where *s* is the number of servers. The system capacity *C* would certainly have to be atleast *s*. If $C = s$, then the maximum number of customers permitted in the system equals the number of servers. In this case there is no need to wait for service.

Assume that all servers operate at the same mean rate μ. So, the service to customers take place on a number of customers simultaneously, but departures will occur one at a time. (The probability of two or more services being completed at exactly the same instant is zero).

If there is only one customer in the system then he will receive service from one of servers at the rate μ. If there are two customers present, they will both receive service, each at the rate μ. So the rate at which this serves works is twice as great as the rate at which a single server works or 2μ. Similarly, if s customers are present, they all will receive service each at the rate μ. So the rate at which this server works is s times as great as the rate at which a single server works or $s\mu$.

State diagram:

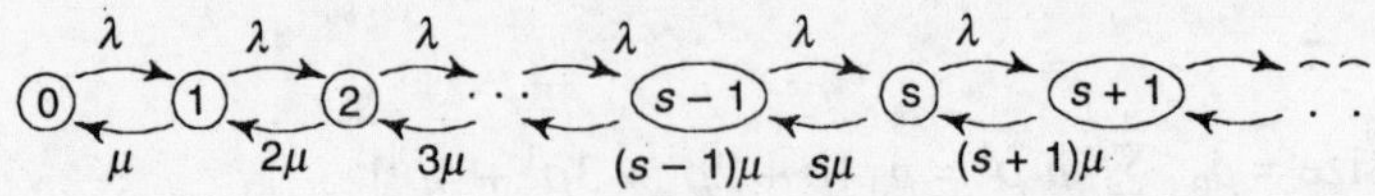

The steady-state equations are

$$0 = -\lambda p_0 + \mu p_1$$
$$0 = \lambda p_0 - (\lambda + \mu)\, p_1 + 2\mu p_2$$
$$0 = \lambda p_1 - (\lambda + 2\,\mu)\, p_2 + 3\mu p_3$$
$$\vdots$$
$$0 = \lambda p_{s-1} - s\,\mu p_s$$

These can be solved in the usual birth-death equations as:

$$p_j = \frac{1}{j!}\left(\frac{\lambda}{\mu}\right)^j p_0, \qquad \text{for } j = 1, 2, \ldots, s.$$

Using the normalization equation to evaluate p_0, we find

$$p_0 = \frac{1}{\sum_{i=0}^{s} \frac{\rho^i}{i!}}, \text{ where } \rho = \lambda/\mu$$

If s is very large relation to ρ, so that the remaining terms in the infinite series are small, so we write p_0 as $e^{-\rho}$, and hence

$$p_j = \frac{1}{j!}\,\rho^j e^{-\rho}$$

is approximately, the distribution of the number of occupied servers.

When s is finite and the approximation is not appropriate, the steady state distribution is

$$p_j = \frac{\frac{\rho^j}{j!}}{\sum_{i=0}^{s} \frac{\rho^i}{i!}}$$

and this distribution is called a **truncated Poisson distribution**. The probability that all servers are occupied is given by

$$p_C = \frac{\dfrac{\rho^C}{C!}}{\sum_{i=0}^{s} \dfrac{\rho^i}{i!}}$$

This formula is known as Erlang's lost call formula or first Erlang function. The mean number of occupied servers or customers in the system is given by $\rho(1 - p_C)$.

The solution discussed above were for "No waiting case". Suppose some waiting capacity is available, the steady-state equations are:

$$0 = -\lambda p_0 + \mu p_1$$

$$0 = \lambda p_{j-1} - (\lambda + j\mu)\, p_j + (j+1)\, \mu p_{j+1} \qquad \text{for } j = 1, 2, \ldots s-1$$

$$0 = \lambda p_{j-1} - (\lambda + s\mu)\, p_j + s\mu p_{j+1} \qquad \text{for } j = s, s+1, \ldots$$

If the total system capacity C is finite, there would be one equation for the last state

$$0 = \lambda p_{C-1} - s\mu p_C$$

hence we get models:

M/M/s: (C/FIFO) Maximum number in the system is limited to C where $C > s$,

$$\lambda_j = \begin{cases} \lambda & 0 \le j < C \\ 0 & j \ge C \end{cases}$$

$$\mu_j = \begin{cases} j\mu & 0 \le j < s \\ s\mu & s \le j \le C \end{cases}$$

we get

$$p_j = \begin{cases} \dfrac{1}{j!}\left(\dfrac{\lambda}{\mu}\right)^j p_0, & 0 \le j < s \\[2ex] \dfrac{1}{s!}\dfrac{1}{s^{j-s}}\left(\dfrac{\lambda}{\mu}\right)^j p_0, & s \le j \le C \end{cases}$$

where

$$p_0 = \left\{ \sum_{j=0}^{s-1} \frac{1}{j!}\left(\frac{\lambda}{\mu}\right)^j + \sum_{j=s}^{C} \frac{1}{s^{j-s}\, s!}\left(\frac{\lambda}{\mu}\right)^j \right\}^{-1}$$

provided $\dfrac{\lambda}{s\mu} < 1$.

(i) Average queue length is given by

$$E(m) = \sum_{n=s}^{C} (n-s)\, p_n = \sum_{n=s}^{C} (n-s) \,.\, \frac{\left(\dfrac{\lambda}{\mu}\right)^n}{s!\, s^{n-s}}\; p_0$$

$$= \frac{1}{s!}\left(\frac{\lambda}{\mu}\right)^s p_0\, \rho \sum_{x=0}^{C-s} x\rho^{x-1}, \qquad \text{for } \rho = \frac{\lambda}{s\mu}$$

$$= \frac{1}{s!}\left(\frac{\lambda}{\mu}\right)^s p_0\, \rho \sum_{x=0}^{C-s} \frac{d}{d\rho}(\rho^x) = \frac{(s\rho)^s}{s!}\, p_0\, \rho\, \frac{d}{d\rho}\left[\sum_{x=0}^{C-s} \rho^x\right]$$

$$= \frac{(s\rho)^s\, p_0\, \rho}{s!}\, \frac{d}{d\rho}\left[\frac{1-\rho^{C-s+1}}{1-\rho}\right]$$

$$= \frac{p_0\,(s\rho)^s\, \rho[1-\rho^{C-s+1} - (1-\rho)(C-s+1)\rho^{C-s}]}{(1-\rho)^2 \,.\, s!}$$

(ii) Average number of customers in the system is given by

$$E(n) = \sum_{n=0}^{C} n\, p_n = \sum_{n=0}^{s-1} n\, p_n + \sum_{n=s}^{C} n\, p_n = \sum_{n=0}^{s-1} n\, p_n + E(m) + s \sum_{n=s}^{C} p_n$$

$$= E(m) + s\left[\sum_{n=0}^{C} p_n - \sum_{n=0}^{s-1} p_n\right] + \sum_{n=0}^{s-1} n\, p_n$$

$$= E(m) + s - p_0 \sum_{n=0}^{s-1} \frac{(s-n)(\rho s)^n}{n!}$$

(iii) Average waiting time in the system can be obtained by using Little's formula, that is

$$E(T) = \frac{E(n)}{\lambda'}, \quad \text{where } \lambda' = \lambda(1 - p_c)$$

Average waiting time in the queue can be obtained by using

$$E(w) = E(T) - \frac{1}{\mu} = \frac{E(T)}{\lambda(1-p_c)}.$$

(M/M/s): (∞/FIFO): This model is the generalization of (*M/M/*1): (∞/FIFO) in the sense that service channels is '*s*' in number instead of a single server. Clearly, since the input is Poisson, and the service is exponential, we have a birth-death process. The mean arrival rate is given by $\lambda_n = \lambda$ for all *n*. If there are more *s* customers in the system, all the servers '*s*' remain busy and each is of mean rate μ, and hence the mean service rate is $s\mu$. Other hand, if there are fewer customers than '*s*' in the system, $n < s$, only *n* of the *s* servers are busy and thus the mean rate is $n\mu$. Hence

$$\mu_n = \begin{cases} n\mu, & 1 \le n < s \\ s\mu, & n \ge s \end{cases}$$

The steady state solution is:

$$p_n = \begin{cases} \dfrac{1}{n!}\left(\dfrac{\lambda}{\mu}\right)^n p_0, & 1 \le n < s \\[2ex] \dfrac{1}{s^{n-s}\, s!}\left(\dfrac{\lambda}{\mu}\right)^n p_0, & n \ge s \end{cases}$$

Making use of the boundary condition $\sum_{n=0}^{\infty} p_n = 1$, we find the value of p_0 as

$$p_0 = \left[\sum_{n=0}^{s-1} \frac{1}{n!} \left(\frac{\lambda}{\mu} \right)^n + \frac{1}{s!} \left(\frac{\lambda}{\mu} \right)^s \frac{s\mu}{s\mu - \lambda} \right]^{-1}$$

(i) Average waiting time in the queue is $\dfrac{\mu \left(\dfrac{\lambda}{\mu} \right)^s p_0}{(s-1)!(s\mu - \lambda)}$

(ii) Average queue length is given by

$$E(m) = \frac{\lambda\mu \left(\dfrac{\lambda}{\mu} \right)^s p_0}{(s-1)!(s\mu - \lambda)^2} = \left(\frac{\lambda}{\mu} \right)^s \cdot \frac{1}{s!} p_0 \cdot \frac{\dfrac{\lambda}{\mu s}}{\left(1 - \dfrac{\lambda}{\mu s} \right)^2}$$

(iii) Average number of customers in the system is:

$$E(n) = E(m) + \frac{\lambda}{\mu}$$

$$= \frac{\lambda\mu \left(\dfrac{\lambda}{\mu} \right)^s p_0}{(s-1)!(s\mu - \lambda)^2} + \frac{\lambda}{\mu}$$

(iv) Expected waiting time that a customer spends in the system is $\dfrac{E(m)}{\lambda} + \dfrac{1}{\mu}$

$$= \frac{1}{(s-1)!} \left(\frac{\lambda}{\mu} \right)^s \frac{\mu}{(s\mu - \lambda)^2} p_0 + \frac{1}{\mu}$$

7.2.1 Example

A super market has two girls running up sales at the counters. If the service time for each customer is exponential with mean 4 minutes and if people arrive in a Poisson fashion at the rate of 10 an hour

(a) What is the probability of having to wait for service?

(b) What is the expected percentage of idle time for each girls?

(c) If a customer has to wait, what is the expected length of his waiting time?

Solution. $s = 2$, $\lambda = \dfrac{10}{60}$ per minute, $\mu = \dfrac{1}{4}$ per minute, $\dfrac{\lambda}{s\mu} = \dfrac{1}{3}$

Therefore the problem is of $(M/M/s)$: $(\infty/FIFO)$ model.

$$p_0 = \left[\sum_{n=0}^{2-1} \frac{1}{n!}\left(\frac{4}{6}\right)^n + \frac{1}{2!}\left(\frac{4}{6}\right)^2 \left(\frac{\frac{1}{2}}{\frac{1}{2}-\frac{1}{6}}\right)\right]^{-1} = \left[\sum_{n=0}^{1} \frac{1}{n!}\left(\frac{4}{6}\right)^n + \frac{1}{2!}\left(\frac{2}{3}\right)^2\left(\frac{3}{2}\right)\right]^{-1}$$

$$= \left[1+\frac{1}{3}+\frac{2}{3}\right]^{-1} = \frac{1}{2}$$

$$p_1 = \frac{\lambda}{\mu}\, p_0 = \frac{2}{3}\times\frac{1}{2} = \frac{1}{3}$$

(a) The probability of having to wait for service is

$$p(n \geq 2) = p(w > 0)$$

$$= \sum_{n=2}^{\infty} p_n = \sum_{n=2}^{\infty} \frac{1}{2!\,2^{n-2}}\left(\frac{2}{3}\right)^n p_0, \quad n \geq 2$$

$$= \sum_{n=2}^{\infty} \frac{2^2}{2.2^n}.\frac{2^n}{3^n}.\frac{1}{2}$$

$$= \sum_{n=2}^{\infty}\left(\frac{1}{3}\right)^n = \sum_{n=0}^{\infty}\left(\frac{1}{3}\right)^n - \frac{1}{3} - 1 = \frac{1}{1-\frac{1}{3}} - \frac{4}{3} = \frac{3}{2} - \frac{4}{3} = \frac{1}{6}$$

(b) The expected idle time for each girl is

$$1 - \frac{\lambda}{s\mu} = 1 - \frac{1}{3} = \frac{2}{3} = 0.67$$

(c) Expected length of customers waiting time in the system

$$E(T) = E(W) + \frac{1}{\mu} = \frac{E(m)}{\lambda} + \frac{1}{\mu} = \frac{1}{(s-1)!}\left(\frac{\lambda}{\mu}\right)^s . \frac{\mu}{(s\mu-\lambda)^2}\, p_0$$

$$= \left(\frac{4}{6}\right)^2 . \frac{\frac{1}{4}}{\left(\frac{1}{2}-\frac{1}{6}\right)^2} \times \frac{1}{2} + 4 = 4.5 \text{ minutes}$$

7.2.2 Example

A Telephone exchange has two long distance operators. It finds that during the peak`load, long distance calls arrive in a Poisson fashion at an average rate of 15 per hour. The length of service on these calls is approximately exponential with mean length 5 minutes.

(a) What is the probability that a subscriber will have to wait for his long distance call during the peak hours of the day.

(b) If the subscribers will wait and one serviced in turn what is the expected waiting time?

Solution.

(a) $s = 2$. $\lambda = \frac{15}{60} = \frac{1}{4}$ per minute, $\mu = \frac{1}{5}$ per minute

$$\rho = \frac{\lambda}{s\mu} = \frac{\frac{1}{4}}{2 \times \frac{1}{5}} = \frac{5}{8}.$$

$$p_0 = \left[\sum_{n=0}^{2-1} \frac{1}{n!}\left(\frac{5}{4}\right)^n + \frac{1}{2!}\left(\frac{5}{4}\right)^2 \left(\frac{\frac{2}{5}}{\frac{2}{5} - \frac{1}{4}} \right) \right]^{-1} = \left[1 + \frac{5}{4} + \frac{25}{12}\right]^{-1} = \left(\frac{13}{3}\right)^{-1} = \frac{3}{13}$$

$$p(w > 0) = \sum_{n=2}^{\infty} \frac{1}{2!2^{n-2}} \left(\frac{5}{4}\right)^n \frac{3}{13} = \sum_{n=2}^{\infty} \frac{2^2}{2.\,2^n} \left(\frac{5^n}{4^n}\right) \cdot \left(\frac{3}{13}\right) = \frac{6}{13} \sum_{n=2}^{\infty} \left(\frac{5}{8}\right)^n$$

$$= \frac{6}{13}\left[\sum_{n=0}^{\infty} \left(\frac{5}{8}\right)^n - 1 - \frac{5}{8} \right] = \frac{6}{13}\left[\frac{1}{1 - \frac{5}{8}} - 1 - \frac{5}{8} \right]$$

$$= \frac{6}{13} \times \frac{25}{24} = \frac{25}{52} \cong 0.48$$

(b) Average waiting time in the queue is

$$\frac{\mu \left(\frac{\lambda}{\mu}\right)^s p_0}{(s-1)!(s\mu - \lambda)^2} = \frac{\frac{1}{5}\left(\frac{5}{4}\right)^2 \times \frac{3}{13}}{\left(\frac{2}{5} - \frac{1}{4}\right)^2} = \frac{125}{39} \cong 3.2 \text{ minutes.}$$

7.2.3 Example

A general insurance company has three claim adjusters in its branch office. People with claims against the company are found to arrive in Poisson fashion at an average rate of 20 per 8 hour day. The amount of time that an adjuster spends with a claimant is found to have negative exponential distribution with mean service time of 40 minutes. Claimants are processed in the order of their appearance.

(a) How many hours a week can an adjuster expect to spend with claimants?
(b) How much time, on the average, does claimant spend in the branch office?

Solution. Here $\lambda = \frac{20}{8} = \frac{5}{2}$ arrivals per hour

$\mu = \frac{1}{40}$ service per minute $= \frac{3}{2}$ per hour, $s = 3$

The service is busy is $\rho = \frac{\lambda}{s\mu} = \frac{5}{9}$

(a) Therefore for 5 working days ($5 \times 8 = 40$ hours), expected weakly time an adjuster spends with claimants $= 5/9 \times 40 = 22.2$ hours.

(b) The average time an adjuster spends in the system is

$$\frac{\mu\left(\frac{\lambda}{\mu}\right)^s p_0}{(s-1)!(s\mu-\lambda)^2}+\frac{1}{\mu}$$

First we calculate p_0:

$$p_0=\left[\sum_{n=0}^{3-1}\frac{1}{n!}\left(\frac{5}{3}\right)^n+\frac{1}{3!}\left(\frac{5}{3}\right)^3\left(\frac{\frac{9}{2}}{\frac{9}{2}-\frac{5}{2}}\right)\right]^{-1}$$

$$=\left[1+\frac{5}{3}+\frac{25}{18}+\frac{125}{72}\right]^{-1}=[4.791]^{-1}=0.20872$$

$$E(n)=\frac{\frac{3}{2}\left(\frac{5}{3}\right)^3 p_0}{2!\left(\frac{9}{2}-\frac{5}{2}\right)^2}+\frac{1}{\frac{3}{2}}\cong 0.847\cong 50\text{ minutes}$$

7.2.4 Example

Given an average arrival rate of 20 per hour, is it better for a customer to get service at single channel with mean service rate of 22 customers or at one of the two channels in parallel, with mean service rate of 11 customers for each of the two channels. Assume that both queues are *M/M/s* type.

Solution. For a single channel, $\lambda = 20$ per hour and

$\mu = 22$ customers per hour. So

$$p_0 = 1-\frac{\lambda}{\mu}=1-\frac{20}{22}=0.09$$

(i) Probability of no customers is 0.09.

(ii) Expected number of customers in the system is $E(n)=\frac{\lambda}{\mu-\lambda}=\frac{20}{22-20}=10$

(iii) Expected number of customers in the queue $=\frac{\lambda^2}{\mu(\mu-\lambda)}=\frac{400}{22\times 2}=9.09$

(iv) Expected waiting time in the queue is $\frac{\lambda}{\mu(\mu-\lambda)}=\frac{20}{44}=0.45$ hrs.

When there are two parallel channels, we have $\lambda = 20$, $\mu = 11$, $s = 2$

$$p_0=\left[\sum_{n=0}^{s-1}\frac{1}{n!}\left(\frac{\lambda}{\mu}\right)^n+\frac{1}{s!}\left(\frac{\lambda}{\mu}\right)^s\frac{s\mu}{s\mu-\lambda}\right]^{-1}$$

$$= \left[\sum_{n=0}^{1} \frac{1}{n!}\left(\frac{20}{11}\right)^n + \frac{1}{2!}\left(\frac{20}{11}\right)^2 \cdot \frac{22}{22-20}\right]^{-1}$$

$$= \left[1 + \frac{20}{11} + \frac{400}{22}\right]^{-1} = (20.99)^{-1} = 0.0476 \text{ hrs.} \cong 27 \text{ minutes.}$$

Expected number of customers in the system

$$E(n) = \frac{\lambda\mu\left(\frac{\lambda}{\mu}\right)^s p_0}{(s-1)!(s\mu-\lambda)^2} + \frac{\lambda}{\mu} = \frac{20\times 11\left(\frac{20}{11}\right)^2 (0.0476)}{(22-20)^2} + \frac{20}{11}$$

$$= \frac{20}{11}\left[\frac{400\times 0.0476}{4} + 1\right] = 8.545$$

Expected number of customers in the queue

$$E(m) = \text{Expected number of customers in the system} - \frac{\lambda}{\mu}$$

$$= 8.545 - \frac{20}{11} = 6.727$$

Expected waiting time in the queue is $\frac{E(m)}{\lambda} = \frac{6.727}{20} = 0.33$ hours = 20 minutes.

On comparison we see that it is better for a customer to wait for 2 channels than a single channel, since the waiting time for a single channel is 0.45 hours, whereas the waiting time for service in a 2 channel is only 0.33 hours.

Next we see examples for multichannel finite population model.

.2.5 Example

A barber shop has two barbers and three chairs for customers. Assume that the customers arrive in a Poisson fashion at a rate of 5 per hour and that each barber service customers according to an exponential distribution with mean of 15 minutes. Further if a customer arrives and there are no empty chairs in the shop, he will leave. What is the probability that the shop is empty? What is the expected number of customers in the shop?

Solution. Here $s = 2$, $C = 3$, $\lambda = \frac{5}{60} = \frac{1}{12}$ customer/minute,

$$\mu = \frac{1}{15} \text{ per minute}$$

$$p_0 = \left[\sum_{n=0}^{2-1} \frac{1}{n!}\left(\frac{\lambda}{\mu}\right)^n + \sum_{n=2}^{3} \frac{1}{2^{n-2}\cdot 2!}\left(\frac{\lambda}{\mu}\right)^n\right]^{-1} = \left[\sum_{n=0}^{1} \frac{1}{n!}\left(\frac{5}{4}\right)^n + \sum_{n=2}^{3} \frac{1}{2!\,2^{n-2}}\left(\frac{5}{4}\right)^n\right]^{-1}$$

$$= \left[1 + \frac{5}{4} + \frac{1}{2!}\left(\frac{5}{4}\right)^2 + \frac{1}{2!}\cdot\frac{1}{2}\left(\frac{5}{4}\right)^3\right]^{-1} = \left[1 + \frac{5}{4} + \frac{1}{2}\times\frac{25}{16} + \frac{1}{4}\times\frac{125}{64}\right]^{-1}$$

$$= (3.52)^{-1} = 0.2841$$

$$p_n = \begin{cases} \dfrac{1}{n!}\left(\dfrac{\lambda}{\mu}\right)^n p_0, & 0 \le n < 2 \\ \dfrac{1}{2!}\dfrac{1}{2^{n-2}}\left(\dfrac{\lambda}{\mu}\right)^n p_0, & 2 \le n \le 3. \end{cases}$$

$$= \begin{cases} \dfrac{1}{n!}\left(\dfrac{5}{4}\right)^n (0.2841), & 0 \le n < 2 \\ \dfrac{1}{2!\,2^{n-2}}\left(\dfrac{5}{4}\right)^n (0.2841), & 2 \le n \le 3. \end{cases}$$

Expected number of customers in the shop (system) is:

$$E(n) = E(m) + s - p_0 \sum_{n=0}^{s-1} \frac{(s-n)(\rho s)^n}{n!}, \rho = \frac{\lambda}{s\mu}$$

$$= \sum_{n=s}^{C} (n-s)\, p_n + s - p_0 \sum_{n=0}^{s-1} \frac{(s-n)\left(\dfrac{\lambda}{\mu}\right)^n}{n!}$$

$$= \sum_{n=2}^{3} ((n-2)\, p_n + 2 - p_0 \sum_{n=0}^{1} \frac{(2-n)(1.25)^n}{n!}$$

$$= p_3 + 2 - p_0\,(2 + 1.25) = p_3 + 2 - (3.25)\, p_0$$

$$= \frac{1}{2(2!)}(1.25)^3\,(0.2841) + 2 - (3.25)\,(0.2841) = 1.226 \text{ customers}$$

7.2.6 Example

A car servicing station has 3 stalls where service can be offered simultaneously. The cars wait in such a way that when a stall becomes vacant, the car at the head of the line pulls up to it. The station can accommodate atmost four cars waiting (seven in the station) at one time. The arrival pattern is Poisson with a mean of one car per minute during the peak hours. The service time is exponential with mean 6 minutes. Find the average number of cars in the service station during peak hours, the average waiting time and the average number of cars per hour that cannot enter the station because of full capacity.

Solution. Here $\lambda = 1$ car per minute, $\mu = \dfrac{1}{6}$ car per minute, $s = 3$, $C = 7$,

$$p_0 = \left[\sum_{n=0}^{3-1} \frac{1}{n!}(6)^n + \sum_{n=3}^{7} \frac{1}{3^{n-3}\,3!}(6)^n\right]^{-1}$$

$$= \left[1 + 6 + \frac{36}{2} + \frac{1}{6}(6)^3 + \frac{1}{3\times 3!}(6)^4 + \frac{1}{3^2\times 3!}(6)^5 + \frac{1}{3^3\times 3!}(6)^6 + \frac{1}{3^4\times 3!}6^7\right]^{-1} = \frac{1}{1141}$$

(i) Expected number of cars in the queue is

$$E(m) = \frac{(s\rho)^s \rho\, p_0}{s!(1-\rho)^2}\,[1 - \rho^{C-s+1} - (1-\rho)(C-s+1)\,\rho^{C-s}]$$

where $\rho = \dfrac{\lambda}{s\mu} = 2.$

So

$$E(m) = \frac{(3\times 2)^3\, 2\, \frac{1}{1141}}{3!(-1)^2}\,[1 - 2^{7-3+1} - (-1)(7-3+1)\,2^{7-3}]$$

$$= (.0631)\,(49) = 3.0919 \text{ cars.}$$

(ii) Expected number of cars in the service station

$$E(n) = E(m) + s - p_0 \sum_{n=0}^{s-1} \frac{(s-n)(\rho s)^n}{n!}, \quad \rho = \frac{\lambda}{s\mu}$$

$$= 3.09 + 3 - p_0 \sum_{n=0}^{2} \frac{(3-n)}{n!}\, 6^n = 6.06 \text{ cars}$$

(iii) Expected waiting time a car spends in the system

$$E(T) = \frac{E(n)}{\lambda(1-p_N)} = \frac{6.06}{1(1-p_7)} = \frac{6.06}{1-.00025} = 6.06 \text{ minutes}$$

where $p_7 = \dfrac{1}{s!\,s^{7-s}}\left(\dfrac{\lambda}{\mu}\right)^7 p_0$ for $7 \le n \le N$

$$= \frac{1}{3!\,3^4}(2)^7\,\frac{1}{1141} = .00025$$

(iv) Expected number of cars per hour that cannot enter the station is $60.\lambda.p_C$

$$= 60 \times 1 \times p_7 = 60 \times \frac{2^7}{3!\,3^4} \times \frac{1}{1141} = 0.015 \text{ car per hour.}$$

Model (General Erlang Queueing Model) (Birth-death Process): Let arrival rate $\lambda = \lambda_n$, service rate $\mu = \mu_n$, depending on n, preceding as in Model I, we arrive

$$P_n = \frac{\lambda_{n-1}\,\lambda_{n-2}\cdots\lambda_0}{\mu_n\,\mu_{n-1}\ldots\mu_0}P_0;\quad P_0 = \frac{1}{1 + \dfrac{\lambda_0}{\mu_1} + \dfrac{\lambda_0\lambda_1}{\mu_1\mu_2} + \ldots}$$

when $\lambda_n = \lambda$, $\mu_n = \mu$, then $P_0 = 1 - \rho$, $P_n = \rho^n(1-\rho)$.

when $\lambda_n = \dfrac{\lambda}{n+1}$, $\mu_n = \mu$, then $P_n = \dfrac{\rho^n e^{-\rho}}{n!}$, $P_0 = e^{-\rho}$

when $\lambda_n = \lambda$, $\mu_n = n\mu$, $P_n = \dfrac{e^{-\rho} \cdot \rho^n}{n!}$, $P_0 = e^{-\rho}$.

7.3 *M*/G/1 QUEUEING SYSTEM

So far we discussed the queueing processes which are either birth and death or non-birth and death processes. They are in either case Markovian and the theory of Markov chains and processes could be applied in their studies. Now we consider the models where the distributions of the inter-arrival time or the service time do not possess the memoryless property. That is not of exponential distribution. The process $N(t)$, giving the state of the system or system size at time t will then be no longer Markovian. However, the analysis of the process can be based on the associated process, which is Markovian. Two techniques are generally used for this purpose. One is due to Kendall and the technique is by means of imbedded Markov chains. The second technique is supplementary variable technique, due to Keilson and Kohavian.

Note that in the queue *M*/*M*/1, *G* represents the service time distribution is general, other things are same as *M*/*M*/1 queue.

We discuss the technique of Imbedded Markov chains.

Consider a single-server queueing system whose arrival process is Poisson with the average arrival rate λ. The service times denoted by B, are independent and identically distributed with mean $\dfrac{1}{\mu}$ and the distribution function is denoted by F_B and its pdf, when it exists, by $f_B = F'_B$. The discipline followed is FIFO. In particular, if we let F_B be the exponential distributed with parameter μ, then we obtain *M*/*M*/1 queueing system. If the service times are assumed to be a constant, then we get the *M*/*D*/1 queueing system.

Let t_n, $n = 1, 2, \ldots$ ($t_0 = 0$) be the nth departure epoch. That is, the instant at which the nth unit completes his service and leaves the system. These points t_n are called the **regeneration points** of the process $\{N(t)\}$. Let X_n be the number of units in the system at time t_n, so that

$$X_n = N(t_n), \qquad n = 1, 2, \ldots$$

The stochastic process $\{X_n, n = 1, 2, \ldots\}$ is a discrete parameter Markov chain, known as the **imbedded Markov chain** of the continuous-parameter stochastic process $\{N(t) : t \geq 0\}$.

Let A_n be the random variable giving the number of units that arrive during the service time of the nth unit, then

$$X_{n+1} = \begin{cases} X_{n-1} + A_{n+1} & \text{if } X_n \geq 1 \\ A_{n+1} & \text{if } X_n = 0 \end{cases}$$

That is, if $X_n = 0$, the next unit to arrive is the $(n + 1)^{\text{st}}$; during its service time if A_{n+1} units arrive, then the $(n + 1)^{st}$ unit departs at time t_{n+1}, leaving A_{n+1} units behind. If $X_n > 0$, then the number of units left behind by the $(n + 1)^{\text{st}}$ units equals $X_{n-1} + A_{n+1}$

Now the service times of all the units have the same distribution so that $A_n \equiv A$ for $n = 1, 2, \ldots$ we have

$$P(A = r) = \int_0^\infty P(\text{A} = r) \mid \text{serivce time of a unit is } t)\, d\text{B}(t) \tag{7.4}$$

where $P\{A = r/\text{service time of a unit is } t\} = \dfrac{e^{-\lambda t}(\lambda t)^r}{r!}$ (7.5)

and so $\quad a_r = P\{A = r\} = \int_0^\infty \dfrac{e^{-\lambda t}(\lambda t)^r}{r!} f_B(t)\, dt$

gives the distribution of A, the number of arrivals during the service time of a unit. Note $\sum_{j=1}^{\infty} a_j = 1$.

The probabilities

$$p_{ij} = P\{X_{n+1} = j/X_n = i\} = \begin{cases} P(A = j - i + 1) = a_{j-i+1}, & i \ge 1, j \ge i - 1 \\ P(A = j) = a_j\ , & i = 0, j \ge 0 \\ 0 & \text{otherwise} \end{cases}$$

The above relation clearly indicate that $\{X_n;\, n \ge 0\}$ is Markov chain having transition probability matrix

$$P = (p_{ij}) = \begin{bmatrix} a_0 & a_1 & a_2 & \cdots \\ a_0 & a_1 & a_2 & \cdots \\ 0 & a_0 & a_1 & \cdots \\ 0 & 0 & a_0 & \cdots \\ \cdot\cdot & \cdot & \cdot & \cdots \end{bmatrix}$$

As every state can be reached from every other state, the Markov chain $\{X_n\}$ is irreducible. Again, as $p_{ii} \ge 0$, the chain is aperiodic. When the traffic intensity $\rho = \lambda/\mu < 1$, the chain is persistent, non-null and hence ergodic. By applying the ergodic theorem of Markov chain (see chapter 6), the limits

$$v_k = \lim_{n \to \infty} p_{jk}(n),\ k = 0, 1, 2, \ldots$$

exists and are independent of the initial state j.
The probabilities $V = (v_0, v_1, \ldots)$, $\Sigma v_j = 1$, are given as the unique solutions of $v = vP$.

Let $G(z) = \sum_{j=0}^{\infty} a_j z^j$ and $V(z) = \sum_{j=0}^{\infty} v_j z^j$ denote the generating functions of the distributions of $\{a_j\}$ and $\{v_j\}$ respectively. We have

$$G(z) = \sum_{j=0}^{\infty} a_j z^j = \sum_{j=0}^{\infty} z^j \left(\int_0^\infty \frac{e^{-\lambda t}(\lambda t)^j}{j!} f_B(t)\, dt \right) = \int_0^\infty e^{-\lambda t} \left[\sum_{j=0}^{\infty} \frac{(\lambda t z)^j}{j!} \right] f_B(t)\, dt$$

$$= \int_0^\infty e^{-\lambda t} \cdot e^{\lambda t z} f_B(t)\, dt. = \int_0^\infty e^{-\lambda t(1-z)}\ f_B(t)\, dt$$

$$= F(\lambda(1 - z)) \tag{7.6}$$

Note that the Laplace transform of a function $f(t)$ is denoted by $F(s)$ and is $F(s) = \int_0^\infty e^{-st} f(t)\, dt$. In this situation, $F(\lambda(1 - z))$ is the Laplace transform of the service time distribution evaluated at $s = \lambda(1 - z)$. Hence

$$E(A) = G'(1) = \left[\frac{d}{dz} G(z)\right]_{z=1} = \left[\frac{d}{dz}(F(\lambda(1-z))\right]_{z=1} = \left\{\left[\frac{dF}{ds}\right]\left[\frac{ds}{dz}\right]\right\}_{s=0}$$

$$= \left[\frac{dF}{ds}\right]_{s=0} (-\lambda) = -\lambda \int_0^\infty \frac{d}{ds}(e^{-st}) f_B(t)dt \text{ as } F(s) = \int_0^\infty e^{-st} f_B(t)dt$$

$$= -\lambda \int_0^\infty -t \cdot e^{-st} f_B(t)\, dt = +\lambda E(B) = \frac{\lambda}{\mu} = \rho \qquad (7.7)$$

Here the reciprocal of the service rate μ of the server equals the average service time $E(B)$ as mentioned earlier.

Now $V = VP$ gives an infinite system of equations. Hence we get

$$v_j = v_0 a_j + \sum_{i=0}^{j+1} v_i a_{j-i+1}$$

$$\Rightarrow \quad \sum_{j=0}^{\infty} v_j z^j = \sum_{j=0}^{\infty} v_0 a_j z^j + \sum_{j=0}^{\infty}\left(\sum_{i=1}^{j+1} v_i a_{j-i+1}\right) z^j$$

$$\Rightarrow \quad V(z) = v_0 \sum_{j=0}^{\infty} a_j z^j + \sum_{i=1}^{\infty}\left(\sum_{j=i-1}^{\infty} v_i a_{j-i+1} z^j\right) \text{ interchanging the order of summation}$$

$$= v_0 \sum_{j=0}^{\infty} a_j z^j + \sum_{i=0}^{\infty}\left(\sum_{k=0}^{\infty} v_i a_k z^{k+i-1}\right)$$

$$= v_0 \sum_{j=0}^{\infty} a_j z^j + \frac{1}{z}\left(\sum_{i=1}^{\infty} v_i z^i \sum_{k=0}^{\infty} a_k z^k\right) = v_0 G(z) + \frac{1}{z}[V(z) - V(0)]\, G(z)$$

$$\Rightarrow \quad V(z) - \frac{1}{z} G(z).V(z) = v_0 G(z) - \frac{1}{z} v_0 G(z)$$

$$\Rightarrow \quad V(z)\left[1 - \frac{1}{z} G(z)\right] = v_0 G(z)\left[1 - \frac{1}{z}\right]$$

$$\Rightarrow \quad V(z) = \frac{v_0 G(z)(z-1)}{z - G(z)} \qquad (7.8)$$

Since $V(1) = \sum_{j=0}^{\infty} v_j = 1 = G(1)$, we can use L'Hospital rule to obtain

$$V(1) = 1 = \lim_{z \to 1} v_0 \frac{G'(z)(z-1) + G(z)}{1 - G'(z)} = \frac{v_0 G(1)}{1 - G'(1)} = \frac{v_0}{1 - G'(1)}$$

provided $G'(1)$ is finite and < 1. Taking $\rho = G'(1)$, it follows that

$$1 = \frac{v_0}{1-\rho} \Rightarrow v_0 - 1 - \rho \qquad (7.9)$$

and, since v_0 is the probability that the server is idle, ρ is the utlization limit. Using (7.6), (7.8) and (7.9), we get,

$$V(z) = \frac{(1-\rho)(z-1)F(\lambda(1-z))}{z - F(\lambda(1-z))} \tag{7.10}$$

This is known as **Pollaczek-Khinchin (P-K) formula**

Remark

The average number of units in the system, in the steady state, is determined by taking the derivative of $V(z)$ with respect to z and taking limit $z \to 1$,

$$E(N) = \lim_{n\to\infty} E(X_n) = \sum_{j=1}^{\infty} jv_j = \lim_{z\to 1} V'(z) = \frac{\rho}{1-\rho}$$

using the derivation of generating function. More generally,

$$E(N) = \rho + \frac{\lambda^2 E(B)^2}{2(1-\rho)}$$

This is known as the Pollaczek-khinchin (*P-K*) mean-value formula.

Note: The result hold true for all scheduling disciplines under the following conditions:

(i) The server is not idle whenever a unit is waiting for service.

(ii) The scheduling is non-preemptive; that is once a unit is under service, it should be completed without interruption.

7.3.1. Model : (M | E_R | 1 : FIFo)

This is a queueing model with Poisson arrivals, Erlang service time with k phases, single channel, first come, first served discipline and infinite population. In this model one unit is served in k phases. The arrival or departure of one unit, therefore means an increase or decrease of k phases in the system and the completion of service of one phase of a unit will mean the decrease of one phase in the system. If at any time there are m units waiting in the queue with one unit in service which has to still pass through s phases, then the total number of phases n in the system at that time will be given by

$$n = mk + s$$

If μ denotes the number of units served per unit time, then $k\mu$ will be the number of phases served per unit time and therefore

$$\lambda_n = \lambda$$
$$\mu_n = \mu k$$

When P_n denotes the steady state probabilities of n phases in the system,

$$\lambda P_{n-k} + k\mu\, P_{n+1} = \lambda P_n + k\mu P_n \qquad n \geq 1$$
$$k\mu P_1 = \lambda P_0$$

The steady state equation gives :

$$0 = -\lambda P_0 + k\mu P_1, \qquad n = 0$$
$$0 = -k\mu P_k + \lambda P_0, \qquad n = k$$
$$0 = -k\mu P_n + k\mu P_{n+1}, \qquad 1 \leq n < k$$

$$\Rightarrow \qquad P_k = \frac{\lambda}{k\mu} P_0, \text{ for } n = k,$$
$$P_{n+1} = P_n, \qquad \text{for } 1 \leq n < k.$$

1. Average number of units in the system

$$E(n) = \frac{k+1}{2k}\cdot\frac{\lambda}{\mu}\frac{\lambda}{\mu-\lambda}+\frac{\lambda}{\mu}$$

2. Average number of units in the queue

$$E(m) = \frac{k+1}{2k}\cdot\frac{\lambda}{\mu}\left(\frac{\lambda}{\mu-\lambda}\right)$$

3. Average time spent by a unit in the system

$$E(T) = \frac{k+1}{2k}\cdot\frac{\lambda}{\mu(\mu-\lambda)}+\frac{1}{\mu}.$$

4. Average waiting time of a unit in the queue

$$E(w) = \frac{k+1}{2k}\cdot\frac{\lambda}{\mu(\mu-\lambda)}$$

For constant service time, taking limit k tends to ∞, we get

$$E(n) = \frac{1}{2}\frac{\lambda^2}{\mu(\mu-\lambda)}+\frac{\lambda}{\mu}$$

$$E(m) = \frac{1}{2}\frac{\lambda}{\mu}\cdot\frac{\lambda}{\mu-\lambda}$$

$$E(T) = \frac{1}{2}\frac{\lambda}{\mu(\mu-\lambda)}+\frac{1}{\mu}$$

$$E(w) = \frac{1}{2}\cdot\frac{1}{\mu}\cdot\frac{\lambda}{\mu-\lambda}$$

when $k = 1$, the Erlang service time distribution reduces to exponential distribution and the values of $E(n)$, $E(m)$, $E(T)$, $E(w)$ are same as that of Model I.

7.3.2 Example

In a factory cafeteria the customers have to pass through three counters. The customers buy coupons at the first counter, select and collect and snacks at the second counter and collect tea at the third. The server at each counter takes on an average 1.5 minutes although the distribution of service time is approximately exponential. If the arrival of customers to the cafeteria is approximately Poisson at an average rate of 6 per hour, calculate

(a) th average time a customer spends waiting in the cafeteria
(b) the average time of getting the service
(c) the most probable time in getting the service

Solution.

Here, number of phases $k = 3$, service time per phase = 1.5 minutes, service time per customer = 1.5 × 3 = 4.5 minutes. Therefore $\mu = \frac{1}{4.5}$ customers/minute = $\frac{40}{3}$ customers/hr, $\lambda = 6$ customers/hour.

(a) Average waiting time

$$E(w) = \frac{k+1}{2k} \cdot \frac{\lambda}{\mu} \cdot \frac{1}{\mu - \lambda} = \frac{3+1}{2 \times 3} \cdot \frac{6 \times 3}{40} \cdot \frac{1}{\frac{40}{3} - 6} = \frac{9}{220} \text{ hrs} = 2.45 \text{ minutes.}$$

(b) Average time of getting the service is the mean of t when it is following the 3rd member of Erlang family.

$$\therefore \text{ Average time spent} = \frac{1}{\mu} = \frac{3}{40} \text{ hours} = 4.5 \text{ minutes}$$

(c) The most probable time spent in getting the service is the modal value of t for the 3rd member of Erlang family

$$\therefore \quad \text{Most probable time spent} = \frac{k-1}{\mu k} = \frac{3-1}{\frac{40}{3} \times 3} = \frac{1}{20} \text{ hour} = 3 \text{ minutes.}$$

7.3.3 Example

A colliery working one shift per day uses a large number of locomotives which break down at random intervals, on average one failing per 8 hour shift. The fitter carries out a standard maintenance schedule on each faulty locomotive. Each of the 5 main parts of the schedule takes an average of $\frac{1}{2}$ hour but the time varies widely. How much time will the fitter have for other tasks and what is the average time a locomotive is out of service?

Solution. Here $k = 5$, $\lambda = \frac{1}{8}$/hour, service time per part = $\frac{1}{2}$ hour. Therefore service time per locomotive = $\frac{5}{2}$ hours. Therefore $\mu = \frac{2}{5}$ hours. Fraction of time the fitter will have for other tasks = Fraction of time for which the fitter is idle

$$= 1 - \frac{\lambda}{\mu} = 1 - \frac{1/8}{2/5} = 1 - \frac{5}{16} = \frac{11}{16}.$$

$\therefore$ Time the fitter will have for other tasks in a day = $\frac{11}{16} \times 8 = 5.5$ hrs.

Average time a locomotive is out of service = Average time spent by the locomotive in the system

$$= \frac{k+1}{2k} \cdot \frac{\lambda}{\mu} \cdot \frac{1}{\mu - \lambda} + \frac{1}{\mu}$$

$$= \frac{5+1}{2 \times 5} \times \frac{1/8}{2/5} \cdot \frac{1}{2/5 - 1/8} + \frac{5}{2} = 3.18 \text{ hours.}$$

7.3.4 Example

In a car manufacturing plant, a loading crane takes exactly 10 minutes to load a car into a wagon and again comes back to the position of loading another car. If the arrivals of cars is a Poisson stream at an average of one every 20 minutes, calculate the average waiting time of a car.

Solution.

Service time is constant here, and equal 6 per hour. Therefore $k = \infty$ and the average waiting time of a car is given by

$$E(w) = \frac{1}{2}\frac{\lambda}{\mu}\frac{1}{\mu-\lambda} = \frac{1}{2}\times\frac{3}{6}\times\frac{1}{6-3} = \frac{1}{12} \text{ hour} = 5 \text{ minutes.}$$

7.3.5 Example

This is the model of $M\,|E_R|\,1$. Consider that the service time υ is Erlang-k $\left(\text{with mean } \frac{1}{\mu}\right)$ having pdf

$$f_B = \frac{(\mu k)^k\, t^{k-1}\, e^{-k\mu t}}{\Gamma(k)}, \quad 0 < t < \infty$$

Lapace transform of Erlang-distribution is

$$F(s) = \left(\frac{\mu k}{s+\mu k}\right)^k$$

From (7.7) and (7.10), we get

$$F(s) = F(\lambda(1-z)) = \left(\frac{\mu k}{\lambda-\lambda z+\mu k}\right)^k = \left(1+\frac{\rho(1-z)}{k}\right)^{-k}\quad \rho = \frac{\lambda}{\mu}$$

and hence

$$V(z) = \frac{(1-\rho)(z-1)}{\left[1+\frac{\rho(1-z)}{k}\right]^k\left[z-\left[1+\frac{(\rho-z)}{k}\right]^{-k}\right]} = \frac{(1-\rho)(z-1)}{z-\left[1+\frac{\rho(1-z)}{k}\right]^k-1}$$

Expanding in powers of z, and comparing coefficients of z^n from both sides, we get υ_n and equivalently p_n.

This model could also be considered as non-birth and death Markovian process, by taking services comprises of k exponential phases each with mean $\frac{1}{k\mu}$.

Various formulae for (M|G|1): (∞/GD) are:

Average number of customers in the system

$$= \frac{\lambda^2\sigma^2+\rho^2}{2(1-\rho)} + \rho, \text{ where } \sigma^2 = \text{ variance of service time} \quad (7.11)$$

$$\text{Average queue length} = \frac{\lambda^2 \sigma^2 + \rho^2}{2(1-\rho)} \tag{7.12}$$

Average waiting time of a customer in the queue

$$= \frac{\lambda^2 \sigma^2 + \rho^2}{2\lambda(1-\rho)} \tag{7.13}$$

Average waiting time that a customer spends in the system

$$= \frac{\lambda^2 \sigma^2 + \rho^2}{2\lambda(1-\rho)} + \frac{1}{\mu} \tag{7.14}$$

(7.10), (7.11) are the $(P - K)$ formulae.

7.3.6 Example

In a heavy machine shop, the overhead crane is 75 percent utilized. Time study observations gave the average service time as 10.5 minutes with a standard deviation of 8.8 minutes. What is the average calling rate for the services of the crane, and what is the average delay in getting service? If the average service time is cut to 8.0 minutes, with standard deviation of 6.0 minutes, how much reduction will occur, an average, in the delay of getting served?

Solution: This is a $(M|G|1) : (\infty/\text{FIFo})$ process.

The average waiting time of a customer in the queue =

$$E(w) = \frac{\lambda^2 \sigma^2 + \rho^2}{2\lambda(1-\rho)}, \rho = \frac{\lambda}{\mu} = 0.75$$

Here

$$\mu = \frac{60}{10.5} = 5.71/\text{hour}$$

$$\lambda = \rho \times \mu = 0.75 \times 5.71 = 4.29 \text{ per hour}$$

Therefore

$$E(w) = \frac{(4.29)^2 \left(\frac{8.8}{60}\right)^2 + (0.75)^2}{2 \times 4.29\,(1-0.75)} \times 60$$

$$= \frac{(18.4041)\,(.02151) + 0.5625}{2.145} \times 60$$

$$= \frac{0.958374}{2.145} \times 60 = 26.807 \text{ minutes}$$

If service time is cut to 8 minutes, then

$$\mu = \frac{60}{8} = 7.5 \text{ per hour}, \rho = \frac{4.29}{7.5} = 0.571$$

That is, the utilization of the crane is reduced to 57.1 percent. Then

$$E(w) = \frac{(4.29)^2\left(\frac{6}{60}\right)^2 + (0.571)^2}{2\times 4.29\,(1-0.571)} \times 60 = \frac{(18.4041)\,(0.01) + (0.326)}{3.68082} \times 60$$

$$= 0.13828 \times 60 = 8.297 \text{ minutes}$$

The average waiting time has a reduction of 18.5 minutes.

7.3.7 Example

Automatic car wash facility operates with only one bay. Car arrive according to a Poisson distribution, with a mean of 4 cars per hour and may wait in the facility's parking lot if the bay is busy. If the service time for all cars is constant and equal to 10 minutes, determine $E(m)$, $E(n)$, $E(T)$, $E(w)$.

Solution. Given $\lambda = 4$ cars/hour, $E(t) = \frac{1}{\mu} = 10$ min. (Given)

That is, $\mu = 6$ cars/hour, variance of service time = 0. So $\sigma = 0$, $\rho = \frac{\lambda}{\mu} = \frac{2}{3}$, Now

$$\text{Average queue length} = E(m) = \frac{\lambda^2\sigma^2 + \rho^2}{2(1-\rho)} = \frac{16\times 0 + \left(\frac{2}{3}\right)^2}{2\left(1-\frac{2}{3}\right)} = \frac{2}{3} = 0.667 \text{ car.}$$

Average number of customers in the system

$$= E(n) = \frac{\lambda^2\sigma^2 + \rho^2}{2(1-\rho)} + \rho = 0.667 + 0.667 = 1.333 \text{ cars}$$

$$\text{Average waiting time of a customer to wait in a system} = E(T) = \frac{E(n)}{\lambda} = \frac{1.333}{4} = 0.333 \text{ hours.}$$

$$\text{Average waiting time of a customer in a queue} = E(w) = \frac{E(m)}{\lambda} = \frac{0.667}{4} = 0.167 \text{ hours.}$$

7.4 SOLVED PROBLEMS

1. A self-service store employs one cashier at its counter. Nine customers arrive on an average every 5 minutes while the cashier can serve 10 customers in 5 minutes. Assuming Poisson distribution for arrival rate and exponential distribution for service rate, find

(i) Average number of customers in the system
(ii) Average number of customers in queue
(iii) Average time a customer spends in the system
(iv) Average time a customer waits before being served.

Solution. Arrival rate $\lambda = \frac{9}{5} = 1.8$ customers/minute

Service rate $\mu = \frac{10}{5} = 2$ customers/minute

(i) Average number of customers in the system

$$= \frac{\lambda}{\mu - \lambda} = \frac{1.8}{2 - 1.8} = 9$$

(ii) Average number of customers in the queue

$$= \frac{\lambda^2}{\mu(\mu - \lambda)} = \frac{(1.8)^2}{2(2 - 1.8)} = 8.1$$

(iii) Average time a customer spends in the system

$$= \frac{1}{\mu - \lambda} = \frac{1}{2 - 1.8} = 5 \text{ minutes}$$

(iv) Average time a customer spends in the queue

$$= \frac{\lambda}{\mu(\mu - \lambda)} = \frac{1.8}{2(2 - 1.8)} = 4.5 \text{ minutes}$$

2. A person repairing radios finds that the time spent on the radio sets has been exponential distribution with mean 20 minutes. If the radios are repaired in the order in which they come in and their arrival is approximately Poisson with an average rate of 15 for 8-hour day, what is the repairman's expected idle time each day? How many jobs are ahead of the average set just brought in?

Solution.

Arrival rate $\lambda = \frac{15}{8 \times 60} = \frac{1}{32}$ radios/minute

Service rate $\mu = \frac{1}{20}$ radios/minute

Number of jobs ahead to the set brought in

= Average number of radios in the system

$$= \frac{\lambda}{\mu - \lambda} = \frac{\frac{1}{32}}{\frac{1}{20} - \frac{1}{32}} = \frac{5}{3}$$

Number of hours for which the repairman remains busy in an 8-hour day

$$= 8\frac{\lambda}{\mu} = 8 \times \frac{\frac{1}{32}}{\frac{1}{20}} = 5 \text{ hours}$$

Time for which repairman remains idel in an 8 hour day = 8–5 = 3 hours.

3. Arrival rate of telephone calls at a telephone booth are according to Poisson distribution, with an average time of 9 minutes between two consecutive arrivals. The length of telephone call is assumed to be exponentially distributed, with mean 3 minutes.

(a) Determine the probability that a person arriving at the booth will have to wait.
(b) Find the average queue length that is formed from time to time.
(c) The telephone company will install a second booth when convinced that an arrival would expect to have to wait at least four minutes for the phone. Find the increase in flow of arrivals which will justify second booth.

Solution.

Arrival rate $\lambda = \frac{1}{9}$ per minute

Service rate $\mu = \frac{1}{3}$ per minute

(a) Probability that a person will have to wait is $\frac{\lambda}{\mu} = \frac{1/9}{1/3} = 0.33$

(b) Average queue length $= \frac{\mu}{\mu - \lambda} = \frac{1/3}{1/3 - 1/9} = 1.5$ persons.

(c) Average waiting time in the queue $= \frac{\lambda_1}{\mu(\mu - \lambda_1)}$

$$\Rightarrow \quad 4 = \frac{\lambda_1}{1/3\,(1/3 - \lambda_1)}$$

$$\Rightarrow \quad \lambda_1 = \frac{12}{7 \times 9} = \frac{4}{21} \text{ arrivals per minute}$$

Therefore increase in flow of arrivals

$$= \frac{4}{21} - \frac{1}{9} = \frac{5}{63} \text{ per minute.}$$

4. On average 96 patients per 24 hour day require the service of an emergency clinic. Also on average, a patient requires 10 minutes of active attention. Assume that the facility can handle only one emergency at a time. Suppose that is costs the clinic Rs 100 per patient treated to obtain an average servicing time of 10 minutes and that each minute of decrease in this average time would cost Rs 10 per patient treated. How much would have to be budgeted by the clinic to decrease the average size of the queue from $\frac{4}{3}$ patients to $\frac{1}{2}$ patient?

Solution.

Arrival rate $\lambda = \frac{96}{24} = 4$ patients/hour,

Service rate $\mu = \frac{1}{10} \times 60 = 6$ patients/hour

Average expected number of patients in the queue $E(m)$

$$= \frac{\lambda}{\mu}\left(\frac{\lambda}{\mu-\lambda}\right) = \frac{4}{6}\cdot\frac{4}{6-4} = \frac{4}{3}$$

This number is to be reduced from $\frac{4}{3}$ to $\frac{1}{2}$. So

$$E(m) = \frac{\lambda}{\mu^1}\cdot\frac{\lambda}{\mu^1-\lambda}. \text{ That is}$$

$$\frac{1}{2} = \frac{4}{\mu^1}\,\frac{4}{\mu^1-4}$$

$$\Rightarrow \quad \mu'^2 - 4\mu' - 32 = 0$$

$$\Rightarrow \quad (\mu'-8)(\mu'+4) = 0$$

$$\Rightarrow \quad \mu' = 8 \text{ or } \mu' = -4$$

$\mu' = -4$ is illogical and so is neglected. Hence $\mu' = 8$ patients/hour.

Therefore average time required by each patient = $\frac{1}{8}$ hour = $\frac{60}{8}$ minutes.

So decrease in time requirement to attend a patient = $10 - \frac{60}{8} = \frac{5}{2}$ minutes.

Hence the budget-required for each patient is Rs. $\left(100 + \frac{5}{2}\times 10\right)$ = Rs 125. In order to decrease the size of the queue, budget requirement per patient should be increased from Rs 100 to Rs 125.

5. A shipping company has a single unloading dock with ships arriving in a Poisson fashion at an average rate of 3/day. The unloading time distribution for a ship with n unloading crews is found to be exponential with average unloading time $\frac{1}{2n}$ days. The company has a large labour supply without regular working hours, and to avoid long waiting times, the company has a policy of using as many unloading crews as there are ships waiting in line or being unloaded. Find

(a) the average number of unloading crews working at any time, and

(b) the probability that more than 4 crews will be needed.

Solution.

Arrival rate $\lambda = 3$ ships/day

Service rate $\mu_n = n\mu$

and mean service rate with one unloading crew

$$\mu_1 = \mu = 2 \text{ ships/day.}$$

(a) Average number of unloading crews

$$= E(n) = \sum_{n=0}^{\infty} n\,p_n = \sum_{n=0}^{\infty} n\frac{\rho^n}{n!}\,e^{-\rho} = \rho e^{-\rho}\sum_{n=1}^{\infty}\frac{\rho^{n-1}}{(n-1)!}$$

$$= \rho e^{-\rho}\left[1+\rho+\frac{\rho^2}{2!}+\frac{\rho^3}{3!}+\ldots\right] = \rho e^{-\rho}\cdot e^{\rho} = \rho = \frac{\lambda}{\mu} = \frac{3}{2}$$

(b) Probability that more that 4 crews will be needed = probability that there are atleast 5 ships in the system = $\sum_{n=5}^{\infty} p_n$

$$= \sum_{n=0}^{\infty} p_n - \sum_{n=0}^{4} p_n = 1 - (p_0 + p_1 + p_2 + p_3 + p_4)$$

$$= 1 - e^{-\rho}\left[1 + \rho + \frac{\rho^2}{2!} + \frac{\rho^3}{3!} + \frac{\rho^4}{4!}\right]$$

$$= 1 - e^{-3/2}\left[1 + \frac{3}{2} + \frac{(3/2)^2}{2!} + \frac{(3/2)^3}{3!} + \frac{(3/2)^4}{4!}\right] = 0.019$$

6. In a busy textile shop in the city, the manager, looks after 96 customers per 24 hour day. For each customer he requires 10 mts to show the different varieties. For each customer it costs Rs. 100/- at an average of servicing the customers at each minute of decreases in this average time would cost Rs. 10 per customer. How much would have to be budgeted to decrease the average size of the queue as manager is the only person to give his service?

Solution. Arrival rate $= \lambda = \frac{96}{24} \times \frac{1}{60} = \frac{1}{15}$ per minute

Service rate $= \mu = 10$ minutes per customer

$= \frac{1}{10}$ per minute per customer

Average number of customers in the queue

$$E(m) = \frac{\lambda^2}{\mu \cdot (r - \lambda)} = \frac{\left(\frac{1}{15}\right)^2}{\frac{1}{10}\left(\frac{1}{10} - \frac{1}{15}\right)} = 1.333$$

If there are customers i.e., manager is idle

$$p_0 = 1 - \rho = 1 - \frac{2}{3} = \frac{1}{3}.$$

To find the new value of μ.

The queue size has to be decreased to $\frac{1}{2}$ customer.

$$E(m) = \frac{\lambda^2}{\mu(\mu - \lambda)} = \frac{1}{2}$$

Substitute the values of λ, we have

$$\lambda = \frac{1}{15}$$

Then $$\frac{\left(\frac{1}{15}\right)^2}{\mu\left(\mu-\frac{1}{15}\right)} = \frac{1}{2}$$

$$\Rightarrow \qquad \mu = \frac{2}{15} \text{ customers per minute}$$

Average rate $= \frac{1}{\mu} = \frac{15}{2}$ minutes

Decreased average rate $= 10 - \frac{15}{2} = 2.5$ minutes

The cost for each customer = Rs. (100 + 2.5 × 10) = Rs. 125/-

7. In a production company, materials arrive at a rate of 30 bags per day. Service time distribution is exponentially distributed with an average of 36 minutes. Find (i) The mean queue size (ii) Probability that queue exceeds 10.

Solution. Arrival rate $= \lambda = \frac{30}{24} \times \frac{1}{60} = \frac{1}{48}$ per minute, Service rate $= \mu = 36$ minutes per bag. Hence

$\mu = \frac{1}{36}$ per minute per bag.

(i) $E(n)$ = Mean queue size

$$= \frac{\lambda}{\mu - \lambda} = \frac{\rho}{1-\rho} = \frac{\frac{1}{48}}{\frac{1}{36} - \frac{1}{48}} = 3 \text{ bags}$$

(ii) Probability the queue size exceeds 10 is given by

$$P(x > 10) = (\rho)^{10} = \left(\frac{\lambda}{\mu}\right)^{10} = \left(\frac{36}{48}\right)^{10} = \left(\frac{3}{4}\right)^{10} = 0.0526.$$

8. In a railway counter customers arrive at the counter at a rate of 12 per hour. There is one clerk serving the customers at a rate of 30 per hour. Assuming the conditions for use of the single channel queueing modal, find

(i) The probability that there is no customer at the counter
(ii) The probability that there are more than 2 customers at the counter
(iii) The probability that there is no customer waiting to be served
(iv) The probability that a customer is being served and nobody is waiting.

Solution. Arrival rate $\lambda = 12$ per hour, Service rate $\mu = 30$ per hour

$$p_n = \left(1 - \frac{\lambda}{\mu}\right)\left(\frac{\lambda}{\mu}\right)^n = \left(1 - \frac{12}{30}\right)\left(\frac{12}{30}\right)^n = (0.6)(0.4)^n \qquad \text{for } n = 0, 1, 2, \ldots$$

(a) $P(\text{no customer at the counter}) = p_0 = 1 - 0.4 = 0.6$

(b) P(more than 2 customers at the counter) $= p_3 + p_4 + \ldots$

$$= 1 - P(\text{atmost 2 customers at the counter})$$
$$= 1 - (p_0 + p_1 + p_2)$$
$$= 1 - \{(0.6) + (0.6)(0.4) + (0.6)(0.4)^2\}$$
$$= 1 - (0.6 + 0.24 + 0.096)$$
$$= 0.064$$

9. In a Bank, every 15 minutes one customer arrives for cashing the cheque. The staff in the only payment counter takes 10 minutes for serving a customer on an average. State suitable assumptions and find

(a) The average queue length

(b) Increase in the arrival rate in order to justify a second counter (when the waiting time of a customer is atleast 15 minutes the management will increase one more counter).

Solution.

(a) Arrival rate= $\lambda = \frac{1}{15} \times 60 = 4$ per hour

Service rate $= \mu = \frac{1}{10} \times 60 = 6$ per hour

$$\text{Average queue length} = \frac{\lambda^2}{\mu(\mu-\lambda)} = \frac{4\times4}{6(6-4)} = 1.33 \text{ customers}$$

(b) The installation of second counter will be justified only if the arrival rate is more than the waiting time. Let λ' be the increased arrival rate. Then expected waiting time in the queue will be

$$E(w) = \frac{\lambda'}{\mu(\mu-\lambda')} \Rightarrow \frac{15}{60} = \frac{\lambda'}{6(6-\lambda')}$$

$$\Rightarrow \qquad \lambda' = 3.6$$

Hence the increase in the arrival rate is $4 - 3.6 = 0.4$ arrivals per minute.

10. At a beauty parlour, with one beautician, ladies arrive according to Poisson distribution with mean arrival rate of 5 per hour and hair design was exponentially distributed with an average design taking 10 minutes. Find

(i) Average number of ladies in the shop and average number waiting to do the hair design.

(ii) % of time an arrival can walk inside the parlour without having to wait.

(iii) % of ladies who has to wait prior to getting into the chair for hair design.

Solution. Given arrival rate = 5 persons per hour $= \frac{5}{60} = \frac{1}{12}$ per minute and service rate $= \mu = \frac{1}{10}$ per minute.

(i) $E(n)$ = Average number of ladies in the shop

$$= \frac{\lambda}{\mu-\lambda} = \frac{\frac{1}{12}}{\frac{1}{10}-\frac{1}{12}} = 4.98 \text{ or } 5 \text{ customers.}$$

$E(m)$ = Average number waiting in the queue

$$= \frac{\lambda^2}{\mu(\mu-\lambda)} = \frac{\left(\frac{1}{12}\right)^2}{\frac{1}{10}\left(\frac{1}{10}-\frac{1}{12}\right)} = \frac{.0069388}{.00167} = 4.15 = 4.$$

(ii) Probability that queue size can be greater than or equal to 1 is

$$P(x \geq 1) = \frac{\lambda}{\mu} = \frac{\frac{1}{12}}{\frac{1}{10}} = \frac{10}{12} = \frac{5}{6} = 0.8333$$

% ladies who have to wait 83.33%.

% of time an arrival can walk without waiting = 100 – 83.33 = 16.667%.

11. In a telephone booth one person was looking after the service and the arrival rate of a customer λ = 5 per hour and service rate has an exponential distribution 8 per hour. Find

(a) The probability that the service is idle

(b) Expected time that a customer is in the queue

(c) The probability that there are atleast two customers in the booth.

Solutions. Arrival rate = λ = 5; Service rate = μ = 8

The model is M/M/1 : FIFO/∞/∞.

(a) $P(\text{server is idle}) = \rho(1-\rho) = \frac{\lambda}{\mu}\left(1-\frac{\lambda}{\mu}\right) = \frac{5}{8}\left(1-\frac{5}{8}\right) = \frac{15}{64}$.

(b) Expected time that the customer is in the queue

$$E(w) = \frac{\lambda}{\mu(\mu-\lambda)} = \frac{5}{8(8-5)} = \frac{5}{8\times 3} = \frac{5}{24} \text{ hrs}$$

(c) Probability that atleast 2 customers in the system

$$P(x \geq 2) = \left(\frac{\lambda}{\mu}\right)^2 = \left(\frac{5}{8}\right)^2 = \frac{25}{64}$$

12. It is assumed that there is only one counter at railway station for issuing the train tickets at which the people arrive in a Poisson distribution at an average rate of 15 per hour. Find out at which average rate must the person who issue the ticket work in order to ensure a probability of 0.90 that the person will not wait longer than 12 mts. [Assume the length of source by the counter person has an exponential distribution].

Solution. Arrival rate = $\lambda = \frac{15}{60}$; service rate = Exponential distribution = μ

$$P(X \geq 12) = 1 - P(X < 12) = 1 - 0.90 = 0.10$$

Since the person has to wait is given by

$$1 - \left(\frac{\lambda}{\mu}\right)e^{-\mu\left(1-\frac{\lambda}{\mu}\right)t}, \quad \text{for } t > 0$$

and $$f(t) = \frac{d}{dt}\left(1 - \frac{\lambda}{\mu} e^{-\mu\left(1-\frac{\lambda}{\mu}\right)}\right) = \frac{\lambda}{\mu} \times \mu\left(1 - \frac{\lambda}{\mu}\right) e^{-\mu\left(1-\frac{\lambda}{\mu}\right)t}$$

Now

$$\int_{12}^{\infty} f(t)dt = 0.10$$

$$\Rightarrow \quad \int_{12}^{\infty} \lambda\left(1 - \frac{\lambda}{\mu}\right) e^{-\mu\left(1-\frac{\lambda}{\mu}\right)t} dt = 0.10$$

$$\Rightarrow \quad \lambda\left(1 - \frac{\lambda}{\mu}\right)\left[\frac{e^{-\mu\left(1-\frac{\lambda}{\mu}\right)t}}{-\mu\left(1-\frac{\lambda}{\mu}\right)}\right]_{12}^{\infty} = 0.10$$

$$\Rightarrow \quad \frac{\lambda}{\mu} e^{-12(\mu-\lambda)} = 0.10$$

$$\Rightarrow \quad \left(\frac{15}{60}\right) e^{-12\left(\mu - \frac{15}{60}\right)} = (0.10)\mu$$

$$\Rightarrow \quad \left(\frac{1}{4}\right) e^{-12\left(\mu - \frac{1}{4}\right)} = (0.10)\mu$$

$$\Rightarrow \quad e^{+(3-12\mu)} = (0.4)\mu$$

$$\Rightarrow \quad \frac{1}{\mu} = 2.48 \text{ mts per second.}$$

13. Trains arrive at the yard every 15 minutes and the service time is 33 minutes. If the line capacity of the yard is limited to 4 trains, find

(i) the probability that the yard is empty

(ii) the average number of trains in the system.

Solution.

$$\lambda = \frac{1}{15} \text{ per minute}, \ \mu = \frac{1}{33} \text{ per minute}, \ C = 4.$$

Hence $\rho = \frac{\lambda}{\mu} = \frac{33}{15} = 2.2$

(i) $p_0 = \frac{1-\rho}{1-\rho^{C+1}} = \frac{\rho - 1}{\rho^{C+1} - 1} = \frac{2.2 - 1}{(2.2)^5 - 1} = 0.0237$

(ii) Average number of trains in the system

$$= \sum_{n=0}^{C} n\, p_n = 0 + p_1 + 2p_2 + 3p_3 + 4p_4 = p_0\,[\rho + 2\rho^2 + 3\rho^3 + 4\rho^4]$$

$$= 0.0237\,[2.2 + 2\,(2.2)^2 + 3(2.2)^3 + 4(2.2)^4] = 3.26.$$

14. **A car park contains 5 cars. The arrival of cars is Poisson at a mean rate of 10 per hour. The length of time each car spends in the car park is negative exponential distribution with mean of 2 hours. How many cars in the car park on average?**

Solution.

$$C = 5;\quad \lambda = \frac{10}{60} = \frac{1}{6};\quad \mu = \frac{1}{2\times 60} = \frac{1}{120}.$$

Hence $\rho = \dfrac{\lambda}{\mu} = \dfrac{120}{6} = 20$

$$p_0 = \left(\frac{1-\rho}{1-\rho^{C+1}}\right) = \left(\frac{\rho-1}{\rho^{C+1}-1}\right) = \left(\frac{20-1}{20^6-1}\right)$$

$$= 2.962 \times 10^{-7}$$

Average number of cars in the park is

$$E(n) = \sum_{n=0}^{5} n\, p_n = p_1 + 2p_2 + 3p_3 + 4p_4 + 5p_5$$

$$= p_0[\rho + 2\rho^2 + 3\rho^3 + 4\rho^4 + 5\rho^5]$$

$$= (2.962 \times 10^{-7})\,[20 + 2\,(20)^2 + 3\,(20)^3 + 4\,(20)^4 + 5\,(20)^5]$$

$$= (2.962 \times 10^{-7})\,[20 + 800 + 24000 + 640000 + 16000000]$$

$$= (2.962 \times 10^{-7})\,(16664820) = 49361196 \times 10^{-7} = 4.94 \simeq 5$$

15. A one person barber shop has six chairs to accommodate people waiting for a hair cut. Assume customers who arrive when all six chairs are full leave without entering the barber shop. Customers arrive at the average rate of 3/hr and spend an average of 15 min. in the barber chair.

(a) What is the probability a customer can get directly into the barber chair upon arrival?

(b) What is the expected number of customers waiting for a hair cut?

(c) What is the effective arrival rate?

Answer: C = maximum number in the system = 7.

$$\lambda = 3/\text{hour};\ \mu = \frac{60}{15} = 4/\text{hour};$$

(a) The probability of a customer to enter directly in to the barber chair is nothing but the probability of no one in the barber shop

$$= p_0 = \frac{1-\rho}{1-\rho^{C+1}} = \frac{1-3/4}{1-(3/4)^8} = 0.2778.$$

(b) Average number of customers in the queue

$$E(m) = \rho^2 \left[\frac{1 - C\rho^{C-1} + (C-1)\rho^C}{(1-\rho)(1-\rho^{C+1})} \right] = 1.39.$$

(c) The effective arrival rate $= \mu\,(1 - p_0)$
$= 4\,(1 - 0.2778) = 2.89/\text{hour}.$

16. At an one man barber shop, the customers arrive according to Poisson process at the average rate of 5 per hour and they are served according to exponential distribution with an average source rate of 10 minutes per person. There are only 5 seats available for the waiting of customers and the customers do not wait if they find no seat available. Find the average number of customer in the system, average queue length and the average time a customer spends in the barber shop. Also find the idle time of the barber.

Solution. Queueing Modol III : M/M/1 : FIFO/C
Arrival rate = $\lambda = 5$; service rate = μ = 10 minutes/person;

(i) Probability of system to be idle is = p_0

$$= \frac{1-\dfrac{\lambda}{\mu}}{1-\left(\dfrac{\lambda}{\mu}\right)^{c+1}} = \frac{1-\dfrac{5}{6}}{1-\left(\dfrac{5}{6}\right)^{5+1}} = 0.2505 = 25\%$$

(ii) Average number of customers in the system

$$L_s = C - \left(\frac{\mu}{\lambda}\right)(1 - p_0) = 5 - \left(\frac{6}{5}\right)(1 - 0.2505) = 4.1006$$

(iii) Average queue length $= L_s - \dfrac{\lambda}{\mu} = 4.1006 - \dfrac{5}{6} = 3.267$

(iv) Average waiting time in the queue

$$W_s = \frac{L_s}{\lambda\,(1-p_N)},\ p_N \text{ the probability of } N \text{ number of customers}$$

$$= \frac{L_s}{\lambda\left(1 - p_0\left(\dfrac{\lambda}{\mu}\right)^N\right)} = \frac{4.1006}{5\left(1 - 0.2506\left(\dfrac{5}{6}\right)^5\right)} = 0.912.$$

17. Trains arrive at the yard every 15 minutes and the service time is 33 minutes. If the line capacity of the yard is limited to 4 trains, find

(i) the probability that the yard is empty
(ii) the average number of trains in the system

Solution. Queueing model : M/M/1 : C/FIFO

$$\lambda = \frac{1}{15} \text{ per minute; } \mu = \frac{1}{33} \text{ per minute, } C = 4$$

(i) $$p_0 = \frac{1-\rho}{1-\rho^{c+1}} = \frac{\rho-1}{\rho^{c+1}-1} = \frac{\frac{33}{15}-1}{\left(\frac{33}{15}\right)^5 - 1} = 0.0237$$

(ii) Average number of trains in the system

$$= \sum_{n=0}^{c} np_n = 0 + p_1 + 2p_2 + 3p_3 + 4p_4 = p_0(\rho + 2\rho^2 + 3\rho^3 + 4\rho^4)$$

$$= 0.0237 \left(\frac{33}{15} + 2\left(\frac{33}{15}\right)^2 + 3\left(\frac{33}{15}\right)^3 + \left(\frac{33}{15}\right)^4 \right) = 3.26$$

18. Assume that the goods trains are coming in a yard at the rate of 30 trains per day and suppose that the interarrival times follow an exponential distribution. The service time for each train is assumed to be exponential with an average of 36 minutes. If the yard can admit 9 trains at a time (there being 10 lines, one of which is reserved for shunting purposes), calculate the probability that the yard is empty and find the average queue length.

Solution. Given

$$\lambda = \frac{30}{60 \times 24} = \frac{1}{48} \text{ and } \mu = \frac{1}{16} \text{ trains per minute}$$

$$\rho = \frac{\lambda}{\mu} = \frac{36}{48} = 0.75, \ C = 9$$

The probability that the yard is empty is given by

$$P_0 = \frac{1-\rho}{1-\rho^{C+1}} = \frac{1-0.75}{1-(0.75)^{10}} = \frac{0.25}{0.90} = 0.28$$

Average queue length is given by

$$E(m) = \frac{\rho^2 [1 - C\rho^{C-1} + (C-1)\rho^C]}{(1-\rho)(1-\rho^{C+1})}$$

$$= \frac{(0.75)^2 (1 - 9(0.75)^8 + 8(0.75)^9)}{(0.25)(0.75)^{10}} = 1.55$$

19. At a railway station, only one train is handled at a time. The railway yard is sufficient only for two trains to wait while other is given signal to leave the station. Trains arrive at the station at an average rate of 6 per hour and the railway station can handle them on an average of 12 per hour. Assuming Poisson arrivals and exponential service distribution, find the steady-state probabilities for the various number of trains in the system. Also find the average waiting time of a new train coming into the yard.

Solution.

Given $\lambda = 6$, $\mu = 12$, so that $\rho = \frac{6}{12} = \frac{1}{2}$, The maximum queue length is 2. That is, the maximum number of trains in the system is 3 (= C).

The probability that there is no train in the system (both waiting and in service) is given by

$$P_0 = \frac{1-\rho}{1-\rho^{C+1}} = \frac{1-\frac{1}{2}}{1-\left(\frac{1}{2}\right)^4} = 0.53$$

Since $P_n = P_0\rho^n$, $P_1 = (0.53)(0.5) = 0.27$, $P_2 = (0.53)(0.5)^2 = 0.13$, $P_3 = (0.53)(0.5)^3 = 0.07$. Hence

$$E(n) = \sum_{n=0}^{C} nP_n = 1 \times (0.27) + 2(0.13) + 3(0.07) = 0.74$$

The average number of trains in the system is 0.71 and each train takes on an average $\frac{1}{12}$ (= 0.085) hours for getting service. As the arrival of new train expects to find an average of 0.74 trains in the system before it

$$E(w) = (0.74)(0.85) \text{ hours} = 0.0629 \text{ hours} = 3.8 \text{ minutes}.$$

20. Patients arrive at a clinic according to a Poisson distribution at a rate of 30 patients per hour. The waiting room does not accommodate more than 14 patients. Examination time per patient is exponential with mean rate 20 per hour.

(a) Find the effective arrival rate at the clinic

(b) What is the probability that an arriving patient will not wait.

(c) What is the expected time until a patient is discharged from the clinic?

Solution. Given $\lambda = 30$/hr, $\mu = 20$/h, $C = 15$,

$$P_0 = \frac{1-\left(\frac{\lambda}{\mu}\right)}{1-\left(\frac{\lambda}{\mu}\right)^{C+1}} = \frac{1-\left(\frac{3}{2}\right)}{1-\left(\frac{3}{2}\right)^{16}} = 0.00076$$

(a) Effective arrival rate $\lambda' = \mu\,(1 - p_0) = 20 \times (1 - 0.0076) = 19.98$ per hour

(b) P (a Patient will not wait) = $P_0 = 0.00076$

(c) $$E(n) = \frac{P_0 \cdot \rho\left[1-(C+1)\rho^C + C\rho^{C+1}\right]}{(1-\rho)(1-\rho^{C+1})} = \frac{(0.00076)\left(\frac{3}{2}\right)\left(1-16\left(\frac{3}{2}\right)^{15}+15\left(\frac{3}{2}\right)^{16}\right)}{\left(1-\frac{3}{2}\right)\left(1-\left(\frac{3}{2}\right)^{16}\right)}$$

$$= 13 \text{ patients (nearly)}$$

$$E(w) = \frac{E(n)}{\lambda'} = \frac{13}{19.98} = 0.65 \text{ h or 39 minutes.}$$

21. A local one person barber shop can accommodate a maximum of 5 people at a time (4 waiting and 1 getting hair cut). Customers arrive according to a Poisson distribution with mean 5 per hour. The barber cuts hair at an average rate of 4 per hour (Exponential service time)

(a) What percentage of time is the barber idle?
(b) What fraction of the potential customers are turned away.
(c) What is the expected number of customers waiting for a hair cut?
(d) How much time can a customer expect to spend in the barber shop?

Solution. Given $\lambda = 5$, $\mu = 4$, $C = 5$

(a) Probability that the barber is idle

$$= P(N = 0) = P_0 = \frac{1 - \dfrac{\lambda}{\mu}}{1 - \left(\dfrac{\lambda}{\mu}\right)^{C+1}} = \frac{1 - \dfrac{5}{4}}{1 - \left(\dfrac{5}{4}\right)^{6}} = 0.08$$

Percentage of time when the barber is idle $\cong 9\%$

(b) P (a customer is away) $= P(C > 5)$

$$= \left(\frac{\lambda}{\mu}\right)^{C} \left[\frac{\left(1 - \dfrac{\lambda}{\mu}\right)}{\left(1 - \left(\dfrac{\lambda}{\mu}\right)^{C+1}\right)}\right] = \left(\frac{5}{4}\right)^{5} \frac{\left(1 - \dfrac{5}{4}\right)}{\left(1 - \dfrac{5}{4}\right)^{6}} = 0.2711$$

Therefore 0.2711 = Probability that the Potential customers turned away.

$$E(m) = E(n) - \sum_{n=1}^{C} p_n = E(n) - (1 - p_0)$$

$$= \frac{\lambda}{\mu - \lambda} - \frac{(C+1)\left(\dfrac{\lambda}{\mu}\right)^{C+1}}{1 - \left(\dfrac{\lambda}{\mu}\right)^{C+1}} - (1 - p_0)$$

$$= \left\{-5 - \frac{6 \times \left(\dfrac{5}{4}\right)^{6}}{1 - \left(\dfrac{5}{4}\right)^{6}}\right\} - (1 - 0.088) = 2.2 \text{ customers}$$

$$E(T) = \frac{E(n)}{\lambda'} = \frac{E(n)}{\mu(1 - p_0)} = \frac{3.1317}{3.6448} = 0.8592 \text{ h} = 51.5 \text{ min.}$$

22. A barber shop has space to accommodate only 10 customers. He can service only one person at a time. If a customer comes to his shop and find it full, he goes to the next shop. Customers randomly arrive at an average rate $\lambda = 10$ per hours and barbers service time is negative exponential with an average of $\frac{1}{\mu} = 5$ minutes per customer. Find P_0, P_n

Solution. Model III

Given $N = 10$, $\lambda = \frac{10}{60}$, $\mu = \frac{1}{5}$, $\rho = \frac{\lambda}{\mu} = \frac{5}{6}$

$$P_0 = \frac{1-\rho}{1-\rho^{11}} = \frac{1-5/6}{1-(5/6)^{11}} = 0.1926$$

$$P_n = \left(\frac{1-\rho}{1-\rho^{11}}\right)\rho^n = (0.1926)(5/6)^n,\ n = 0, 1, 2, \ldots, 10$$

23. Consider a single server queueing system with Poisson input, exponential service times. Suppose the mean arrival rate is 3 calling units per hour, the expected service time is 0.25 hour and the maximum permissible number calling units in the system is two. Derive the steady-state probability distribution of the number of calling units in the system, and then calculate the expected number in the system.

Solution. Given $\lambda = 3$ units per hour, $\mu = 4$ units per hour, $c = 2$. Then traffic intensity $\rho = \frac{\lambda}{\mu} = \frac{3}{4} = 0.75$. The steady-state probability distribution of the number of n customer (calling units) in the system is given by

$$P_n = \frac{(1-\rho)\,\rho^n}{1-\rho^{c+1}},\ \rho \neq 1$$

$$= \frac{(1-0.75)\,(0.75)^n}{1-(0.75)^{2+1}}$$

$$= (0.43)(0.75)^n$$

$$P_0 = \frac{1-\rho}{1-\rho^{c+1}} = \frac{1-0.75}{1-(0.75)^{2+1}} = \frac{0.25}{1-(0.75)^3} = 0.431$$

The expected number of calling units in the system is given by

$$E(n) = \sum_{n=1}^{N} nP_n = \sum_{n=1}^{2} n\,(0.43)(0.75)^n$$

$$= (0.43)\,[0.75 + 2 \times (0.75)^2] = 0.81.$$

24. A petrol bunk has two service stations. The service time follows the exponential distribution with mean 4 minutes and cars arrive for service in a Poisson process at the rate of ten cars per hour. Find the probability that a car has to wait for service. What proportion of time the service station remains idle?

Solution.

Here $s = 2$; $\lambda = 10$ per hour, $\mu = \frac{1}{4}$ per minute $= \frac{60}{4} = 15$ per hour. So $\rho = \frac{\lambda}{s\mu} = \frac{10}{2\times 15} = \frac{1}{3}$.

The proportion of time, the bunk remains busy $= \dfrac{\lambda}{s\mu} = \dfrac{1}{3}$

The proportion of time the bunk remains idle $= 1 - \dfrac{1}{3} = \dfrac{2}{3}$. So percentage of idle period is 67%.

Now

$$p_0 = \left[\sum_{n=0}^{s-1} \frac{1}{n!}\left(\frac{\lambda}{\mu}\right)^n + \frac{1}{s!}\left(\frac{\lambda}{\mu}\right)^s \frac{s\mu}{s\mu - \lambda}\right]^{-1} = \left[\sum_{n=0}^{1} \frac{1}{n!}\left(\frac{2}{3}\right)^n + \frac{1}{2!}\left(\frac{2}{3}\right)^2 \cdot \frac{30}{30-10}\right]^{-1}$$

$$= \left[1 + \frac{2}{3} + \frac{1}{3}\right]^{-1} = \frac{1}{2}$$

So the probability of a car to wait for service is

$$P(n \geq 2) = P(w > 0) = \sum_{n=2}^{\infty} p_n = \sum_{n=2}^{\infty} \frac{1}{2!2^{n-2}}\left(\frac{2}{3}\right)^n p_0 = \frac{1}{6} \simeq 0.167$$

25. Four counters are being run on the frontier of a country to check the passports and necessary papers of the tourists. The tourists choose a counter at random. If the arrival at the frontier is Poisson at the rate of λ and the service time is exponential with parameter $\lambda/2$, what is the steady state average queue at each counter?

Solution.

Here $\quad s = 4, \quad \mu = \lambda/2, \quad \rho = \dfrac{\lambda}{s\mu} = \dfrac{1}{2}$

$$p_0 = \left[\sum_{n=0}^{3} \frac{1}{n!}\left(\frac{\lambda}{\mu}\right)^n + \frac{1}{4!}\left(\frac{\lambda}{\mu}\right)^4 \frac{4\mu}{4\mu - \lambda}\right]^{-1} = \left[\sum_{n=0}^{3} \frac{2^n}{n!} + \frac{1}{4!} \times 2^4 \times \frac{2\lambda}{\lambda}\right]^{-1}$$

$$= \left[1 + 2 + \frac{2^2}{2!} + \frac{2^3}{3!} + \frac{2^5}{4!}\right]^{-1} = \left(\frac{23}{3}\right)^{-1} = \frac{3}{23}$$

The steady state average queue at each counter is

$$E(m) = \frac{\lambda\mu\left(\frac{\lambda}{\mu}\right)^s p_0}{(s-1)!(s\mu - \lambda)^2} = \frac{2^4 \cdot \left(\frac{\lambda^2}{2}\right)}{3!(2\lambda - \lambda)^2} \cdot \frac{3}{23} = \frac{4}{3} \times \frac{3}{23} = \frac{4}{23}.$$

26. A bank has two tellers working on savings accounts. The first teller handles withdrawals only. The second teller handles depositors only. It has been found that the service time distributions of both deposits and withdrawals are exponential with a mean service time of 3 minutes per customer. Depositors and withdrawers are found to arrive in a Poisson fashion throughout the day with mean arrival rate of 16 and 14 per hour. What would be the effect on the average waiting time for depositors and withdrawers if each teller could handle both withdrawals and deposits. What would be the effect if this could only be accomplished by increasing the service time to 3.5 minutes per customer?

Solution.

For withdrawers	**For depositors**
Mean arrival rate $\lambda = 16/\text{hr}$	$\lambda = 14/\text{hr}$
Mean service rate $\mu = \frac{1}{3}$ per minute	$\mu = \frac{1}{3}$ per minute
$= \frac{60}{3} = 20$ per hour	$= \frac{60}{3} = 20$ per hour
Utilisation factor $\rho = \frac{\lambda}{\mu} = \frac{4}{5}$	$\rho = \frac{14}{20} = 0.7$
Waiting time $E(w) = \frac{\rho}{\mu - \lambda} = 0.2$ hour,	$E(w) = \frac{7}{60}$ hour.

When both tellers rendering the same service

$$\lambda = 16 + 14 = 30; \quad \mu = 20/\text{hour and } s = 2$$

So

$$p_0 = \left[\sum_{n=0}^{1} \frac{1}{n!}\left(\frac{30}{20}\right)^n + \frac{1}{2!}\left(\frac{30}{20}\right)^2 \frac{2\times 20}{40-30}\right]^{-1} = 0.1429$$

$$E(w) = \frac{E(m)}{\lambda}, \text{ where } E(w) = \frac{\lambda\mu\left(\frac{\lambda}{\mu}\right)^s p_0}{(s-1)!(s\mu-\lambda)^2}.$$

Hence

$$E(w) = \frac{20\times 30\times\left(\frac{30}{20}\right)^2 \times 0.1429}{(2-1)!(2\times 20-30)^2}\cdot\frac{1}{30}$$

$$= \frac{1.93}{30} = 0.643 \text{ hours.} = 3.86 \text{ minutes.}$$

When the service time is increased to 3.5 minutes per customers, then

$$\mu = 60 \times \frac{1}{3.5} = \frac{120}{7} \text{ per hour,}$$

$$p_0 = \left[\sum_{n=0}^{2-1}\frac{1}{n!}\left(\frac{30}{120/7}\right)^n + \frac{1}{2!}\left(\frac{30}{120/7}\right)^2 \frac{2\times 120/7}{(2\times 120/7)-3}\right]^{-1} = 0.0638$$

and

$$E(m) = \frac{30\times\frac{120}{7}\times\left(\frac{30}{120/7}\right)^2\times(0.0638)}{(2-1)!\left(2\times\frac{120}{7}-30\right)^2} = 5.97$$

Thus $$E(w) = \frac{E(m)}{\lambda} = \frac{5.97}{30} = 0.199 \text{ hours} = 11.94 \text{ minutes.}$$

When the service time is increased to 3.5 minutes per customer instead of 3 minutes, then the average waiting time of a customer in the queue is 11.94 minutes instead of 3.86 minutes.

EXERCISES

1. Find the distribution of the number (Q) in the queue in steady state in a $M/M/1$ queue with rates λ, μ. Find $E(Q)$ and Var (Q). Verify that $E(Q) = \lambda E(W_Q)$, where W_Q is the waiting time of Q. Also verify that Var (W_Q) + Var (υ) = var (W_s), where υ is the service time.

2. Consider a barber shop with two chairs, two barbers, and no room for customers to wait. Say that the state of the system is the number of customers in the shop: 0, 1, 2. If there is an empty chair when a customer arrives, he enters the shop and his haircut beings. If both chairs are occupied when he arrives, he does not enter the shop. As soon as a customer's hair cut is completed, be leaves the shop instantaneously. The barbers do not assist one another when there is only one customer in the shop. On the average a customer arrives every 10 minutes, and each haircut takes an average of 15 minutes

(a) Set up the transition diagram and steady state equations.

(b) Solve the steady-state equations for the distribution of the number of customers in the shop.

(c) Calculate the expected number of busy barbers.

(d) Calculate the expected number of customers turned away per hour.

3. In a college canteen it was observed that there is only one server who takes an average of 4 minutes to serve a customer. If students arrive in the canteen at an average rate of 10 per hour, how much time a student is expected to spend waiting for his turn to place an order? How much students are expected to be in the canteen?

4. Let there be an automobile inspection situation with three inspection stalls. Assume that cars wait in such away that when a stall becomes vacant the car at the head of the line pulls up to it. The station can accommodate atmost 4 cars waiting (seven in station) at one time. The arrival pattern is Poisson with a mean of one car per every minute during peak hours. The service time is exponential with mean 6 minutes. Find the average number of customers in the system during peak hours, the average waiting time and the average number of cars that cannot enter the station became of full capacity.

5. Cars arrive at a petrol bunk having one petrol unit in Poisson with an average of 10 cars per hour. The service time is distributed exponentially with a mean of 3 minutes. Find (i) average number of cars in the system (ii) average waiting time in the queue (iii) average queue, length (iv) the probability that the number of cars in the system is 2.

6. On an average of 96 patients per 24 hour day require the service of an emergency clinic. Also on an average, a patient required 10 minutes of active attention. Assume that the facility can handle only one emergency at a time. Suppose that it cost the clinic Rs 100 per patient treated to obtain an average servicing time of 10 minutes and thus each minute of decrease in this average time would cost Rs 10 per patient treated. How much would have to be budgeted by the clinic to decrease the average size of the queue from $\frac{4}{3}$ to $\frac{1}{3}$?

7. In a railway Marshalling yard, goods train arrives at a rate of 30 trains per day. Assuing that inter-arrival time follows an exponential distribution and the service time distribution is also

exponential with an average of 36 minutes. Calculate
(a) the mean size of the system
(b) the probability that queue size exceeds 10,
(c) if the input of the trains increases to an average 33 per day, what will be the changes in (a) and (b).

8. A Television repairman finds that the time spends on his job has an exponential distribution with mean 30 minutes. If he repairs sets in the order in which they came in and if the arrival of sets is Poisson with an average rate of 10 per 8 hour day, what is his expected idle time per day? How many jobs are ahead of the average set just brought in?
9. A super market has two girls serving at the counters. The customers arrive in a Poisson fashion at the rate of 12 per hour. The service time for each customer is exponential with mean 6 minutes. Find (i) the probability that an arriving customers has to wait for service (ii) the average number of customer in the system (iii) the average time spend by a customer in the super market.
10. Two repairmen are attending five machines in a workshop. Each machine breaks down according to a Poisson distribution with mean 3 per hour. The repair time per machine is exponential with mean 15 minutes.
(a) Find the probability that the two repairmen are idle, that one repairman is idle.
(b) What is the expected number of idle machines not being serviced
11. A post office has 3 windows providing the same services. It receives on an average 30 customers/hr. Arrivals are Poisson distributed and service time exponentially distributed. The post office serves on average 12 customer/hr.
(a) What is the probability that a customer will be served immediately?
(b) What is that probability that a customer will have to wait.
(c) What is the average number of customers in the system?
(d) What is the average total time that a customer must spend in the post office?
12. A two channel waiting line with Poisson arrival has a mean arrival rate of 50 per hour and exponential service with a mean service rate of 75 per hour for each channel. Find
(i) The probability of an empty system
(ii) The probability that an arrival in the system will have to wait.
13. A mechanic services four machines. For each machine, the mean time between service requirements is 10 hours and is assumed to follow an exponential distribution. The repair time tends to follow the same distribution with a mean of two hours. When a machine is down for repairs, the time lost has a value of Rs. 20 per hour. The mechanic costs Rs. 50 per day.
(a) What is the expected number of machines in operation?
(b) What is the expected down-time cost per day?
14. A barber with a one man takes exactly 25 minutes to complete one hair cut. If customers arrive in a Poisson fashion at an average rate of one every 40 minutes, how long on the average must a customer wait for service? Also find the average time a customer spends in the barber shop.
15. In a car manufacturing plant, a loading crane takes exactly 10 minutes to load a car into a wagon and again come back to position to load another car. If the arrival of cars is a Poisson steam at an average of one per every 20 minutes, calculate the average waiting time of a car.
16. In a Jwellery shop there are 5 persons to look after the sales. Arrival of customers is Poisson with 4 per hour and service rate is exponentially distributed with an average service time of 10 minute.
Find (a) Average waiting time of an arrival
(b) Average length of waiting line
(c) Average time an arrival spends in the system.

17. At a port there are 6 unloading berths and 4 unloading crews. When all the berths are full, arriving ships are diverted to an overflow facility 20 kms down the river. Tankers arrive according to a Poisson process with a mean of 1 every 2 hours. An average, it takes 10 hours to unload a tanker, the unloading time following an exponential distribution. Find

(a) How many tankers are there in the port on the average?

(b) How long does a tanker spend at the port on the average?

(c) What is the average arrival rate at the overflow facility?

ANSWERS

(3) $P = 2/3$; $E(m) = \dfrac{\lambda}{\mu(\mu-\lambda)} = \dfrac{\frac{1}{6}}{\frac{1}{4}\left(\frac{1}{4}-\frac{1}{6}\right)} = 8$

$$E(n) = \frac{\lambda}{\mu - \lambda} = \frac{\frac{1}{6}}{\frac{1}{4}-\frac{1}{6}} = 2.$$

(4) $E(m) = 3.09$ cars, $E(n) = 6.06$ cars, $E(T) = 12.3$ minutes.
Expected number of cars per hour that cannot enter the station is 30.3 cars per hour.

(5) $E(n) = 1$; $E(T) = 3$ minutes, $E(m) = 0.5$; $p_2 = 1/8$.

(6) $\mu \simeq 9$ patients/hour.

(7) $\rho = 0.75$; $E(n) = 3$ trains; $\rho(\geq 10) = 0.056$.

(iii) When the input increases to 33 trains per day $\lambda = \dfrac{1}{480}$; $\mu = \dfrac{1}{36}$ trains per minute $E(n) \simeq 5$ trains; $P(\geq 10) = 0.1460$.

(8) $P = 5/8$; $E(n) = 5/3$ jobs.

(9) (i) 0.45 (ii) $E(m) = \dfrac{27}{40}$, $E(n) = 1.87$ or 2 customers
(iii) $E(v) = 9.3$ minutes.

(10)(a) 0.04305, 0.16144 (b) 0.911

(11) (a) 0.29, (b) 0.70 (c) 6 customers, (d) 12 minutes

(12) (i) 0.83 (ii) 0.167

(13) (a) 3 machines (b) Rs. 200 per day

(14) $E(w) = 20.8$ minutes, $E(v) = 45.8$ minutes

(15) $E(w) = 5$ minutes

(16) (a) 12.5 minutes (b) 0.793 (c) 22 minutes

(17) $P_0 = 0.0072$, $E(m) = 0.8203$, $E(n) = 4.35$ tankers Average arrival rate = 0.2930

8

Statistical Inference

Sampling is quite often used in our day-to-day life. For example, in a shop we want to assess the quality of wheat or rice or any other commodity, we take handful of it from the bag and then decide whether we could purchase it or not. For determining population characteristics, instead of enumerating entire population, the individuals in the sample only are observed. Then the sample characteristics are utilised to approximately determine or estimate the population. The error involved in such approximation is known as sampling error.

8.1 INTRODUCTION

8.1.1 Population and Sample

The group of individuals under study is called **population** (or) **universe**. A finite subset of statistical individuals in a population is called a **sample** and the number of individuals in a sample is called the **sample size**.

Sampling is the name given to the method of obtaining n sample from the population.

8.1.2 Parameters and Statistic

Parameters is the name given to mean, variance, mode etc of the population and statistic is the name given to mean, variance, etc of a sample.

Random sampling is one in which each unit of population has an equal chance of being included in it.

8.1.3 Sampling Distribution

The probability distribution of the statistic is known as **sampling distribution**. The standard deviation of the sampling distribution is known as **standard error** (SE).

Standard error plays an important role in the large sample theory and forms the basis of the testing of hypothesis. SE of a statistic may be reduced by increasing the sample size but this results in corresponding increase in cost, labour and time etc. If t is any statistic, then for large samples

$$Z = \frac{t - E(t)}{SE(t)} \sim N(0, 1)$$

Thus if the discrepancy between the observed and the expected value of a statistic is greater than 1.96 times the SE, the hypothesis is rejected at 5% level of significance. Similarly, if

$$|t - E(t)| \leq 1.96 \times \text{SE}(t)$$

the deviation is not regarded significant at 5% level of significance.

8.1.4 Hypothesis

A definite statement about the population parameter is called **hypothesis.**

A hypothesis of no difference is called **Null hypothesis** and denoted by H_0. That is, it is the hypothesis which is tested for possible rejection under the assumption that it is true.

Complementary to null hypothesis H_0 is called an **alternative hypothesis** and is usually denoted by H_1. For example, if we want to test the null hypothesis that the population has a specified mean μ_0 (say), that is H_0: $\mu = \mu_0$, then the alternative hypothesis is

(i) H_1: $\mu \neq \mu_0$
(ii) H_1: $\mu > \mu_0$
(iii) H_1: $\mu < \mu_0$

(i) is known as a two-tailed alternative, (ii) and (iii) are known as right tailed and left tailed alternatives respectively.

Guidelines for Selecting the Null Hypothesis

When the goal of an experiment is to establish an assertion, the negation of the assertion should be taken as the null hypothesis. The assertion becomes the alternative hypothesis.

Notation for the Hypothesis

H_1: The **alternative hypothesis** is the claim we wish to establish
H_0: The **null hypothesis** is the negation of the claim

The two kinds of error and their probabilities are

Type I error: Rejection of H_0 when H_0 is true,
Type II error: Non-rejection of H_0 when H_1 is true
α = probability of making of Type I error (also called the level of significance)
β = probability of making a Type II error.

α, β are referred to as producer's risk and consumer's risk respectively. We summarize the hypothesis H as follows

	Accept H	Reject H
H is true	Correct decision	Type I error
H is false	Type II error	Correct decision

8.1.5 Formulating the Alternative Hypothesis

Example An appliance manufacturer is considering the purchase of a new machine for stamping out sheet metal parts. If μ_0 is the average number of good parts stamped out per hour by his old machine and μ is the corresponding average for the new machine, the manufacturer wants to test the null hypothesis $\mu = \mu_0$ against a suitable alternative. What should the alternative be if he does not want to buy the new machine unless it is more productive than the old one?

Solution. The manufacturer should use the alternative hypothesis $\mu > \mu_0$ and purchase the new machine only if the null hypothesis can be rejected.

8.1.6 Critical Region and Level of Significance

A region in a sample space which amount to rejection of H_0 is termed as critical region. The probability α that a random value of the statistic belongs to the critical region is known as the level of significance. α is fixed in advance before collecting the sample information.

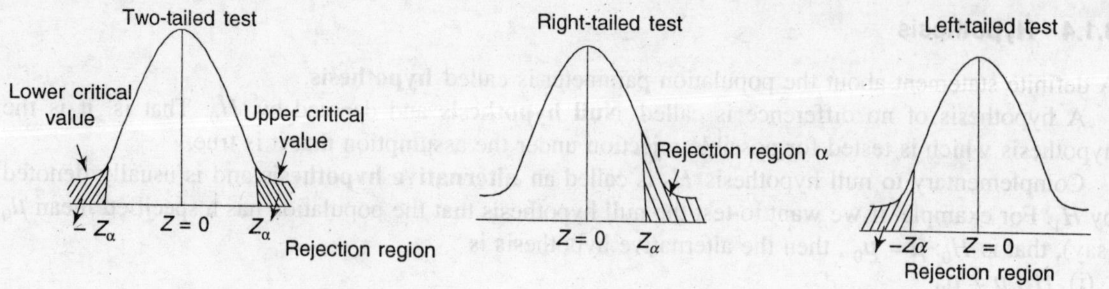

Note 1: Large sample is a sample of size 30 and above. The small sample is a sample of size upto 30.

Note 2 : If H_0 is rejected, then

$$P[|Z| > 1.96] = 0.05 \text{ where } Z \sim N(0, 1)$$

$$\Rightarrow \quad P\left[\left|\frac{X - E(X)}{\sigma_X}\right| > 1.96\right] = 0.05$$

$$\Rightarrow \quad P\left[|X - \mu| > 1.96\ \sigma_X\right] = 0.05$$

where $E(X) = \mu$ and $\text{SE}(X) = \sigma_X^2$. We say that H_0 is rejected at 5% level of significance. Similarly if $|Z| > 2.58$, H_0 is rejected at 1% level of significance.

8.2 TEST OF SIGNIFICANCE FOR LARGE SAMPLES

For large values of *n*, the number of trials, almost all the distributions, (e.g.) Binomial, Poisson, Negative binomial are very closely approximated by normal distribution. In this case we apply the *normal test* which is based upon the *area property* of the normal probability curve. If $X \sim N(\mu, \sigma^2)$, then

$$Z = \frac{X - \mu}{\sigma} = \frac{X - E(X)}{SE(X)} \sim N(0, 1)$$

From the normal probability tables, we have

$$P(-3 \le Z \le 3) = 0.9973$$

$$\Rightarrow \quad P(|Z| \le 3) = 0.9973$$

$$\Rightarrow \quad P(|Z| > 3) = 1 - P(|Z| \le 3) = 0.0027$$

That is, in all probability we should expect a standard normal variate to lie between ±3. Also from normal probability tables, we get

$$P(|Z| \le 1.96) = 0.9595$$

$$P(|Z| \le 2.58) = 0.99$$

$$P(|Z| > 2.58) = 0.01$$

Thus the significant values of Z at 5% and 1% level of significance are 1.96 and 2.58 respectively.

Steps to be Used in Normal Test

(i) Compute the test statistic Z under H_0

(ii) If $|Z| > 3$, H_0 is rejected

(iii) If $|Z| < 3$, we test its significance at certain level.

Thus for two tailed test if $|Z| > 1.96$, H_0 is rejected at 5% level of significance and if $|Z| < 1.96$ may be retained at 5% level of significance. Similarly if $|Z| > 2.58$, H_0 is contradicted at 1% level of significance and if $|Z| < 2.58$, H_0 may be accepted at 1% level of significance.

8.2.1 Procedure for Testing of Hypothesis

(i) Set up null hypothesis H_0: $\mu = \mu_0$
(ii) Set up alternative hypothesis H_1: $\mu \neq \mu_0$

(iii) Test statistic $Z = \dfrac{t - E(t)}{SE(t)}$ under null hypothesis

(iv) *Conclusion* Let the critical value of the test statistic at α level of significance be Z_α.

If $|Z| < Z_\alpha$, that is, the calculated value of Z (in modulus value) is less than Z_α (tabulated value) we accept the null hypothesis at α% level of significance. If $|Z| > Z_\alpha$, we reject the null hypothesis.

Critical value Z_α	Level of significance α 1%	2%	5%
Two tailed test	$\lvert Z_\alpha \rvert = 2.58$	2.33	1.96
Right tailed test	$\lvert Z_\alpha \rvert = 2.33$	1.96	1.645
Left tailed test	$\lvert Z_\alpha \rvert = -2.33$	–1.96	–1.645

We have the following test
(1) Test for single mean
(2) Test for difference of mean
(3) Test for single proportion
(4) Test for difference of proportion

First, we fix the notation. Denote
mean of the population by μ
mean of the sample by $\bar{x}$
standard deviation of population by σ
standard deviation of sample by s
proportion by p.

Note. When σ of the population is known, standard error is $\sigma/\sqrt{n}$ and when σ of the population is not known, then standard error is $s/\sqrt{n}$

8.2.2 Test for Significance of Single Mean ($n > 30$)

Under H_0, the test statistic is

$$Z = \frac{\bar{x} - \mu}{\sigma/\sqrt{n}} \tag{8.1}$$

if n is large, x_i is a random sample of size n drawn from a normal population with mean μ and variance σ^2, sample mean is $\bar{x}$.

8.2.3 Confidence or Fiducial Limits for the Unknown Population Mean μ

95% confidence interval for μ is

$$|Z| \leq 1.96 \text{ or } \left|\frac{\bar{x} - \mu}{\sigma/\sqrt{n}}\right| \leq 1.96$$

i.e

$$\bar{x} - 1.96\,(\sigma/\sqrt{n}) \leq \mu \leq \bar{x} + 1.96\,(\sigma/\sqrt{n})$$

Hence $\bar{x} \pm 1.96\,(\sigma/\sqrt{n})$ are the 95% confidence limits for μ. Similarly $\bar{x} \pm 2.58\,(\sigma/\sqrt{n})$ are known as 99% confidence limits for μ.

8.2.4 Examples

1. The heights of college students in Chennai are normally distributed with standard deviation 6 cm and sample of 100 students had their mean height 158 cm. Test the hypothesis that the mean height of college students in Chennai is 160 cm.

Solution.

Step 1 *Null hypothesis* H_0: $\mu = 160$

Step 2 *Alternative hypothesis* H_1: $\mu \neq 160$

Step 3 *Test statistic for single mean is*

$$Z = \frac{\bar{x}-\mu}{\sigma/\sqrt{n}} = \frac{158-160}{6/\sqrt{100}} = -3.33$$

$$|Z| = 3.33$$

Conclusion $|Z| = 3.33 > 2.58$, we reject the null hypothesis at 1% level of significance.

2. A sample of 900 members has a mean 3.4 cm, standard deviation (SD) 2.61. Is the sample taken from a large population of mean 3.25 and standard deviation 2.61? If the population is normal and its mean is unknown, find the 95% and 98% fiducial limits of true mean.

Solution.

(i) *Null hypothesis* H_0: The sample has been drawn from the population with mean $\mu = 3.25$ cm and SD $\sigma = 2.61$ cm

(ii) *Alternative Hypothesis* H_1 : $\mu \neq 3.25$

(iii) *Test statistic:* Under H_0,

$$Z = \frac{\bar{x}-\mu}{\sigma/\sqrt{n}} = \frac{3.4-3.25}{2.61/\sqrt{900}} = 1.73 < 1.96$$

Conclusion The data don't provide us any evidence against H_0, which may be accepted at 5% level of significance.

95% fiducial limits for μ are $\bar{x} \pm 1.96\,(\sigma/\sqrt{n})$ gives $3.40 \pm 1.96\left(\frac{2.61}{\sqrt{900}}\right)$. That is between 3.5705 and 3.2295. 98% fiducial limits for μ are $\bar{x} \pm 2.33\,(\sigma/\sqrt{n})$ gives 3.6027 and 3.1973.

3. An arithmetic mean of marks in statistics in a sample of 100 students was 51 and $\sigma = 6$. Could this have been a random sample from the population with mean marks 50.

Solution. $n = 100$, $\bar{x} = 51$, $\sigma = 6$

(i) *Null hypothesis* H_0: $\mu = 50$

(ii) *Alternative hypothesis* H_1: $\mu \neq 50$

(iii) *Test statistic*

$$Z = \frac{\bar{x}-\mu}{\sigma/\sqrt{n}} = \frac{51-50}{6/10} = \frac{10}{6} = 1.666$$

$$|Z| = 1.66 < 1.96$$

Conclusion. The random sample comes from a population with mean 50.

8.2.5 Test for Difference of Mean (n > 30)

Let $\bar{x}$ be the mean of a sample size n_1 from a population with mean μ_1 and variance σ_1^2 and let $\bar{x}_2$ be the mean of a sample size n_2 from a population with mean μ_2 and variance σ_2^2. To test the null hypothesis $\mu_1 - \mu_0 = \delta$, where δ is a specified constant. Analogous to the tests concerning one mean, we test the null hypothesis against $\mu_1 - \mu_2 < \delta$, $\mu_1 - \mu_2 > \delta$ and $\mu_1 \neq \mu_2$. The test statistic becomes

$$Z = \frac{(\bar{x}_1 - \bar{x}_2) - \delta}{\sqrt{\dfrac{\sigma_1^2}{n_1} + \dfrac{\sigma_2^2}{n_2}}}$$

Under the null hypothesis H_0: $\mu_1 = \mu_2$, $\delta = 0$, the test statistic

$$Z = \frac{\bar{x}_1 - \bar{x}_2}{\sqrt{\dfrac{\sigma_1^2}{n_1} + \dfrac{\sigma_2^2}{n_2}}}$$

Note.

(1) If the samples are drawn from the same population, $\sigma_1 = \sigma_2 = \sigma$, then

$$Z = \frac{\bar{x}_1 - \bar{x}_2}{\sigma\sqrt{\dfrac{1}{n_1} + \dfrac{1}{n_2}}}$$

(2) σ_1 and σ_2 are not known, then

$$Z = \frac{\bar{x}_1 - \bar{x}_2}{\sqrt{\dfrac{s_1^2}{n_1} + \dfrac{s_2^2}{n_2}}}, \text{ where } s_1, s_2 \text{ are SD of samples}$$

(3) If $\sigma_1 = \sigma_2$, and σ is not known, then

$$\sigma^2 = \frac{n_1 s_1^2 + n_2 s_2^2}{n_1 + n_2} \text{ and}$$

$$Z = \frac{\bar{x}_1 - \bar{x}_2}{\sqrt{\left(\dfrac{n_1 s_1^2 + n_2 s_2^2}{n_1 + n_2}\right)\left(\dfrac{1}{n_1} + \dfrac{1}{n_2}\right)}}$$

95%, 99% confidence limits for the difference of means $|\mu_1 - \mu_2|$ is

$$|\bar{x}_1 - \bar{x}_2| \pm 1.96 \text{ SE } (\bar{x}_1 - \bar{x}_2)$$

and

$$|\bar{x}_1 - \bar{x}_2| \pm 2.58 \text{ SE } (\bar{x}_1 - \bar{x}_2)$$

8.2.6 Examples

1. To test the claim that the resistance of electric wire can be reduced by more than 0.050 ohm by alloying 32 values obtained for standard wire yielded $\bar{x}_1 = 0.136$ ohm, $s_1 = 0.004$ ohm, and 32 values obtained for alloyed wire yielded $\bar{x}_1 = 0.083$ ohm, $s_2 = 0.005$ ohm. At the 0.05 level of significance, does this support the claim?

Solution. Null hypothesis H_0: $\mu_1 - \mu_2 = 0.050$
Alternative hypothesis H_1: $\mu_1 - \mu_2 > 0.050$
Test statistic

$$Z = \frac{(0.136 - 0.083) - 0.050}{\sqrt{\frac{(0.004)^2}{32} + \frac{(0.005)^2}{32}}} = 2.65 > 1.645$$

Conclusion Since $Z = 2.65 > 1.645$, the null hypothesis H_0 must be rejected. That is, the data substantiate the claim.

2. A company claims that its light bulbs are superior to those of its main competitor. If a study showed that a sample of $n_1 = 40$ of its bulbs has a mean lifetime of 647 hours of continuous use with a standard deviation of 27 hours, while a sample of $n_2 = 40$ bulbs made by its main competitor had a mean lifetime of 638 hours of continuous use with a standard deviation of 31 hours, does this substantiate the claim at the 0.05 level of significance?

Solution. Null hypothesis: $\mu_1 - \mu_2 = 0$
Alternative hypothesis: $\mu_1 - \mu_2 > 0$
Test statistic

$$Z = \frac{647 - 638}{\sqrt{\frac{27^2}{40} + \frac{31^2}{40}}} = 1.38$$

Conclusion Since $Z = 1.38 < 1.645$, the null hypothesis H_0 is accepted. That is the observed difference between the two sample means is not significant.

3. The means of two single large samples of 1000 and 2000 members are 67.5" and 68" respectively. Can the samples be regarded as drawn from the sample population of standard deviation 2.5"? (Test at 5% level of significance)

Solution. Null hypothesis: $\mu_1 = \mu_2$, $\sigma = 2.5"$
Alternative hypothesis $\mu_1 \neq \mu_2$
Test statistic

$$Z = \frac{\bar{x}_1 - \bar{x}_2}{\sqrt{\sigma^2\left(\frac{1}{n_1} + \frac{1}{n_2}\right)}} = \frac{67.5 - 68}{2.5\sqrt{\frac{1}{1000} + \frac{1}{2000}}} = -5.1$$

Conclusion Since $|Z| = 5.1 > 3$, we reject the null hypothesis. Hence the samples are certainly not from the sample population with standard deviation 2.5".

8.2.7 Test for Single Proportion

Let X be the number of successes in n trials, p is the probability of success in each trial. Now $q = 1 - p$. Let $p_1 = \dfrac{X}{n}$ is the sample proportion. The test statistic

$$Z = \frac{p_1 - p}{\sqrt{\dfrac{pq}{n}}}$$

95% and 99% confidence interval for p is $p_1 \pm 1.96\sqrt{\dfrac{pq}{n}}$ and $p_1 \pm 2.58\sqrt{\dfrac{pq}{n}}$ respectively. If p is not known, then the probable limits for the proportion in the population are $p_1 \pm \sqrt{\dfrac{p_1 q_1}{n}}$.

8.2.8 Problems

1. A coin is tossed 400 times and it turns up head 216 times. Discuss whether the coin may be unbiased one.

Solution. Null hypothesis: The coin is unbiased: $p = \dfrac{1}{2}$

Alternative hypothesis: $p \neq \dfrac{1}{2}$

Given $n = 400$, $\quad p_1 = \dfrac{216}{n} = \dfrac{216}{400} = 0.54$

$$Z = \frac{0.54 - 0.5}{\sqrt{\dfrac{1}{4} \times \dfrac{1}{400}}} = 1.6 < 1.96$$

Conclusion. Null hypothesis is accepted. That is, the coin is unbiased at 5% level of significance.

2. A dice is thrown 9,000 times and a throw of 3 or 4 is observed 3,240 times. Show that the dice cannot be regarded as an unbiased one and find the limits between which the probability of a throw of 3 or 4 lies.

Solution. If the coming of 3 or 4 is called a success, then probability of getting 3 or 4 is $\dfrac{1}{6} + \dfrac{1}{6} = \dfrac{1}{3}$. $n = 9{,}000$, X = number of success = 3. 240

Null hypothesis H_0: Dice is unbiased one $p = \dfrac{1}{3}$

Alternative hypothesis H_1: $p \neq \dfrac{1}{3}$

Then $q = 2/3$

Test statistic $$Z = \frac{\frac{3240}{9000} - \frac{1}{3}}{\sqrt{\frac{2}{9} \times \frac{1}{9000}}} = 5.36$$

Conclusion. Since $Z = 5.36 > 3$, H_0 is rejected. We conclude the dice is biased. Since H_0 is rejected, $p \neq \frac{1}{3}$. The probable limits for p is

$$p_1 \pm 3\sqrt{\frac{p_1 q_1}{n}} = (0.36) \pm \sqrt{\frac{(0.36)\,(0.64)}{9000}} = 0.375 \quad \text{and} \quad 0.345$$

Hence the probability of getting 3 or 4 almost certainly lies between 0.375 and 0.345.

3. In a sample of 1000 people in Maharashtra 540 are rice eaters and the rest are wheat eaters. Can we assume that both rice and wheat are equally popular in this state at 1% level of significance?

Solution. *Null hypothesis.* H_0 : Both rice and wheat are equally popular in this state. That is $p = \frac{1}{2}$.

Alternative hypothesis : $p \neq \frac{1}{2}$

We know $n = 1000$, $q = \frac{1}{2}$, $p_1 = \frac{540}{1000} = 0.54$. Test statistic

$$Z = \frac{p_1 - p}{\sqrt{\frac{pq}{n}}} = \frac{0.54 - 0.5}{\sqrt{\frac{1}{4} \times \frac{1}{1000}}} = 2.532.$$

Conclusion. Since $|Z| = 2.532 < 2.58$, H_0 is accepted at 1% level of significance.

4. In a large consignment of orange, a sample of 64 were taken, in that 14 were bad. Is it reasonable to assume that 20% of orange were bad.

Solution. Null hypothesis H_0: 20% of oranges are bad.
Alternative hypothesis H_1: 20% of oranges are not bad.

Given $n = 64$, proportion of bad orange $p_1 = \frac{14}{64} = 0.218$

$$p = 20\% \text{ of oranges are bad} = \frac{20}{100} = 0.2$$

$$q = 1 - p = 0.8.$$

Test statistic $$Z = \frac{p_1 - p}{\sqrt{\frac{pq}{n}}} = \frac{0.218 - 2}{\sqrt{\frac{(0.2)\,(0.8)}{64}}} = 0.36$$

Conclusion. Since $|Z| = 0.36 < 1.645$, the null hypothesis is accepted.

8.2.9 Tests for Difference of Proportion

Let two large random samples of n_1 and n_2 items be taken from two different populations and p_1 and p_2 are the proportions of success. To test the difference between the sample proportions is significant or not.

$$Z = \frac{p_1 - p_2}{\sqrt{PQ\left(\frac{1}{n_1} + \frac{1}{n_2}\right)}}, \quad \text{where} \quad P = \frac{n_1 p_1 + n_2 p_2}{n_1 + n_2}, \ Q = 1 - P.$$

95% and 99% confidence interval for difference of proportions are $|p_1 - p_2| \pm 1.96$ SE $(p_1 - p_2)$ and $|p_1 - p_2| \pm 2.58$ SE $(p_1 - p_2)$ respectively.

8.2.10 Examples

1. In a random sample of 1000 persons from town A, 400 are found to be consumers of rice. In a sample of 800 persons from town B, 400 are found to be consumers of rice. Do this data reveal significance difference between town A and B?

Solution. Given $n_1 = 1000, \quad n_2 = 800$

$$p_1 = \frac{400}{1000} = 0.4, \quad p_2 = \frac{400}{800} = 0.5$$

$$P = \frac{p_1 n_1 + p_2 n_2}{n_1 + n_2} = \frac{0.4 \times 1000 + 0.5 \times 800}{1800} = 0.44$$

$$Q = 1 - 0.44 = 0.56$$

Null hypothesis: There is no significant difference between *A* and *B*.

Test statistic

$$Z = \frac{0.4 - 0.5}{\sqrt{(0.44)(0.56)\left(\frac{1}{1000} + \frac{1}{800}\right)}}$$

$$= \frac{-.1}{0.0235} = -4.26$$

Conclusion $|Z| = 4.26 > 3$. So we reject null hypothesis H_0. That is there is a significant difference between *A* and *B*.

2. Before an increase in excise duty on tea, 800 persons out of a sample of 1000 persons were found to be tea drinkers. After an increase in duty, 800 people were tea drinkers in a sample of 1,200 people. Using standard error of proportion, state whether there is a significant decrease in the consumption of tea after the increase in excise duty?

Solution. $n_1 = 1{,}000, \quad n_2 = 1{,}200$

$$p_1 = \frac{800}{1000} = 0.80, \quad p_2 = \frac{800}{1{,}200} = 0.67$$

Null Hypothesis: $H_0 : p_1 = p_2$. That is, there is no significant difference in the consumption of tea before and after the increase in excise duty.

Alternative Hypothesis: $H_1 : p_1 > p_2$

Test statistic $$Z = \frac{p_1 - p_2}{\sqrt{PQ\left(\frac{1}{n_1} + \frac{1}{n_2}\right)}}$$

$$= \frac{0.80 - 0.67}{\sqrt{\left(\frac{16}{22}\right)\left(\frac{6}{22}\right)\left(\frac{1}{1000} + \frac{1}{1200}\right)}} = 6.842$$

where $$P = \frac{n_1 p_1 + n_2 p_2}{n_1 + n_2} = \frac{1000 \times 0.80 + 1200 \times 0.67}{2,200} = \frac{16}{22}$$

and $Q = 1 - P = 1 - \frac{16}{22} = \frac{6}{16}$

Conclusion. Since $|Z| > 1.645$, H_0 is rejected at 5%, level of significance and conclude there is a significant decrease in the consumption of tea after increase in excise duty.

3. In a year there are 956 births in a town *A* of which 52.5% were males, while in town *A* and *B* combined, this proportion in a total of 1,406 births was 0.496. Is there any significant difference in the proportion of male births in the two towns?

Solution. $n_1 = 956, \quad n_2 = 1406 - 956 = 450$

$$p_1 = \frac{52.5}{100} = 0.525, \quad P = \frac{n_1 p_1 + n_2 p_2}{n_1 + n_2} = 0.496$$

$$Q = 1 - P = 1 - 0.496 = 0.504$$

Now

$$P = \frac{956 \times 0.525 + 450\, p_2}{1,406} = 0.496$$

$\Rightarrow$ $p_2 = 0.434$. That is the proportion of males in town *B* is 0.434.

Null hypothesis: $H_0 : p_1 = p_2$

Alternative hypothesis: $H_1 : p_1 \neq p_2$

Test statistic $$Z = \frac{p_1 - p_2}{\sqrt{PQ\left(\frac{1}{n_1} + \frac{1}{n_2}\right)}}$$

$$= \frac{0.525 - 0.434}{\sqrt{(0.496)\,(0.504)\left(\frac{1}{956} + \frac{1}{450}\right)}} = 3.15$$

Conclusion. Since $|Z| > 3$, we reject the hypothesis H_0 so there is a significant difference between *A* and *B*.

8.3 TEST OF SIGNIFICANCE FOR SMALL SAMPLES

When the size of the sample is small its distribution deviates greater from the normal curve. Therefore we can't assume the sampling distribution is approximately normal. Also smaller the sample the greater would be the variation in the values of means of different samples. The standard deviation of the smaller samples tends to be smaller than the true standard deviation of population. In small sample tests, the assumption is that "the population from which the samples are drawn are normal.

Arithmetic Mean. If there are N samples of values $x_1, x_2, \ldots x_N$, then Arithmetic mean or mean

$$\overline{x} = \frac{\sum_{i=1}^{N} x}{N}.$$

8.3.1 Degree's of Freedom (df)

It is the number of values in a sample, which may be assigned arbitrarily. If $\overline{x}$ is the mean of a sample of n observations, then only $(n-1)$ of the n values are assigned arbitrarily, then the remaining value is automatically determined dependent upon others. For example, there are five values x_1, x_2, x_3, x_4, x_5 and the total is 30. We can assign four values, the fifth value is determined obviously since the total can't exceed 30. The degrees of freedom are four and one is last as one condition is imposed, that is the total number of five values is 30. If more restrictions are placed in our choice and in this case the degrees of freedom will be last by the number of restrictions. The number of restrictions is known as *independent constraints*. So degrees of freedom are nothing but the reduction of number of constraints from the number of observations in a sample.

8.3.2 Students *t* Distribution

If $\overline{x}$ is the mean of a random sample of size n taken from a normal population having the mean μ and the variance σ^2 and

$$S^2 = \frac{\sum_{i=1}^{n} (x_i - \overline{x})^2}{n-1}, \text{ then } t = \frac{\overline{x} - \mu}{S/\sqrt{n}}$$

is a random variable having the t-distribution with $(n-1)$ degree's of freedom k. The probability density function is

$$f(t) = \frac{1}{\sqrt{k}\,\beta\left(\frac{1}{2}, \frac{k}{2}\right)} \frac{2}{\left(1+\frac{t^2}{k}\right)^{\frac{k+1}{2}}}, \quad -\infty < t < \infty$$

Note. Observe that $Z_{0.025} = 1.96 = t_{0.025}$ for degrees of freedom $k = \infty$. Hence when $k \to \infty$, t distribution approaches the standard normal distribution. In fact the values of t_α for 29 or more degrees of freedom are close to the corresponding values of Z_α. Hence the standard normal distribution provides a good approximation to the t distribution for samples of size 30 or more.

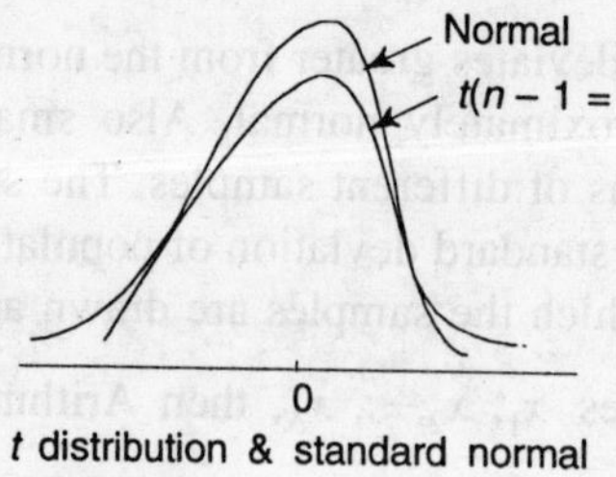

t distribution & standard normal distribution

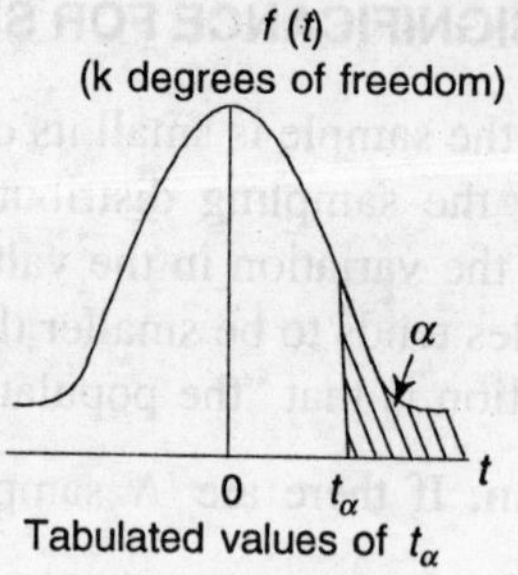

Tabulated values of t_α

Properties

(i) The variable of *t*-distribution varies from $-\infty$ to ∞.

(ii) The *t*-distribution is symmetrical and has mean 0.

(iii) 95% and 99% confidence limits for μ are $\bar{x} \pm t_{0.05}\dfrac{S}{\sqrt{n}}$, $\bar{x} \pm t_{0.01}\dfrac{S}{\sqrt{n}}$ respectively.

Applications. The '*t*' distribution is useful to test

(i) when the two sample means differ significantly

(ii) whether the sample mean $\bar{x}$ differ significantly with population mean

(iii) whether an observed correlation or regression coefficient is significant.

8.3.4 Problems

1. A random sample of size 16 has 53 as mean. The sum of the squares of the deviation taken from the mean is 150. Can the sample be taken from the population having 56 as mean? Obtain 95% confidence limits of population mean.

Solution.

1. *Null hypothesis:* $H_0 : \mu = 56$
2. *Alternative hypothesis:* $H_1 : \mu \neq 56$.

   Given $n = 16$, $\bar{x} = 53$, $\Sigma(x_i - \bar{x})^2 = 150$
3. *Test Statistic:*

$$t = \frac{\bar{x} - \mu}{S/\sqrt{n}} = \frac{53 - 56}{3.16/4} = -3.79$$

where
$$S = \sqrt{\frac{\sum(x_i - \bar{x})^2}{n-1}} = \sqrt{\frac{150}{15}} = \sqrt{10} = 3.16.$$

4. *Degrees of Freedom:* $n - 1 = 16 - 1 = 15$ *df*
5. *Conclusion:* Compare the calculated value with the tabulated value of *t* for 15 *df* and at certain level of significance. The tabulated value $t_{0.05}$ for 15 *df* is 1.753. Since the calculated value > tablulatead value, the null hypothesis is rejected that is, the sample is not taken from the population having mean 56.

95% confidence limits for μ is $\bar{x} \pm \dfrac{S}{\sqrt{n}} t_{0.05} = 53 \pm \dfrac{3.161}{\sqrt{16}}(1.753) = 54.38, 51.61$

2. The mean weekly sales of soap bars in departmental stores was 146.3 bars per store. After an advertising campaign the mean weekly sales in 22 stores for a typical week increased to 153.7 and showed a standard deviation of 17.2. Was the advertising campaign successful?

Solution.

1. *Null hypothesis:* The advertising campaign is not successful

$$H_0: \mu = 146.3$$

2. *Alternative hypothesis:* $H_1: \mu > 146.3$ (Right tail)

Given $n = 22, \mu = 146.3, \bar{x} = 153.7$, sample SD $= s = 17.2$.

3. *Test statistic:* Under the null hypothesis

$$t = \frac{\bar{x}-\mu}{S/\sqrt{n}} = \frac{\bar{x}-\mu}{s} \cdot \sqrt{n-1} = \frac{153.7-146.3}{17.2} \times \sqrt{21}$$

$= 19716$, since SD $= s = \sqrt{\dfrac{\sum(x-\bar{x})^2}{n}} = 17.2$ gives

$$ns^2 = \Sigma(x-\bar{x})^2 = S^2(n-1). \text{ That is, } \frac{S^2}{n} = \frac{s^2}{n-1}$$

4. *Conclusion:* Tabulated value of t for 21 *df* at 5% level of significance for single tailed test is 1.323. Since calculated value > tabulated value, H_0 is rejected. That is H_1 is accepted at 5% level of significance.

3. A random sample of 10 boys had the following I.Q.'s 70, 120, 110, 101, 88, 83, 95, 98, 107, 100. Do these data support the assumption of a population mean I.Q. of 100? Find a reasonable range in which most of the mean IQ values of samples of 10 boys lie?

Solution.

1. *Null hypothesis H_0:* The data support the assumption of population mean of 100. That is $\mu = 100$.
2. *Alternative hypothesis H_1:* $\mu \neq 100$.
3. $\bar{x}$ and S^2 to be calculted: $\bar{x} = \dfrac{\sum x}{n} = \dfrac{972}{10} = 97.2.$

x	$x - \bar{x}$	$(x-\bar{x})^2$
70	27.2	739.84
120	22.8	519.84
110	12.8	163.84
101	3.8	14.44
88	–9.2	84.64
83	–14.2	201.64
95	–2.2	4.84
98	0.8	0.64
107	9.8	96.04
100	2.8	7.84
Total 972		1833.60

$$S^2 = \frac{1833.60}{9} = 203.73.$$

The test statistic $t = \dfrac{\bar{x}-\mu}{S/\sqrt{n}} = \dfrac{97.2-100}{4.5136} = \dfrac{-2.8}{4.5136} = -0.62$

Conclusion: $|t| = .62$. Since tabulated value $t_{0.05}$ for $9df$ for two tailed test is 1.833, and calculated value < tabulated value, H_0 is accepted at 5% level of significance.

95% confidence limits within which the mean IQ values of samples of 10 boys will lie are

$$\bar{x} \pm t_{0.05}\left(\frac{S}{\sqrt{n}}\right) = 97.2 \pm (1.833)\ (4.514) = 105.5,\ 88.93$$

Hence the 95% confidence interval is [88.93, 105.5].

8.3.5 *t*-test for Difference of Means

To test two independent samples x_i $(i = 1, 2, \ldots n)$, $y_j = (j = 1, 2, \ldots n)$ have been drawn from the population with same mean (or) the two sample means $\bar{x}$ and $\bar{y}$ differ significantly or not

$$H_0 : \mu_x = \mu_y.$$

Test statistic
$$t = \frac{\bar{x}-\bar{\mu}}{S\sqrt{\dfrac{1}{n_1}+\dfrac{1}{n_2}}}$$

where $\bar{x}$ = mean of first sample

n_1 = size of first sample

$\bar{y}$ = mean of second sample

n_2 = size of second sample

$$S^2 = \frac{1}{n_1+n_2-2}\left[\sum_{i-1}^{n_1}(x_i-\bar{x})^2 + \sum_{j=1}^{n_2}(y_j-\bar{y})^2\right]$$

t is a distribution with $n_1 + n_2 - 2$ degrees of freedom.

Assumptions

(i) Parent populations, from which the samples are drawn are normally distributed.

(ii) The population variance are equal and unknown that is, $\sigma_x = \sigma_y = \sigma^2$ (say)

(iii) The two samples are random and independent.

8.3.6 Examples

1. Samples fo two types of electric light bulbs were tested for length of life and following data were obtained:

	Type I	Type II
Sample No.	$n_1 = 8$	$n_2 = 7$
Sample means	$\bar{x}_1$ = 1,234 hrs	$\bar{x}_2$ = 1,036 hrs
Sampled S.D.	s_1 = 36 hrs	s_2 = 40 hrs

Is the difference in the means sufficient to warrant that type I is superior to type II regarding length of life?

Solution. Null hypothesis $H_0 : \mu_x = \mu_y$

Alternative hypothesis H_1: $\mu_x > \mu_y$. That is type I is superior than type II.

$$S^2 = \frac{1}{n_1 + n_2 - 2}\left[\sum_{i=1}^{n_1}(x_i - \bar{x}_1)^2 + \sum_{j=1}^{n_2}(y_j - \bar{x}_2)^2\right]$$

$$= \frac{1}{13}[8 \times 36^2 + 7 \times 40^2] = (40.73)^2$$

where $$s_1^2 = \frac{\sum(x_i - \bar{x}_1)^2}{n_1} = 36^2$$

and $$s_2^2 = \frac{\sum(y_j - \bar{x}_2)^2}{n_2} = 40^2$$

gives $$\sum_i (x_i - \bar{x}_1)^2 = 8 \times 36^2 \quad \text{and} \quad \sum_j (y_j - \bar{x}_2)^2 = 7 \times 40^2$$

Test statistic

$$t = \frac{\bar{x}_1 - \bar{x}_2}{\sqrt{S^2\left(\frac{1}{n_1} + \frac{1}{n_2}\right)}} = \frac{1234 - 1036}{40.73}\sqrt{\frac{n_1 n_2}{n_1 + n_2}}$$

$$= \frac{1234 - 1036}{40.73}\sqrt{\frac{7 \times 8}{15}} = 9.39$$

The tabulated value for 13*df* with 5% level of significance for right tail test is 1.350. (This is the value of $t_{0.10}$ for 13*df* from two tail tables).

Conclusion. Calculated value > tabulated value. Hence H_0 is rejected and H_1 is accepted.

2. In a certain experiment to compare two types of rat foods *A* and *B*, the following results of increase in weights were observed in rats.

rat No.		1	2	3	4	5	6	7	8	Total
Increase in wts	Food A	49	53	51	52	47	50	52	53	407
in lbs	Food B	52	55	52	53	50	54	54	53	423

Assuming that the two samples of rats are independent, can we conclude that food *B* is better than food *A*?

Solution. *Null hypothesis* H_0: $\mu_x = \mu_y$
Alternative hypothesis H_1: $\mu_x < \mu_y$

$$\bar{x} = \frac{\sum x}{n} = \frac{407}{8} = 50.875$$

$$\bar{y} = \frac{\sum y}{n} = \frac{423}{8} = 52.875$$

Food A

x	$x-\bar{x}$	$(x-\bar{x})^2$
49	−1.875	3.5156
53	2.125	4.5156
51	0.125	0.0156
52	1.125	1.2656
47	−3.875	15.0156
50	−0.875	0.7656
52	1.125	1.2656
53	2.125	4.5156
		30.875

Food B

y	$y-\bar{y}$	$(y-\bar{y})^2$
52	−0.875	0.7656
55	2.125	4.5856
52	−0.875	0.7656
53	0.125	0.0156
50	−2.875	8.2656
54	1.125	1.2656
54	1.125	1.2656
53	0.125	0.0156
		16.8748

$$S^2 = \frac{1}{n_1+n_2-2}\left[\sum(x-\bar{x})^2+(y-\bar{y})^2\right]$$

$$= \frac{1}{14}[30.875 + 16.875]$$

$$= 3.41$$

Test statistic
$$t = \frac{\bar{x}-\bar{y}}{\sqrt{S^2\left(\frac{1}{n_1}+\frac{1}{n_2}\right)}} = \frac{52.875-50.875}{\sqrt{3.41\left(\frac{1}{8}+\frac{1}{8}\right)}} = 2.17$$

Tabulated value $t_{0.05}$ for $(8 + 8 - 2) = 14df$ for one tail test is 1.345. Since $|t| = 2.17 > 1.345$, H_0 is rejected at 5% level of significance and accept H_1. Hence we conclude food B is better than food A.

8.3.7 *t*-test for Significance of Correlation Coefficient

Null hypothesis H_0: $\rho = 0$. That is population correlation coefficient is zero. The test statistic

$$t = \frac{r}{\sqrt{1-r^2}}\sqrt{n-2}$$

r is observed correlation coefficient of n samples and t distribution is with $(n-2)$ degrees of freedom.

8.3.8 Examples

1. A random sample of 27 pairs of observations from a normal population gave a correlation coefficient of 0.6. Is this significant of correlation in population.
   Solution. $H_0 : \rho = 0$. That is the observed sample correlation coefficient is not significant of any correlation in the population. Give $r = 0.6$, $n = 27$.

   Test stafistic
   $$t = \frac{r}{\sqrt{1-r^2}}\sqrt{n-2} = \frac{0.6}{\sqrt{1-0.36}}\sqrt{25} = 3.75$$

   Tabulated value $t_{0.05}$ for $25df$ is 1.708.
   Conclusion. $|t| > t_{0.05}$, it is highly significant. Hence H_0 is rejected. Thus the variables are correlated in the population.

2. A coefficient of correlation of 0.2 is derived from a random sample of 625 pairs of observations. Is this value r significant? (ii) What are the 95% confidence limits to the correlation coefficient in the population.

Solution. *Null hypothesis* H_0: $\rho = 0$

Test statistic

$$t = \frac{r\sqrt{n-2}}{\sqrt{1-r^2}} = \frac{(0.2)\sqrt{625-2}}{\sqrt{1-0.4}} = 5.09$$

Since $df = 625 - 2 = 623$, the significant values of t are same as the case of normal distribution, $t_{0.05} = 1.96$.

Conclusion. $|t| > t_{0.05}$, H_0 is rejected. That is the sample correlation is significant of correlation in population.

3. Find the least value of r in a sample of 18 pairs of observations from a bivariate normal population, significant at 5% level of significance.

Solution. Here $n = 18$, From the table $t_{0.05}$ for $(18 - 2) = 16$ df is 1.746. Under H_0: $\rho = 0$,

$$t = \frac{r\sqrt{n-2}}{\sqrt{1-r^2}}$$

In order that the calculated value of t is significant at 5% level of significance

$$\left|\frac{r\sqrt{n-2}}{\sqrt{1-r^2}}\right| > t_{0.05}$$

$$\Rightarrow \quad \left|\frac{r\sqrt{16}}{\sqrt{1-r^2}}\right| > 1.746 \Rightarrow 16r^2 > (1.746)^2(1-r^2)$$

$$\Rightarrow \quad (19.06)r^2 > 3.0625 \Rightarrow r^2 = \frac{3.0625}{19.06} \quad \Rightarrow \quad r^2 = 0.1606$$

hence $|r| = 0.4008$

4. The table gives the ages of 10 women and mens. Compute the correlation coefficient and test the significance difference.

Men Age x:	23	27	28	29	30	31	33	35	36	39
Women Age y:	18	22	23	24	25	26	28	29	30	32

Solution. $\bar{x} = 31.1$, $\bar{y} = 25.7$, H_0: $\rho = 0$

Correlation coefficient
$$r = \frac{\frac{\sum xy}{n} - \bar{x}\,\bar{y}}{\sqrt{\left(\frac{\sum x^2}{n} - \bar{x}^2\right)\left(\frac{\sum y^2}{n} - \bar{y}^2\right)}}$$

$$r = \frac{\frac{8171}{10} - (31.1)(25.7)}{\sqrt{\left[\frac{9875}{10} - (31.1)^2\right]\left[\frac{6763}{10} - (25.7)^2\right]}} = 0.995$$

Test statistic

$$t = \frac{0.99}{\sqrt{1-(0.99)^2}}\sqrt{10-2} = 28.17$$

Conclusion. $|t| = 28.17 > t_{0.05}$ at 8 $df = 1.86$. Hence H_0 is rejected at 5% level of significance.

8.3.9 *F*-test for Equality of Population Variances

Suppose we want to test (i) whether two independent sample $x_i (i = 1 \text{ to } n_1)$ and y_j ($j = 1$ to n_2) have been drawn from the normal populations with the same variance σ^2. (or) (ii) whether the two independent estimates of the population variance are homogeneous or not.

Under the null hypothesis $H_0 : \sigma_x^2 = \sigma_y^2 = \sigma^2$

Test statistic F is given by

$$F = \frac{S_1^2}{S_2^2}$$

where $$S_1^2 = \frac{1}{n_1 - 1}\sum_{i=1}^{n_1}(x_i - \bar{x})^2$$

and $$S_2^2 = \frac{1}{n_2 - 1}\sum_{j=1}^{n_2}(y_j - \bar{y})^2$$

The probability density function of F is given by

$$f(F) = \frac{(r_1/r_2)^{r_1/2}}{\beta\left(\frac{r_1}{2}, \frac{r_2}{2}\right)} \times \frac{F^{r_1/2 - 1}}{\left(1 + \frac{r_1 F}{r_2}\right)^{\frac{r_1 + r_2}{2}}}$$

Note. F is a random variable having the F-distribution with $n_1 - 1$ and $n_2 - 1$ degrees of freedom. The numerator is always greater than denominator. Otherwise take $F = \frac{S_2^2}{S_1^2}$. The critical region for testing the null hypothesis $\sigma_x^2 = \sigma_y^2$ against the alternative hypothesis $\sigma_x^2 > \sigma_y^2$ is $F > F_\alpha$ where F_α is tabulated value. Similarly, the critical region for testing the null hypothesis against the alternative hypothesis $\sigma_x^2 < \sigma_y^2$ is $F < F_{1-\alpha}$ calculated by making use of the relation $F_{1-\alpha}(v_1, v_2) = \frac{1}{F_\alpha(v_2, v_1)}$, where F_α is the appropriate critical value of F for $n_2 - 1$ and $n_1 - 1$ degrees of freedom. For two sided alternative $\sigma_x^2 \neq \sigma_y^2$, the critical region is $F < F_{1-\alpha/2}$ or $F > F_{\alpha/2}$ where $F = \frac{S_1^2}{S_2^2}$ and the degrees of freedom are $n_1 - 1$ and $n_2 - 1$.

8.3.10 Examples

1. It is desired to determine whether there is less variability in the silver plating down by company 1 than that done by company 2. If independent random sample of size 12 of the two companies work yield $S_1 = 0.035$ mil and $S_2 = 0.062$ mil, test the null hypothesis $\sigma_1^2 = \sigma_2^2$ against the alternative hypothesis $\sigma_1^2 < \sigma_2^2$ at the 0.05 level of significance.

Solution.

1. *Null hypothesis* H_0: $\sigma_1^2 = \sigma_2^2$

Alternative hypothesis H_1: $\sigma_1^2 < \sigma_2^2$

F-test statistic

$$F = \frac{S_2^2}{S_1^2} = \frac{(0.062)^2}{(0.035)^2} = 3.14$$

Conclusion Since $F = 3.14 > F_{0.05} = 2.82$ for (11, 11) degrees of freedom, the null hypothesis is rejected and H_1 is accepted.

2. Two random samples were drawn from two normal population and there values are

A	66	67	75	76	82	84	88	90	92		
B	64	66	74	78	82	85	87	92	93	95	97

Test whether the two population have same variance at 5% level of significance?

Solution. H_0: $\sigma_A^2 = \sigma_B^2$

H_1: $\sigma_A^2 \neq \sigma_B^2$

$\bar{A} = 80$; $\bar{B} = 83$

												Total
$(A - \bar{A})^2$	196	169	25	16	4	16	64	100	144			734
$(B - \bar{B})^2$	361	289	81	25	1	4	16	81	100	144	196	1298

$$S_1^2 = \frac{734}{8} = 91.75; \; S_2^2 = \frac{1298}{10} = 129.8$$

Test statistic

$$F = \frac{S_2^2}{S_1^2} = \frac{129.8}{91.75} = 1.4147$$

Conclusion. $|F| = 1.4147 <$ tabulated value $F_{0.05}$ at $(n_2 - 1, n_1 - 1) = (10, 8)$ *df* is 3.07. So H_0 is accepted at 5% level of significance.

3. Two random samples have the following data:

Sample	Size	Sample mean	Sum of squares of deviation from mean
1	10	15	90
	12	14	108

Test whether the sample come from same normal population.

Solution. A normal population has two parameter viz μ and variance σ^2. To test the two independent samples have been drawn from the same population we have (i) to test the equality of population means (ii) to test the equality of population variances.

Null hypothesis. The two samples have drawn from the same normal population H_0: $\mu_1 = \mu_2$; $\sigma_1^2 = \sigma_2^2$. The equality of means will be tested by applying t-test and equality of variance is tested by applying F-test.

Given $$n_1 = 10,\ n_2 = 12,\ \bar{x}_1 = 15,\ \bar{x}_2 = 14,$$

$$\Sigma(x_1 - \bar{x}_1)^2 = 90,\ \Sigma(x_2 - \bar{x}_2)^2 = 108$$

F-test $$S_1^2 = \frac{\sum(x_1 - \bar{x}_1)^2}{n_1 - 1} = \frac{90}{9} = 10$$

$$S_2^2 = \frac{\sum(x_2 - x_2)^2}{n_2 - 1} = \frac{108}{11} = 9.82$$

Since $S_1^2 > S_2^2$, $$F = \frac{S_1^2}{S_2^2} = \frac{10}{9.82} = 1.018$$

Tabulated value $F_{0.05}$ at (9, 11) is 2.90 > 1.018, the null hypothesis that the equality of variance for two population is accepted. Since $\sigma_1^2 = \sigma_2^2$, we apply t-test for testing

$$H_0 : \mu_1 = \mu_2$$

t-test: under $$H_0: \mu_1 = \mu_2,\ H_1 : \mu_1 \neq \mu_2,$$

$$t = \frac{\bar{x}_1 - \bar{x}_2}{S\sqrt{\frac{1}{n_1} + \frac{1}{n_2}}}$$

where $$S^2 = \frac{1}{n_1 + n_2 - 2}[\Sigma(x_1 - \bar{x}_1)^2 + \Sigma(x_2 - \bar{x}_2)^2] = 9.9$$

Test statistic $$t = \frac{15 - 14}{\sqrt{9.9\left(\frac{1}{10} + \frac{1}{12}\right)}} = \frac{-1}{\sqrt{9.9\ \frac{11}{60}}} = -0.742$$

$t_{0.05}$ for 20 df is 2.086, since the tabulated value > calculated value, H_0 is accepted.

Note. If F-test is failed, then $\sigma_1^2 \neq \sigma_2^2$, hence t-test should not be used.

8.3.11 Chi-square Distribution

If $X_i = (i = 1, 2, ..., n)$ are n independent normal variates with mean μ_i and variance $\sigma_1^2 = 1, 2, ..., n$, then

$$\chi^2 = \sum_{i=1}^{n}\left(\frac{x_i - \mu_i}{\sigma_i}\right)^2$$

is a chi-square variate with n degrees of freedom.

Application of Chi-square distribution

(i) To test the 'goodness of fit'

(ii) To test the independence of attributes

(iii) To combine various probabilities obtained from independent experiments to given a single test of significance.

8.3.12 Chi-square Test for Population Variance

Suppose we want to test whether a random sample x_i has drawn from a normal population with a specified variance $\sigma^2 = \sigma_0^2$ (say). Then under null hypothesis $\sigma^2 = \sigma_0^2$,

$$\chi^2 = \frac{ns^2}{\sigma^2} \text{ with } (n-1)df$$

The above formula is applied when the population from which sample is drawn is normal. If the sample size $n > 30$, we have $Z = \sqrt{2\chi^2} - \sqrt{2n-1}$.

Example Test the hypothesis that $\sigma = 10$ given that $s = 15$ for a random sample of size 50 from a normal distribution.

Solution. Null hypothesis $H_0 : \sigma = 10$;
Alternative hypothesis H_1: $\sigma \neq 10$

$$n = 50,\ s = 15,\ \chi^2 = \frac{ns^2}{\sigma^2} = \frac{50 \times 15^2}{100} = 112.5$$

Since n is large,

$$Z = \sqrt{2 \times 112.5} - \sqrt{2 \times 50 - 1} = 5.05$$

$|Z| > 3$. H_0 is rejected at all levels of significance.

8.3.13 Goodness of Fit

If O_i $(i = 1, 2, \ldots, n)$ is a set of observed (experimental) frequencies and $E_i = (i = 1, 2, \ldots, n)$ is the corresponding set of expected (theoretical) frequencies, the Pearson's Chi-square is given by

$$\chi^2 = \sum_{i=1}^{n} \frac{(O_i - E_i)^2}{E_i} \text{ and}$$

$$\sum_{i=1}^{n} O_i = \sum_{i=1}^{n} E_i$$

In a set of n observations the degrees of freedom for χ^2 are $(n-1)$ since one *df* being lost because of the linear constraint $\sum_i O_i = \sum_i E_i = N$ on the frequencies.

8.3.14 Conditions for the Validity of χ^2-Test—Goodness of Fit

(i) The sample observations should be independent

(ii) Constraints on the frequency are linear

i.e. $\sum_i O_i = \sum_i E_i$

(iii) N the total frequency should be greater than 50.

(iv) No theoretical cell frequency should be less than 5. If any theoretical cell frequency is less than 5, then for application of χ^2-test it is added with the proceeding or succeeding frequency so that the added frequency is more than 5 and finally adjust the *df* lost in adding.
If r independent linear constraints are imposed, then it can be subtracted from *df*.
In addition, if any of the population parameter(s) is (are) calculated from the given data and used for computing the expected frequencies then in applying χ^2-test of goodness of fit, subtract one *df* for each parameter calculated.

8.3.15 Examples

1. Accidents that occurred during the various days of the week. Test whether the accidents are uniformly distributed over the week.

Days	M	T	W	Th	F	S
No. accidents	14	18	12	11	15	14

Solution. *Null hypothesis.* The accidents are uniformly distributed over the week under the null hypothesis, the expected frequency

$$E = \frac{\text{Total number of observed frequency}}{\text{No. of days}}$$

$$= \frac{84}{6} = 14$$

O	E	$\frac{(O-E)^2}{E}$
14	14	0
18	14	$\frac{16}{4}$
12	14	$\frac{4}{14}$
11	14	$\frac{9}{14}$
15	14	$\frac{1}{14}$
14	14	0
		Total = 2.143

Degrees of freedom:

n = number of observations – number of independent restriction
= 6 – 1 = 5

Conclusion. Calculated value $\chi^2 = 2.143$. The tabulated value $\chi^2_{0.05}$ for 5 *df* is 11.07. Since tabulated value $\chi^2_{0.05} > \chi^2$, H_0 is accepted.

2. The theory predicts the proportion of beans in the four groups A, B, C, D should be 9 : 3 : 3 : 1. In an experiment among 1600 beans, the numbers in the four groups were 882, 313, 287, 118. Does the experimental result support the theory?

Solution. *Null hypothesis* H_0: The theory fits well into the experiments.
Under the null hypothesis, the expected (theoretical) frequencies can be computed as follows.
Total number of beans = 1600. These are divided in the ratio 9 : 3 : 3 : 1. Therefore

$$E(882) = \frac{9}{16} \times 1600 = 900; \; E(313) \frac{3}{16} \times 1600 = 300$$

$$E(287) = \frac{3}{16} \times 1600 = 300; \; E(118) = \frac{1}{16} \times 1600 = 100$$

Therefore

$$\chi^2 = \sum \frac{(O-E)^2}{E} = \frac{(882-900)^2}{900} + \frac{(313-300)^2}{300}$$

$$+ \frac{(287-300)^2}{300} + \frac{(118-100)^2}{100}$$

$$= 0.3600 + 0.5633 + 0.5633 + 3.2400 = 4.7266$$

Degrees of freedom = 4 – 1 = 3.
Conclusion. $\chi^2_{0.05}$ for $3df = 7.815 >$ calculated value, the null hypothesis is accepted at 5% level of significance.

3. A survey of 320 families with 5 children each revealed the following distribution.

No. of boys	5	4	3	2	1	0
No. of girls	0	1	2	3	4	5
No. of families	14	56	110	88	40	12

Is this result consistent with the hypothesis that male and female births are equally probable?
Solution. *Null hypothesis:* Male and female births are equally distributed.
Observed frequency is given. To calculate the expected frequency we use the *binomial distribution.*

$$p = \text{The probability of male birth} = \frac{1}{2}$$

$$q = \text{The probability of female birth} = \frac{1}{2}$$

$$f(r) = Np(r) = \text{Probability of r male births}$$
$$= 320 \times 5\,C_r p^r q^{5-r}, \qquad r = 0, 1, 2, 3, 4, 5$$
$$= 320 \times 5C_r \left(\frac{1}{2}\right)^r \left(\frac{1}{2}\right)^{5-r}$$
$$= 320 \times 5C_r \left(\frac{1}{2}\right)^5$$
$$= 320 \times \frac{1}{32} \times 5C_r$$
$$= 10 \times 5C_r.$$

Now
$$f(0) = 10 \times 5C_0 = 10$$
$$f(1) = 10 \times 5C_1 = 10 \times 5 = 50$$

$$f(2) = 10 \times 5C_2 = 100$$
$$f(3) = 10 \times 5C_3 = 100$$
$$f(4) = 10 \times 5C_4 = 50$$
$$f(5) = 10 \times 5C_5 = 10$$

Observed frequency	Expected frequency	$(0-E)^2$	$\frac{(0-E)^2}{E}$
14	10	16	1.60
56	50	36	0.72
110	100	100	1.00
88	100	144	1.44
40	50	100	2.00
12	10	4	0.40
			Total 7.16

$$\chi^2 = 7.16$$

Degrees of freedom: $n - 1 = 6 - 1 = 5\ df$

Conclusion $\lambda^2_{0.05}$ = for $5df$ is 11.07 > calculated value, so the null hypothesis is accepted at 5% level of significance.

4. A set of 5 coins is tossed 3200 times and the number of heads appear in each test is noted. The test is as follows

No. of heads:	0	1	2	3	4	5
Frequency:	80	570	1100	900	500	50

Test the hypothesis that coins are unbaised

Solution. *Null hypothesis H_0:* The coins are unbiased.

$$p = \text{Probability of getting head} = \frac{1}{2}$$

$$q = \text{Probability of getting tail} = \frac{1}{2}$$

$$n = 5$$

$$P(X = x) = nc_x\, p^x\, q^{n-x},\ x = 0, 1, 2, 3, 4, 5$$

The expected frequency is the frequency that we are expected for $x = 0, 1, 2, 3, 4, 5$.

$$P(X = x) = 5C_x \left(\frac{1}{2}\right)^x \left(\frac{1}{t}\right)^{5-x} = \frac{1}{32} \times 5C_x$$

$$f(x) = NP(X = x) = 3200 \times \frac{1}{32} \times 5C_x = 100 \times 5C_x$$

$$f(0) = 100 \times 5C_0 = 100$$
$$f(1) = 100 \times 5C_1 = 500$$
$$f(2) = 100 \times 5C_2 = 1000$$
$$f(3) = 100 \times 5C_3 = 1000$$
$$f(4) = 100 \times 5C_4 = 500$$
$$f(5) = 100 \times 5C_5 = 100$$

O	E	$(O-E)^2$	$\frac{(O-E)^2}{E}$
80	100	400	4
570	500	4000	9.8
1100	1000	10,000	10
900	1000	10,000	10
500	500	0	0
500	100	2500	25
			58.8

Degrees of freedom = $n - 1 = 6 - 1 = 5$

Conclusion: $\chi^2_{0.05}$ for $5df = 11.07 <$ calculated value. So the null hypothesis is rejected.

5. Fit a Poisson distribution to the following data and test the goodness of fit

x:	0	1	2	3	4	5	6
frequency $f(x)$:	275	72	30	7	5	2	1

Solution. Mean of the given distribution is $\bar{x} = \frac{\sum x}{n} = 0.482.$

The frequency distribution of r successes is given by Poisson law as

$$f(r) = NP(X = r) = \frac{392 \times e^{-m}\,(m)^r}{r!}, \text{ where}$$

$$N = \Sigma f = 392 \text{ and } m = \bar{x} = 0.482. \text{ Now}$$

$$f(0) = \frac{392 \times e^{-0.482}\,(0.482)^0}{0!} = 242.1$$

$$[e^{-0.482} = \text{antilog}(-0.482 \log e)$$
$$= \text{antilog}\,(-0.482 \times 2.7183)$$
$$= 0.6176]$$

$$f(1) = (0.482) \times 242.1 = 116.69$$

$$f(2) = \frac{(0.482)^2}{2} \times 242.1 = 28.12$$

$$f(3) = \frac{(0.482)^3}{6} \times 242.1 = 4.518$$

$$f(4) = \frac{(0.482)^4}{24} \times 242.1 = 0.544$$

$$f(5) = \frac{(0.482)^5}{5!} \times 242.1 = 0.052$$

$$f(6) = \frac{(0.482)^6}{6!} \times 242.1 = 0.004$$

Hence theoretical Poisson frequencies correct to four decimals are:

x:	0	1	2	3	4	5	6	Total
expected frequency:	242.1	116.7	28.1	4.5	0.5	0.1	0	392

Observed frequency	Expected frequency	$(O-E)^2$	$\frac{(O-E)^2}{E}$
275	242.1	1082.41	4.471
72	116.1	1998.09	17.121
30	28.1	3.61	0.128
7, 5, 2, 1 } = 15	4.5, 0.5, 0.1, 0 } = 5.1	98.01	19.217
392	391.4		40.937

According to 8.3.14, the cell frequencies less than 5 are added with the preceding frequencies.

Constraints are:

(i) $\Sigma O \neq \Sigma E$

(ii) The parameter m is estimated from the given data and is used to compute expected frequency.

(iii) The last four cells whose frequencies are less than 5 are added together so that 3 df are lost.

Degrees of freedom: n = No. of observations – No. of constraints

$= 7 - 5 = 2$

Conclusion $\chi^2_{0.05}$ for $2df$ is 5.99 < calculated value. So it is highly significant. We conclude that the Poisson distribution is not a good fit to the given data.

8.3.16 Test of Independence of Attributes

Consider two attributes A and B, A divided into r classes, $A_1, A_2, \ldots A_r$ and B divided into s classes $B_1, B_2, \ldots, B_s$. The various cell frequencies can be expressed in the $r \times s$ contingency table, where (A_i) is the number of persons possessing the attribute A_i $(i = 1$ to $r)$, (B_j) is the number of persons possessing the attributes B_j, $(j = 1$ to $s)$ and (A_iB_j) is the number of persons possessing both the attributes A_i and B_j $(i = 1$ to $r, j = 1$ to $s)$. Also

$$\sum_{i=1}^{r} A_i = \sum_{j=1}^{s} B_j = N \text{ (total frequency)}.$$

Null hypothesis: H_0: Two attributes A and B are independent.

B \ A	A_1	A_2	A_3	...	A_r	Total
B_1	A_1B_1	A_2B_1	A_3B_1	...	A_rB_1	(B_1)
B_2	A_1B_2	A_2B_2	A_3B_2	...	A_rB_2	(B_2)
B_3	A_1B_3	A_2B_3	A_3B_3	...	A_rB_3	(B_3)
⋮	⋮	⋮	⋮	⋮	⋮	
B_s	A_1B_s	A_2B_s	A_3B_s	...	A_rB_s	(B_s)
Total	(A_1)	(A_2)	(A_3)	...	(A_r)	N

Expected frequency $E(A_iB_j) = \dfrac{(A_i)(B_j)}{N}$ $i = 1$ to r, $j = 1$ to s

In a $r \times s$ contingency table in calculating expected frequencies, the row totals, the column totals, and the grand total remain fixed. The fixation of 'r' column totals and 's' row totals imposes $(r + s)$ constraints on the cell frequencies. But since $\sum_{i=1}^{r}(A_i) = \sum_{j=1}^{s}(B_j) = N$, the total number of independent constraints is only $(r + s - 1)$. Further since the total number of the cell frequencies is $r \times s$, the required degrees of freedom is $rs - (r + s - 1) = (r - 1)(s - 1)$.

8.3.17 Examples

1. Two sample polls of votes for two candidates A and B for a public office are taken, one from among residents of rural areas. The results are given below. Examine whether the nature of the area is related to voting preference in this election.

Votes for / Area	A	B	Total
Rural	620	380	1000
Urban	550	450	1000
Total	1170	830	2000

Solution. *Null hypothesis:* The nature of area is independent of the voting preference in the election

$$E(620) = \frac{1170 \times 1000}{2000} = 585$$

$$E(380) = \frac{830 \times 1000}{2000} = 415$$

$$E(550) = \frac{1170 \times 1100}{2000} = 585$$

$$E(450) = \frac{830 \times 1000}{2000} = 415$$

$$\chi^2 = \Sigma \frac{(0 - E)^2}{E} = \frac{(620 - 585)^2}{585} + \frac{(380 - 415)^2}{415}$$

$$+ \frac{(550 - 585)^2}{585} + \frac{(450 - 415)^2}{415} = 10.089$$

Conclusion. Tabulated value $\chi^2_{0.05}$ for $(2 - 1)(2 - 1) = 1$ df is 3.841 < calculated value. So the null hypothesis is rejected at 5% level.

2. The following table gives the number of good and bad parts produced by each of 3 shifts in a part. Is there any association between the shifts and quality of parts produced by the factory.

Shift	Good	Bad	Total
Day	900	130	1030
Evening	700	170	870
Night	400	200	600
Total	2000	500	2500

Solution. *Null hypothesis:* There is no association between the shifts and quality

$$E(900) = \frac{2000 \times 1030}{2500} = 824; \; E(130) = \frac{500 \times 1030}{2500} = 696$$

$$E(700) = \frac{2000 \times 870}{2500} = 206; \; E(170) = \frac{500 \times 870}{2500} = 174$$

$$E(400) = \frac{2000 \times 600}{2500} = 480; \; E(200) = \frac{500 \times 600}{2500} = 120$$

$$\chi^2 = \sum \frac{(O-E)^2}{E}$$

$$= \frac{(900-824)^2}{824} + \frac{(130-696)^2}{696} + \frac{(700-206)^2}{206}$$

$$+ \frac{(170-174)^2}{174} + \frac{(400-480)^2}{480} + \frac{(200-120)^2}{120}$$

$$= 101.77$$

Conclusion The tabulated value $\chi^2_{0.05}$ at 2 df = 5.991 < calculated value. So the null hypothesis is rejected at 5% level of significance.

8.4 SOLVED PROBLEMS

1. A normal population has a mean of 0.1 and standard deviation of 2.1. Find the probability that mean of a sample of size 900 will be negative.

Solution. The standard normal variate corresponding to $\bar{x}$ is given by

$$Z = \frac{\bar{x} - \mu}{\sigma/\sqrt{n}} = \frac{\bar{x} - 0.1}{2.1/3.0} = \frac{\bar{x} - 0.1}{0.07}$$

$$\Rightarrow \qquad \bar{x} = (0.1) + Z(0.07)$$

where $Z \sim N(0, 1)$. The required probability p that the sample mean is negative is

$$p = P(\bar{x} < 0) = P((0.1) + Z(0.07) < 0)$$

$$= P\left(Z < \frac{-0.1}{0.07} = -1.43\right)$$

$$= P(Z > 1.43)$$

$$= 0.5 - P(0 < Z < 1.43)$$

$$= 0.5 - 0.4236 = .0764$$

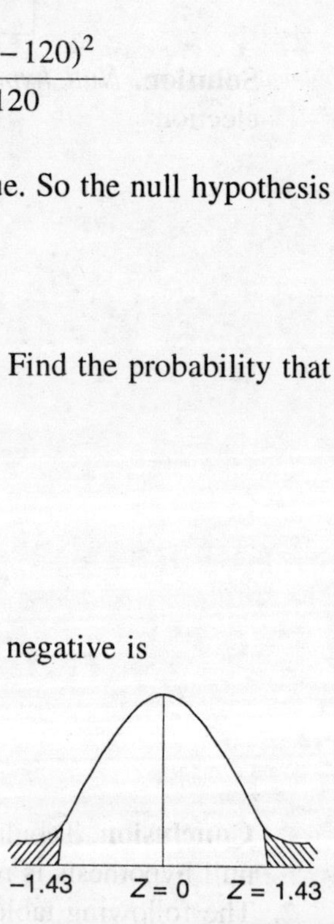

2. The guaranteed average life of a certain type of electric light bulbs is 1000 hrs with a standard deviation of 125 hrs. It is decided to sample the output so as to ensure that 90% of the bulbs do not fall short of the guaranteed average by more than 2.5%. What must be the minimum size of the sample?

Solution. $\mu = 1000$, $\sigma = 125$. Since we do not want the sample mean to be less than the guaranteed average mean ($\mu = 1000$) by more than 2.5%. We have

$$\bar{x} = 1000 - 2.5\% \text{ of } 1000$$

$$= 1000 - \frac{2.5}{100} \times 1000 = 975$$

Let n be the given sample size. Then

$$Z = \frac{\bar{x} - \mu}{\sigma/\sqrt{n}} > \frac{975 - 1000}{125/\sqrt{n}} = \frac{-25}{125/\sqrt{n}} = \frac{-\sqrt{n}}{5}$$

That is $Z > \frac{-\sqrt{n}}{5}$. By given condition

$$P\left(Z > \frac{-\sqrt{n}}{5}\right) = 0.90$$

$$\Rightarrow \quad P\left(0 < Z < \frac{\sqrt{n}}{5}\right) + P(Z = 0 \text{ to } \infty) = 0.90$$

$$\Rightarrow \quad P\left(0 < Z + \frac{\sqrt{n}}{5}\right) = 0.90 - 0.50 = 0.40$$

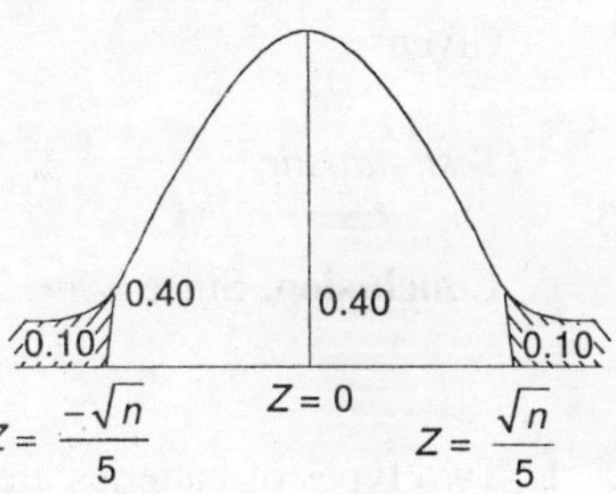

From normal tables,

$$\frac{\sqrt{n}}{5} = 1.28 \Rightarrow n = 25 \times (1.28)^2 = 41.$$

3. There are two brands of car tyres A and B in the market. A sample of 100 tyres of brand A has average life of 37500 kms with SD 2500 kms. Another sample of 75 tyres of brand B has an average life of 39,000 kms with SD 3000 kms. Can we conclude that there is a significant difference in population means?

Solution. *Null hypothesis* H_0: $\mu_1 = \mu_2$. There is no difference in the means of the population from which the samples are drawn.

Alternative hypothesis H_1: $\mu_1 \neq \mu_2$; $\mu_1 < \mu_2$

Given: $n_1 = 100$ $\quad n_2 = 75$

$\bar{x}_1 = 37500$ $\quad \bar{x}_2 = 39000$

$s_1 = 2500$ $\quad s_2 = 3000$

Test statistic

$$Z = \frac{|\bar{x}_1 - \bar{x}_2|}{\sqrt{\frac{s_1^2}{n_1} + \frac{s_2^2}{n_2}}} = \frac{|37500 - 39000|}{\sqrt{\frac{(2500)^2}{100} + \frac{(3000)^2}{75}}} = 3.511 > 3$$

Conclusion. At 5% level of significance, $Z_\alpha = 1.645$, so H_0 is rejected and H_1 is accepted.

4. Define a confidence interval.
We can find out the values of the population parameters by means of sample statistic. But this can be done approximately. With the help of standard error we can specify the maximum and minimum—within which the parameters of the population are expected to lie with a specified degree of confidence. Such an interval is known as confidence interval.
5. 1000 light bulbs with a mean life of 120 days are installed in new factory. Their length of life is normally distributed with SD 20 days. How many bulbs will expire in less than 90 days.

Solution. $$Z = \frac{90-120}{20} = -1.5$$

$$P(Z < -1.5) = 0.668 \text{ (by table)}$$

No. of bulbs expected to expire in less than 90 days

$$= 1000 \times .0668 = 66.8.$$

6. A sample of 400 male students is found to have a mean height of 67.47″, can it be reasonably regarded as a sample from a large population with mean hight 67.39″ and standard error 1.30″.
Solution. *Null hypothesis* H_0: $\bar{x} = \mu$
Alternative hypothesis H_1: $\bar{x} \neq \mu$

Given $$n = 400,\ \bar{x} = 67.47'',\ \mu = 67.39'',\ \sigma = 1.30$$

Test statistic $$Z = \frac{\bar{x}-\mu}{\sigma/\sqrt{n}} = \frac{67.47-67.39}{.065} = 1.2$$

Conclusion. Since $|Z| < 2.58$, H_0 is accepted.

EXERCISES

1. Two types of batteries are tested for their length of life and the following results were obtained
Battery A: $N_1 = 10$; $\bar{x}_1 = 500$ hrs; $\sigma_1^2 = 100$
Battery B: $N_2 = 10$, $\bar{x}_2 = 560$ hrs; $\sigma_2^2 = 121$
Compute '*t*' and test whether there is a significant difference in the two means?
2. Two independent sample of 8 and 7 items respectively had the following values:

Sample 1:	9	13	11	11	15	9	12	14
Sample 2:	10	12	10	14	9	8	10	

Is the difference between the means of samples significant?
3. A certain stimulus administered to each of the 12 patients resulted in the following increase of blood pressure

5 2 8 –1 3 0 –2 1 5 0 4 6

Can it be concluded that the stimulus will, in general, be accompanied by an increase in blood pressure?
4. In two large populations, there are 30% and 25% respectively of fair haired people. Is this difference likely to be hidden in samples of 1,200 and 900 respectively from the two populations?
5. In a simple sample of 600 men from a certain large city, 450 are found to be smokers. In one of 900 from another large city, 450 are smokers. Do the data indicate that the cities are significantly different with respect to the prevalence of smoking among men?
6. A biased coin was thrown 400 times and head resulted 240 times. Find the standard error of the observed proportion of heads and deduce that the probability of getting a head in a single throw of the coin lies almost certainly between 0.53 and 0.67.

7. The following table gives the classification of 100 workers according to sex, and the nature of work. Test whether nature of work is independent of the sex of the worker

	Skilled	Unskilled	Total
Male	40	20	60
Female	10	30	40
Total	50	50	

(5% value of chi-square with 1 *df* is 3.841).

8. A sample of 1000 members is found to have a mean of 3.67 cm. Can it be reasonably regarded as a simple sample from a population with mean 3.25 cm and SD 2.35 cm? How will you proceed if the sample variance is also given.

9. In an experiment on immunization of cattle from tuberculosis the following results were obtained.

	Affected	Unaffected
Inoculated:	12	28
Not Inoculated:	13	7

Examine the effect of vaccine in controlling the incidence of the disease.

10. The students of the same age-group from two different schools were compared for variability in their mathematical skill. A random sample of 25 pupils from one school had a variance of 16 marks while a random sample of 22 pupils from the other school has a variance of 8 marks. Examine if the difference in variability is significant.

ANSWERS

(1) $t = 12.1$ (calculated value)
(2) $t = 1.21$
(3) $t = 2.89$ (H_0 is rejected)
(4) $Z = 2.56 > 1.96$
(5) $Z = 9.6$, it is highly significant
(6) $SE = 0.2445$
(10) $F = 1.06$ not significant.

9

Design of Experiments and Quality Control

9.1 ANALYSIS OF VARIANCE

The analysis of variance is a powerful statistical tool for tests of significance. The alternative procedure to t-test for testing whether three or more samples are drawn from the same population is provided by the technique of analysis of variance. The basic purpose of the analysis of variance is to test that homogeneity of several means.

9.1.1 Definition

According to Prof R.A. Fisher, Analysis of Variance (ANOVA) is the '*separation of variance ascribable to one group of causes from the variance ascribable to other groups*'. By this technique the total variation in the sample data is expressed as the sum of its non-negative components where each of these components is a measure of the variation due to some specific independent source or factor or cause.

Assumptions. For the validity of the F-test in ANOVA the following assumptions are made.

(i) The observations are made

(ii) Parent population from which observations are taken is normal

(iii) The variance of populations are equal.

The important technique of analysis of variance has been discussed separately as

(i) One-way classification

(ii) Two-way classification

In one-way classification the data are classified according to only one factor.

In two-way classification the data are classified according to two different factors.

9.1.2 One-Way Classification

Suppose we have N observations x_{ij} ($i = 1, 2, \ldots, k, j = 1, 2, \ldots n_i$) are divided into k classes of sizes $n_1, n_2, \ldots, n_k$,in such a way that each observations belongs to exactly one class. The data will then be as in the Table 9.1

Where x_{ij} is j^{th} observation in the ith class, n_i is the number of observation in the ith class, and $n_1 + n_2 + \ldots + n_k = N$

The total variation in the observations x_{ij} can be split into the following two components.

(i) The variation *between the classes* or the variation due to different bases of classification, commonly known as *treatments.*

(ii) The variation *within the classes* i.e. inherent variation of the random variable within the observations of a class.

Table 9.1

Class	Observations	Means	Sum of Squares
1	$x_{11}\ x_{12}\ \cdots\ x_{1n_1}$	$\bar{x}_1$	$\sum_{j=1}^{n_1}(x_{1j}-\bar{x}_1)$
2	$x_{21}\ x_{22}\ \cdots\ x_{2n_2}$	$\bar{x}_2$	$\sum_{j=1}^{n_2}(x_{2j}-\bar{x}_2)$
3	$x_{31}\ x_{32}\ \cdots\ x_{3n_3}$	$\bar{x}_3$	$\sum_{j=1}^{n_3}(x_{3j}-\bar{x}_3)$
.	…………………	.	.
.	…………………	.	.
.	…………………	.	.
i	$x_{i1}\ x_{i2}\ \cdots\ x_{in_i}$	$\bar{x}_i$	$\sum_{j=1}^{n_i}(x_{ij}-\bar{x}_i)$
..	…………………	.	
k	$x_{k1}\ x_{k2}\ \cdots\ x_{kn_k}$	$\bar{x}_k$	$\sum_{i=1}^{n_k}(x_{kj}-\bar{x}_k)$

9.1.3 Mathematical Analysis of the Model

If we write

$$x_i. = \bar{x}_i \quad = \text{mean of the } i^{\text{th}} \text{ class}$$

$$= \sum_{j=1}^{n_i} \frac{x_{ij}}{n_i} \quad (i = 1, 2, \ldots k)$$

and $\sigma = \sum_i \sum_j x_{ij}$,

$$x\,..= \bar{\bar{x}} = \text{overall mean} = \frac{1}{N}\sum_{i=1}^{k}\sum_{j=1}^{n_i} x_{ij} = \frac{1}{N}\sum_{i=1}^{k} n_i \bar{x}_i$$

Then

$$\sum_{i=1}^{k}\sum_{j=1}^{n_i}(x_{ij}-\bar{\bar{x}})^2 = \sum_{i=1}^{k}\sum_{j=1}^{n_i}(x_{ij}-\bar{x}_i+\bar{x}_i-\bar{\bar{x}})^2$$

$$= \sum_i\sum_j (x_{ij}-\bar{x}_i)^2 + \sum_i n_i(\bar{x}_i-\bar{\bar{x}})^2 + 2\Big[\sum_i(\bar{x}_i-\bar{\bar{x}})\sum_j(x_{ij}-\bar{x}_i)\Big]$$

But $\sum_{j=1}^{n_i}(x_{ij}-\bar{x}_i) = 0$, since the algebraic sum of the deviations of the observations from their mean is zero. Therefore

$$\sum_i^k\sum_j^{n_i}(x_{ij}-\bar{\bar{x}})^2 = \sum_{i=1}^{k}\sum_{j=1}^{n_i}(x_{ij}-\bar{x}_i)^2 + \sum_i n_i(\bar{x}_i-\bar{\bar{x}})^2 \tag{9.1}$$

$S_{T^2} = \sum_i\sum_j (x_{ij}-\bar{\bar{x}})^2$ is known as **total sum of squares** T.S.S.

$$S_{E^2} = \sum_i \sum_j (x_{ij} - \bar{x}_i)^2$$ is called **within sum of squares** or **Error sum of squares** (SSE)

$$S_{t^2} = \sum_i n_i (\bar{x}_i - \bar{\bar{x}})^2$$ is called **S.S. due to treatments** (SST)

Hence (9.1) gives

$$\text{TSS} = \text{SSE} + \text{SST}$$

Degrees of Freedom for Various SS

TSS is computed from the N quantities of the form $(x_{ij} - \bar{\bar{x}})$ will carry $(N - 1)df$, one df is lost because of the linear constraint

$$\sum_{i=1}^{k} \sum_{j=1}^{n_i} (x_{ij} - \bar{\bar{x}}) = 0.$$

Similarly the treatment sum of squares

$$\sum_{i=1}^{k} n_i (\bar{x}_i - \bar{\bar{x}})^2$$ will have $(k - 1)df$ since, $\sum_{i=1}^{k} n_i (\bar{x}_i - \bar{\bar{x}}) = 0$

and the error SS $= \sum_i \sum_j (x_{ij} - \bar{x}_i)^2$ will have $(N - k)df$ since it is based on N quantities which are subjected to k linear constraints

$$\sum_{j=1}^{n_i} (x_{ij} - \bar{x}_i) = 0, \; i = 1, 2, \ldots, k$$

Hence the df for various SS are also additive, since $N - 1 = (N - k) + (k - 1)$

9.1.4 Mean Sum of Squares (MSS)

The sum of squares divided by its degrees of freedom gives the corresponding variance or the mean sum of squares MSS. Thus

$$\frac{S_{t^2}}{(k-1)} = \frac{SST}{(k-1)} = s_{t^2}$$ is the MSS due to treatments and

$$\frac{S_{E^2}}{(N-k)} = \frac{SSE}{N-k} = s_{E^2}$$ is the MSS due to error.

Null hypothesis. We want to test the equality of the population means i.e., the homogeneity of different means. So

$$H_0 : \mu_1 = \mu_2 = \ldots = \mu_k = \mu \qquad \text{(population means are equal)}$$

Test statistic for H_0 is provided by the variance ratio

$$F = \frac{s_{t^2}}{s_{E^2}} \tag{9.2}$$

If H_0 is true, then F should take the value 1, otherwise it should be greater than unity. In order to find out if an observed value of F is significantly greater than unity, we have to obtain the sampling distribution of the statistic F defined in (9.2).

If an observed value of F obtained from (9.2) is greatrd than the tabulated value of F for $(k-1, N-k)$ df at specified level of significance, then H_0 is rejected at that level.

To test for the equality of mean curing times we complete the *Analysis of variance* (ANOVA) table by including the mean square errors and the value of F.

ANOVA Table For One-Way Classification

Sources of variation	Degrees of freedom	Sum of squares	Mean sum of squares	Variance F
Treatment (Between classes)	$k-1$	S_{t^2}	$s_{t^2} = \dfrac{S_{t^2}}{k-1}$	$\dfrac{s_{t^2}}{s_{E^2}}$
Error within classes	$N-k$	S_{E^2}	$s_{E^2} = \dfrac{S_{E^2}}{N-k}$	
Total	$N-1$	S_{T^2}		

Note. For practical problems, we can use the simplified formulae.

$$\text{TSS} = \sum_i\sum_j (x_{ij} - \bar{\bar{x}})^2 = \sum_i\sum_j x_{ij}^2 - \frac{(\sum\sum x_{ij})^2}{N}$$

$$= \sum_i\sum_j x_{ij}^2 - \frac{T^2}{N}$$

where T is the grand total of all observations and $N = n_1 + n_2 + \ldots + n_k$. $\sum\sum x_{ij}^2$ is known as Row sum of squares (RSS) and the expression $\dfrac{T^2}{N}$ is called the **correction factor** (CF)

$$\text{SSE} = \sum_i\sum_j x_{ij}^2 - \sum_i\left(\frac{T_i^2}{n_i}\right)$$

Between classes

$$\text{SST} = \sum_i n_i(\bar{x}_i - \bar{\bar{x}})^2 = \sum_i \frac{T_{i^2}}{n_i} - \frac{T^2}{N}.$$

Hence to simplify the arithmetic we subtract origin from all observations. This will not effect the value of the sums of squares. This may also divided by a suitable constant C. This will reduce the size of SS. To get the actual value, multiply finally sum of squares by C^2.

9.1.5 Example

The following data shows the monthly salaries of employees selected at random from 4 companies A, B, C and D.

	Salaries in Rs. (Employees)					
A	560	590	530	600	510	570
B	490	510	500	480	470	
C	610	580	590	620		
D	580	590	620	640	550	

Test whether the salaries differ from company to company.

Solution.

Null hypothesis: H_0: The salaries do not differ significantly from company to company.

To simplify the arithmetic we subtract 550 from each observation and divide by 10.

The given data is reduced to

								T_i
A	1	4	−2	5	−4	2	$n_1 = 6$	$T_1 = 6$
B	−6	−4	−5	−7	−8		$n_2 = 5$	$T_2 = -30$
C	6	3	4	7			$n_3 = 4$	$T_3 = 20$
D	3	4	7	9	0		$n_5 = 5$	$T_4 = 23$
								$T = 19$

N = number of samples = 20

To calculate $\sum x_{ij}^2$ we prepare the square table.

							T_i
A	1	16	4	25	16	4	66
B	36	16	25	49	64		190
C	36	9	16	49			110
D	9	16	49	81	0		135
							$\sum x_{ij}^2 = 521$

Now correction term $= \dfrac{T^2}{N} = \dfrac{19^2}{20} = 18.05$

(i) Total SS $= \sum_{ij} x_{ij}^2 - \dfrac{T^2}{N} = 521 - 18.05 = 502.95$

(ii) Between class = SST $= \sum \dfrac{T_i^2}{n_i} - \dfrac{T^2}{N}$

$$= \frac{(6)^2}{6} + \frac{(-30)^2}{5} + \frac{(20)^2}{4} + \frac{(23)^2}{5} - 18.05$$

$$= 6 + 180 + 100 + 105.8 - 18.05$$

$$= 391.8 - 18.05 = 373.75$$

(iii) Within classes = SSE = TSS − SST = 129.20

Total number of observation is $N = 20$. Number of classes is $k = 4$. Hence the total number of degrees of freedom for TSS $= N - 1 = 19$.

Number of df associated with SST $= k - 1 = 4 - 1 = 3$.

Number of df associated with SSE $= N - k = 20 - 4 - 16$.

Since we have divided by 10, we should multiply by $(10)^2$ each class.

ANOVA Table is

Source of variation	df	sum of squares	Mean SS	F
Between companies	3	$373.75 \times 100 = 37375$	$\frac{37375}{3} = 12455$	$\frac{12455}{807.5} = 15.41$
Within companies	16	$129.2 \times 100 = 12920$	$\frac{12920}{16} = 807.50$	
Total	19	$502.95 \times 100 = 50295$		

Conclusion: $F_{0.05}$ at (3, 16)df is 3.24 < calculated value 15.9. Hence H_0 is rejected at 5% level of significance. Hence the salaries differ from company to company between classes and within classes.

9.1.6 Example

The following tables shows the lives in hours of four batches of electric bulbs.

Batches								
1	1600	1610	1650	1680	1700	1720	1800	
2	1580	1640	1640	1700	1750			
3	1460	1550	1600	1620	1640	1660	1740	1820
4	1510	1520	1530	1570	1600	1680		

Perform an analysis of variance of these data and show that a significance test does not reject their homogenity.

Solution.

Null hypothesis: Various batches are homogeneous

Shifting the origin to $C = 1640$ and then dividing by 10, the above data reduces to

Batches									T_i	$\sum_j x_{ij}^2$
1	−4	−3	1	4	6	8	16	—	28	398
2	−6	0	0	6	11	—	—	—	11	193
3	−18	−9	−4	−2	0	2	10	18	−3	853
4	−13	−12	−11	−7	−4	4	—	—	−43	515
								Total	−7	1959

$$T = \sum T_i = -7. \text{ So corrector factor } = \frac{T^2}{N} = \frac{(-7)^2}{26} = \frac{49}{26} = 1.884615;$$

$$\text{(i) TSS} = \sum_{ij} x_{ij}^2 - \frac{T^2}{N} = 1959 - 1.884615 = 1957.115385$$

$$\text{Between sum of squares} = \text{SST} = \sum_{i=1}^{4}\left(\frac{T_i^2}{n_i}\right) - \frac{T^2}{N}$$

$$= \frac{(28)^2}{7} + \frac{(11)^2}{5} + \frac{(-3)^2}{8} + \frac{(-3)^2}{6} - 1.884615$$
$$= 445.492 - 1.884615$$
$$= 443.60706$$

within sum of square = TSS – SST = 1957.115385 – 443.607 = 1513.50833.

ANOVA Table

Source of variation	df	Sum of squares	Means SS	F
Between batches	3	443.60706	$\frac{443.60706}{3} = 147.86903$	$\frac{147.86903}{68.79583} = 2.1493$
Within batches	22	1513.50833	$\frac{1513.50833}{22} = 68.79583$	
Total	25	1957.11539		

Conclusion. $F_{0.05}$ for (3, 22) *df* is 3.05 > 2.1493. Hence H_0 is accepted at 5% level of significant. Hence we conclude that four batches may be regard as homogeneous.

9.1.7 Two Way Classification

Let N observations be classified according to two classifications A and B in such a way that each observations belong to exactly one A class and exactly one B class. If there are m A classes and n B classes, then we have $N = mn$. Let

x_{ij} = the observation belong to the ith A class (A_i) and j^{th} B class (B_j)

Then the data can be written in the following form

	B_1	B_2	...	B_n	Means	Raw Total
	O	O	...	O		
A_1	x_{11}	x_{12}	...	x_{1n}	$\bar{x}_{1.}$	$T_{1.}$
A_2	x_{21}	x_{22}	...	x_{2n}	$\bar{x}_{2.}$	$T_{2.}$
A_3	x_{31}	x_{32}	...	x_{3n}	$\bar{x}_{3.}$	$T_{3.}$
A_i	x_{i1}	x_{i2}	...	x_{in}	$\bar{x}_{i.}$	$T_{i.}$
A_m	x_{m1}	x_{m2}	...	x_{mn}	$\bar{x}_{m.}$	$T_{m.}$
Means	$\bar{x}_{.1}$	$\bar{x}_{.2}$	...	$\bar{x}_{.n}$	$\bar{\bar{x}}$	
Column Total	$T_{.1}$	$T_{.2}$	...	$T_{.n}$		N

If we write

$$\bar{x}_{i.} = \frac{1}{n}\sum_{j=1}^{n} x_{ij}, \ (i = 1, 2, \ldots m)$$

$$\bar{x}_{.j} = \frac{1}{m}\sum_{i=1}^{m} x_{ij}, \ (j = 1, 2, \ldots n)$$

$$\bar{\bar{x}} = \text{overall mean} = \frac{1}{nm}\sum_i \sum_j x_{ij}$$

$$= \frac{1}{n}\sum_j \left(\frac{1}{m}\sum_i x_{ij}\right) = \frac{1}{n}\sum_j \bar{x}_{.j}$$

Also

$$\bar{\bar{x}} = \frac{1}{m}\sum_i \bar{x}_{i.}$$

Then

$$\sum_i \sum_j (x_{ij} - \bar{\bar{x}})^2 = n\sum_i (\bar{x}_{i.} - \bar{\bar{x}})^2 + m\sum_i (\bar{x}_{.j} - \bar{\bar{x}}) + \sum_i \sum_j (x_{ij} - \bar{x}_{i.} - \bar{x}_{.j} + \bar{\bar{x}})^2$$

or

$$S_{T^2} = S_t + S_{V^2} + S_{E^2}$$

where

$$S_{T^2} = \sum_i \sum_j (x_{ij} - \bar{\bar{x}})^2 \text{ is the total SS = TSS.}$$

$$S_{t^2} = n\sum_i (\bar{x}_{i.} - \bar{\bar{x}})^2 \text{ is the SS due to treatments = SST}$$

$$S_{V^2} = m\sum_j (\bar{x}_{.j} - \bar{\bar{x}})^2 \text{ is the SS due to varieties}$$

$$S_{E^2} = \sum_i \sum_j (x_{ij} - \bar{x}_{i.} - \bar{x}_{.j} + \bar{\bar{x}})^2 \text{ is the error or residual SS.}$$

Null hypothesis. We set up the null hypothesis that the treatments as well as varieties are homogeneous.

That is: H_0: There is no significant difference between the column means as well as between row means.

Degrees of freedom S_{T^2} is computed from $N = nm$ quantities $(x_{ij} - \bar{\bar{x}})$ which are subjected one linear constraint $\sum_i \sum_j (x_{ij} - \bar{\bar{x}}) = 0$ will carry $(N - 1)df$

S_{t^2} will be based on $(k - 1)$ *df* since $\sum_i (\bar{x}_{i.} - \bar{\bar{x}}) = 0$,

S_{V^2} will have $(n - 1)df$, since $\sum_j (\bar{x}_{.j} - \bar{x}) = 0$.

S_{E^2} will have $(N - 1) - (m - 1) - (n - 1) = (n - 1)(m - 1)$df

Mean S.S Mean SS treatments $= \dfrac{S_{t^2}}{m-1} = s_{t^2}$

Means SS varieties $= \dfrac{S_{V^2}}{n-1} = s_{V^2}$

$$\text{Error Mean SS} = \frac{S_{E^2}}{(n-1)(m-1)} = s_{E^2}$$

Anova Table for Two-way Classified Data

Source of variation	Sum of squares	df	MSS	Variance ratio F
Treatments (Between A class)	$S_{t^2} = \sum_i n(\bar{x}_i. - \bar{\bar{x}})^2$	$m-1$	$s_{t^2} = \frac{S_{t^2}}{m-1}$	$F_t = \frac{s_{t^2}}{s_{E^2}}$
Varieties (Between B class)	$S_{V^2} = m\sum_j (\bar{x}._j - \bar{\bar{x}})^2$	$n-1$	$s_{V^2} = \frac{S_{V^2}}{n-1}$	$F_V = \frac{s_{V^2}}{s_{E^2}}$
Error (Residual)	$S_{E^2} = \sum_i\sum_j (x_{ij} - \bar{x}_i. - \bar{x}._j - \bar{\bar{x}})^2$	$(n-1)(m-1)$	$s_{E^2} = \frac{S_{E^2}}{(n-1)(m-1)}$	
Total	$\sum_i\sum_j (x_{ij} - \bar{\bar{x}})^2$	$nm-1$		

The tabulated value for F_t is $F_{(m-1),\,(n-1)(m-1)}$ at certain level of significance and the tabulated value for F_V is $F_{(n-1),\,(n-1)(m-1)}$ at certain level of significance.

Remark. For numerical computation of the various sum of squares the following formulae are used.

(i) $\text{TSS} = \sum_i\sum_j x_{ij}^2 - \frac{T^2}{mn}$

(ii) Treatment SS (Between A class) = SST = $\frac{1}{n}\sum_{i=1}^{m} T_i.^2 - \frac{T^2}{mn}$

(iii) Varieties SS (Between B class) = $\frac{1}{m}\sum_{j=1}^{m} T._j^2 - \frac{T^2}{mn}$

(iv) Residual SS = TSS – (SST + varieties SS)

9.1.8 Example

To compare three varieties of wheat and 4 Blocks of three plots of land each were used. The following table shows the yield of the 12 plots in kg. Analyse the data.

Varieties	Blocks B_1	B_2	B_3	B_4
V_1	3	4	6	4
V_2	4	6	5	3
V_3	2	3	4	5

Solution.

Null hypothesis H_0: There is no significant difference between varieties and blocks of the plots *w*.
Alternative hypothesis H_1: There is a significant difference between varieties and blocks of the plots used.

We prepare table for calculating T_i . and $T_{.j}$, T and $\sum xij^2$ as follows.

(i)

	B_1	B_2	B_3	B_4	T_i
V_1	3	4	6	4	17
V_2	4	6	5	3	18
V_3	2	3	4	5	14
$T_{\times j}$	9	13	15	12	49 = T

(ii) Table of squares

	B_1	B_2	B_3	B_4	Total
V_1	9	16	36	16	77
V_2	16	36	25	9	86
V_3	4	9	16	25	54
Totals	29	61	77	50	217

Calculation of sums of square: $\sum xij^2 = 217$.

$$\text{Correction Term} = T^2 = \left(\frac{(49)^2}{3\times 4}\right) = 200.1$$

$$\text{Total SS} = \sum_{i,\,j} x_{ij}^{\ 2} - \frac{T^2}{mn} = 217 - 200.1 = 16.9$$

$$\text{SS between varieties } A = \sum_i \left(\frac{T_{i.}^{\ 2}}{n}\right) - \frac{T^2}{mn} = \frac{1}{4}[17^2 + 18^2 + 14^2] - 200.1 = 2.15$$

$$\text{SS. between blocks } B = \frac{1}{m}\sum_j T_{\cdot j}^2 - \frac{T^2}{mn} = \frac{1}{3}[9^2 + 13^2 + 15^2 + 12^2] - 200.1 = 6.23.$$

Residual SS = TSS – (SS between varieties A + SS between blocks B)
= 16.9 – (2.15 + 6.23) = 8.52

Since $m = 3$, $n = 4$, the total number of $df = mn - 1 = 12 - 1 = 11$. *Df* between varieties is $m - 1 = 2$, *Df* between blocks is $n - 1 = 3$. The *df* for the residual SS are $11 - (2 + 3) = 6$. Also $(m - 1)(n - 1) = 2 \times 3 = 6$.

ANOVA Table

Source	DF	SS	MSS	Variance ratio F
Between varieties	2	$S_t^2 = 2.15$	$s_t^2 = \frac{2.15}{2.0} = 1.075$	$F_t = \frac{s_t^2}{s_E^2} = \frac{1.075}{1.42} = .757\ F_{(2,\,6)}$
Between blocks	3	$S_v^2 = 6.23$	$s_v^2 = \frac{6.23}{3} = 2.07$	$F_v = \frac{s_v^2}{s_E^2} = \frac{2.07}{1.42} = 1.463\ F(3,\,6)$
Residual	6	$S_E^2 = 8.52$	$s_E^2 = \frac{8.52}{6} = 1.42$	
Total	11	16.9		

Conclusion: $F_{(2, 6)}$ at 5% level of significance is 5.14, $F_{(3, 6)}$ at 5% level of significance is 4.75. Hence $F_t < F_{(2, 6)}$ and $F_v < F_{(3, 6)}$. So H_0 is accepted at 5% level of significant. That is, there is no significant difference between varieties and blocks of the plots used. The yields would still vary due to the difference in soil fertility. Such factors beyond human control is spoken of as *experimental error*. It is due to

(i) the inherent variability in the experimental material to which treatments are applied
(ii) the lack of uniformity in the methodologies of conducting the experiment
(iii) the lack of representativeness of the sample to the population under study.

Remark. In order to test the significance of the difference between two treatments, we first obtain an estimate of the experimental error and then apply some test of significance.

Replication. Replication means the execution of an experiment more than once. In other words, the repetition of treatments under investigation is known as *replication.*

Precision. The reciprocal of the variance of the mean is termed as the precision. Thus for an experimental replicated r terms, the precision is given by

$$\frac{1}{V(\bar{x})} = \frac{r}{\sigma^2}$$

where σ^2 is the error variance per unit.

Uniformity Trials. A trial in which the field (experimental material) is divided into small units (plots) and the same treatment is applied on each of the units and their yields are recorded.

Principles of an Experimental Design

(i) Replication
(ii) Randomisation
(iii) Local control

Randomisation. It is a process of assigning the treatment to various experimental units in a purely chance manner. It eliminates bias in any form.

Note that randomisation without replication is not sufficient. When randomisation of treatments to various units is accompanied by an adequate number of replications then only we apply the tests of significance (t-test or F-test).

Local control. The process of reducing the experimental error by dividing the relatively heterogeneous experimental area (field) into homogeneous blocks.

9.2 DESIGN OF EXPERIMENTS

The design of experiments may be defined as "the logical construction of the experiment in which the degree of uncertainty with which the inference is drawn may be well defined."

9.2.1 Terminology in Experimental Design

Experiment. An experiment is a device or a means of getting an answer to the problem under consideration. It is classified as

(a) Absolute (b) Comparative

Absolute experiments consist in determining the absolute value of some characteristic like (i) obtaining the average intelligence quotient I.Q. of a group of people (ii) finding the correlation coefficient between two variables in a bivarnate distribution etc.

Comparative experiments are designed to compare the effect of two or more objects on some people characteristic e.g., comparison of different manures or fertilizers, different kinds of varieties of a crop, different cultivation processes or different pieces of land in a field, or different diets or medicines in a dietary or medical experiments etc.

Experimental Unit. The smallest division of the experimental material to which we apply the treatments and on which we make observations on the variable under study is termed as *experimental unit. e.g.* In field experiment the plot of *land* is the experimental unit.

Blocks. The whole experimental unit are divided sometimes into relatively homogeneous subgroups. These subgroups are called *blocks.*

Yield. The measurement of the variable under study on different experimental units (e.g. plots in field experiments) are termed as *yields.*

Experimental Error. Let us suppose that a large homogeneous field is divided into different plots (of equal shape and size and different treatments are applied to these plots. If the yields from some of the treatments are more than those of the others, the experimenter is faced with the problem of deciding if the observed differences are really due to treatment effects or they are due to uncontrolled factors. Experience tells that if the same treatment. Local control reduces the experimental error and increases the efficiency of the design.

Size of the Plot. The size of the plot depends on a number of factors such as the total experimental area available, the number of treatments, the number of replication of each treatment, and so on.

Fairfield Smith's Variance law

$$V_x = \frac{V_1}{x^b}$$

where V_x is the variance of the yield per unit area from plots of size x units, V_1 is the variance among plots of size unity and b, the regression coefficient is a soil characteristic indicating the relationship between adjacent units. The limiting value of b are 1 and 0.

$b = 1$ means that the experimental unit is composed of a random selection of x individuals. In this case we get

$$V_x = \frac{V_1}{x}$$

so that the precision of the experiment increases with an increase in the plot size.

Shape of Blocks and Plots. The shape and size of the blocks will usually depend upon the shape and size of the plots. In order to control the experiments error it is desirable to divide the whole experimental area into different blocks such that within each block there is as much homogeneity as possible but between blocks there is maximum variation. Further each block is to be divided into as many plots as the number of treatments. For maximum precision the plots should be rectangular in shape with their long sides parallel to the direction of the fertility gradient and the blocks should be arranged one after the other along the fertility gradient as shown in the figure.

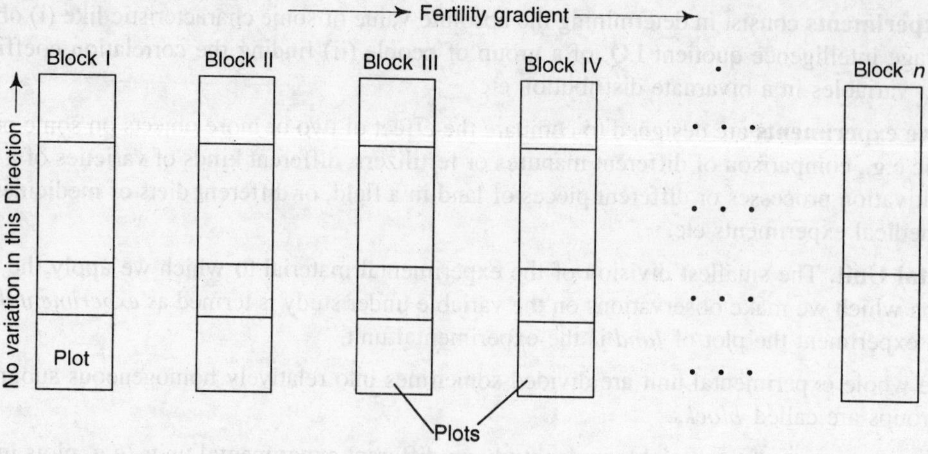

9.3 COMPLETELY RANDOMISED DESIGN (C.R.D)

The completely randomised design is the simplest of all the designs, based on principles of **randomisation** and **replication**. In this design treatments are allocated at random to the experimental units over the entire experimental material. Suppose we have v treatments, the ith treatment being replicated r_i times, $i = 1, 2 \dots v$. Then the whole experimental material is divided into $n = \sum r_i$ experimental units and the treatments are distributed completely at random over the units subject to the condition that the ith treatment occurs r_i times.

Suppose we have k varieties of wheat to be experimented on. For the purpose of experimentation, we select a large piece of land such that it is *homogeneous* with respect to fertility. This implies that the fertility is the same everywhere in the land and there are no gradients or ridges of fertility. We divide the land in *N plots* of equal size. The varieties $V_1, V_2, \dots V_k$ of wheat are then allotted to $n_1, n_2, \dots n_k$ plots respectively, so that $n_1 + n_2 + \dots + n_k = N$. The allotment of varieties to plots is done by a process of randomisation. The plots may be numbered from 1 to k. Then

n_1 plots may be selected by numbers for V_1

n_2 plots may be selected by numbers for V_2

. . .

n_k plots may be selected by numbers from V_k

The varieties are sown on their plots and yields if the plots are measured. These constitute the data for statistical analysis.

Let x_{ij} = yield of the ith variety on the jth plot allotted to it, $j = 1, 2, \dots, n_i$, $i = 1, 2, \dots k$.

Variety	yields			
V_1	x_{11}	x_{12}	$\dots$	x_{1n1}
V_2	x_{21}	x_{22}	$\dots$	x_{2n2}
V_3	x_{31}	x_{32}	$\dots$	x_{3n3}
$\vdots$	$\vdots$	$\vdots$	$\vdots$	
V_i	x_{i1}	x_{i2}	$\dots$	x_{ini}
$\vdots$	$\vdots$			
V_k	x_{k1}	x_{k2}	$\dots$	x_{knk}

The yields are classified according to the varieties. It is a case of one way classification.

ANOVA Table for C.R.D

Source of Variance	df	S.S	M.S.S.	Varianoo ratio F
Treatments Between varieties	$k-1$	$S_{T^2} = \sum_i n_i (\bar{x}_i . - \bar{\bar{x}})^2$	$s_{T^2} = \frac{S_{T^2}}{k-1}$	$F_T = \frac{s_{T^2}}{s_{E^2}}$ ~ $F(k-1, n-k)$
Within Varieties Error	$N-k$	$S_{E^2} = \sum_{ij} (x_{ij} - \bar{x}_i .)^2$	$s_{E^2} = \frac{S_{E^2}}{N-k}$	
Total	$N-1$	$\sum_{i,j} (x_{ij} - \bar{\bar{x}}_i)^2$		

Remarks. The following formulae for the calculation of various SS are:

$$\text{TSS} = \sum_{i,j} (x_{ij} - \bar{\bar{x}}_i)^2 = \sum_{i,j} xij^2 - \frac{T^2}{N}, N = \sum_{i=1}^{k} n_i$$

$$\text{SST} = \sum n_i (\bar{x}_i . - \bar{\bar{x}})^2 = \sum_{i=1}^{k} \left(\frac{T_i}{n_i}\right)^2 - \frac{T^2}{N}$$

$$\text{SSE} = \text{TSS} - \text{SST} = \sum_{i,j} x_{ij}^2 - \sum_i n_i \bar{x}_i .^2$$

Null hypothesis: H_0: The treatment effects are same.

9.3.1 Example

To test the effect of three diets, three groups of 10 animals were allotted to each diet. The design used was CRD as there was the strong evidence that the animals were homogeneous. During the course some animals died. The data obtained finally is as below.

Diets	Increase in weight in gms					
A	5	6	3	4	8	
B	2	0	1	3		
C	4	9	10	12	8	11

Analyse the data.

Solution. H_0: The diets *A*, *B*, *C* do not differ significantly.

To simplify calculations we subtract 5 from each observations

Diet							T_i
A	0	1	−2	−1	3		1
B	−3	−5	−4	−2			−14
C	−1	4	5	7	3	6	$\frac{24}{11}$

$$n_1 = 5, n_2 = 4, n_3 = 6, N = n_1 + n_2 + n_3 = 15.$$

$$\text{Correction Term} = \frac{T_i^2}{N} = \frac{121}{15} = 8.07$$

Table of Squares

Diet							T_i
A	0	1	4	1	9		15
B	9	25	16	4			54
C	1	16	25	49	9	36	136
							205

$$\sum x_{ij}^2 = 205$$

(i) Total SS = $\sum x_{ij}^2 - \frac{T_i^2}{N} = 205 - 8.07 = 196.93$

(ii) Between diet SS = $\sum \frac{T_i^2}{n_i} - \frac{T^2}{N}$

$$= \frac{1^2}{5} + \frac{(-14)^2}{4} + \frac{(24)^2}{6} - 8.07$$

$$= \frac{1}{5} + \frac{196}{4} + \frac{576}{6} - 8.07 = 137.1$$

(iii) Within diet SS = 196.93 – 137.13 = 59.8
The *df* for treatments (between diets) is $k - 1 = 2$
The *df* for Error is $N - k = 12$.

The Analysis of Variance Table

Source of Variance	DF	SS	MSS	F
Between diets	2	$S_{T^2} = 137.13$	$\frac{137.13}{2} = 68.56$	$F_T = \frac{S_{T^2}}{S_{E^2}} = 13.76$
Within diets	12	$S_{E^2} = 59.8$	$\frac{59.8}{12} = 4.98$	
Total	14	196.93		

Conclusion. $F_{(2, 14)}$ for 5% level of significance is 3.74. As calculated value $F_T > F_{(2, 14)}$, H_0 is rejected at 5% level of significance.

9.3.2 Example

A set of data involving four "tropical feed stuffs *A, B, C, D*" tried on 20 chicks is given below. All the twenty chicks are treated alike in all respects except the feeding treatments and each feeding treatment is given to 5 chicks. Analyse the data. Weight gain of baby chicks fed on different feeding materials composed of tropical feed stuffs are:

						T_i
A	55	49	42	21	52	219
B	61	112	30	89	63	355
C	42	97	81	95	92	407
D	169	137	169	85	154	714
					T	1695

$$\sum_i \sum_j x_{ij}^2 = 55^2 + 49^2 + 42^2 + \ldots + 85^2 + 154^2 = 181445$$

$$\text{Correction factor C.F} = \frac{T^2}{N} = \frac{(1695)^2}{20} = 143651.25$$

$$\text{TSS} = \sum_i \sum_j x_{ij}^2 - \frac{T^2}{N} = 181445 - 143651.25 = 37793.75$$

$$\text{SST} = \sum_{i=1}^{k} \left(\frac{T_i}{n_i}\right)^2 - \frac{T^2}{N}$$

$$= \frac{219^2}{5} + \frac{355^2}{5} + \frac{407^2}{5} + \frac{714^2}{5} - 143651.25 = 26234.95$$

Error SS = TSS – SST = 37793.75 – 26234.95 = 11558.80

Df for treatment SS is 4 – 1 = 3 and *df* for error SS is 19 – 3 = 16.

Anova Table

Source of variation	SS	df	MSS = $\frac{ss}{dt}$	Variance ratio F
Treatments	26234.95	3	8744.98	$F_T = \frac{8744.98}{722.42} = 12.105$
Error	11558.80	16	722.42	
Total	37793.75	19		

Conclusions: Tabulated value $F_{(3, 16)}$ at 5% level of significant is 3.06 < calculated value F_T. Hence H_0 is rejected at 5% level of significance.

Advantages

(i) CRD results in the maximum use of the experimental units since all the experimental material can be used.

(ii) The design is flexible.

(iii) It provides the maximum number of degrees of freedom for the estimation of the error variance which increases the sensitivity or the precision of the experiment for small experiments.

Disadvantages. Since randomisation is not restricted in any direction to ensure that the units receiving one treatment are similar to those receiving the other treatment, the whole variations among the experimental units is included in the residual variance. This makes the design less efficient and results in less sensitivity in detecting significant effects.

9.4 RANDOMISED BLOCK DESIGN RBD

Suppose we have *k* varieties of wheat to be experimented. For the purpose of experimentation we select a large piece of land which is not homogeneous and the fertility gradient is only in one direction.

In CRB the land is homogeneous so far as fertility is concerned.

It is possible to divide the land available into a number of blocks each of which is homogeneous in fertility (atleast more homogeneous than the whole piece). In RBD the land is divided into *r* blocks

$B_1, B_2, ..., B_r$ of same size. Each Block is then divided into k plots of same size. k treatments are allotted to k plots in a block by a process of randomisation.

Thus each variety is allotted to 1 plot in each of the r blocks. Within a block treatments are allotted at random.

Let x_{ij} = yield of the ith variety from the ith plot in the jth block.

These observations can be tabulated as:

Varieties	Blocks							
	B_1	B_2	B_3	...	B_j	...	B_r
V_1	x_{11}	x_{12}	x_{13}		...		x_{1r}
V_2	x_{21}	x_{22}	x_{23}		...		x_{2r}
V_i	...	...	...	...	x_{ij}	...	x_{ir}
.							
V_k	x_{k1}	x_{k2}	x_{k3}	...	x_{kj}	...	x_{kr}

The yields are classified according to blocks and varieties. This is the case of two way of classification.

Anova Table

Source Variance	DF	S.S	MSS	F
Between Varieties	$k-1$	$S_{T^2} = \sum_i r(\bar{x}_i. - \bar{\bar{x}})^2$	$s_{T^2} = \dfrac{S_{T^2}}{k-1}$	$F_T = \dfrac{s_{T^2}}{s_{E^2}}$
Between Blocks	$r-1$	$S_{B^2} = \sum_j k(\bar{x}._j - \bar{\bar{x}})^2$	$s_{B^2} = \dfrac{S_{B^2}}{r-1}$	$F_B = \dfrac{s_{B^2}}{s_{E^2}}$
Error or Residual	$(k-1)(r-1)$	$S_{E^2} = \sum_{ij}(x_{ij} - \bar{x}_i. - \bar{x}._j + \bar{\bar{x}})^2$	$s_{E^2} = \dfrac{S_{E^2}}{(k-1)(r-1)}$	

If F_T is significant it indicates that the varieties differ in their performance. If F_2 is significant it shows that the blocks differ in fertility.

Remark. In RBD the land is divided into blocks which are more homogeneous. Further the effects of the differences between the blocks are separated in the final analysis and eliminated from comparisons between varieties.

9.4.1 Example

In a RBD 3 varieties of rice were tested in 4 blocks. The following table shows the yields obtained (in kg)

Varieties	Blocks			
	B_1	B_2	B_3	B_4
V_1	19	20	22	19
V_2	16	18	13	23
V_3	22	25	19	29

Analyse the data.

Solution. Null hypothesis H_0: Varieties as well as blocks are homogeneous.
Alternative hypothesis H_1: Varieties and blocks are not homogeneous.
To make the computation simple subtract 21 from each observation.

(i)

	B_1	B_2	B_3	B_4	T_i
V_1	−2	−1	1	−2	−4
V_2	−5	−3	−8	2	−14
V_3	1	4	−2	8	11
$T_{\cdot j}$	−6	0	−9	8	−7

$$\text{Correction Term } \frac{T^2}{N} = \frac{(-7)^2}{12} = \frac{49}{12} = 4.08, \text{ where } m = 3,\ n = 4$$

(ii) Table of Squares

	B_1	B_2	B_3	B_4	Totals
V_1	4	1	1	4	10
V_2	25	9	64	4	102
V_3	1	16	4	64	85
Total	30	26	69	72	197

$$\sum xij^2 = 197$$

Now

(a) $\text{Total SS} = \sum xij^2 - \dfrac{T^2}{N} = 197 - 4.08 = 192.92$

(b) $\text{Between varieties SS} = \sum_{i=1}^{4} \dfrac{T_{i.}^{\ 2}}{n_i} - \dfrac{T^2}{N} = \dfrac{16}{4} + \dfrac{196}{4} + \dfrac{121}{4} - 4.08 = 79.17$

(c) $\text{Between blocks SS} = \sum_{j}^{3} T_{.j^2} - \dfrac{T^2}{N}$

$$= \frac{(-6)^2}{3} + \frac{0^2}{3} + \frac{(-9)^2}{3} + \frac{8^2}{3} - 4.08$$

$$= \frac{36}{3} + \frac{81}{3} + \frac{64}{3} - 4.08 = 56.25$$

(d) Residual SS = Total SS − (Between varieties SS + Between blocks SS)
= 192.92 − (79.17 + 56.25) = 57.5

Degrees of freedom. Total number of $Df = mn - 1 = 12 - 1 = 11$
No. of Df for between varieties SS $= m - 1 = 2$
No. of Df for between blocks SS $= n - 1 = 3$
No. of Df for residuals SS $= (m - 1)(n - 1) = 6$

Anova Table

Source Variation	DF	SS	MSS	F
Between Varieties	$m - 1 = 2$	$S_{T^2} = 79.17$	$s_{T^2} = 39.585$	$F_T = \dfrac{s_{T^2}}{s_{E^2}} = 4.13$
Between blocks	$n - 1 = 3$	$S_{B^2} = 56.25$	$s_{B^2} = 18.75$	$F_B = \dfrac{s_{B^2}}{s_{E^2}} = 1.25$
Residuals	$(m - 1)(n - 1) = 6$	$S_{E^2} = 57.5$	$s_{E^2} = 9.58$	
Total	$mn - 1 = 11$	192.92		

Conclusion. Tabulated value $F_{(2, 6)}$ at 5% level of significant is 5.14 > calculated F_T and tabulated value $F_{(3, 6)}$ at 5% level of significance is 4.76 > Calculated F_B. Hence H_0 is accepted.

9.4.2 Efficiency of RBD Relative to CRD

The efficiency E of RBD with m treatments and r replicates relative to the same experiment conducted as a CRD is given by

$$E = \frac{r(m-1)\, s_{E^2} + (r-1)\, s_{B^2}}{(mr-1)\, s_{E^2}}$$

Advantages

(i) *Accuracy.* This design is more efficient or accurate than CRD for most types of experimental work. The elimination of "between SS from Residual SS usually results in a decrease of Error Mean SS.

(ii) *Flexibility.* In RBD no restrictions are placed on the number of treatments or the number of replicates. In general at least two replicates are required to carry out the test of significance.

The error of any treatment can be isolated and any number of treatment may be omitted from the analysis without complicating it.

Disadvantages: RBD is not suitable for large number of treatments or for cases in which complete blocks contains considerable variability.

9.5 LATIN SQUARE DESIGN (LSD)

In RBD whole of the experimental area is divided *relatively homogeneous blocks* and treatments are allocated at random to units within each block i.e. randomisation was subjected to one restriction, i.e. within blocks. A useful method of eliminating fertility variations consists in an experimental layout which will control variation in two perpendicular directions. Such a layout is a Latin Square Design (LSD).

Layout of Design. In this design the number of treatments is equal to the number of replications. Thus in case of m treatments, there have to be $m \times m = m^2$ experimental units. The whole of experimental area is divided into m^2 experimental units (plots) arranged in a square so that each row as well as each column contains m units (plots). The m treatments are then allocated at random to these rows and columns in such a way that *every treatment occurs once and only once in each row and in each column.* Such a layout is known as $m \times m$ Latin Square Design (LSD) and is extensively used in agricultural experiments.

Statistical Analysis of m × m LSD for one observation per Experimental Unit.

Let y_{ijk} $(i, j, k = 1, 2, ..., m)$ denote the response from the unit in the ith row, j^{th} column and receiving the kth treatment. If a single observation is made per experimental unit, then the linear additive model is

$$y_{ijk} = \mu + \alpha_i + \beta_j + \tau_k + \varepsilon_{ijk}\ (i, j, k) \in S,$$

where S represents the set of m^2 values, μ is the constant mean effect, α_i, β_j, τ_k are the effects due to i^{th} row, j^{th} column, k^{th} treatment component assumed to be normally distributed with mean zero and variance σ^2. If we write

$$T = y_{...} = \text{Total of all the } m^2 \text{ observations}$$
$$R_i = y_{i..} = \text{Total of the } m \text{ observations in the } i^{th} \text{ row}$$
$$C_j = y_{.j.} = \text{Total of the } m \text{ observations in the } j^{th} \text{ column}$$
$$T_k = y_{..k} = \text{Total of the observations from } k^{th} \text{ treatment}$$

then

$$\text{TSS} = \text{Total sum of squares}$$
$$= \sum_{i,j,k \in S} (y_{ijk} - \bar{y}_{...})^2 = \sum y_{ijk}^2 - \frac{T^2}{N},\ N = m^2$$
$$S_{R^2} = \text{SSR} = \text{Sum of squares due to rows}$$
$$= m \sum_i (\bar{y}_{i..} - \bar{y}_{...})^2 = \frac{1}{m} \sum_i R_i^2 - \frac{T^2}{N}$$
$$S_{C^2} = \text{SSC} = \text{Sum of squares due to columns}$$
$$= m \sum (\bar{y}_{.j.} - \bar{y}_{...})^2 = \frac{1}{m} \sum_j C_j^2 - \frac{T^2}{N}$$
$$S_{T^2} = \text{SST} = \text{Sum of squares due to treatments}$$
$$= m \sum (\bar{y}_{..k} - \bar{y}_{...})^2 = \frac{1}{m} \sum T_k^2 - \frac{T^2}{N}$$
$$S_{E^2} = \text{SSE} = \text{Sum of squares due to error} = \text{TSS} - (\text{SSR} + \text{SSC} + \text{SST})$$

Anova Table for m × m LSD

Source of variation	df	SS	MSS	Variance ratio F
Rows	$m-1$	S_{R^2}	$s_{R^2} = \dfrac{S_{R^2}}{m-1}$	$F_R = \dfrac{s_{R^2}}{s_{E^2}}$
Columns	$m-1$	S_{C^2}	$s_{C^2} = \dfrac{S_{C^2}}{m-1}$	$F_C = \dfrac{s_{C^2}}{s_{E^2}}$
Treatments	$m-1$	S_{T^2}	$s_{T^2} = \dfrac{S_{T^2}}{m-1}$	$F_T = \dfrac{s_{T^2}}{s_{E^2}}$
Error	$(m-1)(m-2)$	S_{E^2}	$s_{E^2} = \dfrac{S_{E^2}}{(m-1)(m-2)}$	
Total	$m^2 - 1$	TSS		

The variance ratio F_R, F_C, F_T follow F-distribution with $(m-1)$, $(m-1)(m-2)$ df under the null hypothesis H_α, H_β, H_τ respectively.

Let $F_\alpha = F_\alpha\{(m-1), (m-1)(m-2)\}$ be the tabulated value of F at $\alpha\%$ level of significance. Thus if $F_R > F_\alpha$ we reject H_α and if $F_R \le F_\alpha$ we accept H_α. Similarly we test for H_β and H_τ where

Null hypothesis H_α: $\alpha_1 = \alpha_2 = \ldots = \alpha_m = 0$

H_β: $\beta_1 = \beta_2 = \ldots = \beta_m = 0$

H_τ: $\tau_1 = \tau_2 = \ldots = \tau_m = 0$

9.5.1 Example

An experiment was carried out to determine the effect of claying the ground on the field of barley grains amount of clay used were:

A: No clay
B: Clay at 100 per acre
C: Clay at 200 per acre
D: Clay at 300 per acre

The yields were in plots of 8 meters by 8 meters layout:

Row \ Column →	I	II	III	Iv	Row total (R_i)
I	*D* 29.1	*B* 18.9	*C* 29.4	*A* 5.7	83.1
II	C 16.4	*A* 10.2	*D* 21.2	*B* 19.2	67
III	*A* 5.4	*D* 38.8	*B* 24.0	*C* 37.0	105.2
IV	B 24.9	C 41.7	A 9.5	D 28.9	105.0
Column Total (C_j)	75.8	109.6	84.1	90.8	360.3

Perform the ANOVA and calculate the critical difference for the treatments mean yields.

Solution. H_0: Variations due to row, column, and treatment are not differ significantly. The four treatments totals are

(A) $= 5.7 + 10.2 + 5.4 + 9.5 = 30.8$
(B) $= 18.9 + 19.2 + 24 + 24.9 = 87.0$
(C) $= 29.4 + 16.4 + 37.0 + 41.7 = 124.5$
(D) $= 29.1 + 21.2 + 38.8 + 28.9 = 118.0$

$N = 16$, Grand total $T = 360.3$

$$\text{CF} = \frac{T^2}{N} = \frac{(360.3)^2}{16} = 8113.50$$

$$\text{TSS} = \sum y_{ijk}^2 - \frac{T^2}{N} = (29.1)^2 + (18.9)^2 + (29.4)^2 + \ldots$$

$$+ (41.7)^2 + (9.5)^2 + (28.9)^2 - 8113.50$$

$$= 100667.59 - 8113.50 = 1954.09$$

$$SSR = \frac{1}{m}\left[\sum R_i^2\right] - CF$$

$$= \frac{1}{4}\,[(83.1)^2 + (67)^2 + (105.2)^2 + (105)^2] - 8113.50$$

$$= \frac{33486.65}{4} - 8113.50 = 258.1625$$

$$SSC = \frac{1}{4}[(75.8)^2 + (109.6)^2 + (84.1)^2 + (90.8)^2] - 8113.50$$

$$= \frac{33075.25}{4} - 8113.50 = 155.3125$$

$$SST = \frac{1}{4}[(A)^2 + (B)^2 + (C)^2 + (D)^2] = \frac{1}{4}[(30.8)^2 + 87^2 + (124.5)^2 + (118)^2] - 8113.50$$

$$= \frac{37941.89}{4} - 8113.50 = 1371.9725$$

$$\text{Error SS} = TSS - SSR - SSC - SST = 168.6425$$

ANOVA Table for LSD

Source of Variation	df	SS	MSS	Variance ratio
Rows	3	258.1625	86.054	$F_R = \frac{86.054}{28.107} = 3.06$
Columns	3	155.3125	51.770	$F_C = \frac{51.77}{28.107} = 1.841$
Treatments	3	1371.9725	457.324	$F_T = \frac{457.324}{28.107} = 16.27$
Error	6	168.6425	28.1070	

Conclusion. The tabulated $F_{0.05}$ at (3, 6) *df* is 4.76

Note $F_R = 3.06 < 4.76$; $F_C = 1.84 < 4.76$; $F_T = 16.27 > 4.76$

Hence we conclude that the variation due to rows and columns is not significant but the treatments at different levels of clay, have significant effect on the yield.

To determine which of the treatment pairs differ significantly we have to calculate the critical difference (CD)

S.E of difference between any two treatments means = $s_E\sqrt{\frac{2}{m}} = \sqrt{\frac{2\times 28.1070}{4}} = 3.748$

Therefore
$$CD = 3.748 \times t_{0.05} \text{ (for error } df)$$
$$= 3.748 \times t_{0.05} \text{ (at 6 } df)$$
$$= 3.748 \times 2.447 = 9.1713$$

We row arrange the treatments in their decreasing order of magnitude as:

Treatment	Mean yield	$\|D\|$ = difference
C	$\frac{124.5}{4} = 31.12$	31.12 – 29.5 = 1.62
D	$\frac{118}{4} = 29.5$	29.5 – 21.726 = 7.774
B	$\frac{87}{4} = 21.726$	21.726 – 7.7 = 14.026
A	$\frac{30.8}{4} = 7.7$	

Hence we conclude

(i) By comparing the difference between the mean yields for different treatments with the critical difference we find that the difference between mean yields of C and D is not significant and they may therefore regarded alike as regards their effect on yield. Similar argument holds for the pair D and B.

(ii) The difference between C and B is 9.4 exceeds the CD. Hence the treatments C and B are significantly different from each other as regards their effect on yields. Similarly we argue for other pairs.

9.5.2 Example

Four types of health drinks A, B, C, D were tried on the school children. In order to study the effects of the age groups of the children and localities, four schools from four different localities, were selected and students were divided into four age groups. The Latin square design was arranged. The gain in weights in same units are recorded below. Test whether the localities age groups and the varieties of food have any significant effect on the gain in weight.

Age groups

	5-8	8-11	11-14	14-17	Total
	A 2	B 1.8	C 2.1	D 1.5	7.4
	D 1.3	A 1.4	B 1	C 1.2	4.9
	B 1.7	C 1.6	D 1.1	A 1.9	6.3
Total	5.9	5.8	6.2	6.1	24

Solution. H_0: The localities, age groups and the treatments are similar and they do not differ significantly.

Treatment totals are: $A = 7.3, B = 6, C = 5.8, D = 4.9$

Total number of items $= 16$

Correction factor $= \frac{T^2}{N} = \frac{24^2}{16} = 36$

$$\sum_{i,j,k} y_{ijk}^{\;2} = 2^2 + (1.8)^2 + (2.1)^2 + \ldots + (1.5)^2 = 38.32$$

$$\text{TSS} = \sum_{i,j,k} y_{ijk}^{\;2} - \frac{T^2}{N} = 38.32 - 36 = 2.32$$

$$\text{SSR} = \frac{1}{4}\{(7.4)^2 + (4.9)^2 + (6.3)^2 + (5.4)^2\} - \frac{T^2}{N} = 0.905$$

$$\text{SSC} = \frac{1}{4}\{(5.9)^2 + (5.8)^2 + (6.2)^2 + (6.1)^2\} - \frac{T^2}{N} = 0.025$$

$$\text{SST} = \frac{1}{4}(A^2 + B^2 + C^2 + D^2)$$

$$= \frac{1}{4}[(7.3)^2 + 6^2 + (5.8)^2 + (4.9)^2\} - \frac{T^2}{N} = 0.735$$

$$\text{SSE} = \text{TSS} - (\text{SSR} + \text{SSC} + \text{SST})$$
$$= 2.32 - (0.905 + 0.025 + 0.735) = 0.655$$

ANOVA Table

Source of Variation	Sum of Squares	Degrees of freedom	MSS	Variance Ratio F
Between Rows (Locality)	0.905	3	$\frac{0.905}{3} = .302$	$\frac{.302}{.109} = 2.77$
Between Columns (Age)	0.025	3	$\frac{.025}{3} = .008$	$\frac{.008}{.109} = .07$
Between Treatments (Health drinks)	0.735	3	$\frac{0.735}{3} = .245$	$\frac{.245}{.109} = 2.25$
Error	0.655	6	$\frac{.655}{6} = .109$	
Total	2.32	15		

Conclusion. Tabulated value $F_{0.05}$ at $(3, 6)df$ is $4.76 > F_R, F_C, F_T$. Hence H_0 is accepted at 5% level of significance.

9.6 STATISTICAL QUALITY CONTROL (SQC)

9.6.1 Quality

Quality can be defined in two different ways. Goods are said to be of good quality if they satisfy the consumer requirement. That is, goods are said to be of good quality if they meet the expected functional use. So quality means the best for customers under specified conditions.

9.6.2 Statistical Quality Control

Statistical quality control is simply a statistical method determining the extent to which quality goals are being met without necessarily checking every item produced and for indication whether or not the variations which occur are exceeding normal expectations.

Variation in quality occur due to two causes.

1. *Chance Variation:* This type of variation in quality occur because of climate changes, strike etc. The variations in the quality of the products due to chance causes behave in a random manner. Level of variation will continue unless something special occurs.
2. *Assignable Variation:* Such variations can be the result of the several factors such as a change in the raw material, a new operation, improper machine settings, broken (or) worn out parts, mechanical faults in plant etc. Out of these two variations chance variation in the process is preferred. If only chance variation is present, the output of a process will form a pattern which is stable over time and predictable.

9.6.3 Process Control

This is concerned with the control of the quality of the goods while they are in the *process of production.* In process control we could deduct the *defect* and *faults* of the process and could correct the errors.

The aim of process control is to distinguish between these two types of variation and identify the corrective action.

Product control. This is concerned with finished goods that are acceptable or not acceptable. Otherwise with the acceptance of the lot. In this the finished goods are tested as acceptable or non-acceptable products. Process control is concerned with the operations of machines, raw materials, manpower and poor quality of materials, during the process while the product control is concerned more with the quality of the finished product.

Variables

Variables are those quality characteristics of a product or item which are measurable and can be expressed in their respective units of measurements.

Example

(i) Weights of packed materials
(ii) Tensile strength of rods
(iii) Life of an electric bulb
(iv) Diameter of wires etc.

Attributes

Attributes are the characteristics of products which are not measurable.

(i) Blemishes in piece of cloth
(ii) Spots in white cloths of same size
(iii) Bubbles of air in windscreen

A Defective is an item (or) unit which contains one (or) more defects, errors, and faults but a defect is an individual flow and error.

9.6.4 Control Charts

A control chart is a statistical device principally used for the study and control of respective processes.

There are mainly two types of control charts namely

(i) control charts for variables
(ii) control charts for attributes

Under the category of control charts for variables we have

1. The mean chart (or) $\overline{X}$-chart
2. The Range chart (or) R-chart
3. σ-chart

Under the category of control charts for attributes we have

1. p-chart (for proportion of defectives)
2. np-chart (for number of defectives)
3. c-chart (for number of defectives per unit)

Various Charts of Quality Control

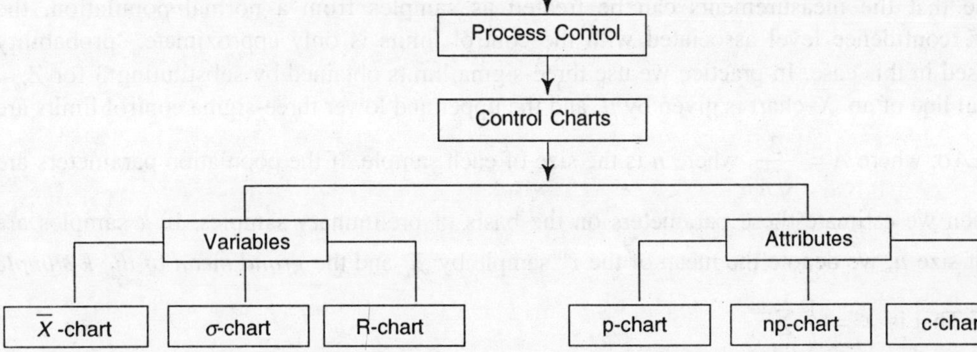

p-chart is used for monitoring fraction defective. np-chart is used for monitoring number of defective items in a sample of given size and c-chart is used for monitoring the defects in a single part or assembled unit.

$\overline{X}$-chart is used for monitoring process average. R-chart is used for monitoring process variability

A control chart consists of three horizontal lines namely

(a) A central line (C.L)
(b) Upper central line (UCL)
(c) Lower central line (LCL)

The lines on either side of the central line are known as central limits. In practice, the control charts are drawn using a graph sheet.

Model of the chart

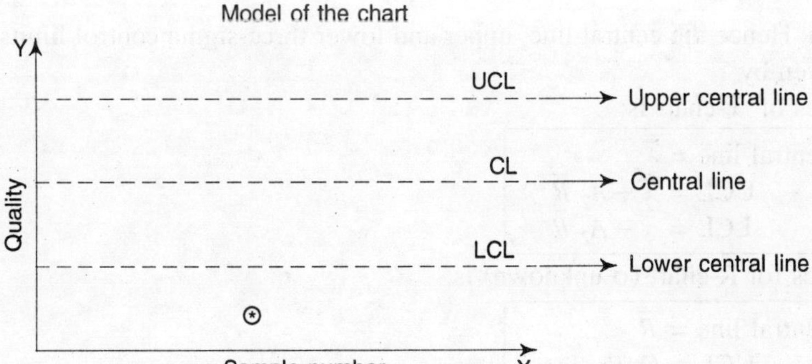

⊛-Denotes out of control situations.

9.6.5 Control Chart for Measurements

If the process mean and standard deviation, μ and σ are known, and it is reasonable to treat the measurements as samples from a normal population, we can assert with probability $1 - \alpha$ that the mean of a random sample of size n will fall between $\mu - Z_{\alpha/2}\left(\frac{\sigma}{\sqrt{n}}\right)$ and $\mu + Z_{\alpha/2}\left(\frac{\sigma}{\sqrt{n}}\right)$. These two limits on $\bar{x}$ provide upper and lower control limits and under the given assumptions, they enable the quality control engineer to determine whether or not to make an adjustment in the process.

If μ and σ are unknown, then it is necessary to estimate their values from a large sample (or samples) taken while the process is "in control". For this reason and because there may be no assurance that the measurements can be treated as samples from a normal population, the $(1 - \alpha)100\%$ confidence level associated with the control limits is only approximate, 'probability limits' are used in this case. In practice we use three-sigma limits obtained by substituting 3 for $Z_{\alpha/2}$.

The central line of an $\bar{X}$-chart is given by μ, and the upper and lower three-sigma control limits are given by $\mu \pm A\sigma$, where $A = \frac{3}{\sqrt{n}}$ where n is the size of each sample. If the population parameters are unknown, then we estimate these parameters on the basis of preliminary samples. If k samples are used, each of size n, we denote the mean of the ith sample by $\bar{x}_i$ and the *grand mean of the k sample means* by $\bar{\bar{x}}$. That is $\bar{\bar{x}} = \frac{1}{k}\sum_{i=1}^{k} \bar{x}_i$.

The process variability σ can be estimated either from the standard deviations or the ranges of the k-samples.

If R_i denote the range of the ith sample, then mean sample range

$$\bar{R} = \frac{1}{k}\sum_{i=1}^{k} R_i$$

Since $\bar{\bar{x}}$ provides an unbiased estimate of the population mean μ, the central line for the $\bar{x}$ chart is given by $\bar{\bar{x}}$. The statistic $\bar{R}$ does not provide an unbiased estimate of σ, but multiplying $\bar{R}$ by the constant A_2, we obtain an unbiased estimate of $\frac{3\sigma}{\sqrt{n}}$. The constant multiplier A_2 is tabulated for various values of n. Hence the central line, upper and lower three-sigma control limits UCL, LCL for an $\bar{X}$-chart are given by

Control chart values of $\bar{x}$-chart is:

$$\begin{aligned} \text{Central line} &= \bar{\bar{x}} \\ \text{UCL} &= \bar{\bar{x}} + A_2\,\bar{R} \\ \text{LCL} &= \bar{\bar{x}} - A_2\,\bar{R} \end{aligned}$$

Control chart values for R-chart (σ unknown) is:

$$\begin{aligned} \text{Central line} &= \bar{R} \\ \text{UCL} &= D_4\bar{R} \\ \text{LCL} &= D_3\bar{R} \end{aligned}$$

D_3, D_4 are calculated from tables of central chart depending on the size of the sample.

If range of samples are not given directly, then compute R_i as biggest value B_i of the variable in the sample —smallest value S_i of the variable in the sample.

Conclusion. If all the sample points fall within the control limits, the process is under statistical control.

9.6.6 Example

A food company puts mango juice into cans advertised as containing 10 ounces of the juice. The weights of the juice drained from cans immediate after filling for 20 samples are taken by random method. Construct an $\bar{X}$ chart to control the weights of mango juice for the filling.

Sample No	H Cans each sample			
	Can1	Can 2	Can 3	Can 4
1	15	12	13	20
2	10	8	8	14
3	8	15	17	10
4	12	17	11	12
5	18	13	15	4
6	20	16	14	20
7	15	19	23	17
8	13	23	14	16
9	9	8	18	5
10	6	10	24	20
11	5	12	20	15
12	3	15	18	18
13	6	18	12	10
14	12	9	15	18
15	15	15	6	16
16	18	17	8	15
17	13	16	5	4
18	10	20	8	10
19	5	15	10	12
20	6	14	12	14

Solution.

Calculation for $\bar{X}$-chart

Sample No	Weight of each can				$\sum x_i$	$\bar{x}_i$	R = Sample range
	Four cans in each samples n = 4						highest-least
1	15	12	13	20	60	15	8
2	10	8	8	14	40	10	6
3	8	15	17	10	50	12.5	9
4	12	17	11	12	52	13	6
5	18	13	15	4	50	12.5	14
6	20	16	14	20	70	17.5	6
7	15	19	23	17	74	18.5	8
8	13	23	14	16	66	16.5	10
9	9	8	18	5	40	10	13
10	6	10	24	20	60	15	18
11	5	12	20	15	52	13	15
12	3	15	18	18	54	13.5	15
13	6	18	12	10	46	11.5	12

(Contd.)

(Contd.)

14	12	9	15	18	54	13.5	9
15	15	15	6	16	52	13.0	10
16	18	17	8	15	58	14.5	10
17	13	16	5	4	38	9.5	12
18	10	20	8	10	48	12	12
19	5	15	10	12	42	10.5	10
20	6	14	12	14	46	11.5	8

$$\sum \bar{x}_i = 263, \quad \sum R = 211$$

Now

(i) $\bar{\bar{x}} = \dfrac{\sum \bar{x}_i}{20} = \dfrac{263}{20} = 13.15 = \text{Central line}$

(ii) $\bar{R} = \dfrac{\sum R}{20} = \dfrac{211}{20} = 10.55$

(iii) $A_2 = 0.729$ (From table for $n = 4$)

$$\text{UCL} = \bar{\bar{x}} + A_2 \bar{R}$$
$$= 13.15 + 0.729 \times 10.55 = 20.84$$
$$\text{LCL} = \bar{\bar{x}} - A_2 \bar{R}$$
$$= 13.15 - 0.729 \times 10.55 = 5.46$$

$\bar{x}$ chart is:

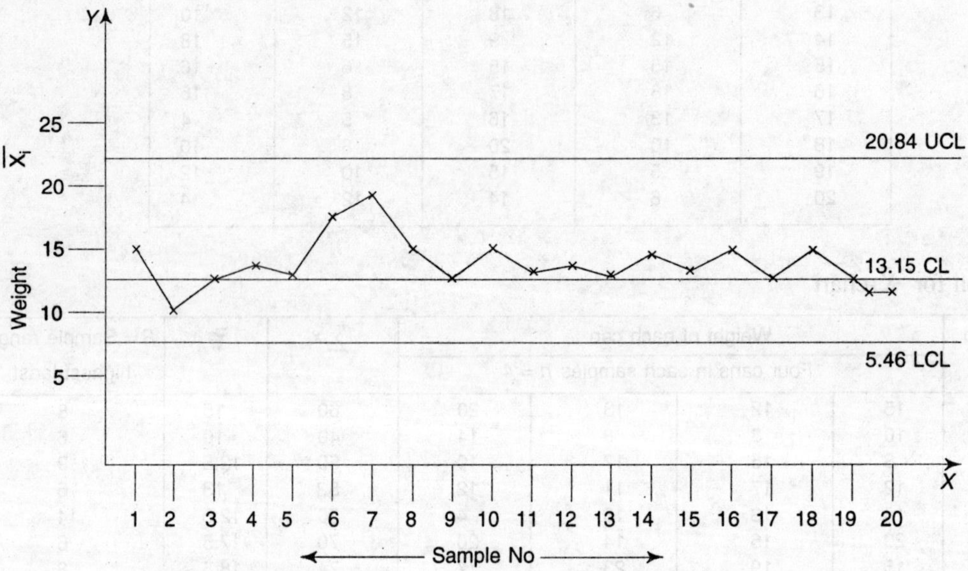

Conclusion. Since all the points are falling within control limits the process in a state of control and hence there is nothing to worry.

9.6.7 Example

You are given the values of sample mean (X) and the range (R) for ten samples of sizes for each. Draw mean and Range charts and comment on the state of control of the process.

Sample No	1	2	3	4	5	6	7	8	9	10
$\overline{X}$	43	49	37	44	45	37	51	46	43	47
R	5	6	5	7	7	4	1	6	4	6

You may use the following control chart constants (for $n = 5$, $A_2 = 0.58$, $D_3 = 0$, $D_4 = 2.115$)

Solution.

Calculation for $\overline{X}$-chart

$$\overline{\overline{x}} = \frac{43+49+37+44+45+37+51+46+43+47}{10}$$

$$= \frac{442}{10} = 44.2 \text{ (central line)}$$

where $\qquad \text{UCL} = \overline{\overline{x}} + A_2\overline{R}$ and $\text{LCL} = \overline{\overline{x}} - A_2\,\overline{R}$

$$\overline{R} = \frac{5+6+5+7+7+4+1+6+4+6}{10} = \frac{51}{10} = 5.1$$

$$\text{UCL} = 44.2 + (0.58) \times (5.1) = 44.2 + 2.958 = 47.158 = 47$$

$$\text{LCL} = 44.2 - 2.958 = 41.242 = 41.2$$

$\overline{X}$-chart

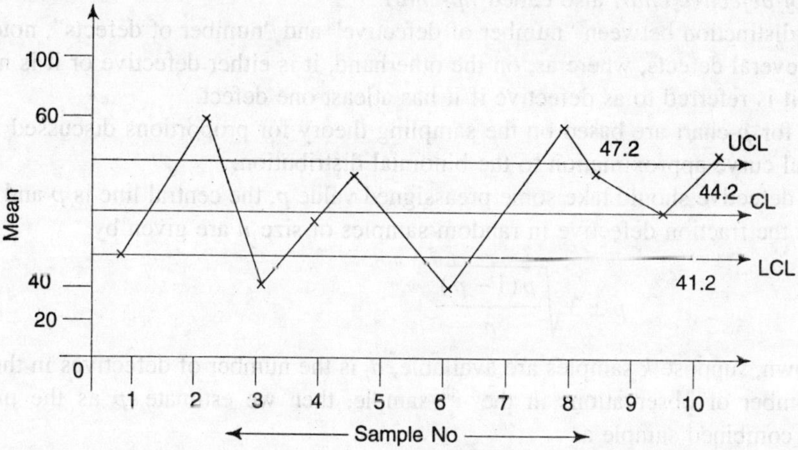

From the given table we see that means corresponding to sample numbers 2, 7 are greater than UCL and the means corresponding to the sample numbers 3, 6 are less than LCL value.

Four of the 10 samples points fall outside the control limits. Hence the system is not in a control.

Calculation of R-chart

The central line $\qquad \overline{R} = 5.1$

$$\text{UCL} = D_4\overline{R} = 2.115 \times 5.1 = 10.7865 = 10.8$$

$$\text{LCL} = D_3\overline{R} = 0 \times 5.1 = 0$$

R-chart

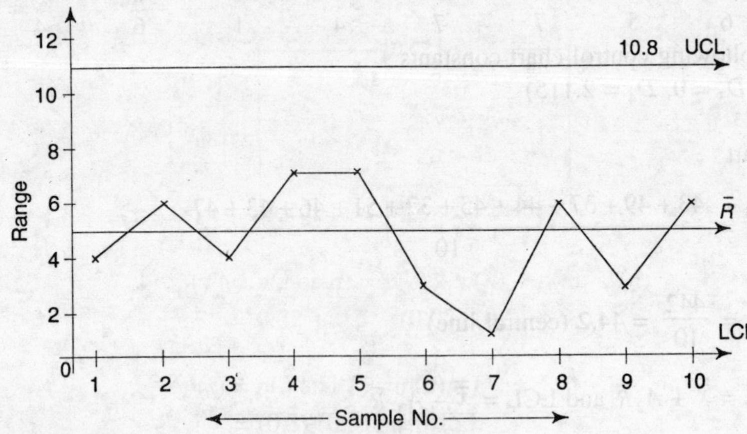

Comment. The system is in control since all the points line within the control limits.

9.6.8 Control Charts for Attributes

Here we discuss two fundamental kinds of control charts used in connection with attribute sampling, the *fraction-defective chart,* also called a *p-chart*, the *number-of-defects chart* also called a *c-chart* and the *number-of-defective chart* also called *np-chart.*

To clarify the distinction between "number of defective" and "number of defects", note that a unit tested can have several defects, where as, on the otherhand, it is either defective or it is not. In many applications a unit is referred to as defective if it has atleast one defect.

Control limits for *p*-chart are based on the sampling theory for proportions discussed in chapter 8 and on the normal curve approximation to the binomial distribution.

If the fraction defective should take some preassigned value *p*, the central line is *p* and three sigma control limits for the fraction defective in random samples of size *n* are given by

$$p \pm 3\sqrt{\frac{p(1-p)}{n}}$$

If *p* is not known, suppose *k* samples are available, d_i is the number of defectives in the ith sample, and n_i is the number of observations in the ith sample, then we estimate *p* as the proportion of defectives in the combined sample as

$$\bar{p} = \frac{d_1 + d_2 + \ldots + d_k}{n_1 + n_2 + \ldots + n_k}$$

and *control chart values of p-chart*

$$\text{Central line} = \bar{p}$$

$$\text{UCL} = \bar{p} + 3\sqrt{\frac{\bar{p}(1-\bar{p})}{n}}$$

$$\text{LCL} = \bar{p} - 3\sqrt{\frac{\bar{p}(1-\bar{p})}{n}}$$

If LCL in some case may be negative for p small, then it is customary to consider LCL as zero. Also there is one more complication that can arise if p is small. Generally speaking, the use of the above control limits for p-charts is unrealistic whenever n and p are such that the underlying binomial distribution cannot be approximated by a normal curve, since it is possible only when np and $n(1 - p)$ are greater than 15. In such cases it is best to use Poisson approximation to binomial.

Equivalent to the p-chart for the fraction defective is the control chart for the number of defectives. If p is estimated by $\bar{p}$, the control-chart values for a number-of-defectives chart, np-chart are as follows.

$$\text{Central line} = n\bar{p}$$
$$\text{UCL} = n\bar{p} + 3\sqrt{n\bar{p}(1-\bar{p})}$$
$$\text{LCL} = n\bar{p} - 3\sqrt{n\bar{p}(1-\bar{p})}$$

There are situations where it is necessary to control the number of defects in a unit of product. For example, in the production of newsprint one may wish to control the number of defects per roll. This situation is like the one in Poisson distribution. Thus, if c is the number of defects per manufactured unit, c is taken to be a value of a random variable having the Poisson distribution.

The central line for number-of-defects chart (c-chart) is the parameter m of the corresponding Poisson distribution and the three-sigma control limits are based on SD of this distribution ($\sqrt{m}$). If m is not known, if k is the number of units of product available for estimating m, and if c_i is the number of defects in the ith unit, then m is estimated by

$$\bar{c} = \frac{1}{k}\sum_{i=1}^{k} c_i \quad \text{(mean number of defects}$$

and the *control-chart values for the c-chart are:*

$$\text{Central line} = \bar{c}$$
$$\text{UCL} = \bar{c} + 3\sqrt{\bar{c}}$$
$$\text{LCL} = \bar{c} - 3\sqrt{\bar{c}}$$

9.6.9 Example

Assume that 20 litres of milk bottles are selected at random from a process. The number of air bubbles (defects) observed from the bottles is given by

Bottle No.	1	2	3	4	5	6	7	8	9	10	11	12	13	14	15	16	17	18	19	20	
Defects	4	5	7	3	3	5	6	2	4	8	3	5	4	3	4	5	3	7	6	13	100

Ascertain whether the process is in a state of statistical control.

Solution.

1. $\bar{c}$ = Average number of defects

$$= \frac{100}{20} = 5 = \text{central line}$$

2. $\text{UCL} = \bar{c} + 3\sqrt{\bar{c}} = 5 + 3\sqrt{5} = 5 + 6.708 = 11.708$

3. $\text{LCL} = \bar{c} - 3\sqrt{\bar{c}} = 5 - 6.708 = -1.708 = 0$ (since LCL cannot be negative)

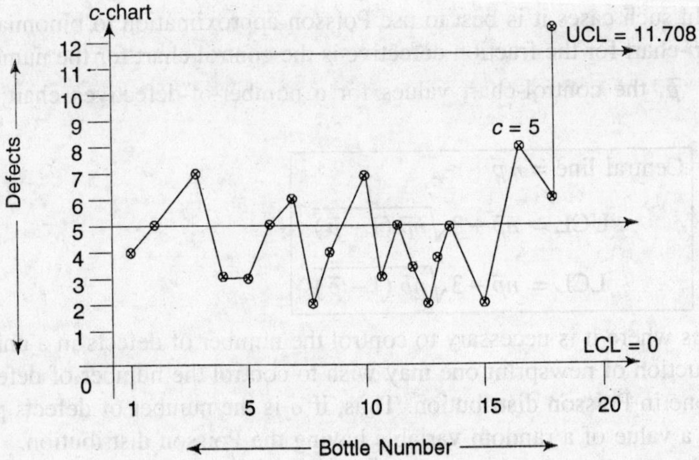

Conclusion It is clear from the chart that only one point in respect of last sample falls out side the control limit.

9.6.10 Example

Twenty pieces of cloth out of different rolls contained respectively 1, 4, 3, 2, 5, 4, 6, 7, 2, 3, 2, 5, 7, 6, 4, 5, 2, 1, 3 and 8 imperfections. Ascertain whether the process is in a state of statistical control.

Solution.

$$\bar{c} = \frac{1+4+3+\ldots+8}{20} = \frac{80}{20} = 4 = \text{central line}$$

$$\text{UCL} = \bar{c} + 3\sqrt{\bar{c}} = 4 + 3\sqrt{4} = 4 + 6 = 10$$

$$\text{LCL} = \bar{c} + 3\sqrt{\bar{c}} = 4 + 3\sqrt{4} = 4 - 6 = -2 \text{ (or) } 0$$

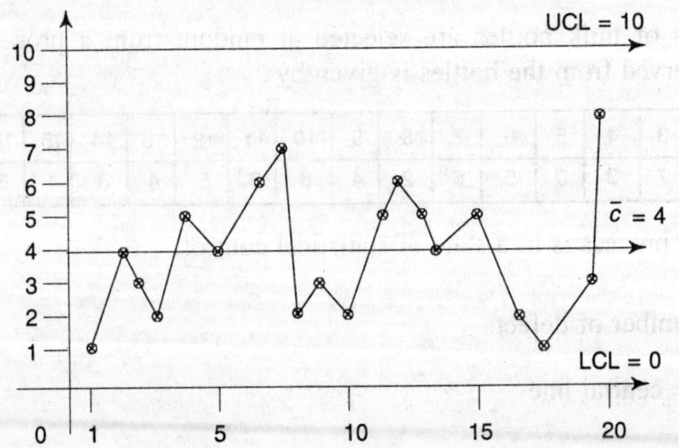

Conclusion Since all the points lies within the central limits the system is in control.

9.6.11 Example

Samples of 100 tubes each are drawn randomly from the output of a process that produces several thousand units daily. Sample items are inspected for quality and defective tubes are rejected. The results of 15 samples are shown below.

Sample No	1	2	3	4	5	6	7	8	9	10	11	12	13	14	15
No. of defective tubes	8	10	13	9	8	10	14	6	10	13	18	15	12	14	9

On the basis of information given above prepare a control chart for fraction defective. What conclusion do you draw from the control chart?

Solution.

Now $$\bar{p} = \frac{\text{Total defective}}{\text{No. of items inspected}}$$

$$= \frac{169}{15\times100} \text{ each sample contains 100 items}$$

$$= 0.113 = \text{central line}$$

Sample No.	No. of defects	Fraction defective
1	8	.08
2	10	.10
3	13	.13
4	9	.09
5	8	.08
6	10	.10
7	14	.14
8	6	.06
9	10	.10
10	13	.13
11	18	.18
12	15	.15
13	12	.12
14	14	.14
15	9	.09
Total	169	

$$\text{UCL} = \bar{p} + 3\sqrt{\frac{\bar{p}(1-\bar{p})}{n}} = \sqrt{.113\,(1-.113)}$$

$$= .113 + .095 = .208 = .21$$

$$\text{LCL} = .113 - .095 = .018 = .02$$

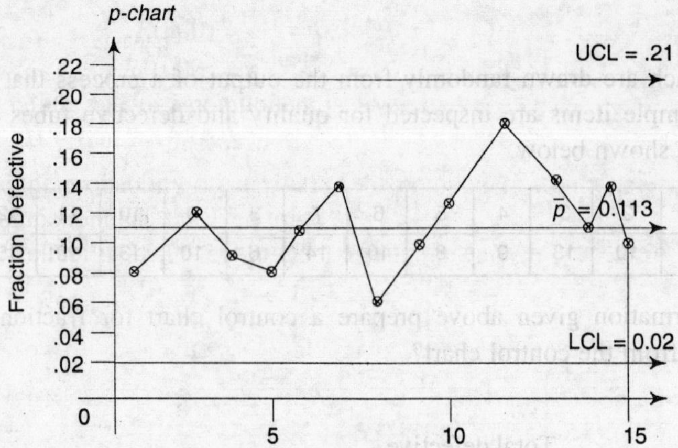

Conclusion. Since all the points lies within the control limits, the system is in control.

Control chart for the standard deviation σ—σ-chart

Central line is $y = \bar{\sigma}$, σ^2—variance, σ is standard deviation

$$\text{UCL} = B_4\bar{\sigma},$$

$$\text{LCL} = B_3\bar{\sigma}$$

B_3, B_4 are table values.

If σ is not known, in this case

$$\text{UCL} = B_4\bar{s}$$

$$\text{LCL} = B_3\bar{s}$$

where $\bar{s} = \dfrac{\text{sum of the sample S.D's}}{\text{No. of samples}}$

Construction of σ chart is same as R-chart.

9.6.12 Example

Consider the Example 9.6.11. Construct a *np*-chart and test whether the process is in a state of statistical control.

Solution.

$$\text{UCL} = n\bar{p} + 3\sqrt{n\bar{p}\,(1-\bar{p})} = n\left[\bar{p} + 3\sqrt{\frac{\bar{p}\,(1-\bar{p})}{n}}\right].$$

$$= 100(.21) = 21$$

$$n\bar{p} = 100 \times .113 = 11.3 = \text{CL}$$

$$\text{LCL} = n\bar{p} - 3\sqrt{n\bar{p}\,(1-\bar{p})}$$

$$= n\left[\bar{p} - 3\sqrt{\frac{\bar{p}\,(1-\bar{p})}{n}}\right] = 100 \times (.02) = 2$$

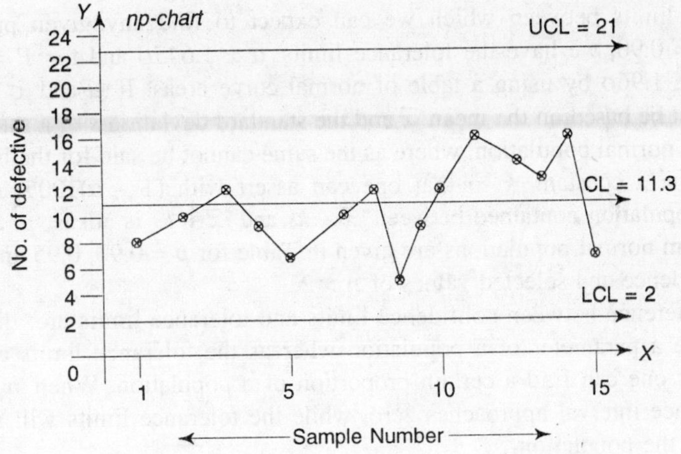

Conclusion Since all the points lie within the central limits, the system is in control

9.6.13 Advantages of Statistical Quality Control

1. Reduction of costs
2. Greater efficiency
3. Easy to apply
4. Early deduction of faults
5. Adherence to specification
6. To determine the effect of change in process
7. Statistical quality control ensures overall co-ordination

9.6.14 Limitations of SQC

The techniques of QC should not be used mechanically rather they should be matched to the process being studied. The application of standard procedures without adequate study of the process is extremely dangerous and has in the past led to statistical methods being discredited.

Statistical method applied on a production process are only an information service and as such must be conditioned by the process to which they are applied.

The responsibility for quality and process decisions rests with the manager incharge of the process and not with satisfaction. The charts do not reduce the managers responsibility.

9.7 TOLERANCE LIMITS

In every phase of industrial quality control there is the problem of comparing some quality characteristic or measurement of a finished product against given specifications. These specifications, usually called *tolerance limits* are determined by the customer or by the design engineer. If a product is made without prior specifications, or if modifications are made, it is desirable to know within what limits the process can hold a quality characteristic. This gives the concept of *'natural tolerance limits'*. If information regarding the distribution in question are available, then it is simple to find natural tolerance limits. For example, if long experience with a product enables us to assume that a certain dimension is normally distributed with the mean μ and the standard deviation σ, then it is easy to construct tolerance limits between which we can expect to find any given proportion P of the population. For $P = 0.90$, we have the tolerance limits $\mu \pm 1.645\sigma$ and for $P = 0.95$ we have the tolerance limits $\mu \pm 1.96\sigma$ by using a table of normal curve areas. If μ and σ are not known, the tolerance limits must be based on the mean $\bar{x}$ and the standard deviation s of a sample. The limits $\mu \pm 1.96\sigma$ includes 95% normal population, where as the same cannot be said for the limits $\bar{x} \pm 1.96s$. It is possible to determine a constant k so that one can assert with $(1 - \alpha)100\%$ confidence that the proportion of the population contained between $\bar{x} - ks$ and $\bar{x} + ks$ is atleast p. Such values of k of random samples from normal populations are given in Table for $p = 0.90$, 0.95 and 0.99 with 95% or 99% levels of confidence and selected values of $n > 2$.

The essential difference between confidence limits and tolerance limits are: the confidence limits are used to estimate a parameter of a population whereas the tolerance limits are used to indicate between what limits one can find a certain proportion of a population. When n becomes large, the length of a confidence interval approaches zero, while the tolerance limits will approach the corresponding values for the population.

9.7.1 Example

The cardboard industry is considering new standards for the cardboard used in boxes. One test involves placing weight on the box unit it bursts. The burst strength, in pounds per square inch, for 40 boxes are

210	234	216	232	262
183	227	197	248	218
256	218	244	259	263
185	218	196	235	223
212	237	275	240	217
263	240	247	253	269
231	254	248	261	268
262	247	292	238	215

Obtain a 95% tolerance bound that will be less than proportion 0.95 of the population of burst strength.

Solution. $\bar{x} = \dfrac{\sum x}{n} = 237.32$; $s = 25.10$. From table $k = 2.126$, so $L = \bar{x} - ks = 237.32 - 2.126$ $(25.10) = 183.96$, which is rounded to 183. We are 95% confident that atleast a proportion 0.95 of the population of burst strengths, for cardboard boxes, is above 183 psi.

9.7.2 Acceptance Sampling

Inspection of product is an integral part of all production process. The situation in which products, grouped in lots are sampled and the results of the sample are use to draw inferences concerning the quality of the product is called acceptance sampling.

Since the cost of inspection is rarely negligibie (sometimes it is nearly as high as or higher than the cost of production), it is sometimes desirable to inspect each item in a lot. Thus, acceptance inspection usually involves *sampling;* specifically speaking, a random sample is selected from each lot and the lot is accepted if the number of defectives found in the sample does not exceed a given *acceptance number.* This procedure is equivalent to a test of *null hypothesis* that the proportion defective p

in the lot equals some specified value p_0 against the alternative that it equals p_1, where $p_1 > p_0$. In acceptance sampling the value p_0 is called the *acceptable quality level* or AQL and p_1 is called the *lot tolerance percent defective* or *LTPD* or rejectable quality level (RQL). The probability of a Type I error, α, can be interpreted as an upper limit to the proportion of 'good' lots (lots with $p \leq p_0$) that will be rejected, and is called *producer's risk.* The probability of a Type II error β gives an upper bound to the proportion of 'bad' lots (lots with $p \geq p_1$) that will be accepted; it is called the *consumer's risk.*

Single Sampling Plan

It is simply a specification of the sample size and the acceptance number to be used, and its choice is usually based on a specified AQL and LTPD in association with given producers (or) consumers risk. It is best described by its operating characteristic or OC curve, which gives the probability of acceptance for each value that can be assumed by the lot proportion defective p. If a sample of size n is taken from a lot containing N units, and if the acceptance number is c, the probability of accepting a lot containing the proportion of defectives p can be calculated by hypergeometric distribution $L(p) = \sum_{x=0}^{C} h(x; n, Np; N)$.

9.7.3 Operating Characteristic Curve (OC) Curve

It is a graphic representation of the relationship between the probability of acceptance $P_\alpha(p)$ for variations in the lot quality p. For any given fraction, defected p in a submitted lot, the OC curve show, the probability p_α that such a lot will be accepted by the given sampling plan (or) in otherwords, the OC curve shows the long run percentage of submitted lots that would be accepted if great many lot of any stated quality were submitted for inspection.

In drawing the OC curve the following two terms are important.

(a) In order to measure the *customer's risk,* we must define maximum percentage of defective items in lots which the consumer wishes to accept. That is LTPD.

(b) In order to measure the *producer's risk* we define minimum percentage of defective items in a lot, that is AQL.

The producer's risk is the probability that a lot having the AQL will be rejected and the consumer's risk is the probability that a lot having LTPD will be accepted.

The shape of Typical OC curve is:

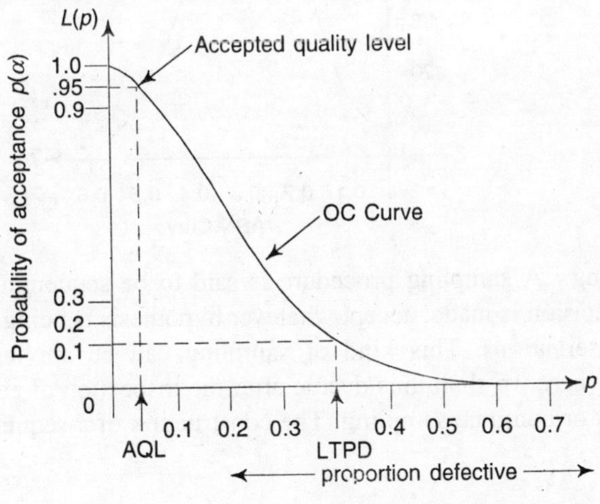

9.7.4 Double Sampling Plan

A double sampling plan involves the selection of a random sample of size n_1 from a lot; if the sample contains c_1, or fewer defectives, the lot is accepted; if it contain c_1' or more defectives $c_1' > c_1$, the lot is rejected; otherwise, a second sample of six n_2 is taken from the lot, and the lot is accepted unless the total number of defectives in the combined sample of size $n_1 + n_2$ exceeds c_2.

Similarly we define a multiple-sampling plan.

The advantage of double or multiple sampling is that there is a high probability that a very good lot will be accepted or a very poor lot will be rejected on the basis of the first sample, thus reducing the required amount of inspection.

To observe how a double or multiple sampling plan works, we introduce the *average sample number* (ASN) curve. Consider the double sampling plan

Sample	Sample size	Combined Samples		
		Size	Acceptance number	Rejection number
First	15	15	1	5
Second	30	45	5	6

If the incoming lot quality is $p = 0.05$, the probability that a second sample will be required is the same as the probability that there will be 2, 3 or 4 defectives in a sample of size 15. According to Binomial distribution table, the probability equals

$$\sum_{k=0}^{4} 15c_k (0.05)^k (.95)^{15-k} - \sum_{k=0}^{1} 15c_1(0.05)^k (.95)^{n-k} = 0.170$$

Thus, on the average, it will require a sample of size (15 + (0.170)30 = 20.1 to decide whether to accept or reject an incoming lot of quality $p = 0.05$.

A graph showing the relation between the average sample size (also called *average sample number*) and the incoming lot quality is called an ASN curve.

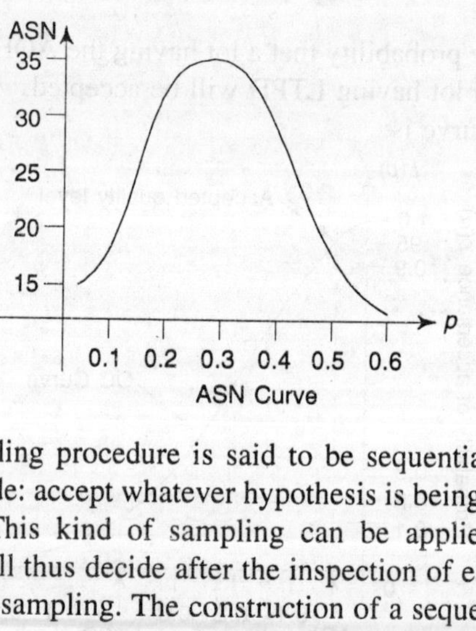

ASN Curve

Sequential Sampling A sampling procedure is said to be sequential if, after each observation, one of the following decision is made: accept whatever hypothesis is being tested, reject the hypothesis and make another observations. This kind of sampling can be applied only in connection with acceptance sampling, where we shall thus decide after the inspection of each successive item whether to accept a lot, reject it or continue sampling. The construction of a sequential sampling plan consists of finding two sequences of numbers a_n and r_n where n is the number of observations, so that the lot is accepted as soon as the number of defectives is less than or equal to a_n for some n, the lot is rejected as soon as the number of defectives is greater than or equal to r_n for some n, and sampling continues as long as the number of defectives in a sample of size n falls between a_n and r_n. If an acceptance plan is to have p_0 and p_1 as its AQL and LTPD, the producer's risk α, consumer's risk β, then a_n and r_n can be computed as follows.

$$a_n = \frac{\log\left(\frac{\beta}{1-\alpha}\right) + n\log\left(\frac{1-p_0}{1-p_1}\right)}{\log\left(\frac{p_1}{p_0}\right) - \log\left(\frac{1-p_1}{1-p_0}\right)}$$

$$r_n = \frac{\log\left(\frac{1-\beta}{\alpha}\right) + n\log\left(\frac{1-p_0}{1-p_1}\right)}{\log\left(\frac{p_1}{p_0}\right) - \log\left(\frac{1-p_1}{1-p_0}\right)}$$

If a_n is not an integer, then it is replaced by the largest integer $\leq a_n$ and if r_n is not an integer than r_n is replaced by the smallest integer $\geq r_n$.

9.8 SOLVED PROBLEMS

1. Calculate the probability of lot acceptance for a sampling plan with sample size $n = 5$ and acceptance number $c = 0$ for a lot of fraction defective p = 0.1, 0.3, and 0.5.

Solution.

$$P(X = x) = nC_x\, p^x q^{n-x}$$
$$P(\text{accept})/p(X = 0) = 5C_0\, p^0\, q^5 = q^5 = 1$$
$$P(\text{accept})/p(X = 0.1) = (0.9)^5 = .590$$
$$P(\text{accept})/p(X = 0.3) = (0.7)^5 = .168$$
$$P(\text{accept})/p(X = 0.5) = (0.5)^5 = .031$$
$$P(\text{accept}) = 1 \text{ when } p = 0$$
$$= 0 \text{ when } p = 1$$

2. Discuss the multiple sampling plan by an illustration.

Solution.

Sample	Sample size	Combined Samples		
		Size	Acceptance No.	Rejection No.
I	20	20	—	3
II	20	40	1	4
III	20	60	3	5
IV	20	80	3	6
V	20	100	5	7
VI	20	120	6	8
VII	20	140	7	8

In the first step, the lot is rejected if there are 3 or more defective; other wise the sampling continues.

In the second step, the lot is accepted if the combined sample contains atmost 1 defective, and it is rejected if there are 4 or more defectives; otherwise the sampling continues. This process goes on, if necessary, until in the final step the lot is accepted if there are at most 7 defectives in the combined sample of size 140, and otherwise it is rejected.

EXERCISES

1. A single sampling plan calls for a sample of size 150.
   (a) use the normal approximation to find the acceptance number c if the AQL is to be 2.5% and the producer's risk is to be $\alpha = 0.05$
2. A sequential sampling plan is to have $p_0 = 0.01$, $p_1 = 0.10$, $\alpha = 0.05$, $\beta = 0.20$. Determine the acceptance and rejection numbers for $n = 1, 2, \ldots 50$.
3. Calculate the average sample size required for selected value of p and sketch the ASN curve for the following double-sampling plan

Sample	Sample size	Combined samples		
		Size	Acceptance No.	Rejection No.
First	10	10	0	3
Second	35	45	3	4

4. Draw the mean and range charts and control of the process.

Sample number:	1	2	3	4	5	6	7	8	9	10
$\bar{x}$	43	49	37	44	45	37	51	46	43	47
R	5	6	5	7	7	4	8	6	4	6

5. Draw suitable control chart for the following data pertaining to the number of foreign coloured threads in 15 pieces of cloth of 7, 2, 3, 20, 21, 5, 4, 3, 10, 8, 0, 9, 6, 7 and 20.
6. Thirty-five successive samples of 100 castings each, taken from a production line, contained, respectively, 3, 3, 5, 3, 5, 0, 3, 2, 3, 5, 6, 5, 9, 1, 2, 4, 5, 2, 0, 10, 3, 6, 3, 2, 5, 6, 3, 3, 2, 5, 1, 0, 7, 4 and 3 defectives. If the fraction defective is to be maintained at 0.02 construct a p chart for these data and state whether or not this standard is being met.
7. Twenty pieces of cloth out of different rolls contain respectively 1, 4, 3, 2, 5, 4, 6, 7, 3, 2, 5, 7, 6, 4, 5, 2, 1, 3 and 8 imperfections. Ascertain whether the process is in a state of statistical control.

ANSWERS

(4) $\bar{x}$ chart: UCL = 47.56; LCL = 40.84
R chart: UCL = 12.26; LCL = 0

(5) $\bar{c} = 9$, UCL = 18, LCL = 0

(7) $\bar{c} = 4$, UCL = 10, LCL = −2

10

Reliability Engineering

10.1 CONCEPTS OF RELIABILITY

The theory of reliability deals with the failure law of systems. Whenever the system is put in operation, some time or other it is bound to fail either due to external causes or due to an internal structural defect. One possible way to improve the performance of the system is to introduce additional components to help the system operate successfully in case of failure of one or more components. Any system with additional components is called **redundant system**. However, there are other situations where it is possible to have a repair facility in which the components that have failed can be repaired and use it again.

10.1.1 Definition

Let the random variable X be the life time or the time to failure of a component. The probability that the component survives until some time t is called the **reliability** $R(t)$ of the component. Thus

$$R(t) = P(X \geq t) = 1 - F(t), \quad t \geq 0$$
$$= 1, \quad t < 0.$$

where F is the distribution function of X. $F(t)$ is also called **unreliability**.

The component is assumed to be working properly at time $t = 0$ [that is, $R(0) = 1$] and no component can work forever without failure.

Denote by $f_X(t)$ the probability density function of X, when t exists.

Consider a fixed number of identical component N_0 under test. After time t, $N_f(t)$ components have failed and $N_s(t)$ components have survived with $N_f(t) + N_s(t) = N_0$. The probability of survival may be written as:

$$p = \frac{N_s(t)}{N_0}$$

If $N_s(t)$ gets smaller and $R(t)$ decreases:

$$R(t) = \frac{N_s(t)}{N_0} = \frac{N_0 - N_f(t)}{N_0} = 1 - \frac{N_f(t)}{N_0}$$

Here the total number of components N_0 is constant, while the number of failed components N_f increases with time. Taking derivatives on both sides, we get

$$R'(t) = -\frac{1}{N_0} N'_f(t)$$

$N'_f(t)$ denotes the rate at which components fail. When $N_0 \to \infty$, $\frac{N'_f(t)}{N_0}$ is the failure density function $f_X(t)$. So

$$R'(t) = -f_X(t)$$

The conditional probability that the component does not survive for an (additional) interval of duration x given that it has survived until time t can be written as

$$\frac{P(t < X < t + x)}{P(X > t)} = \frac{F(t + x) - F(x)}{R(t)}$$

10.1.2 Definition

Hazard rate or failure rate function of the distribution at time t is

$$h(t) = \frac{f_X(t)}{R(t)} = -\frac{R'(t)}{R(t)}$$

Thus $h(t)\,\Delta t$ represents the conditional probability that a component survive to age t will fail in the interval $(t, t + \Delta t]$. Consider $h(x) = \frac{f_X(x)}{R(x)}$. Integrating both sides we get

$$\int_0^t h(x)dx = \int_0^t \frac{f(x)}{R(x)}\,dx = \int_0^t -\frac{R'(x)}{R(x)}\,dx = -\int_{R(0)}^{R(t)} \frac{dR}{R}$$

or $-\log h(x) = \int_0^t h(x)dx$, using the boundary condition $R(0) = 1$. Therefore

$$R(t) = \exp\left[-\int_0^t h(x)\,dx\right]$$

The cumulative failure rate $H(t) = \int_0^t h(x)dx$ is called **cumulative hazard**. Hence $R(t) = e^{-H(t)}$. Note that if the lifetime is exponentially distributed, then $H(t) = \lambda t$ and we obtain the exponential reliability function.

The following points gives the difference between $f(t)$ and $h(t)$. (i) The quantity $f(t)\Delta t$ is the unconditional probability that the component will fail in the interval $(t, t + \Delta t]$, whereas the quantity $h(t)\Delta t$ is the conditional probability that the component will fail in the same time interval, given that it has survived until time t. (ii) $h(t) \geq f(t)$, as $R(t) \leq 1$. (iii) $f(t)$ is a probability density function, but $h(t)$ is not.

10.1.3 Definition Conditional Reliability $R_t(y)$

Conditional Reliability $R_t(y)$ to be the probability that the component survives an additional interval of duration y given that it has survived until time t. Thus

$$R_t(y) = \frac{R(t + y)}{R(t)}$$

10.1.4 Theorem

If a component has a survival probability over an additional period of length y which is same as its present age, then the failure rate or hazard rate is constant.

Proof: Given $R_t(y) = R(y)$ for all $y, t \geq 0$. That is $\dfrac{R(t+y)}{R(t)} = R(y)$, which implies $R(t+y) = R(y)\,R(t)$. Rearranging, we get

$$\frac{R(y+t)-R(y)}{R(t)} = \frac{(R(t)-1)\,R(y)}{t}$$

Taking limit $t \to 0$ and noting that $R(0) = 1$, we get

$$R'(y) = R'(0)\,R(y)$$

$$\Rightarrow \quad \frac{R'(y)}{R(y)} = R'(0)$$

$$\Rightarrow \quad \frac{R'(y)}{R(y)} = -\lambda, \text{ if we take } R'(0) = -\lambda$$

$$\Rightarrow \quad R(y) = e^{-\lambda y} \qquad y > 0$$

$$\Rightarrow \quad Z(t) = h(t) = \frac{-R'(y)}{R(y)} = \lambda \text{ (a constant)}$$

Note: If the conditional survival probability increases with age, then hazard rate will be a decreasing function of age and the corresponding distribution is known as DFR (decreasing failure rate distribution). Similarly, we define IFR, (increasing failure rate distribution). If the failure rate is a function of age, then it is called **hazard function.**

10.2 SYSTEM RELIABILITY

Suppose that a system consists of n independent components $1, 2, \ldots i \ldots n$ and that the system is capable of just two modes of performances. The performance of each component can be denoted by a random variable X_i, which takes on the value $x_i = 1$ if the component performs satisfactorily for the desired time and $x_i = 0$ if the component fails during this time.

The 'performance' of the system is measured by the binary random variable $\phi(X_1, X_2, \ldots X_n)$ where

$$\phi(X_1, X_2, \ldots X_n) = \begin{cases} 1 & \text{if system performs satisfactorily during time } [0, t] \\ 0 & \text{if system fails during time } [0, t] \end{cases}$$

The function ϕ is called **structure function of the system**. The system structures generally considered are described below.

10.2.1 Series System

The series system is the simplest and most common of all configurations. For a series system, the system fails if any component of the system fails. That is, the system performs satisfactorily if and only if all the components perform satisfactorily. The structure function for a series system is given by

$$\phi(X_1, X_2, \ldots X_n\} = X_1\, X_2 \ldots X_n = \min\,\{X_1, X_2, \ldots X_n\}$$

This follows because each X_i is either 1 or 0.

Hence the structure function takes on the value 1 if each X_i equals 1 or if the minimum of the X_i equals 1. Assume that a device function for some time after which it fails and then remains in that non-functioning state. The duration of time T, for which it function is its **life time.**

Reliability of Series Systems

The reliability of a system can be expressed as

$$R = P\{\phi(X_1, X_2,\ldots,X_n) = 1\}$$

Thus, for a series system the reliability is

$$\begin{aligned} R_s = P\{X_1\, X_2 \ldots X_n = 1\} &= P\{X_1 = 1,\, X_2 = 1, \ldots X_n = 1\} \\ &= P\{X_1 = 1\}\; P\{X_2 = 1\} \ldots P\{X_n = 1\} \\ &= p_1\, p_2 \ldots p_n \end{aligned}$$

Let the lifetime of the system is T and that of each component is T_i.

The system reliability at time t is $R(t) = P\{T > t\}$ and the reliability of ith component at time t is

$R_i(t) = P\{T_i > t\}$ Hence

$$\begin{aligned} R_s(t) = P(T > t) &= P\{\phi(X_1, X_2,\ldots X_n) > t)\} \\ &= \prod_{i=1}^{n} P(X_i > t) = \prod_{i=1}^{n} R_i(t) \end{aligned} \tag{10.1}$$

The hazard rate function of the system is

$$h(t) = -\frac{R'(t)}{R(t)} = \frac{\sum_{i=1}^{n} (-R_i'(t)) \prod_{j \neq i}^{n} R_j(t)}{\prod_{j=i}^{n} R_j(t)}$$

$$= \sum_{i=1}^{n} \frac{-R_i'(t)}{R_i(t)} = \sum_{i=1}^{n} h_i(t)$$

where $h_i(t)$ is the hazard rate function of the component i. Thus, if the lifetime distribution of each component is an IFR, then that of the series system is also IFR.

10.2.2 Parallel System

A parallel system of n components is defined to be a system that fails if all components fail, or alternatively, the system performs satisfactorily if atleast one of the n components performs satisfactorily (with all components operating simultaneously). This property of parallel systems is often called "**redundancy**" (i.e. additional components existing within the system to help the system operate successfully in case of failure of one or more components). The structure function for a parallel system is

$$\begin{aligned} \phi(X_1, X_2,\ldots X_n) &= 1 - (1 - X_1)\,(1 - X_2)\ldots(1 - X_n) \\ &= \max\,(X_1, X_2,\ldots X_n) \end{aligned}$$

This follows because each X_i is either 1 or 0. The structure function takes on the value 1 if atleast one of the X_i equals 1 or if the largest X_i equals 1.

Reliability of Parallel Systems

The structure function for a parallel system is

$$\phi(X_1, X_2,\ldots X_n) = \max(X_1, X_2\ldots X_n)$$

and the reliability is

$$\begin{aligned} R_p(p_1, p_2,\ldots p_n) &= P(\max (X_1, X_2,\ldots X_n) = 1) \\ &= 1 - P \text{ (all } X_i = 0) \\ &= 1 - P(X_1 = 0, X_2 = 0,\ldots X_n = 0) \\ &= 1 - (1 - p_1)(1 - p_2)\ldots(1 - p_n) \end{aligned}$$

Thus for $X = (X_1, X_2,\ldots X_n)$,

$$\begin{aligned} R_p(t) &= P(X > t) = P(\max(X_1, X_2,\ldots X_n) > t) \\ &= 1 - \prod_{i=1}^{n} P(X_i \le t) = 1 - \prod_{i=1}^{n} (1 - R_i(t)) \end{aligned} \tag{10.2}$$

For parallel system, the life time distribution of the system is not necessarily IFR, even if that of each component is. The reliability of a parallel system of n independent components, each with an exponential failure law (with parameter λ) is

$$R_P(t) = 1 - (1 - e^{-\lambda t})^n \tag{10.3}$$

10.2.3 *k* Out of *n* System

Some systems are assembled such that the system operates if k out of n component function properly. Note that the series system is a k out of n system with $k = n$, and the parallel system is k out of n system, with $k = 1$. The structure function is given by

$$\phi(X_1, X_2,\ldots X_n) = \begin{cases} 1 & \text{if } \sum_{i=1}^{n} X_i \ge k \\ 0 & \text{if } \sum_{i=1}^{n} X_i < k \end{cases}$$

and the reliability is given by

$$R(p_1, p_2,\ldots p_n) = P\left(\sum_{i=1}^{n} X_i \ge k\right)$$

If $p_1 = p_2 = \ldots. = p_n = p$, then $\sum_{i=1}^{n} X_i$ has a binomial distribution with parameters n and p, so that

$$R(p_1, p_2,\ldots p_n) = \sum_{i=k}^{n} nC_i\, p^i (1 - p)^{n-i}$$

10.2.4 Procedure for Finding the Exact Reliability

First we fix some terminology in graph theory.

10.2.5 Definition

A graph consists of a set of points called **vertices** or **nodes**, with certain pairs of nodes being joined by lines called **edges** or **branches**.

e.g.

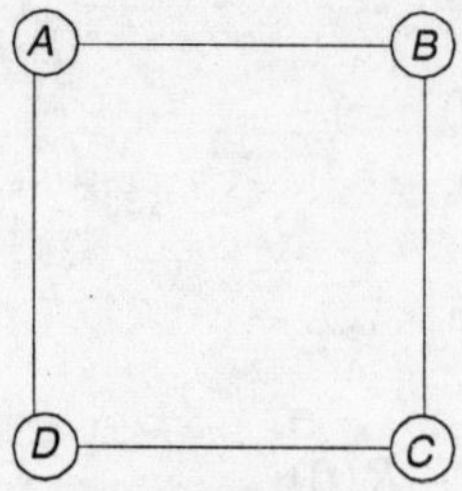

Undirected Graph

Fig. 10.1

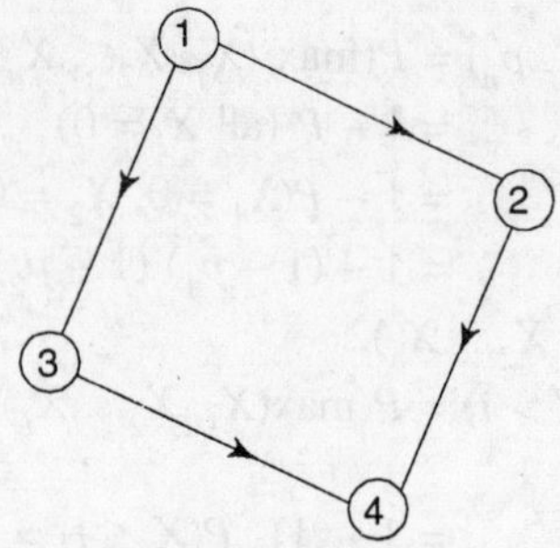

Directed Graph

Fig. 10.2

If the directions of the edges are given, then it is called **directed graph**; otherwise it is an **undirected graph**. A **network** is considered to be a graph with a flow of some type in its branches. A **chain** between two vertices i and j is a sequence of branches connecting these two vertices. When the direction of travel along the chain is also specified, it is called a **path**. A **cycle** is a chain connecting a node to itself. A **source** is a vertex in a network such that the flow moves away from that node. A **sink** is a vertex where the flow moves towards that node.

10.2.6 Definition

A minimal path is a minimal set of components that by functioning, ensures the successful operation of the system. A minimal cut is a minimal set of components that, by failing, ensures the failure of the system.

For example, consider the system that can be represented by the network

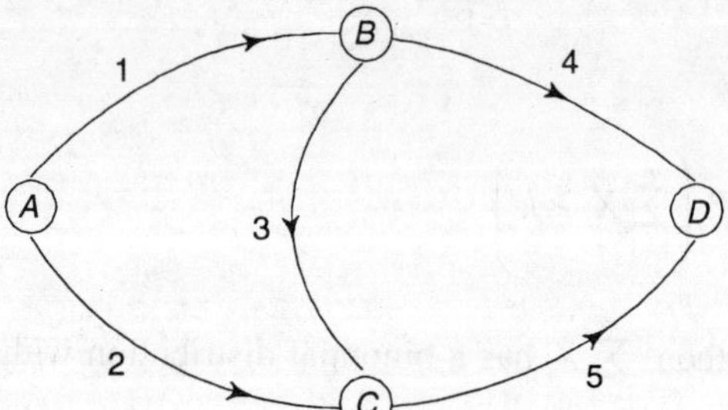

Fig. 10.3

This system consists of five components that are connected. The system will operate successfully if there exists a flow from source A to sink D through the directed graph., i.e. 1, 4 operate successfully or 2, 5 operate successfully or 1, 3, 5 operate successfully.

If the performance of the ith component is denoted by the binary random variable X_i, then it takes the value 1 with probability p_i and 0 with probability $(1 - p_i)$. For the system given in Fig. 10.3, the minimal paths and cuts are

minimal path	minimal cut
X_1X_4	X_1X_2
$X_1X_3X_5$	X_4X_5
X_2X_5	$X_2X_3X_4$
	X_1X_5

Using all the minimal paths, we can obtain the exact system reliability. The system will operate if all the components in atleast one of the minimal paths operate. Hence paths operate as a parallel system, so that

$$\begin{aligned}\phi(X_1, X_2, X_3, X_4, X_5) &= \max(X_1X_4, X_1X_3X_5, X_2X_5)\\ &= 1-(1-X_1X_4)(1-X_1X_3X_5)(1-X_2X_5)\\ &= X_1X_4 + X_1X_3X_5 + X_2X_5 - X_1X_3X_4X_5 - X_1X_2X_4X_5 - X_1X_2X_3X_5 + X_1X_2X_3X_4X_5\end{aligned}$$

as $X_i^2 = X_i$

Now

$$\begin{aligned}R(p_1, p_2, p_3, p_4, p_5) &= P\{\phi(X_1, X_2, X_3, X_4, X_5) = 1\}\\ &= E(\phi(X_1, X_2, X_3, X_4, X_5))\\ &= E(X_1X_4 + X_1X_3X_5 + X_2X_5 - X_1X_3X_4X_5\\ &\quad - X_1X_2X_4X_5 - X_1X_2X_3X_5 + X_1X_2X_3X_4X_5)\\ &= p_1p_4 + p_1p_3p_5 + p_2p_5 - p_1p_3p_4p_5 - p_1p_2p_4p_5 - p_1p_2p_3p_5 + p_1p_2p_3p_4p_5.\end{aligned}$$

Using all the minimal cuts also, we can obtain the exact system reliability. Since the system will fail if and only if all the components in atleast one of the minimal cut fail, the system reliability can be expressed as

$$\begin{aligned}R(p_1, p_2, p_3, p_4, p_5) &= 1 - P(\phi(X_1, X_2, X_3, X_4, X_5) = 0)\\ &= 1 - P((X_1 = 0, X_2 = 0) \cup (X_4 = 0, X_5 = 0) \cup (X_2 = 0, X_3 = 0, X_4 = 0)\\ &\quad \cup (X_1 = 0, X_5 = 0))\end{aligned} \qquad (10.4)$$

We know

$$\begin{aligned}P(A\cup B\cup C\cup D) &= P(A) + P(B) + P(C) + P(D) - P(A\cap B) - P(A\cap C) - P(A\cap D) - P(B\cap C)\\ &\quad - P(B\cap D) - P(C\cap D) + P(A\cap B\cap C) + P(A\cap B\cap D) + P(A\cap C\cap D)\\ &\quad + P(B\cap C\cap D) - P(A\cap B\cap C\cap D)\end{aligned} \qquad (10.5)$$

Using (10.5) in (10.4), we get

$$\begin{aligned}R(p_1, p_2, p_3, p_4) &= 1 - q_1q_2 - q_4q_5 - q_2q_3q_4 - q_1q_5 + q_1q_2q_3q_4 + q_1q_2q_5 + q_2q_3q_4q_5\\ &\quad + q_1q_4q_5 - q_1q_2q_3q_4q_5\end{aligned}$$

where $q_i = 1 - p_i$. The other method to obtain the same result by taking

$$\begin{aligned}\phi(X_1, X_2, X_3, X_4, X_5) = \min\,[&1-(1-X_1)(1-X_2),\ 1-(1-X_4)(1-X_5),\ 1-(1-X_2)(1-X_3)(1-X_4),\\ &1-(1-X_1)(1-X_5)]\end{aligned}$$

Taking expectations on both sides gives the required result.

Computation of Mean Time to Failure

Let X denote the life time of a component so that its reliability $R(t) = P(X > t)$ and $R'(t) = -f(t)$. Then the expected life or the mean time to failure (MTTF) of the component is given by

$$E(X) = \int_0^\infty t f(t)\, dt$$

$$= -\int_0^\infty tR'(t)dt$$

$$= [-tR(t)]_0^\infty + \int_0^\infty R(t)dt \qquad \text{by integration by parts}$$

$$= \int_0^\infty R(t)\, dt \qquad R(t) \to 0, \quad \text{as } t \to \infty.$$

$$E(X^k) = \int_0^\infty t^k f(t)\, dt$$

$$= -\int_0^\infty t^k R'(t)\, dt$$

$$= [-tk\, R(t)]_0^\infty + \int_0^\infty kt^{k-1} R(t)\, dt$$

$$= \int_0^\infty kt^{k-1} R(t)\, dt.$$

$$\text{Var}(X) = E(X^2) - (E(x))^2 = \int_0^\infty 2t\, R(t)\, dt - \left[\int_0^\infty R(t)\, dt\right]^2$$

If $\qquad R(t) = e^{-\lambda t}$, then $E(X) = \int_0^\infty e^{-\lambda t}\, dt = \frac{1}{\lambda}$ and

$$\text{Var}(X) = \int_0^\infty 2te^{-\lambda t}\, dt - \frac{1}{\lambda^2} = \frac{1}{\lambda^2}.$$

For series system,

$$E(X) = \int_0^\infty R(t)\, dt \le \min\left(\int_0^\infty R_i(t)\, dt\right)$$

$$= \min_i E(X_i)$$

For parallel system

$$E(X) = \int_0^\infty R(t)\, dt \ge \max_i\left(\int_0^\infty R_i(t)\, dt\right)$$

$$= \max_i E(X_i)$$

If X_i is exponentially distributed with parameter λ, then for parallel system $R_X(t) = 1 - (1 - e^{-\lambda t})^n$ and

$$E(X) = \int_0^\infty R_X(t)dt$$

$$= \int_0^\infty (1 - (1 - e^{-\lambda t})^n)\, dt$$

$$= \frac{1}{\lambda}\int_0^1 \frac{1-u^n}{1-u}\, du, \text{ where} \qquad u = 1 - e^{-\lambda t}$$

$$dt = \frac{1}{\lambda(1-u)}\, du$$

Since the integrand is the sum of finite geometric series

$$E(X) = \frac{1}{\lambda}\int_0^1 \left(\sum_{i=1}^{n} u^{i-1}\right) du$$

$$= \frac{1}{\lambda}\sum_{i=1}^{n-1}\int_0^1 u^{i-1}\, du$$

$$= \frac{1}{\lambda}\sum_{i=1}^{n}\left[\frac{u^i}{i}\right]_0^1$$

$$= \frac{1}{\lambda}\sum_{i=1}^{n}\frac{1}{i}$$

$$= \frac{\log n}{\lambda}$$

Mean Time between Failures ***(MTBF)***

"Mean time to failure" should be used in the case of simple components which are not repaired when they fail but are replaced by good components.

"Mean time between failures" should be used with repairable equipment or systems. We can use MTBF for both non-repairable and repairable components.

MTTF is the mean time to first failure, whereas MTBF is the mean time between two successive component failures i.e., between the nth and $(n + 1)^{\text{th}}$ failures in a system when n is relatively large. The relationship between MTTF & MTBF is

$$\frac{1}{\text{MTBF}} = \sum_{j=1}^{m}\frac{1}{E(T_j)}$$

where $E(T_j)$ is the MTTF of the jth component. It is obvious that MTBF is a function of time. When a system is first operated will all new components, then MTTF and MTBF are same. After that MTBF will fluctuate until and after many failures and replacements and it will stabilize at the value given by

$$\frac{1}{\text{MTBF}} = \sum_{j=1}^{m}\frac{1}{E(T_j)}.$$

10.2.7. Example

An airborne electronic system has a radar, a computer and an additional unit with MTBF of 83, 167, and 500 *h*, respectively. Find the system MTBF and the reliability for a 5*h* operating time.

Solution. MTBF for a series system can be calculated by converting the individual MTBFs into failure rate, adding these to get a system failure rate and converting this to a system MTBF.

Unit	MTBF h	Failure rate/1000h.
Radar	83	$\frac{1000}{83}$ = 12 (approx)
Computer	167	$\frac{1000}{167}$ = 6 (approx)
Additional	500	$\frac{1000}{500}$ = 2
		Total = 20

System $$\text{MTBF} = \frac{1000}{20} = 50\text{h}.$$

For a 5*h* operating time

$$R_{\text{sys}}(5) = e^{-\frac{1}{50}\times 5} = 90\% \text{ (approx)}$$

while for Radar $$R = e^{-5/83} = 94\% \text{ (approx)}$$

Computer $R = 97\%$ and additional unit $R = 99\%$.

10.2.8 Example

Consider a system consisting of n components such that the failure of the ith component occurs in accordance with a Poisson process of intensity a_i. Find the reliability of the system under different system structures. Here T_i is exponential with mean $\frac{1}{a_i}$. That is $R_i(t) = e^{-a_i t}$

(i) *Series System*

$$R(t) = \prod_{i=1}^{n} R_i(t) = e^{-\Sigma a_i t}$$

T is exponential with mean $\frac{1}{\Sigma a_i}$.

(ii) *Parallel System*

$$R(t) = 1 - \prod_{i=1}^{n} (1 - e^{a_i t})$$

$$= \Sigma e^{a_i t} - \Sigma\, e^{-(a_i + a_j)t} + \ldots\ldots \pm e^{-(\Sigma a_i)\, t}$$

and $$E(T) = \Sigma \frac{1}{a_i} - \Sigma \frac{1}{a_i + a_j} + \ldots\ldots \pm \frac{1}{\Sigma a_i}$$

(iii) k out of n System: Let $a_i = a$ for all i. Then

$$R(t) = P(T > t)$$

$$= \sum_{r=k}^{n} {}^nC_r \,(e^{-at})^r\,(1 - e^{-at})^{n-r}$$

and

$$E(T) = \int_0^\infty R(t)\,dt$$

$$= \sum_{r=k}^{n} {}^nC_r \int_0^\infty e^{-atr}\,(1 - e^{-at})^{n-r}dt$$

$$= \sum_{r=k}^{n} {}^nC_r \int_0^\infty y^{r-1}\,(1 - y)^{n-r}\,\frac{dy}{a}, \quad \text{put} \quad y = e^{-at}$$

$$= \frac{1}{a}\sum_{r=k}^{n} {}^nC_r\,\beta\,(r, n - r + 1)$$

$$= \frac{1}{a}\sum_{r=k}^{n} \frac{n!}{r!(n-r)!}\,\frac{\Gamma(r)\,\Gamma(n-r+1)}{\Gamma(n+1)}$$

$$= \frac{1}{a}\sum_{r=k}^{n} \frac{n!}{r!(n-r)!}\,\frac{(r-1)!(n-r)!}{n!} \quad \text{since } \Gamma(n) = (n-1)!$$

$$= \frac{1}{a}\sum_{r=k}^{n} \frac{(r-1)!}{(r-1)!\,r}$$

$$= \frac{1}{a}\sum_{r=k}^{n} \frac{1}{r}.$$

10.2.9 Example

Let us consider a system consisting of two units 1 and 2, connected in series. Let X and Y be the random variables denoting failure times of units 1 and 2 respectively. Suppose X and Y are independently and exponentially distributed. Let ϕ be the time to failure for the system. Find the distribution function F_ϕ and density function f_ϕ.

Solution. $Z > z$ if and only if $X > z$ and $Y > z$,

Thus

$$P\,(Z > z) = P\,(X > z)\,P(Y > z)$$
$$= (1 - F_X(z))\,(1 - F_Y(z))$$

Now

$$F_Z(z) = P(Z \le z)$$
$$= 1 - P(Z > z)$$

$$= 1 - \{[1 - F_X(z)][1 - F_Y(z)]\}$$
$$= F_X(z) + F_Y(z) - F_X(z)\, F_Y(z)$$

When X and Y are exponentially distributed, we have

$$F_X(z) = e^{-\lambda z},\ z \geq 0$$
$$F_Y(z) = e^{-\mu z},\ z \geq 0$$

Thus

$$P(Z > z) = 1 \text{ for } z < 0$$
$$= e^{-\lambda z}\, e^{-\mu z} = e^{-(\lambda + \mu)z},\ z \leq 0$$

Hence
$$F_Z(z) = 1 - P(Z > z)$$
$$= 0 \text{ for } z < 0$$
$$= 1 - e^{-(\lambda+\mu)z}, \text{ for } z \geq 0$$

and
$$f_Z(z) = 0 \quad \text{for } z < 0$$
$$= (\lambda + \mu)e^{-(\lambda+\mu)z}, \text{ for } z \geq 0.$$

10.2.10 Example

Suppose the system consists of two units connected in parallel. Find the distribution and density function of ϕ.

Solution. When the units of the system are connected in parallel, the random variable ϕ is given by $\phi = \max(X, Y)$. The probability mass is $F_Z(z)$ and is $F_Z(z) = F_{XY}(z, z)$. If X and Y are independent, then $F_Z(z) = F_X(z)\, F_Y(z)$. Differentiating with respect to z we get

$$f_Z(z) = \frac{\partial F_{XY}(z, z)}{\partial x} + \frac{\partial F_{XY}(z, z)}{\partial y}$$

$$= \int_{-\infty}^{z} f_{XY}(z, y)dy + \int_{-\infty}^{z} f_{XY}(x, z)\, dx.$$

Series-parallel Systems

The formulas (10.1) and (10.2) for the reliability computation of series and parallel system can be used in combination to compute the reliability of a system having both series and parallel parts.

Consider a series-parallel system of n serial stages where stage i consists of n_i identical components in parallel. Let the reliability of each component at stage i be R_i. Assuming all components are independent, the system Reliability

$$R_{sp} \text{ is } \prod_{i=1}^{n} [1 - (1 - R_i)^{n_i}].$$

Reliability of systems with more general inter-connections cannot be computed with the above formula.

Example: Series-parallel system

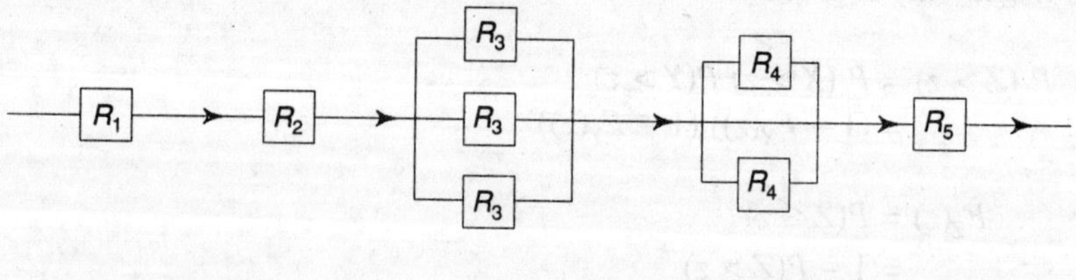

Parallel Redundancy: To replicate components with small reliabilities.

Standby Redundancy: Assume that the system has one component operating and $(n-1)$ unpowered spares. The failure rate of an operating component is λ, and a unpowered spare does not fail. Furthermore, the switching equipment is failure-free. Let T_i be the lifetime of the ith component from the point it is put into operation until its failure. Then the system lifetime T is given by

$$T = \sum_{i=1}^{n} T_i$$

with $E(T) = \frac{n}{\lambda}$ and Var $(T) = \frac{n}{\lambda^2}$. The reliability expression for a standby redundant system with a total of n independent components, each of which has an exponentially distributed lifetime with parameter λ is given by

$$R(t) = \sum_{k=0}^{n-1} \frac{(\lambda t)^k}{k!} e^{-\lambda t},\ t \geq 0 \tag{10.6}$$

Triple Modular Redundancy (TMR)

In such a system there are three components, two of which are required to be in working order for the system to function properly. TMR increases reliability over the simplex system only if the simplex reliability is greater than 0.5. Otherwise this type of redundancy decreases reliability.

10.2.11 Example

Consider evaluating the reliability R of the five component systems shown in Fig. on next page.

The system is said to be functioning properly only if all the components on atleast one path from S to Z are functioning parallel.

Define for $i = 1, 2, \ldots\ldots 5$, event X_i = "component i is functioning properly", R_i = reliability of component $i = P(X_i)$. Let X = system is functioning properly, R = system reliability = $P(X)$. It is clear that X is the union of four events.

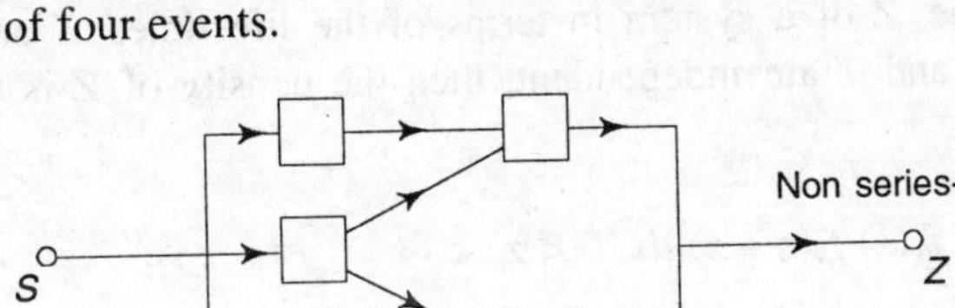

Non series-parallel system

$$X = (X_1 \cap X_4) \cup (X_2 \cap X_4) \cup (X_2 \cap X_5) \cup (X_3 \cap X_5).$$

These four events are not mutually exclusive. Using total probability theorem, we have

$$\begin{aligned} P(X) &= P(X/X_2)\ P(X_2) + P(X/\bar{X}_2)P(\bar{X}_2) \\ &= P(X/X_2)\ R_2 + R(X/\bar{X}_2)\ (1 - R_2) \end{aligned} \tag{10.7}$$

To compute $P(X/X_2)$, we observe that since component C_2 is functioning, the component C_1 and C_3 is irrelevant. Thus, under this condition, the system is equivalent to two components C_4 and C_5 in parallel. Therefore, we get

$$P(X/X_2) = 1 - (1 - R_4)\ (1 - R_5) \tag{10.8}$$

To compute $P(X/\bar{X}_2)$, as component C_2 is known to have malfunctioned, the resulting equivalent system is a series-parallel one whose reliability is computed:

$$P(X/\bar{X}_2) = 1 - (1 - R_1R_4)(1 - R_3R_5) \tag{10.9}$$

Substitute (10.8) and (10.9) in (10.7) we get

$$R = 1 - R_2(1 - R_4)(1 - R_5) - (1 - R_2)(1 - R_1R_4)(1 - R_3R_5).$$

10.2.12 Example

Consider a parallel-redundant system with n components, each having a constant failure rate μ. If k denotes the number of components operating properly, then the death process with linear rate describes the behavior of this system. The distribution of time to failure is then given by $p_o(t)$ and system Reliability is $R(t) = 1 - p_0(t)$. By (6.40)

$$p_k(t) = {}^nC_k\,(e^{-\mu t})^k\,(1 - e^{-\mu t})^{n-k}.$$

For $k = 0$, we have

$$p_o(t) = (1 - e^{-\mu t})^n. \text{ Hence}$$

$$R(t) = 1 - (1 - e^{-\mu t})^2$$

which is same as (10.3).

10.2.13 Example

Consider a standby-redundant system with n components, each with a constant failure rate μ. Then $p_0(t)$ gives the distribution of the time to failure of such a system, which is same as (10.6).

10.2.14 Example

Consider a system with two statistically identical components, each with an exponential failure law with parameter λ. Only one component is required to be operative for the system. The other component is kept it as spare one. When the operative component is failed, then it is replaced by the spare. Assume that the spare component does not fail and that the failure detection and switching equipment is perfect. We can characterize the life time Z of a system in terms of the life times X and Y of individual components of $Z = X + Y$. If X and Y are independent, then the density of Z is the convolution of densities of X and Y. That is

$$f_Z(z) = \int_0^z f_X(x)\,f_Y(z - x)\,dx,\ 0 < z < \infty$$

Further, if X and Y have exponential distributions with parameter λ; that is $f_X(x) = \lambda e^{-\lambda x}$, $x > 0$, and $f_Y(y) = \lambda e^{-\lambda y}$ $y > 0$. Then

$$f_Z(z) = \int_0^z \lambda e^{-\lambda x} \cdot \lambda e^{-\lambda(z-x)}\,dx$$

$$= \lambda^2 e^{-\lambda z} \int_0^z dx$$

$$= \lambda^2\,e^{-\lambda z}\,z.$$

$$F_Z(t) = \int_0^t f_Z(z)\,dz$$

$$= \int_0^t \lambda^2 z e^{-\lambda z} dz$$

$$= \lambda^2 \left[\int_0^t z \, d\left(\frac{e^{-\lambda z}}{-\lambda} \right) \right]$$

$$= \lambda^2 \left[\left(\frac{z \cdot e^{-\lambda z}}{-\lambda} \right)_0^t - \int_0^t \frac{e^{-\lambda z}}{-\lambda} dz \right]$$

$$= \lambda^2 \left[\frac{t}{-\lambda} e^{-\lambda t} - \frac{e^{-\lambda t}}{\lambda^2} + \frac{1}{\lambda^2} \right]$$

$$= 1 - e^{-\lambda t} - \lambda t e^{-\lambda t}$$

The reliability of the two-component stand by redundant system is

$$R(t) = 1 - F_Z(t)$$

$$= 1 - (1 - e^{-\lambda t} - \lambda t e^{-\lambda t}) = (1 + \lambda t) e^{-\lambda t}.$$

Reliability of two-component standby systems.

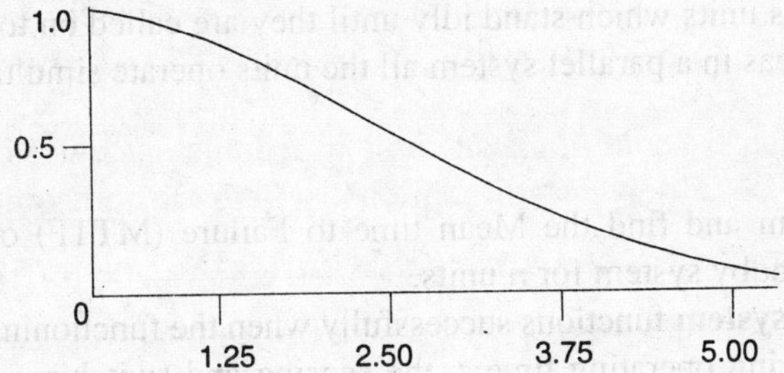

In this example we assumed the failure rates of the two components were the same. If the failure rates are distinct and if X and Y are exponentially distributed with parameter λ_1, λ_2, $\lambda_1 \neq \lambda_2$, then $Z = X + Y$ has a two-stage hypo exponential distribution with parameters, λ_1 and λ_2. Now

$$f_Z(z) = \int_0^z f_X(x) f_Y(z - x) \, dx$$

$$= \int_0^z \lambda_1 e^{-\lambda_1 x} \cdot \lambda_2 e^{-\lambda_2 (z - x)} \, dx$$

$$= \lambda_1 \lambda_2 e^{-\lambda_2 z} \int_0^z e^{(\lambda_2 - \lambda_1) x} \, dx$$

$$= \lambda_1 \lambda_2 e^{-\lambda_2 z} \left[\frac{e^{-(\lambda_1 - \lambda_2) x}}{-(\lambda_1 - \lambda_2)} \right]_0^z$$

$$= \lambda_1 \lambda_2 e^{-\lambda_2 z} \left[\frac{e^{-(\lambda_1 - \lambda_2) z}}{-(\lambda_1 - \lambda_2)} + \frac{1}{\lambda_1 - \lambda_2} \right]$$

$$= \frac{\lambda_1 \lambda_2}{\lambda_1 - \lambda_2} e^{-\lambda_2 z} + \frac{\lambda_1 \lambda_2}{\lambda_2 - \lambda_1} e^{-\lambda_1 \cdot z}$$

10.2.15 Example

Consider the triple modular redundancy (TMR) system ($k = 2$, $n = 3$). It is 2 out of 3 system. So its reliability is given by

$$R_{TMR}(t) = \sum_{i=2}^{3} 3C_i\, p^i (1-p)^{3-i}$$
$$= 3C_2 p^2 (1-p) + 3C_3\, p^3$$
$$= 3p^2 - 2p^3$$
$$= 3R^2(t) - 2R^3(t)$$

If $R_{TMR}(t) = e^{-\lambda t}$, then

$$R_{TMR}(t) = 3e^{-2\lambda t} - 2e^{-3\lambda t}.$$

10.2.16 Example

What is the difference between standby system and parallel system?

Solution. A standby system has units which stand idly until they are called on to operate by a sensing and switching subsystem whereas in a parallel system all the units operate simultaneously.

10.2.17 Example

Define two-unit standby system and find the Mean time to Failure (MTTF) of a two-unit standby system. Also generalize the standby system for n units.

Solution. A two-unit standby system functions successfully when the functioning unit does not fail or if the functioning unit fails during operating time t, the sensing and switching subsystems functions properly and the standby unit (not having failed but remains idle) functions properly for the remainder of the time.

If the failure rate for both the units is λ, then

$$R(t) = e^{-\lambda t} + (\lambda t)\, e^{-\lambda t}$$

And the MTTF of the system is

$$E(T) = \int_0^\infty R(t)dt = \int_0^\infty e^{-\lambda t}\, dt + \int_0^\infty (\lambda t) e^{-\lambda t}\, dt$$
$$= \frac{1}{\lambda} + \frac{\lambda}{\lambda^2}$$
$$= \frac{2}{\lambda} = 2\left(\frac{1}{\lambda}\right)$$

So the MTTF of the system is the double that for a single unit.

10.2.18. Multiunit-standby System

Assume the 100% reliability of the sensing-switching unit and the standby units while idling. Equation for two-unit standby system is

$$R(t) = e^{-\lambda t} + \lambda t\, e^{-\lambda t}$$

which can be expanded to the situation having n identical units with one functioning and the remaining $(n-1)$ units in standby to give

$$R(t) = e^{-\lambda t}\left[1 + \lambda t + \frac{(\lambda t)^2}{2!} + \ldots + \frac{(\lambda t)^{n-1}}{(n-1)!}\right].$$

and, in this case the MTTF of the system is: $E(T) = \dfrac{n}{\lambda} = n\left[\dfrac{1}{\lambda}\right]$.

10.3 RELIABILITY AND AVAILABILITY OF MARKOVIAN SYSTEMS

Availability of a device is defined as the probability that the device is operating satisfactorily at any time, where the total time includes operating time, active repair time, logistic time and administrative time.

The classification of availability are :

* Instantaneous (or point) Availability
* Average Up-time Availability (or Mean Availability)
* Steady State Availability
* Inherent Availability
* Achieved Availability and Operational Availability.

Assume that under failure of a component, it is repaired and restored to be "as good as new".

We define the **instantaneous availability or Point Availability** $A(t)$ of a component (or a system) as the probability that the component is properly functioning at time t. Note that in the absence of a repair or a replacement, availaibility $A(t)$ is equal to the reliability of the component. The point availability is $A(t) = R(t) + \int_0^t R(t-x)\, m(x)\, dx$ where $m(x)$ is the renewal density function of the system.

Let T_i be the duration of the ith functioning period of a component. Let T_i's be identically distributed with cumulative distribution function $W(t)$ and density function $w(t)$. Let D_i be the system down time for the ith repair or replacement. Let D_i's be identically distributed with distribution function $G(t)$ and density function $g(t)$. Let $X_i = \{T_i + D_i\}$, $i = 1, 2, \ldots$.
be random variables, which is mutually independent. Then $\{X_i, i = 1, 2, \ldots..)$ is a renewal process with the density function $f(t)$ and is the convolution of w and g. Thus we have $L(f(t)) = L(w(t)) \cdot L(g(t))$. Hence, by (6.48), we have

$$L(m(t)) = \frac{L(w(t))\, L(g(t))}{1 - L(w(t))\, L(g(t))} \tag{10.10}$$

10.3.1 Theorem

If the failure time distribution and repair-time distribution $g(t)$ are given, then the instantaneous availability $A(t)$ as a function of time is given by

$$A(t) = L^{-1}\left\{\frac{1 - L(w(t))}{s[1 - L(w(t))\, L(g(t))]}\right\}$$

Proof: The functioning of a component at time t can be taken as the combination of two events.

(i) The component has not failed from the beginning (no renewals in the period $(0, t]$) with Reliability $R(t)$.

(ii) The last repair occurred at time x, $0 < x < t$, and the component has continued to work since that time with the probability $\int_0^t R(t-x)\, m(x)\, dx$

So,

$$A(t) = R(t) + \int_0^t R(t-x)\, m(x)\, dx.$$

Taking Laplace transform on both sides, we get

$$L(A(t)) = L(R(t)) + L\left(\int_0^t R(t-x)\, m(x)\, dx\right)$$

$$= L(R(t)) + L(R(t)) \cdot L(m(t))$$

$$= L(R(t))\, [1 + L(m(t))]$$

$$= L(R(t)) \left[1 + \frac{L(w(t))\, L(g(t))}{1 - L(w(t))\, L(g(t))}\right] \text{ using } (10.10)$$

$$= \frac{L(R(t))}{1 - L(w(t))\, L(g(t))}$$

Since $R(t) = 1 - W(t)$ and as $\frac{d}{dt} W(t) = w(t) \Rightarrow L(W'(t)) = L(w(t)) \Rightarrow sL(W(t)) - W(0) = L(w(t)) \Rightarrow sL(W(t)) = L(w(t)) \Rightarrow L(W(t)) = \frac{L(w(t))}{s}$, we have

$$L(R(t)) = L(1 - W(t))$$

$$= L(1) - L(W(t)) = \frac{1}{s} - \frac{L(w(t))}{s} = \frac{1 - L(w(t))}{s}$$

Substituting, we get

$$L(A(t)) = \frac{1 - L(w(t))}{s(1 - L(w(t))\,.\,L(g(t))}.$$

10.3.2 Definition

Limiting Availability or simply **availability**, A is the probability of functioning at any specific time of a component. That is

$$A = \lim_{t \to \infty} A(t)$$

The limiting reliability is

$$\lim_{t \to \infty} R(t) = \lim_{t \to \infty} 1 - W(t) = 0$$

whereas the limiting availability is non-zero.

We know $$e^{-st} = 1 - st + \frac{(st)^2}{2!} + \ldots$$

$$\simeq 1 - st, \text{ for small values of } s.$$

By definition of Laplace transform

$$L(w(t)) = \int_0^\infty e^{-st}\, w(t)dt$$

$$\simeq \int_0^\infty (1 - st)\, w(t)dt$$

$$= \int_0^\infty w(t)dt - s \int_0^\infty t\, w(t)dt$$

$$= 1 - \frac{s}{\lambda}, \qquad \text{since } \int_0^\infty w(t)dt = 1$$

where $1/\lambda$ is MTTF. Also

$$L\,(g(t)) = \int_0^\infty e^{-st}\, g(t)dt$$

$$= \int_0^\infty g(t)dt - s \int_0^\infty tg(t)dt$$

$$= 1 - \frac{s}{\mu}$$

where $\frac{1}{\mu}$ is the mean time to repair (MTTR). We know by final value theorem of Laplace transform

$$\lim_{t \to \infty} A(t) = \lim_{s \to 0} sL(A(t)),$$

$$A = \lim_{s \to 0} sL\,(A(t))$$

$$= \lim_{s \to 0} \left[\frac{1 - L(w(t))}{1 - L(w(t))\, L(g(t))} \right]$$

$$= \lim_{s \to 0} \left[\frac{1 - \left(1 - \frac{s}{\lambda}\right)}{\left(1 - \frac{s}{\lambda}\right)\left(1 - \frac{s}{\mu}\right)} \right]$$

$$= \lim_{s \to 0} \frac{1/\lambda}{\frac{1}{s}\left(\frac{s}{\lambda} + \frac{s}{\mu} - \frac{s^2}{\lambda\mu}\right)}$$

$$= \frac{\frac{1}{\lambda}}{\frac{1}{\lambda} + \frac{1}{\mu}}.$$

$$= \frac{MTTF}{MTTF + MTTR}$$

This shows that the limiting availability depends on the mean time to failure and mean time to repair, and not on the nature of distributions of failure times and repair times.

10.3.3 Definition

Steady state availability is the steady-state probability that the system is in state i, the state with the system functioning properly.

10.3.4 Example

Consider a component with a constant failure rate λ. Upon a failure, it is repaired with an exponential repair-time distribution of parameter μ. Thus, the MTTF is $1/\lambda$ and MTTR is $1/\mu$. If the failure distribution is also exponential, show that

$$\lim_{t \to \infty} A(t) = \frac{\mu}{\lambda + \mu}.$$

Solution.

Given $w(t) = \lambda e^{-\lambda t}$; $g(t) = \mu e^{-\mu t}$. Now $L(w(t)) = \frac{\lambda}{s + \lambda}$; $L(g(t)) = \frac{\mu}{s + \mu}$.

$$L(A(t)) = \frac{1 - \frac{\lambda}{(s+\lambda)}}{s\left[1 - \frac{\lambda\mu}{(s+\lambda)(s+\mu)}\right]}$$

$$= \frac{s+\mu}{s(s+(\lambda+\mu))} = \frac{1}{\lambda+\mu}\left[\frac{(\lambda+\mu)(s+\mu)}{s(s+(\lambda+\mu))}\right]$$

$$= \frac{1}{\lambda+\mu}\left[\frac{\mu}{s} + \frac{\lambda}{s+(\lambda+\mu)}\right]$$

$$= \frac{\frac{\mu}{\lambda+\mu}}{s} + \frac{\frac{\lambda}{\lambda+\mu}}{s+(\lambda+\mu)}$$

$$\Rightarrow \qquad A(t) = L^{-1}\left(\frac{\frac{\mu}{\lambda+\mu}}{s}\right) + L^{-1}\left(\frac{\frac{\lambda}{\lambda+\mu}}{s+(\lambda+\mu)}\right)$$

$$= \frac{\mu}{\lambda+\mu} L^{-1}\left(\frac{1}{s}\right) + \frac{\lambda}{\lambda+\mu} L^{-1}\left(\frac{1}{s+(\lambda+\mu)}\right)$$

$$= \frac{\mu}{\lambda+\mu} + \frac{\lambda}{\lambda+\mu} e^{-(\lambda+\mu)t}$$

as $L^{-1}\left(\frac{1}{s}\right) = 1$ and $L^{-1}\left(\frac{1}{s+a}\right) = e^{-at}$. So the limiting availability is

$$\lim_{t\to\infty} A(t) = \frac{\mu}{\lambda+\mu}$$

when $\mu \to 0$

$$A(t) = e^{-\lambda t} = R(t).$$

Thus for a system which is not under repair, the availability and reliability are same.

Average Uptime Availability (or Mean Availability)

The mean availability is the proportion of time during a mission or time period that the system is available for use. It represents the mean value of the instantaneous availability function over the period (0, T] and is given by

$$\overline{A(t)} = \frac{1}{t}\int_0^t A(u)\, du.$$

Inherent Availability A_I

It is the steady state availability when considering only the corrective downtime of the system. For a single component, this can be computed by

$$A_I = \frac{MTTF}{MTTF + MTTR}$$

For a system,

$$A_I = \frac{MTBF}{MTBF + MTTR}$$

where *MTBF* is the mean time between failures.

Achieved Availability A_A

It is very similar to inherent availability with the exception that preventive maintenance down times are also included. Specifically, it is the steady state availability when considering corrective and preventive downtime of the system. It can be computed by looking at the mean time between maintenance actions,

MTBM and the mean maintenance downtime, $\overline{M}$, and

$$A_A = \frac{MTBM}{MTBM + \overline{M}}$$

Operational Availability A_0

It is a measure of the average availability over a period of time and it includes all experienced sources of downtime, such as administrative downtime, logitic downtime, etc. Operational availability is the ratio of the system uptime and total time. That is

$$A_o = \frac{\text{Uptime}}{\text{Operating cycle}}$$

where the operating cycle is the overall time period of operating being investigated and uptime is the total time the system was functioning during the operating cycle. When there is no specified logistic downtime or preventive maintenance, A_o returns the mean availability of the system.

10.4 MAINTAINABILITY, PREVENTIVE MAINTENANCE

All systems, except for non-recoverable ones such as unmanned satellites, Mars, etc which are used on a continuous or intermittent basis for any period of time, are subject to maintenance at some time or other. Properly maintained equipment can function at an effectively higher reliability than unmaintained equipment.

We classify maintenance as

(i) Off-schedule (imperfect)

(ii) Scheduled (preventive)

Off Schedule Maintenance

The system is not serviced after each mission but is allowed to proceed on the next mission, provided the system functions satisfactorily. Failed units are not replaced or repaired until the system fails.

Preventive Maintenance

It is performed at regular intervals for the purpose of keeping the system in a condition consistent with built-in levels of reliability, performance and safety. That is, it is performed to prevent component and system failure rates from exceeding design levels. It achieves its purpose by inspections, servicing, and minor and major overhauls during which the actions fit into three classes.

(1) Regular care of normally operating devices, components which require attention (lubrication, refueling, cleaning, etc).

(2) Checking for, with replacement or repair of failed redundant components.

(3) Replacement or overhaul of components or devices which are nearing wear out.

Periodic maintenance is performed every T hours, starting at time zero. Each device and component is checked. Each one which has failed is replaced by a new component. For useful life, the system is restored to "as good as new" condition after each maintenance operation.

For a periodic maintenance, the time period t can be written as

$$t = jT + \tau, \qquad j = 1, 2, \ldots, 0 \le \tau < T$$

For a time period $t = T$, we have

$$R_{PM}(t = T) = R(T).$$

For $j = 2$, $\tau = 0$, the system has to operate for the first T hours without failure. After all failed elements are replaced, T hours more of failure-free operations are necessary. Thus

$$R_{PM}(t = 2T) = [R(T)]^2$$

If $0 < \tau < T$, then τ hours more of failure-free operations are necessary, then reliability is

$$R_{PM}(t = 2T + \tau) = (R(T))^2 R(\tau)$$

In general,

$$R_{PM}(t = jT + \tau) = (R(T))^j R(\tau), \quad j = 1, 2, \ldots\ldots, 0 \le \tau < T$$

The MTTF for periodic maintenance is

$$\text{MTTF}_{PM} = \int_0^{\infty} R_{PM}(t)\, dt$$

$$= \sum_{j=0}^{\infty} \int_{jT}^{(j+1)T} R_{PM}(t)\, dt$$

If $t = jT + \tau$, then $dt = d\tau$ and the limits of the integral are zero and T. Hence

$$\text{MTTF}_{PM} = \sum_{j=0}^{\infty} \int_0^{T} R_{PM}(t)\, dt$$

$$= \sum_{j=0}^{\infty} \int_0^{T} (R(T))^j R(\tau) d\tau$$

$$= \frac{\int_0^{T} R(\tau)\, d\tau}{1 - R(T)}$$

Markovian Models in Reliability

The assumption that the failure process and the repair process, whenever repairs are considered are Markovian is often made. In such cases, we write Chapman-Kolmogorov equation of the Markov process of the resulting system. From the solutions of these equations, one can find the system reliability.

10.4.1 Example

Consider a two component-parallel system with a single repairman available for maintenance. Suppose that the failure and repair process are both Markovian with rates a and b respectively. Find the reliability of the system.

Solution. Here the process $\{X(t); t \ge 0\}$ has 3 states $i = 0, 1, 2$, the state i denoting the number of failed units at the instant t. The state 2 is an absorbing state. The process may be represented by a birth and death process having 3 states 0, 1, 2, state 2 being absorbing state.

Let $p_i = P\{X(t) = i\}$; the Kolmogorov equations are

$$p_0'(t) = -2ap_o(t) + bp_1(t)$$

$$p_1'(t) = -(a + b)\, p_1(t) + 2ap_0(t)$$

$$p_2'(t) = ap_1(t)$$

Let the initial conditions be $p_0(0) = 1$, $p_i(0) = 0$ for $i = 1, 2$. The above equations can be solved by taking their Laplace transform and using the initial conditions. First two equations can be written as

$$sL(p_o(t)) - p_0(0) = -2a\, L(p_o(t)) + bL\,(p_1(t))$$

$$sL(p_1(t)) - p_1(0) = -(a + b)\, L(p_1(t)) + 2aL\,(p_0(t))$$

which implies

$$(s + 2a)L(p_0(t)) - bL(p_1(t)) = 1$$

and

$$(s + a + b)\, L\, (p_1(t)) - 2aL\, (p_0(t)) = 0$$

Solving we get

$$L(p_1(t)) = \frac{2a}{s^2 + (3a+b)s + 2a^2}$$

$$L(p_0(t)) = \frac{a+b+s}{s^2 + (3a+b)s + 3a^2}$$

$L(p_1(t))$ can be written as

$$L(p_1(t)) = \frac{2a}{(s-s_1)(s-s_2)}, \text{ where}$$

$$s_i = \frac{-(3a+b) \pm \sqrt{(3a+b)^2 - 8a^2}}{2}, \ i = 1, 2$$

$$= \frac{-(3a+b) \pm \sqrt{(a+b)^2 + 4ab}}{2}$$

Thus

$$p_1(t) = L^{-1}\left[\frac{2a}{(s-s_1)(s-s_2)}\right]$$

$$= \frac{2a}{s_1 - s_2} L^{-1}\left(\frac{1}{s-s_1} - \frac{1}{s-s_2}\right)$$

$$= \frac{2a}{s_1 - s_2}\left[e^{s_1 t} - e^{s_2 t}\right]$$

Similarly we find

$$L(p_0(t)) = \frac{a+b+s}{(s-s_1)(s-s_2)}$$

$$= \frac{1}{s_1 - s_2}\left(\frac{a+b+s_1}{s-s_1} - \frac{a+b+s_2}{s-s_2}\right)$$

This implies

$$p_0(t) = \frac{1}{s_1 - s_2}\left[a + b + s_1)\, e^{s_1 t} - (a + b + s_2)\, e^{s_2 t}\right]$$

The system reliability is given by

$$R\,(t) = p_0(t) + p_1(t)$$

$$= \frac{(3a+b+s_1)\, e^{s_1 t} - (3a+b+s_2)\, e^{s_2 t}}{s_1 - s_2}$$

10.5 SOLVED PROBLEMS

1. How we calculate reliability by making use of binomial distribution.

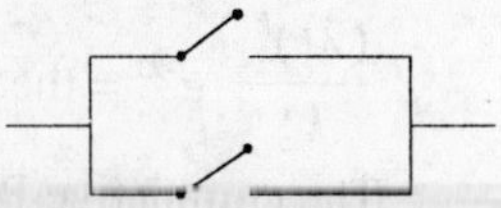

Solution. Consider two switches as in Fig.

Each switch functions properly (p) or it does not q.

Thus $p + q = 1$. The ways in which these functions

Both functions properly	p^2
One function properly and other does not	$pq + qp = 2pq$
Both not functioning	q^2

This covers all possible way of functioning. Thus

$$p^2 + 2pq + q^2 = 1$$

That is, $$(p + q)^2 = 1.$$

If we consider a system with $n = 3$ components, the possible ways of functioning are:

Three functions properly	p^3
Two functions properly one does not	$p^2q + pqp + qp^2 = 3p^2q$
One function properly and two does not	$pq^2 + qpq + q^2p = 3pq^2$
Three are not functioning	q^3

Adding gives $(p + q)^3 = 1$. Generalizing for system having n components, we get $(p + q)^n = 1$.

Now

$$P_n(k) = nC_k\, p^{n-k}\,(1 - p)^k$$

So the probability of an event not occurring k times in n trials, if the probability of the event in each trial is p.

2. How we make use of Poisson distribution for calculating reliability?

Solution. If the sample size is indefinite, it is not possible to use Binomial distribution. Equivalent to $(p + q)^n = 1$, we write it as $e^{-x}\, e^{x} = 1$.

That is $$e^{-x}\left(1 + x + \frac{x^2}{2!} + \ldots + \frac{x^n}{n!} + \ldots\right) = 1.$$

We interpretate as:

If x is the average number of occurrences of an event, then

e^{-x} = the probability that the event will not occur

xe^{-x} = the probability that the event will occur exactly once.

$\frac{x^2}{2!}\, e^{-x}$ = the probability that the event will occur exactly twice.

$\frac{x^n}{n!}\, e^{-x}$ = the probability that the event will occur exactly n times.

In reliability, the event is failure and the average number of failures in time t is given by λt when failure rate is constant. Consequetly, $e^{-\lambda t}$ is $R(t)$ for a system having a constant failure rate λ.

$(\lambda t)e^{-\lambda t}$ = the probability of exactly one failure occurring in time t.

$\frac{(\lambda t)^2}{2!}\, e^{-\lambda t}$ = the probability of exactly two failures occuring in time t.

$\frac{(\lambda t)^k}{k!} e^{-\lambda t}$ = the probability of exactly k failures occurring in time t.

The cumulative Poisson distribution (the probability of k or fewer failures) is given by

$$\sum_{k=0}^{\infty} \frac{e^{-\lambda t} (\lambda t)^k}{k!}$$

It is used to determine the probability of some specified number of failures during a given period or to calculate the number of spares required when units operate in useful life.

3. Consider Example 5 below with a second transmitter added in parallel with the first. Determine the new reliability.

Solution.

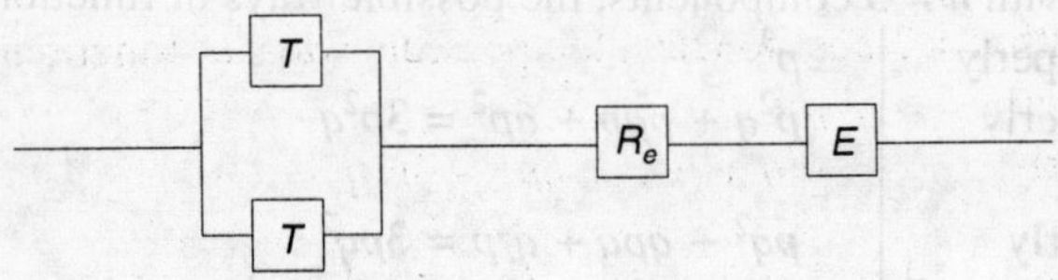

$$R_{\text{trans}} = 1 - (1 - 0.85)^2 = 0.9775$$

$$R_{\text{sys}} = (0.9775)\,(0.99)\,(0.94) = 0.9097.$$

4. Consider a three-element system (each element having a reliability of 0.90). Calculate the reliabilities of the four arrangements given below. Draw appropriate conclusions.

(a) *Series*

$R = (0.90)^3 = 0.729$

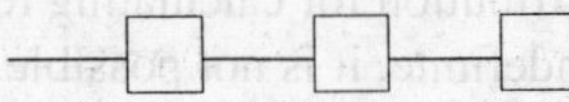

(b) *Parallel-series*

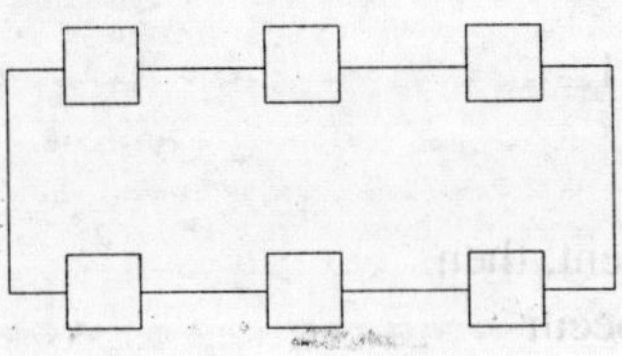

$$R = (1 - (1 - (0.90)^3)^2 = 0.927$$

(c) *Mixed-Parallel*

$$R = [1 - (1 - (0.90)^2]\,[1 - (1 - 0.90)^2] = 0.954$$

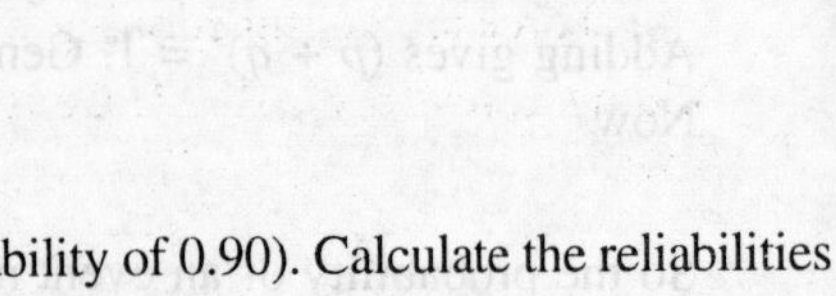

Series-Parallel:

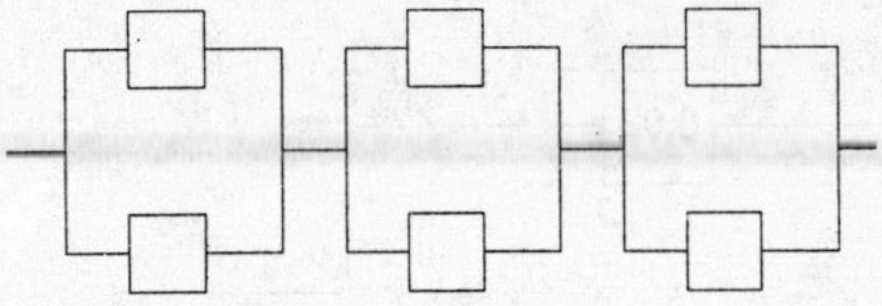

$$R = [1 - (1 - 0.90)^2]^3 = 0.970.$$

Conclusion: The reliability of a system with *m parallel* paths of *n* elements each (reliability of a single element is *p*) is

$$R = 1 - (1 - p^n)^m \qquad \text{(parallel-series)}$$

The reliability of a system of *n* series units with *m* parallel elements in each unit is

$$R = [1 - (1 - p)^m]^n \qquad \text{(series-parallel)}$$

For parallel-series system, there are only *m* possible paths and for series-parallel system there are m^n possible paths for system success.

We conclude that the reliability (*p*) of the units increases, the difference between parallel-series and series-parallel reliabilities decreases.

5. A system consists of four independent components in series, each having a realiability of 0.970. What is the reliability of the system?

Solution. $R = \prod_{i=1}^{4} R_i(t) = (0.970)^4 = 0.885.$

Consider a communication system consisting of a transmitter, a receiver, and an encoder. Failure of one causes system failure. During 8 hr of communication the unit operates for various periods.

6. Determine the reliability for maintaining communications without failure for 8 hr.

unit	Operating period *h*	Failure rate failures/*h*	λt	*Rt*
Transmitter	6	0.0267	0.16	0.85
Receiver	8	0.00125	0.01	0.99
Encoder	4	0.015	0.06	0.94
		Total	0.23	

Solution. $R_{sys} = e^{-\Sigma\lambda t} = e^{-0.23} = 0.79.$

7. Calculate the system reliability

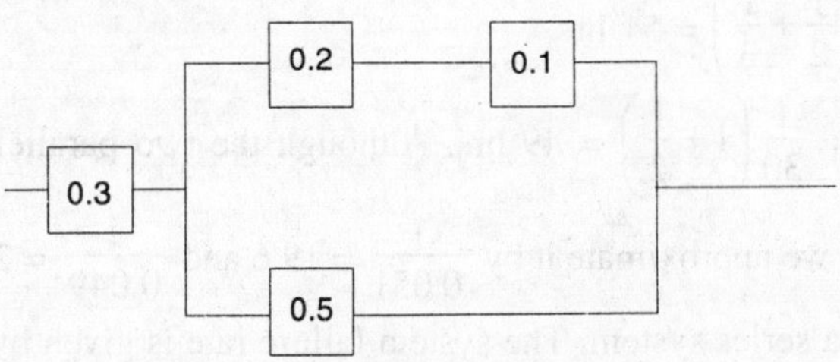

Solution. Let us replace each parallel path by an equivalent single path and ultimately reduce the system to one consisting of only series elements. Let *b* and *c* are reduced to equivalent element *e*, where

$$p(e) = p(b \text{ and } c) = p(b)\,p(c) = (0.2)\,(0.1) = 0.02$$

The system now takes the form

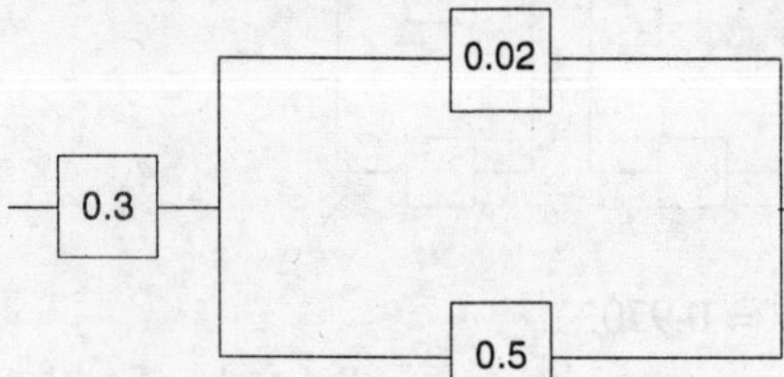

Let
$$p(f) = p(e \text{ or } d) = p(e) + p(d) - p(e) \cdot p(d)$$
$$= 0.02 + 0.5 - (0.02)\,(0.5) = 0.51$$

Now
$$p(s) = p(a \text{ and } f) = p(a)p(f)$$
$$= (0.3)\,(0.51) = 0.153.$$

8. Find the mean time to failure of the system, assuming the exponential model and that the reliabilities are given for 10 hours of operation

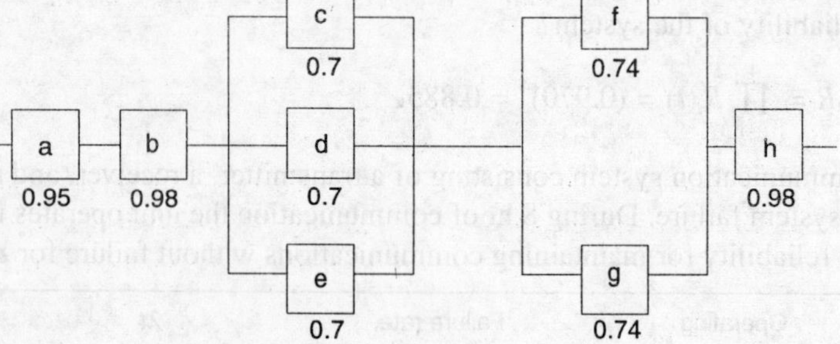

Solution. The failure rate of component 'a' is given by $0.95 = e^{-10\alpha} \Rightarrow \alpha = 5.1 \times 10^{-3}$ failures per hours. Similarly we can calculate the failure of each component is

Component	a	b	c	d	e	f	g	h
failure rate	5.1	2	35.7	35.7	35.7	30	30	2

The mean time between failures (MTBF) for the parallel system c, d, e is

$$\frac{1}{35.7}\left(1 + \frac{1}{2} + \frac{1}{3}\right) = 51 \text{ hrs.}$$

MTBF for f and $g = \frac{1}{30}\left(1 + \frac{1}{2}\right) = 49$ hrs. Although the two parallel assemblies do not have constant failure rates we approximate it by $\frac{1}{0.051} = 19.6$ and $\frac{1}{0.049} = 20$ per thousand hrs. Treat the entire system as a series system. The system failure rate is given by 5.1 + 2 + 19.6 + 20 + 2 = 48.7 failures per thousand hours. The mean time to failure of the system is $\frac{1}{48.7} = 0.02053 =$ 20 hrs (approx).

9. Find the hazard rate function $h(t)$ of the exponential distribution with pdf ae^{-at}, a > 0.
Solution. The hazard rate function is given by

$$z(t) = h(t) = \frac{f(t)}{R(t)} = \frac{-R'(t)}{R(t)}, \text{ where}$$

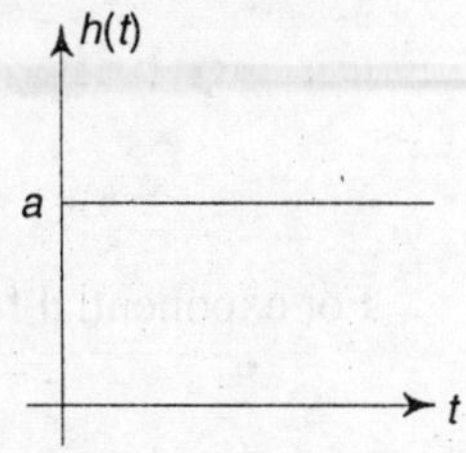

$$R'(t) = -f(t) = -ae^{-at} \text{ and}$$

hence
$$R(t) = \int R'(t)dt = \frac{-ae^{-at}}{-a} = e^{-at}. \text{ So}$$

$$h(t) = \frac{ae^{-at}}{e^{-at}} = a.$$

Hence the hazard rate is a constant.

10. Find the hazard rate function of the weibull distribution given by $a\alpha t^{n-1} e^{-at^{\alpha}}$, a > 0, $\alpha > 0$.

Solution.
$$h(t) = \frac{f(t)}{R(t)} = \frac{a\alpha t^{\alpha-1} e^{-at^{\alpha}}}{-\int a\alpha t^{\alpha-1} e^{-at^{\alpha}} dt}$$

Let $\int a\alpha t^{\alpha-1} e^{-at^{\alpha}} dt$. Put $t^{\alpha} = x$, then $\alpha t^{\alpha-1} dt = dx$. Now

$$\int a\alpha t^{\alpha-1} e^{-at^{\alpha}} dt = \int ae^{-ax} dx = \frac{-ae^{-ax}}{a} = -e^{-ax} = -e^{-at^{\alpha}}$$

Hence

$$h(t) = \frac{a\alpha t^{\alpha-1} e^{-at^{\alpha}}}{-(-e^{-at^{\alpha}})} = a\alpha t^{\alpha-1}.$$

11. Define the probability that the system does not fail in service before time t, $R_T(t)$.
Solution.
$$R_T(t) = R(t) \text{ for } t < T$$
$$= R(T) \text{ for } t = T$$

where $R(T)$ is the reliability of the system after an operating period of T hours.
After performing the first maintenance operation, the system becomes as good as new and therefore the reliability after the first T hours of operation is

$$R_T(t) = R(T)\, R(t - T)\; T < t$$
$$= (R(T))^2 \text{ for } t = 2T$$

Similarly, after two maintenance operations

$$R_T(t) = R(T)^2\, R(t - 2T),\; 2T < t$$
$$= R(T)^3 \text{ for } t = 3T$$

so on. In general at any interval $jT < t < (j + 1)T$ the reliability of the system is

$$R_T(t) = R(T)^j\, R(t - jT)$$

12. Find the mean time to in-service failure.

Solution.
$$\text{MTTSF} = \int_0^{\infty} R_T(t)dt$$

$$= \sum_{j=0}^{\infty} [R(T)]^j \int_{jT}^{(j+1)T} R(t-jT)dt$$

$$= \frac{\int_0^T R(x)\,dx}{1-R(T)}$$

For exponential failure time distribution,

$$\text{MTTSF} = \frac{\int_0^T (e^{\lambda x})^n\,dx}{1-e^{-n\lambda T}} = \frac{1}{n\lambda}.$$

That is, there is no effect of maintenance on the reliability of series systems.

13. If the system has m identical units in parallel with $m - 1$ units as redundant units, then

$$R(T) = 1 - (1 - p(T))^m$$

$$R_T(t) = \{1 - (1 - p(T)^m\}^j \{1 - [1 - P(t - jT)]^m\}$$

$$\text{MTTSF} = \frac{\int_0^T \{1-(1-p(x))^m\}\,dx}{(1-p(T))^m} = \frac{\int_0^T \{1-(1-e^{\lambda x})^m\}}{(1-e^{-\lambda T})^m}$$

If there is no maintenance $T = \infty$ and hence

$$\text{MTTSF} = \frac{1}{\lambda}\sum_{r=1}^{m}\frac{1}{r}.$$

which is nothing but MTTF of an unmaintained in unit parallel system.
For a single unit system

$$\text{MTTSF} = \frac{1}{\lambda}$$

and therefore maintenance has no effect on a single unit system.

14. Define the constant hazard model.

Solution. The constant hazard model is of the form $h(t) = \lambda$, where λ is a constant and is independent of time.

An item with constant hazard rate have the following reliability and associated functions.

$$f(t) = \lambda e^{-\lambda t}$$

$$R(t) = e^{-\lambda t}$$

$$F(t) = 1 - e^{-\lambda t}$$

$$\text{MTTF} = \int_0^{\infty} e^{-\lambda t}\,dt = \frac{1}{\lambda}.$$

15. Find the Mean time to failure of a linear hazard model.

Solution. The linear hazard model takes the form $h(t) = bt$, $t > 0$, where b is a constant.

Now

$$R(t) = \exp\left[-\int_0^{\infty} bt\,dt\right] = \exp\left(\frac{-bt^2}{2}\right)$$

$$f(t) = bt \cdot \exp\left(\frac{-bt^2}{2}\right)$$

$$\text{MTTF} = \int_0^\infty e^{\frac{-bt^2}{2}}\, dt = \frac{\Gamma\left(\frac{1}{2}\right)}{2\sqrt{\frac{b}{2}}} = \sqrt{\frac{\pi}{2b}}.$$

16. Find the MTTF of a non linear hazard model.

Solution. A non linear hazard model takes the form $h(t) = at^b$, where a and b are constants.

$$R(t) = \exp\left[\frac{-at^{b+1}}{b+1}\right]$$

$$f(t) = at^b \cdot \exp\left[\frac{-at^{b+1}}{b+1}\right]$$

$$\text{MTTF} = \frac{\Gamma\left(\frac{1}{b+1}\right)}{(b+1)\left[\frac{a}{(b+1)}\right]^{\frac{1}{b+1}}}$$

17. Write the hazard rate for gamma model.

Solution. $$h(t) = \frac{\lambda}{(a-1)!}\,(\lambda t)^{a-1}\,\frac{e^{-\lambda t}}{R(t)},$$

a is a positive integer and λ is a positive constant

$$R(t) = \sum_{j=0}^{a-1} \frac{(\lambda t)^j}{j!} e^{-\lambda t}$$

$$f(t) = \frac{\lambda}{(a-1)!}\,(\lambda t)^{a-1}\, e^{-\lambda t}$$

$$\text{MTTF} = \frac{a}{\lambda}.$$

18. The mean time to failure of a particular type of component is 800 h. What is the probability that a similar component will fail in an operating time of

(a) 200 h (b) 400 h (c) 800 h (d) 1000 h.

Solution. Reliability,

$$R(t) = e^{-t/\text{MTTF}}$$

Therefore $$R(200) = e^{-\frac{200}{800}} = e^{-0.25} = 0.7788.$$

$$R(400) = e^{-\frac{400}{800}} = 0.6065$$

$$R(800) = e^{-1} = 0.3679$$
$$R(1000) = e^{-1000/800} = e^{-1.25} = 0.2865.$$

19. The early failure rate of a component is given by

$$h(t) = ae^{-bt}$$

Determine the probability of survival of the component from age T for a mission time t h, given that the component has survived up to age T.

Solution. Reliability of the component during the early failure period is

$$R(t) = \exp\left[-\int_0^t h(t)\,dt\right]$$
$$= \exp\left[-\int_0^t ae^{-bt}\,dt\right] = \exp\left[-\frac{a}{b}(1-e^{-bt})\right]$$

The probability of component survival upto age T and $T + t$ are:

$$R(T) = \exp\left[\frac{-a}{b}(1-e^{bt})\right]$$
$$R(T + t) = \exp\left[\frac{-a}{b}(1-e^{-b(T+t)})\right]$$

The probability of survival of the component in the interval T to $T + t$ hours is given by

$$R(T, t) = \frac{R(T+t)}{R(T)} = \frac{\exp\left[\frac{-a}{b}(1-e^{-b(T+t)})\right]}{\exp\left[\frac{-a}{b}(1-e^{-bT})\right]}$$

$$= \exp\left[-\left(\frac{a}{b}\right)e^{-bT}(1-e^{-bt})\right].$$

20. Find the hazard rate for the Weibull distribution and graph $h(t)$ versus time for $\alpha = 1, 2, 3$.

Solution. We know the pdf of Weibull distribution is $f(t) = \frac{\alpha}{\beta} t^{\alpha-1} e^{-\frac{t^\alpha}{\beta}}$, $0 \le t < \infty$, $\alpha > 0$, $\beta > 0$ and cumulative distribution function is

$$F(t) = 1 - e^{-\frac{t^\alpha}{\beta}}$$

Now

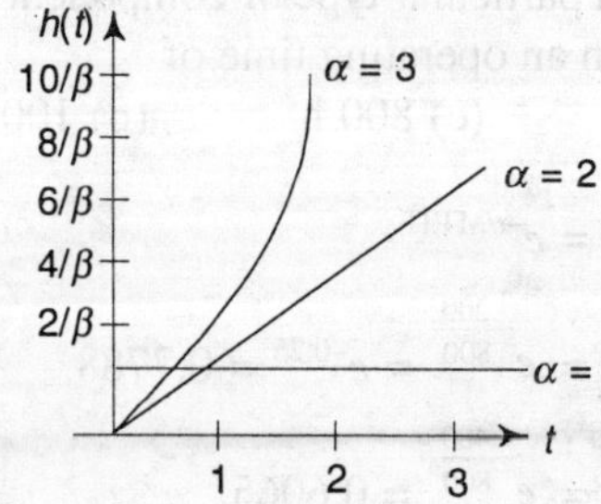

$$h(t) = \frac{f(t)}{1-F(t)} = \frac{\left(\frac{\alpha}{\beta}\right) t^{\alpha-1}\, e^{-t^{\alpha}/\beta}}{1-(1 \;\; e^{-t^{\alpha}/\beta})} = \frac{\alpha}{\beta}\, t^{\alpha-1}$$

when the shape parameter α is equal to 1, then

$$h(t) = \frac{1}{\beta}$$

when $\alpha = 2$, $h(t) = \frac{2}{\beta} t$; when $\alpha = 3$, $h(t) = \frac{3}{\beta} t^2$ a second order function of time t.

21. Find the reliability of the system shown in figure given that $p_A = .90$, $p_B = .95$, $p_C = .95$, $p_D = .92$, $p_E = .97$, $p_F = .92$, $p_G = .97$

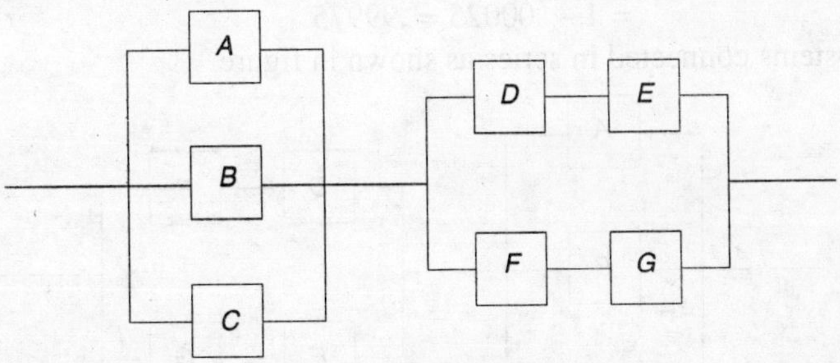

Solution. Since the reliabilities of the pairs of components (D, E), (F, G) are identical, it follows that the reliabilities of these two series subsystems are equal.

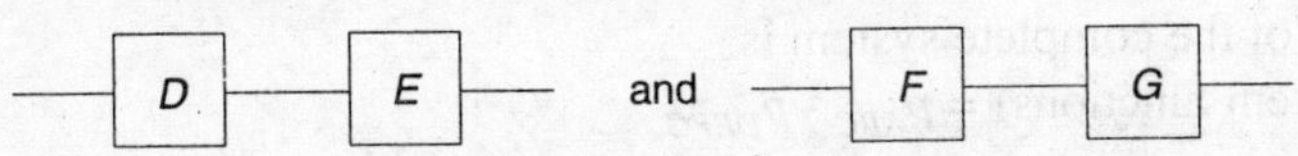

The reliability of these series is

$$p_{DE} = p_{FG} = (0.92)(0.97) = .8924.$$

The reliability of the parallel subsystem containing these two series subsystems

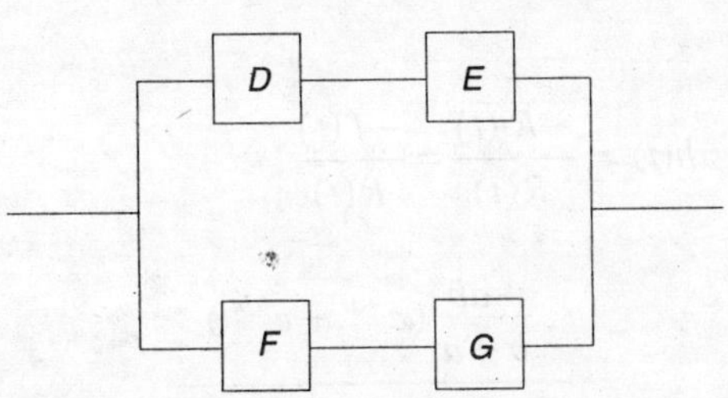

$$\begin{aligned} p_{DEFG} &= 1-(1-p_{DE})(1-p_{FG}) \\ &= 1-(1-.8924)(1-.8924) \\ &= 1-.0115778 = .9884 \end{aligned}$$

Next we compute the reliability of the parallel subsystem consisting of components A, B, C,

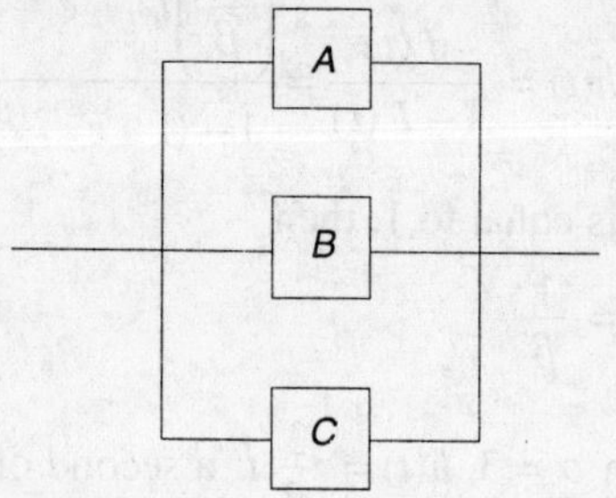

$$p_{ABC} = 1 - (1 - p_A)(1 - p_B)(1 - p_C)$$
$$= 1 - (1 - .90)(1 - .95)(1 - .95)$$
$$= 1 - .00025 = .99975$$

The two subsystems connected in series as shown in figure

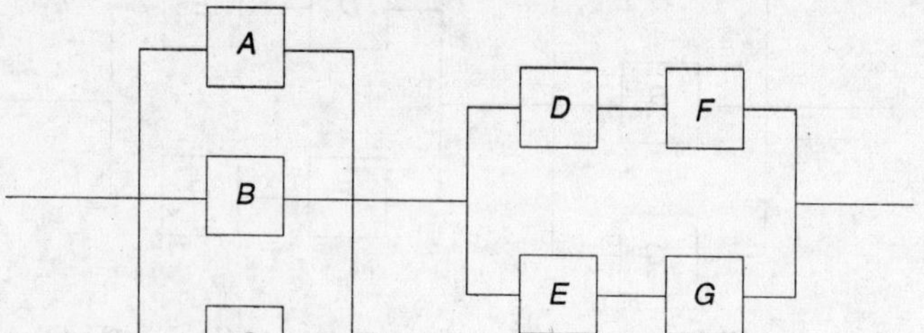

The reliability of the complete system is

$$P(\text{system functions}) = p_{ABC} \cdot p_{DEFG}$$
$$= (.99975)\,(.9884) = .9881.$$

22. Find the hazard rate of the distribution

$$f(t) = \frac{ab}{b-a}\,[e^{-at} - e^{-bt}],\ t > 0, \quad a \neq b.$$

Solution.

$$h(t) = \frac{-R'(t)}{R(t)} = \frac{-f(t)}{R(t)}$$

$$= \frac{\dfrac{-ab}{b-a}\,(e^{-at} - e^{-bt})}{\left(\dfrac{ab}{b-a}\right)\left(\dfrac{e^{-at}}{-a} + \dfrac{e^{-bt}}{b}\right)}$$

$$= \frac{ab\,(e^{-at} - e^{-bt})}{be^{-at} - ae^{-bt}}$$

23. The density function is given by

$$f(t) = \frac{100}{(t+20)^3}, \ t \ge 0, \text{ derive } R(t) \text{ and } R(1).$$

Solution. Let
$$f(t) = \frac{100}{(t+20)^3}, \ t \ge 0$$

$$R(t) = \int_t^{\infty} f(t)dt$$

$$= \int_t^{\infty} \frac{100}{(t+20)^3} dt = \int_{\frac{1}{z}-20}^{0} 100z(-dz),$$

by putting
$$z = \frac{1}{t+20}, \ dz = \frac{-dt}{(t+20)^2}$$

$$= -100 \int_{\frac{1}{z}-20}^{0} zdz \quad = -100\left(\frac{z^2}{2}\right)_{1/z-20}^{0}$$

$$= -100\left(-\frac{1}{2}\left(\frac{1}{z} - 20\right)^2\right) = 50t^2$$

Hence
$$R(1) = 50.$$

24. In a computer system, MTTF of a particular chip is 1000 h. What is the probability that a similar chip will fail in the operating time.

(a) 100 h (b) 200 h (c) 400 h (d) 800 h (e) 1000 h.

Solution. $R(t) = e^{-t/\text{MTTF}}$, MTTF = 1000 h.

(a) $R(100) = e^{-\frac{100}{1000}} = e^{-\frac{1}{10}} = 0.9048$

(b) $R(200) = e^{-\frac{200}{1000}} = e^{-\frac{2}{10}} = 0.8187$

(c) $R(400) = e^{-\frac{400}{1000}} = e^{-\frac{4}{10}} = 0.6703$

(d) $R(1000) = e^{-\frac{-1000}{1000}} = e^{-1} = 0.3678$

EXERCISES

1. Given three components with respective reliabilities $R_1 = 0.8$, $R_2 = 0.75$ and $R_3 = 0.98$. Compute the reliabilities of the two systems, show below.

(i)

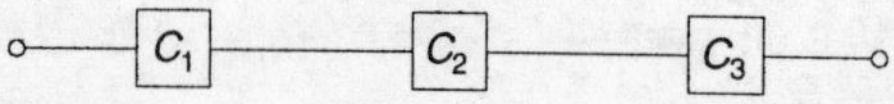

(ii)

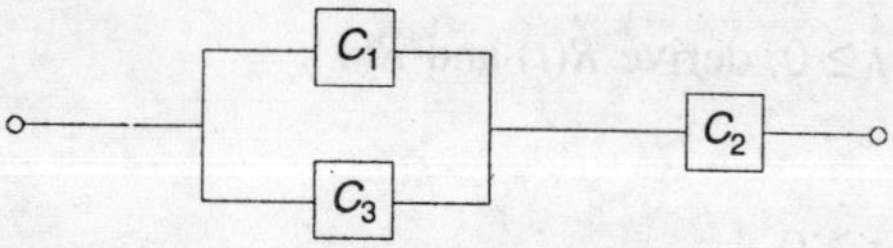

2. The failure rate of a certain component is $h(t) = \lambda_0 t$, where $\lambda_0 > 0$ is a given constant. Determine the reliability $R(t)$ of the component.
3. Weibull distribution. It is a failure distribution with density given by $f(t) = \lambda \alpha t^{\alpha-1} e^{-\lambda t^{\alpha}}$ and distribution function is $F(t) = 1 - e^{-\lambda t\alpha}$ with hazard rate $h(t) = \lambda \alpha t^{\alpha-1}$ and cumulative hazard is a power function $H(t) = \lambda t^{\alpha}$, where α is a shape parameter.
If the lifetime X in hours of a component is modeled by a Weibull distribution with $\alpha = 2$, and starting with a longer number of components, it is observed that 15% of the components that have lasted 90 hours fail before 100 hours. Determine the parameter λ.
[*Hint:* Given $P(X < 100/X > 90) = 0.15$, so

$$P(X < 100 \mid X > 90) = \frac{P(90 < X < 100)}{P(X > 90)} = \frac{F_X(100) - F_X(90)}{1 - F_X(90)}$$

$$= \frac{e^{-\lambda (90)^2} - e^{-\lambda (100)^2}}{e^{-\lambda (90)^2}} \Bigg]$$

4. A system consists of 100 units in series, each unit having a reliability of 0.99. What is the reliability of the system.
5. A system has n non-redundant components; their failure rates λ_i are constant, but all are different. Find the reliability and MTTF of the system.
6. Find the MTBF of the system if five components in series have constant failure rates of 1.4, 1.7, 1.9, 1.2 and 1.6 failures per 1000 h respectively.
7. A given component has a constant failure rate of 0.0050 failure/h. Determine
   (a) Reliability of one unit for an operating time of 300 hr.
   (b) Reliability of two units in series for an operating time of 150 hr.
   (c) Reliability of four units in series for an operating time of 75 hr.
8. A system consists of five identical components connected in parallel. What must be the reliability of each component, if the reliability of the system must be 98%.
9. A system has 5 identical components connected in series. It is desired that the system reliability is 0.95. Determine how good each component should be ?
10. A parallel system is composed of 8 identical independent components. If the system reliability is 0.95, how poor the components be.
11. Find the MTTF of the system, assuming the exponential model and that the reliabilities are given for 10 hours of operation.

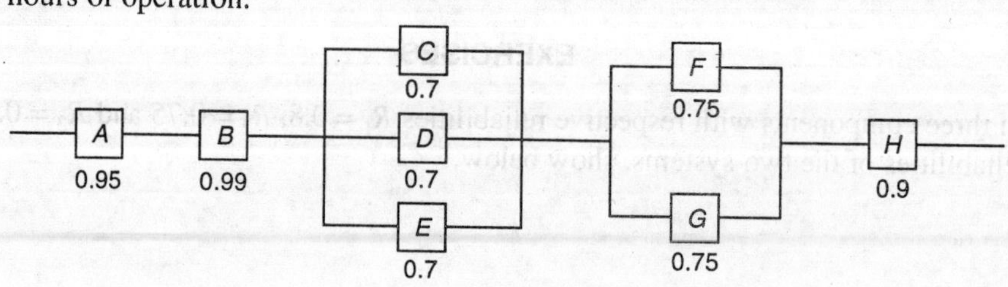

12. Consider 4 identical units connected in parallel having the reliability 0.7. The successful operator of the system depends on the performance of any one of the unit. Determine the system reliability.

13. A system with three independent components works correctly if atleast one component is functioning properly. Failure rates of the individual components are $\lambda_1 = 0.01$, $\lambda_2 = 0.002$, $\lambda_3 = 0.04$ (assume exponential lifetime distributions).
(*a*) Determine the probability that the system will work for 100 hours.
(*b*) Determine the density functions of the lifetime X of the system.

14. Consider a series system consisting of n independent components. Assuming that the lifetime of the ith component is Weibull distributed with parameters λ_i and α, show that the system lifetime also has a Weibull distribution.

15. Consider a job consisting of three tasks. Tasks 1 and 2 are non-interfering and hence can be executed in parallel. Task 3 cannot be started until both task 1 and 2 have completed. If T_1, T_2, T_3 respectively, denote the times of execution of three tasks, then the time of execution of the entire job is given by

$$T = \max\{T_1, T_2\} + T_3 = M + T_3.$$

[**Hint.** Assume T_1, T_2 are continuous random variables with uniform distribution over $[t_1 - t_0, t_1 + t_0]$, T_3 is a continuous random variable uniformly distributed over $[t_3 - t_0, t_3 + t_0]$. Assume that T_1, T_2, T_3 are mutually independepent. We have to find the probability that

$T > t_1 + t_3$. $F_M(m) = P[M \le m] = P[\max(T_1, T_2) \le m]$

$$= \phi(T_1 \le m, \text{ and } T_2 \le m) = P(T_1 \le m)\, P(T_2 \le m)$$

$$= F_{T_1}(m) \,.\, F_{T_2}(m) = \begin{cases} 0, & m < t_1 - t_0 \\ \dfrac{(m - t_1 + t_0)^2}{4t_0^{\,2}}, & t_1 - t_0 \le m < t_1 + t_2 \\ 1, & m \ge t_1 + t_0 \end{cases}$$

$$F_{T_1}(t) = \begin{cases} 0, & t < t_1 - t_0 \\ \dfrac{t - t_1 + t_0}{2t_0^{\,2}}, & t_1 - t_0 \le t < t_1 + t_0 \\ 1, & \text{otherwise} \end{cases}$$

and
$$f_M(m) = \begin{cases} \dfrac{m - t_1 + t_0}{2t_0^{\,2}}, & t_1 - t_0 < m < t_1 + t_0 \\ 0, & \text{otherwise} \end{cases}$$

$$P(T > t_1 + t_3) = \iint_A f_{M,T_3}(m, t)\, dm\, dt$$

$$= \iint_A f_M(m)\, f_{T_3}(t)\, dm\, dt$$

$$= \frac{2}{3}.$$

16. Consider the triple modular redundant (TMR) system ($n = 3$, $m = 2$). If we define a "short" mission by the mission time $t \le t_0$, then prove that TMR type of redundancy improves reliability only for short missions, not for long missions.
[**Hint.** Reliability of a single component is given by

$R(t) = e^{-\lambda t}$, then $R_{TMR}(t) = 3R^2(t) - 2R^3(t)$

$$= 3e^{-2\lambda t} - 2e^{-3\lambda t}.$$

For $0 \le t < t_0$, $R_{TMR}(t) \ge R(t)$, and for $t_0 \le t < \infty$, $R_{TMR}(t) \le R(t)$. Then if t_0 is the solution of the equation,

$$3e^{-2\lambda t_0} - 2e^{-3\lambda t_0} = e^{-\lambda t_0} \Rightarrow t_0 = \frac{0.7}{\lambda}\Bigg]$$

17. Obtain an expression for the reliability of a two component stand by redundant system.

$$\left[\textbf{Ans. } R(t) = 1 - F(t) = 1 - \sum_{k=0}^{1} \frac{(\lambda t)^k}{k!} e^{-\lambda t} = (1 + \lambda t)e^{-\lambda t}, t \ge 0\right]$$

18. Prove that the TMR/simplex has a higher reliability then either a simplex or an ordinary TMR system for all $t \ge 0$.

It detects a single component failure, discards the failed component, and reverts to one of the non failing simplex components.

[**Hint.** Let X, Y, Z, denotes the times to failure of three components. Let W denote the residual time to failure of the selected on failure component. Let X, Y, Z be mutually independent, and exponentially distributed with parameter λ. Min {X, Y, Z} is exponentially distributed with parameters 3λ. Let L = Min {X, Y, Z} + W. Then L has 2-stage hypoexponential distribution with parameters 3λ and λ. Then

$$F_L(t) = 1 - \frac{3\lambda}{2\lambda} e^{-\lambda t} + \frac{\lambda}{2\lambda} e^{-3\lambda t} = 1 - \frac{3e^{-\lambda t}}{2} + \frac{e^{-3\lambda t}}{2}.$$

So $$R(t) = 1 - F_L(t) = \frac{3e^{-\lambda t}}{2} - \frac{e^{-3\lambda t}}{2}.$$

The following diagram compares TMR/Simplex, TMR.

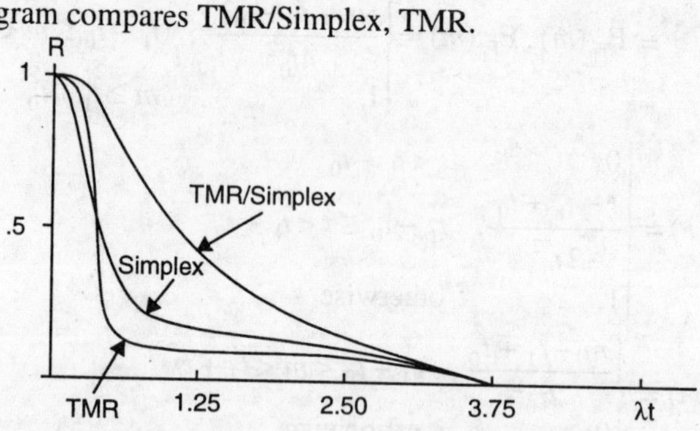

19. Compare the TMR/Simplex reliability with two-component and three component redundant systems having stand by redundancy. Graph the expressions on the same plot.

20. Compare reliability expressions for the simplex system, two component parallel redundant system, and two component stand by redundant system. Assume that the failure rate of an active component is constant at $\frac{1}{1000}$ per hour, the failure rate of a spare is zero, and that the switching mechanism is fault free.

Answers

9. $P^5 = 0.95 \Rightarrow 5 \log P = \log (0.95) \Rightarrow P = \log (0.95)^{1/5}$.

10. $(1 - P(x))^8 = 0.05 \Rightarrow P(x) = 1 - (0.05)^{1/8}$

11. MTBF for the parallel system C, D, E = 51 hrs.

MTBF for F, G = 52 hrs.

Model Questions

PART A (20 × 1 = 20 MARKS)

1. If A and B are independent events then $P(A \cup B) = P(A) + P(B)$.
2. If A and B are any two events, then $P(A \cup \overline{B}) = 1 - P(A)\, P(B/A)$.
3. If A and B are any two events, then $P(A \cap B) \geq P(A) + P(B) - 1$.
4. Distinguish between mutually exclusive and independent of two events.
5. If X and Y are two random variables, then Var $(X + Y)$ = ______.
6. For Poisson distribution with parameter λ, $E(X)$ = ______ and Var (X) = ______.
7. If the correlation coefficient $\rho_{XY} = 0$, then X and Y are ______.
8. Suppose X and Y are independent random variables, then $E\left(\frac{X}{Y}\right)$ = ______.
9. State two properties of cumulative distribution function.
10. If X and Y are independent random variables then $P(X \leq x, Y \leq y)$ = ______.
11. State the Chebyshev inequality.
12. Define the convergence in mean square.
13. State the weak law of large numbers.
14. State two properties of the Auto-correlation function.
15. Give an example for non-stationary Random process.
16. Define the cross-sectional density function.
17. Define the Gaussian process.
18. Define covariance functions of the random process.
19. When do you say that a random process is second order stationary?
20. Define Poisson process and state one of its properties.

PART B (5 × 12 = 60 MARKS)

21. (a) (i) Let A and B be two independent events defined on some probability space and let $P(A) = 1/3$, $P(B) = 3/4$. Find $P(A \cup B)$, $P(A/A \cup B)$ and $P(B/A \cup B)$.

 (ii) A box of fuses contain 20 fuses, of which 5 are defective. If 3 of the fuses are selected at random and removed from the box in succession without replacement. What is the probability that all three fuses are defective?

OR

 (b) (i) Suppose there are three boxes containing 2 white and 3 black balls; 3 white and 2 black, and 4 white and 1 black balls respectively. There is equal probability of each box being chosen. One ball is drawn from the box chosen at random. What is the probability that a white ball is drawn?

 (ii) Show that $P(A \cap B) \leq P(A) \leq (A \cup B) \leq P(A) + P(B)$.

22. (a) (i) The p.d.f. of the R.V. X is given by

$$f(x) = \begin{cases} \dfrac{10}{x^2} & \text{for } x > 10 \\ 0, & \text{for } x \le 10 \end{cases}$$

What is the cumulative distribution function of X? Find $P(X > 20)$ and $P(20 < X < 30)$.

(ii) The joint p.d.f. of X and Y is given by

$$f(x, y) = \begin{cases} \dfrac{e^{\frac{-x}{y}} e^{-y}}{y}, 0 < x < \infty, 0 < y < \infty \\ 0 \qquad \text{otherwise} \end{cases}$$

Find $E\left(\dfrac{X}{Y} = y\right)$?

(OR)

(b) (i) If X has an exponential distribution, prove that $P(X > s + t) = P(X > s)\, P(X > t)$.

(ii) The joint p.d.f. of X and Y is given by

$$f(x, y) = \begin{cases} \dfrac{1}{2} ye^{-xy}, 0 < x < \infty, 0 < y < 2 \\ 0 \qquad , \text{otherwise} \end{cases}$$

What is $E\left(e^{\frac{x}{2}} / Y = 1\right)$?

23. (a) (i) Let X be a continuous random variable with p.d.f.

$$f(x) = \begin{cases} 1, 0 < x < 1 \\ 0, \text{otherwise} \end{cases}$$

Find the p.d.f. of the random variable $Y = X^3$ and hence find $E(Y)$.

(ii) Explain why there can be no random variable X for the moment generating function $M_X(t) = \dfrac{t}{1-t}$.

OR

(b) (i) Show that the distribution for which the characteristic function is $e^{-|t|}$ has the density function

$$f(x) = \frac{1}{\pi} \frac{1}{1 + x^2} \quad \text{for } -\infty < x < \infty.$$

(ii) A random variable X has the density function $f(x) = e^{-x}$, $x \ge 0$. Show that Chebyshev's inequality is $P[|x - 1| > 2] < \dfrac{1}{4}$ and show that the actual probability is e^{-3}.

24. (a) (i) Find the spectral density function $S(w)$ of the auto-correlation function of random process is given by

$$R_{XX}(\tau) = \begin{cases} 1 - \dfrac{|\tau|}{T} & \text{for } |\tau| < T \\ 0 & \text{otherwise.} \end{cases}$$

(ii) Consider the process $X(t) = A \cos wt + B \sin wt$, where A and B are uncorrelated random variables each with mean 0 and variance 1 and w is a positive constant. Show that the process $X(t)$ is covariance stationary.

Or

(b) (i) Prove that the auto-correlation function $R_{XX}(\tau)$ takes maximum value at $\tau = 0$.

(ii) Prove that the random process $X(t) = A_0 \cos (w_0 t + \theta)$ where, A_0 and w_0 are constant and θ is uniformally distributed in the interval on $[-\pi, \pi]$ is wide sense stationary.

25. (a) (i) The auto-correlation function of a random process $X(t)$ is given by $R(\tau) = ae^{-b^2\tau^2}$ where 'a' and 'b' are constants. Evaluate the spectral density function.

(ii) Find the auto-correlation function of a random telegraph signal.

Or

(b) (i) Consider the random phenomenon called the random telegraph signal that takes the value either +1 or –1 by changing sign in a random fashion. Assume that the signal is +1 at time $t = 0$, and that the probability of the number of changes of sign in time t, denoted by k, follows the Poisson process:

$$P(X(t) = k) = \frac{(\lambda t)^k}{\angle k} e^{-\lambda t},\ k = 0, 1, 2, \ldots, \lambda > 0.$$

Show that the average number of sign change in time t is equal to $e^{-2\lambda t}$

(ii) State two properties of cross-spectral density function.

PART A (20 × 1 = 20 MARKS)

State True or False:

1. $P(\bar{A} \cap B) = P(B) - P(A \cap B)$.
2. If $P(A) > P(B)$ then $P(A/B) < P(B/A)$.
3. If A and B are independent then A and $\bar{B}$ are also independent.
4. If A and B are mutually exclusive events then $P[A/\bar{B}] = P(A)/[1 - P(B)]$.
5. $(A \cap B) \cap (\bar{B} \cap C) = \phi$.
6. State the properties of probability density function of a one dimensional random variable.
7. If $F(x)$ is the cumulative distribution function of a random variable X, then $x_1 < x_2$ implies $F(x_1) = F(x_2)$. True or false.
8. Define the conditional probability density of Y given X.
9. Define Binomial random variable.
10. If X and Y are independent then $E(XY) = E(X) \cdot E(Y)$. True or false.
11. Define correlation coefficient between two random variables X and Y.
12. If $\phi_X(t)$ is the characteristic function of a random variable X, then $\phi_X(0) = \ldots$..
13. Define the moment generating function of a one dimensional discrete random variable.
14. Define convergence in mean square.
15. Almost certain convergence implies convergence in probability—True or false.
16. Define Stationary Random Process.

17. When do you say that a random process is second order stationary?
18. Define a Gaussian process.
19. What is low pass noise?
20. Define Bandpass Noise.

PART B (5 × 12 = 60 MARKS)

21. (a) A and B throw alternately with a pair of dice. A wins if he throws 6 before B throws 7 and B wins if he throws 7 before A throws 6. If A begins find the probability of A winning.

Or

(b) Show that for any three events, A, B, C

$$P(A \cup B \cup C) = P(A) + P(B) + P(C) - P(A \cap B) - P(A \cap C) - P(B \cap C) + P(A \cap B \cap C).$$

22. (a) A continuous random variable can assume values between 2 and 5 and has density function $f(x) = k(1 + x)$. Find the value of k, $P(x < 4)$ and $P(x > 4)$.

Or

(b) The joint probability density function of the two dimensional variables (X, Y) is given by

$$f(x, y) = \begin{cases} \frac{8}{9}xy; 1 \le x \le y \le 2 \\ 0 \quad ; \text{elsewhere.} \end{cases}$$

(i) Find the marginal density functions of X and Y.
(ii) Find the conditional density functions of Y given $X = x$, and the conditional density functions of X given $Y = y$.

23. (a) A random variable X has the density function $f(x) = e^{-x}$; $x \ge 0$. Show that Chebyshev's inequality gives $P[|X - 1| > 2] < \frac{1}{4}$ and show that the actual probability is e^{-3}.

Or

(b) Prove that the random process $X(t) = A \cos (w_0 t + \theta)$ is wide sense stationary.

24. (a) Calculate the power spectral density of a stationary random process for which the auto-correlation is $R_{XX}(\tau) = \sigma^2 e^{-\alpha|\tau|}$.

Or

(b) Given a stationary random process $X(t) = 10 \cos [100\, t + \theta]$ where θ is a random variable uniformly distributed in the interval $[-\pi, \pi]$. Show that $X(t)$ is ergodic in the auto-correlation function.

25. (a) The occurrence of a pulse follows Poisson process. Calculate the auto-correlation function such a random telegraph signal.

Or

(b) A circuit has an impulse response given by

$$h(t) = \frac{1}{T}; 0 \le t \le T = 0; \text{elsewhere}$$

Evaluate $S_{YY}(w)$ interms of $S_{XX}(w)$.

PART A

1. If $P(A) = \frac{3}{4}$, $P(B) = \frac{5}{8}$ prove that $P(A \cap B) \ge \frac{3}{8}$.

2. Define mutually exclusive events.
3. If $B \subset A$ then prove that $P(B) \leq P(A)$.
4. If $P(A) = \frac{1}{3}$, $P(B) = \frac{3}{4}$ and $P(A \cup B) = \frac{11}{12}$ find $P(A/B)$ and $P(B/A)$.
5. If $p(x) = \begin{cases} 0.1\,x, & x = 1, 2, 3, 4 \\ 0, & \text{elsewhere} \end{cases}$ find $p\{((1/2 < x < 5)/2x > 1\}$.
6. Verify whether $f(x) = \begin{cases} |x| \text{ in } -1 \leq x \leq 1 \\ 0 \text{ elsewhere} \end{cases}$, can be the p.d.f. of a continuous random variable.
7. Given the following bivariate probability distribution, obtain marginal distributions of X and Y:

Y \ X	–1	0	1
0	$\frac{1}{15}$	$\frac{2}{15}$	$\frac{1}{15}$
1	$\frac{3}{15}$	$\frac{2}{15}$	$\frac{1}{15}$
2	$\frac{2}{15}$	$\frac{1}{15}$	$\frac{2}{15}$

8. Find a constant $b > 0$ so that function $f(x) = \begin{cases} \frac{e^{3x}}{4}, & 0 \leq x \leq b \\ 0 & \text{elsewhere} \end{cases}$ is a valid density function.
9. If the p.d.f. of X is $f_X(X) = e^{-X}$, $X > 0$ find the p.d.f. of $Y = 2X + 1$.
10. Let X and Y be two random variables such that $Y \leq X$ then prove that $E(Y) \leq E(X)$.
11. If the joint p.d.f. X and Y is given by

$$F_{XY}(X, Y) = \begin{cases} \frac{1}{Y}\, 0 < X < 1, 0 < Y < 1 \\ 0 \text{ elsewhere} \end{cases} \text{ find } E[X/Y].$$

12. A random variable X has density function e^{-x} for $x \geq 0$. Show that $P[|x - 1| > 2] < 1/4$ using Chebyshev's inequality.
13. What is the difference between a random variable and a random process?
14. When are two random processes said to be orthogonal?
15. Define Ergodic random process.
16. What is cross power density spectrum?
17. State the properties of a Gaussian process.
18. Define band-limited white noise.
19. How is filtering done in electrical system?
20. Prove that two independent Poisson processes is not a Poisson process.

PART B (5 × 12 = 60 MARKS)

21. (a) A throw is made with three dice. What is the probability that the sum of the numbers thrown will be 12?

    (b) One card is drawn from a regular deck of 52 cards. What is the probability that of the card being either red or a king?

Or

22. (a) It is known that in a city 3 out of 5 people are in favour of a certain proposal. What is the chance that if 5 people are chosen at random there will be a majority amongst them against the proposal?

(b) If A, B, C are mutually independent events then $A \cup B$ and C are also independent.

23. (a) A random variable X has the following probability function:

Values of x, x:	0	1	2	3	4	5	6
$p(x)$:	k	$3k$	$5k$	$7k$	$9k$	$11k$	$13k$

(i) Find k

(ii) Evaluate $P(x < 4)$, $P(x \geq 5)$, and

(iii) $P(3 < x < 6)$.

(b) If $f(x, y) = \begin{cases} e^{-(x+y)} & x \geq 0,\ y \geq 0 \\ 0 & \text{elsewhere} \end{cases}$ is the joint p.d.f. of random variables X and Y. Find

(i) $P(X < 1)$,

(ii) $P(X < 1 \cap Y < 3)$, and

(iii) $P\left(X < \frac{1}{Y} < 3\right)$.

Or

24. (a) If the joint probability function of X and Y is $f(x, y) = \begin{cases} x + y & 0 < x < 2, 0 < y < 1 \\ 0 & \text{elsewhere.} \end{cases}$ Show that X and Y are stochastically dependent.

(b) Verify that the following is a distribution function

$$F(x) = \begin{cases} 0, & x < -a \\ \frac{1}{2}\left(\frac{x}{a} + 1\right), & -a < x < a \\ 1, & x > a. \end{cases}$$

25. (a) Two random variables X and Y have the following Joint probability density function

$$f(x, y) = \begin{cases} 2 - x - y; & 0 \leq x \leq 1, 0 \leq y \leq 1 \\ 0 & \text{otherwise.} \end{cases}$$

Find:

(i) Marginal probability density functions of X and Y.

(ii) Var (X) and Var (Y).

(iii) Covariance between X and Y.

(b) Find the characteristic function of the Poisson distribution.

Or

26. (a) Find the density function $f(x)$ corresponding to the characteristic function defined as

$$\phi(t) = \begin{cases} 1 - |t|, & \text{for } |t| \leq 1 \\ 0 & \text{, for } |t| > 1. \end{cases}$$

(b) A fair coin is tossed 10 times. Find the probability of getting 3 or 4 or 5 heads using Central limit theorem.

27. (a) Prove that the random process $X(t) = A \cos (w_c t + \theta)$ is not stationary if it is assumed that A and w_c are constants and θ is uniformly distributed variable on the interval $(0, \pi)$.

(b) If $[X(t)]$ is a wise sense stationary process with auto-correlation $R(\tau) = Ae^{-\alpha(\tau)}$, determine the second order moment of the random variable $X(8) - X(5)$.

Or

28. (a) The auto-correlation function of a wide sense stationary random process is given by $R_{XX}(\tau) = \dfrac{4\tau^2 + 100}{\tau^2 + 4}$. Find the mean and variance of this process.

(b) The auto-correlation function of the random telegraph signal process is given by $R(\tau) = a^2 e^{-2r|(t)|}$. Determine the power density spectrum of the random telegraph signal.

29. (a) Assume that a normal stationary process $X(t)$ with zero mean and auto-correlation $R_{XX}(\tau)$ is passed through a system which gives the output $Y(t) = X^2(t)$. Show that the auto-correlation function $Y(t)$ is given by $R_{YY}(\tau) = R_{XX^2}(0) + 2 R_{XX^2}(\tau)$.

(b) If customers arrive at a counter in accordance with a Poisson process with a mean rate of 2 per minute find the probability that the interval between two consecutive arrivals is

(i) more than 1 minute

(ii) between 1 minute and 2 minutes

(iii) 4 minutes or less.

Or

30. (a) If $X(t)$ is a stationary random process with derivative $X'(t)$ then for a given t, prove that the random processes $X(t)$ and $X'(t)$ are orthogonal and uncorrelated.

(b) Prove that if the narrow band noise $n(t)$ is Gaussian then its inphase component $n_1(t)$ and quadrature component $n_Q(t)$ are jointly Gaussian.

PART A

1. Given $P(A) = \frac{1}{3}$, $P(B) = \frac{1}{4}$, $P(A \cup B) = \frac{1}{6}$, find $P(\bar{A} \cup \bar{B})$.

2. If A and B are independent events, prove that A and $\bar{B}$ are also independent events.

3. Verify that the following function $f(x)$ is a p.d.f.

$$f(x) = 1 - |1 - x| \text{ for } 0 < x < 2.$$

4. If a random variable X has the p.d.f. $f(x) = \begin{cases} \frac{1}{4}, \text{if } -z < x < 2, \\ 0 \text{ otherwise,} \end{cases}$ find $P(|x| > 1)$.

5. State whether the following statement is true or false. "The mean of a binomial distribution is 3 and the variance is 4". Explain.

6. If X has a Poisson distribution and $P(x = 0) = P(x = 1) = k$, find the value of k?

7. Establish the relation between hazard rate and mean time to failure.

8. An equipment consists of three components A, B and C which one connected in parallel and the respective reliabilities are $R_A = 0.92$, $R_B = 0.94$ and $R_C = 0.90$. Calculate the equipment reliability.

9. Define:

(a) Sample

(b) Standard Error for mean $\bar{x}$.

10. Find 95% confidence limits for the mean of a normal distributed population from which the following sample was taken:

15, 17, 10, 18, 16, 9, 7, 11, 13, 14.

11. Define null and alternative hypothesis.

12. What are the assumptions about *t*-test?

13. Define:
(a) One-tailed test statistics
(b) Two-tailed test statistics

14. What do you understand by analysis of variance.

15. Explain:
(a) Type-I error
(b) Type-II error.

16. The following data gives the means of two samples taken from a population. Examine whether there is any significant difference between the two samples: $n_1 = 1000$; $n_2 = 2000$; $\bar{x}_1 = 65.5$; $\bar{x}_2 = 68$; $\sigma = 2.5$.

17. Describe:
(a) *p*-chart
(b) R-chart.

18. The number of defectives in a sample lot of 45 is 9. What is the fraction defective?

19. Describe the operating characteristics curve.

20. A manufacturer draws 10 items hourly from his production line and subjects them to an attribute inspection. By the end of the day a total of 120 units were inspected and 6 defective units found. What is the fraction defective p for a day?

PART B (5 × 12 = 60 MARKS)

Answer All the Questions

21. (a) The contents of 3 Urns are

Urn 1 : 1 white, 3 red, 2 black balls
Urn 2 : 3 white, 1 red, 1 black balls
Urn 3 : 3 white, 3 red, 3 black balls

Two balls are chosen from a randomly selected urn
(i) What is that probability that the balls are 1 white and 1 red?
(ii) What is the probability that the above event comes from urn 2?

(b) Given the moment generating function $M_x(t) = e^{3t+8t^2}$ of the R.V. X, find the moment generating function of the R.V. $Z = \frac{1}{4}(X - 3)$, and use it to determine the mean and variance of Z.

Or

(c) A continuous random variable has the probability density function defined by

$$f(x) = \begin{cases} kxe^{-\lambda x}, \text{ if } x \geq 0, \text{ and } \lambda > 0 \\ 0 \qquad \text{otherwise} \end{cases}$$

Determine the constant k and find mean and variance.

(d) If the joint probability density of X and Y is given by

$$f(x, y) = \begin{cases} \frac{2}{3}(x+2y), \text{ for } 0<x<1, 0<y<1 \\ 0 \quad , \text{elsewhere} \end{cases}$$ Find the conditional mean and the conditional variance of X given $Y = 1/2$.

22. (a) (i) If X is an exponentially distributed random variable, prove that $P(X > t + s/X > s) = P(x > t)$

(ii) If X is normally distributed with mean 8 and standard deviation 4, find (1) $P(5 \leq X \leq 10)$, (2) $P(x \geq 15)$.

(b) (i) Derive the reliability of two unit series system and hence obtain the mean time to failure (*MTTF*).

(ii) Calculate the reliability of a 2 out of 5 units having an identical constant failure rate of components $\lambda = 0.01$/hour for a mission time of 10 hours.

Or

(c) If X is a geometrically distributed random variable with p.m.f. $P(X = x)$ and c.d.f. $F(X)$.

(i) Find the value of the ratio $\dfrac{P(X=x)}{1-F(X-1)}$.

(ii) P.T. $P(X \geq n + m/X \geq n) = P(X \geq m)$.

(d) (i) Find the mean and variance of the Weibull distribution.

(ii) Given p.d.f. of the R.V. X as

$$f(x) = \begin{cases} \frac{1}{5}e^{\frac{-x}{5}}, x>0 \\ 0 \quad , \text{otherwise} \end{cases}$$

Find (1) $P(x < 6/x > 3)$
(2) $P(x > 6)$.

23. (a) A survey of 100 randomly selected residents of a certain city showed that $\frac{1}{5}$ of them were born in that city. Construct a 0.95 confidence interval estimate of the proportion p of residents in that city who were born here?

(b) Test runs for the operating time before overheating randomly selected motors of a certain type showed a sample mean of $\bar{x} = 39.7$ hours and a sample standard deviation of $s = 2.5$ hours. Find a 0.95 confidence interval estimate for the standard deviation σ in the length of time a motor of this type will operate before overheating.

Or

(c) Independent random samples of size $n_1 = 16$ and $n_2 = 25$ from normal populations with $\sigma_1 = 4.8$ and $\sigma_2 = 3.5$ have the means $\bar{x}_1 = 18.2$ and $\bar{x}_2 = 23.4$. Find a 90% confidence interval for $\mu_1 - \mu_2$.

(d) An industrial designer wants to determine the average amount of time it takes an adult to assemble an 'easy-to-assemble" toy. Use the following data (in minutes), a random sample, to construct a 95% confidence interval for the mean of the population sampled:

17, 13, 18, 19, 17, 21, 29, 22, 16, 28, 21, 15
26, 23, 24, 20, 8, 17, 17, 21, 32, 18, 25, 22
16, 10, 20, 22, 19, 14, 30, 22, 12, 24, 28, 11.

24. (a) a quality-control engineer suspects that the proportion of defective units among certain manufactured items has increased from the set limit of 0.01. To test his claim, he randomly selected 100 of these items and found that the proportion of defective units in the sample was .02. Test the engineer's hypothesis at the 0.05 level of significance.

(b) The following are the cholesterol contents, in milligrams per package, which four laboratories obtained for 6—once package of three very similar diet foods:

	Diet food-*A*	Diet food-*B*	Diet food-*C*
Laboratory 1	3.4	2.6	2.8
Laboratory 2	3.0	2.7	3.1
Laboratory 3	3.3	3.0	3.4
Laboratory 4	3.5	3.1	3.7

Perform a two-way analysis of variance, testing the null hypothesis concerning the laboratories and the diet foods at the level of significance $\alpha = 0.05$.

Or

(c) In comparing the variability of family income in two metropolitan areas, a survey yielded the following data: $n_1 = 100$, $n_2 = 100$, $s_1^2 = 25$ and $s_2^2 = 10$, where n_1, n_2 are the sample sizes and s_1^2, s_2^2 are the sample variance of incomes in thousands of dollars for the two areas, respectively. Assuming that the populations are normal, test the hypothesis H_0: $\sigma_1^2 = \sigma_2^2$ versus H_1: $\sigma_1^2 > \sigma_2^2$ at the $\alpha = 0.05$ level of significance

(d) The following are the number of mistakes made in five successive weeks by four technicians working for a medical laboratory:

Technician I	13	16	12	14	15
Technician II	14	16	11	19	15
Technician III	13	18	16	14	18
Technician IV	18	10	14	15	12

Test at the level of significance $\alpha = 0.05$ whether the differences among the four sample means can be attributed to chance

25. (a) Ten electric motors were examined for defects. The numbner of defects associated with motors 1, 2, 3, 4, 5, 6, 7, 8, 9, and 10 were 6, 5, 11, 9, 9, 6, 7 and 8 respectively. Develop the C-chart.

(b) Manufacturer inspects 100 Televisions daily, picked randomly from the production and determines the number of defectives by visual examination. He wishes to use the data accumulated during a 10- day period to construct an attributed control chart for number of defectives. The results of the daily inspection of 100 Televisions is shown in the following table. Does the attribute control chart reveal that the process is under control?

Sample number:	1	2	3	4	5
Sample size:	100	100	100	100	100
Number of defective *np*:	6	14	16	10	2
Sample number:	6	7	8	9	10
Sample size:	100	100	100	100	100
Number of defective *np*:	18	5	12	7	8

Or

(c) Ten inspection sample lots of five amplifiers each one drawn from production. The following table lists the power output obtained for each amplifier. Plot an $\bar{X}$ chart and R chart.

Inspection sample lot number:	1	2	3	4	5
$\bar{X}$:	11.0	12.0	12.8	14.0	13.6
R:	4	4	6	4	6
Inspection sample lot number:	6	7	8	9	10
$\bar{X}$:	12.8	11.8	12.6	13.0	11.8
R:	5	5	6	4	6

(d) Construct a control chart for the proportion of defectives obtained in repeated random samples of size 100 from a process which is considered to be under control when the proportion of defective p is equal to 0.20. Draw the control line and the upper lower control limits on graph paper.

PART A

1. What is the probability that a leap year selected at random contains 53 Sundays?
2. The probability that x solves a problem is p_1 and the probability that y solves it is p_2, what is the probability that the problem will be solved when x and y work independently?
3. Let A and B two events such that $P(A) = \frac{3}{4}$ and $P(B) = \frac{5}{8}$, show that $\frac{3}{8} \le P(A \cap B) \le \frac{5}{8}$.
4. If $A \subset B$, $P(A) = \frac{1}{4}$ and $P(B) = \frac{1}{3}$, find $P\left(\frac{A}{B}\right)$ and $P\left(\frac{B}{A}\right)$.
5. Define the probability mass function of a discrete random variable.
6. If the p.d.f. of a random variable is $f(x) = \frac{x}{2}$ in $0 \le x \le 2$, find $P\{x > \frac{1.5}{x} > 1\}$.
7. The following table represents the joint probability distribution of the discrete random variables X and Y:

Y \ X	1	2	3
1	$\frac{1}{12}$	$\frac{1}{6}$	0
2	0	$\frac{1}{9}$	$\frac{1}{5}$
3	$\frac{1}{18}$	$\frac{1}{4}$	$\frac{2}{15}$

(a) Evaluate marginal distribution of X and Y.
(b) Find $F(X + Y < 4)$.

8. The joint p.d.f. of (X, Y) is $f(x, y) = 6e^{-2x-3y}$, $x \ge 0$, $y \ge 0$ find the marginal density of X and conditional density of Y given X.
9. Let X be a continuous random variable with p.d.f. $f(x) = \begin{cases} \frac{x}{12} & \text{in } 1 < x < 5 \\ 0 & \text{elsewhere} \end{cases}$ find the probability density function $Y = 2X - 3$.

10. If the random variable X follows $N(0, 2)$ and $Y = 3X^2$ find the mean and variance of Y.

11. Define moment generating function.

12. What is the difference between convergence every where and almost every where of a random sequence $\{X_n\}$?

13. If $\{X(s,t)\}$ is a random process. What is the nature of $X(s, t)$ when (a) s is fixed, and (b) t is fixed.

14. Examine whether the Poisson process $\{X(t)\}$ given by $P\{X(t) = \gamma\} = e^{-\lambda t} \dfrac{(\lambda t)^r}{\underline{|r}}$, $(r = 0, 12\ldots)$ is of variance stationary.

15. Define power spectrum.

16. Is all wide sense stationary process are strict sense stationary process?

17. Define band limited white noise.

18. Define shot noise process.

19. Write the expressions for in-phase and quadrature components of a narrow-band noise signal $n(t)$.

20. Define a homogeneous Poisson process.

PART B (5 × 12 = 60 MARKS)

21. (a) Three men toss in succession for a prize to be given to one who tosses heads first. Find their respective chances of winning the prize.

(b) A box contains 4 bad and 6 good tubes. Two are drawn out from the box at a time. One of them is tested and found to be good. What is the probability that the other one is also good?

Or

22. (a) If $P(A) = P(B) = P(AB)$. Show that $P(A\bar{B} + \bar{A}B) = 0$.

(b) A six faced die is so biased that it is twice as likely to show an even number as an odd number when thrown. If it is thrown twice, what is the probability that sum of the two numbers thrown is even.

23. (a) The probability function of an infinite discrete distribution is given by $P(X = j) = \dfrac{1}{2^j}$ $(j = 1, 2, \ldots. \infty)$. Verify that the total probability is 1 and find the mean and variance of the distribution.

(b) The distribution function of a R.V.X. is given by $F(X) = 1 - (1 + x)e^{-x}$ for $x \geq 0$. Find the density function, mean and variance.

Or

24. The length of time in (minutes) that a certain lady speaks on the telephones is found to be random phenomenon with a probability function specified by the function

$$f(x) = \begin{cases} Ae^{\frac{-x}{5}}, & \text{for } x \geq 0 \\ 0 & \text{otherwise} \end{cases}$$

(a) Find the value of A that makes $f(x)$ a p.d.f.

(b) What is the probability that the number of minutes that she will talk over the phone is (i) more than 10 minutes, (ii) less than 5 minutes, (iii) between 5 and 10 minutes?

25. (a) Find the characteristic function with p.d.f. $f(x) = \alpha/2e^{-\alpha|x|}$, $-\infty < x < \infty$. Hence find its mean and variance.

(b) If $Y = X^2$ where X is a Gaussian random variable with zero mean and variance σ^2, find the p.d.f. of the random variable Y.

Or

26. A distribution with unknown mean μ has variance equal to 1.5. Use central limit theorem to find how large a sample be taken from the distribution in order that the probability will be at least 0.95 that the sample mean will be within 0.5 of the population mean.

27. (a) Prove that a finite order stationary random process has a constant mean.

(b) Consider a random process $X(t)$ whose mean is zero and $R_{XX}(\tau) = A\ \delta(\psi)$. Show that the process $X(t)$ is mean ergodic.

Or

28. (a) The power spectral density of a random process is given by

$$\delta_{XX}(w) = \begin{cases} \pi & |w| < 1 \\ 0 & \text{elsewhere} \end{cases}$$

Find its auto-correlation function.

(b) Given that the auto-correlation function for a stationary ergodic process with no periodic components is $R_{XX}(\tau) = 25 + \dfrac{4}{1+6\tau^2}$. Find the mean value and variance of the process $\{X(t)\}$.

29. Prove that if a Gaussian random process X(t) is applied to a linear, time invariant system, the random process $Y(t)$ developed at the output of the system is also Gaussian.

Or

30. In a white Gaussian noise $X(t)$ of zero mean and spectral density $\dfrac{N_0}{2}$ is applied to a law pass RC filter shown in figure, determine the auto-correlation function of the output.

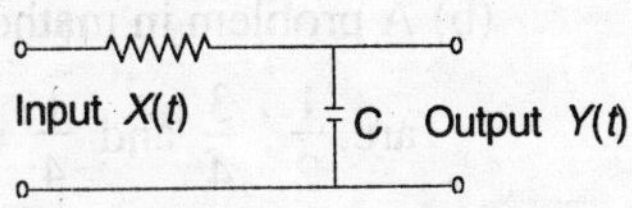

PART A

1. A and $\bar{A}$ are mutually exclusive events of A, prove that $P(\bar{A}) \leq 1$.
2. Two dice are thrown. Find the probability that sum of the number got is 8.
3. $P(A) = \dfrac{3}{4}$, $P(B) = \dfrac{2}{3}$. If A and B are independent find $P(A \cup B)$.
4. If A and B are independent events, prove that $P(A \cup B) = 1 - P(\bar{A})\, P(\bar{B})$.
5. The probability distribution of a random variable X is given below.

X:	0	1	2	3
$P(X)$:	k	$3k$	$5k$	k

Find $P(X \geq 2)$.

6. Determine the value of C if $f(x) = C(1 + x)$ in (2, 5) is a probability density function.
7. Define cumulative distribution function.
8. Define joint probability distribution function for continuous random variable.

9. Given

$$f_{xy}(X, Y) = \begin{cases} CX(X-Y), & 0 < X < 2, -X < Y < X \\ 0 & , \text{elsewhere} \end{cases}$$

evaluate 'C'.

10. If the random variable X has the p.d.f.

$$f(x) = \begin{cases} \frac{x}{\sqrt{2}} & \text{in } 1 < x < 5 \\ 0 & \text{elsewhere} \end{cases}$$

find the p.d.f. of $Y = 2X - 3$.

11. If the joint p.d.f. of (X, Y) is given by

$$f(x, y) = 24\,y\,(1 - x),\ 0 \le y \le x \le 1, \text{ find } E(XY).$$

12. What is the advantage of characteristic function over moment generating function?

13. Define stochastic process.

14. State the condition that a random process is a strongly stationary process.

15. Prove that Auto-correlation function is an even function.

16. State the reproductivity property of the Poisson distribution.

17. Define Gaussian process.

18. Define short time average and sample time average of $X(t)$.

19. Define power spectrum.

20. Define matched filter.

PART B (5 × 12 = 60 MARKS)

21. (a) If two dice are thrown, what is the probability that the sum is neither 7 nor 11.

(b) A problem in mathematics is given to the three students A, B and C whose chances of solving it are $\frac{1}{2}, \frac{3}{4}$ and $\frac{1}{4}$ respectively. What is the probability that the problem will be solved?

Or

22. (a) Prove that A, B and C are random events in a sample space and if A, B, C are pair-wise independent and A is independent of $(B \cup C)$ then prove that A, B and C are mutually independent.

(b) Find the probability of getting atleast 60 heads when 100 fair coins are tossed

23. (a) A random variable X has the following probability distribution

X:	0	1	2	3	4	5	6	7
$P(X)$:	0	k	$2k$	$2k$	$3k$	k^2	$2k^2$	$7k^2 + k$

Find

(i) the value of k

(ii) $P\left(1.5 < X < \frac{4.5}{X} > 2\right)$

(iii) The smallest value of λ for which $P(X \le \lambda) > \frac{1}{2}$.

(b) The diameter of an electric cable say X is assumed to be continuous random variable with p.d.f. $f(x) = 6x(1 - x),\ 0 \le x \le 1$

(i) Check that above is p.d.f.

(ii) Determine a number such that

$$P(X) < b = P(X) > b.$$

Or

24. (a) Joint distribution of X and Y are given by $f(x, y) = 4xy\, e^{-(x^2 + y^2)}$, $x > 0, y > 0$. Test whether X and Y are independent. For the above distribution find the conditional density of X given $Y = y$.

(b) X and Y are two random variables having Joint density function $f(x, y) = \frac{1}{27}(x + 2y)$ where x and y can assume only the integar values 0, 1 and 2, find the conditional distribution of Y for $X = x$.

25. (a) If X and Y are two independent random variables with $f_x(X) = e^{-x} \cup (X)$, $f_y(Y) = e^{-y} \cup (Y)$ and $Z = (X - Y) \cup (X - Y)$, prove that $E(z) = \frac{1}{2}$

(b) If X and Y are independent random variables having density function $f_1(X) = \begin{cases} 2e^{-2x} & X \geq 0 \\ 0 & X < 0 \end{cases}$ and

$f_2(Y) = \begin{cases} 3e^{-3y} & Y \geq 0 \\ 0 & Y < 0 \end{cases}$ and find the density function of their sum $U = X + Y$.

Or

26. (a) Show that the distribution for which characteristic function is $e^{(-w)}$ has the density function $f(X) = \frac{1}{\pi}\frac{1}{1+X^2}$, $-\infty < X < \infty$.

(b) A random sample of size 100 is taken from a population whose mean is 60 and S.D. is 20. Using central limit theorem with what probability can we assert that the mean of the sample will not differ from $\mu = 60$ by more than 4.

27. (a) If $X(t) = z \cos t$ where z is a random variable uniformly distributed in (0, 1). Prove that the mean of the short time average of $\{X(t)\}$ is given by

(i) $\frac{1}{2}\frac{\sin T/2}{T/2}\cos(t - T/2)$ and

(ii) its auto-correlation is

$$\frac{1}{3}\frac{(\sin T/2)^2}{T/2}\cos(t_1 - T/2)\cos(t_2 - T/2).$$

(b) Prove that the Random Telegraphic signal process is a covariance stationary process.

Or

28. (a) If $[X_1(t)]$ and $\left[X_1'(t)\right]$ are jointly stationary, prove that $R_{X'X'}(u) = -\frac{d^2}{du^2}R_{XX}(u)$ (where dashes denotes differentiation).

(b) The process $[X_1(t)]$ is stationary with $E(X(t)) = 1$ and $R(u) = 1 + e^{-2|u|}$, find the mean and variance of the R.V. $Y = \int_0^1 X(t)\, dt$.

Or

29. (a) If the member function of a Gaussian process are uncorrelated then they are independent.

(b) Prove that the difference of two independent Poisson processes is not a Poisson process.

Or

30. (a) Find the power spectrum of the process whose auto-correlation is given by $R(u) = e^{-\alpha u^2}$

(b) Find the unit impulse response $h(t)$ of the Wiener filter system with $S_{NN}(w) = N_0$ and $S_{FF}(w) = \dfrac{3\alpha^2 N_0}{\alpha^2 + w^2}$ and if the signal and noise are uncorrelated.

Appendix A

DISCRETE DISTRIBUTIONS

Distribution	Parameters	p.m.f $f_X(x)$	Mean $E(x)$	Variance var (x)
Bernoulli	p $0 \le p \le 1$	$f_X(0) = p;$ $f_X(1) = q = 1 - p$	p	$p(1-p)$
Binomial	n, p $n \ge 1, 0 \le p \le 1$	$nC_x\, p^x q^{n-x}, q = 1 - p$ $x = 0, 1, \ldots n$	np	$np(1-p)$
Geometric	p $0 < p \le 1$	$p(1-p)^{x-1}$ $x = 1, 2, \ldots$	$\frac{1}{p}$	$\frac{1-p}{p^2}$
Negative Binomial	p $0 < p \le 1$	${}^{n-1}C_{x-1}\, p^x q^{n-x}$ $q = 1 - p;\ x = 1, 2, \ldots$ $n = x, x+1, \ldots$	$\frac{1}{p}$	$\frac{x(1-p)}{p^2}$
Poisson	$\lambda, \lambda > 0$	$\frac{e^{-\lambda} \lambda^x}{x!}$ $x = 0, 1, 2, \ldots$	λ	λ

CONTINUOUS DISTRIBUTION

Distribution	Parameters	p.d.f $F_X(x)$	Mean $E(x)$	Variance var (X)
Erlang	$r, l,$ $r \ge 1, l > 0$	$\frac{\lambda e^{-\lambda x} (\lambda x)^{r-1}}{(r-1)!}$	$\frac{r}{\lambda}$	$\frac{r}{\lambda}$
Exponential	l $l > 0$	le^{-lx}	$\frac{1}{\lambda}$	$\frac{1}{\lambda^2}$
Gamma	a, l $a > 0, l > 0$	$\frac{\lambda e^{-\lambda x} (\lambda x)^{\alpha-1}}{\Gamma_{(\alpha)}}, x > 0$	$\frac{\alpha}{\lambda}$	$\frac{\alpha}{\lambda^2}$
Normal	m, s^2	$\frac{1}{\sigma\sqrt{2\pi}} e^{-\frac{(x-\mu)^2}{2\sigma^2}}$	m	s^2
Weibull	$l > 0, a > 0$	$\lambda \alpha x^{\alpha-1} e^{-\lambda x^{\alpha}}, x > 0$	$\left(\frac{1}{\lambda}\right)^{\frac{1}{\alpha}} \Gamma_{\left(1+\frac{1}{\alpha}\right)}$	$\left[\left(\frac{1}{\lambda}\right)^{\frac{2}{\alpha}} \Gamma_{\left(1+\frac{2}{\alpha}\right)}\right] - (E(x))^2$

Appendix B

LAPLACE TRANSFORM OF $f(t) = L(f(t)) = F(s) = \int_0^{\infty} e^{-st} f(t)\,dt$

Function $f(t)$, $t > 0$	$L(f(t)) = F(s)$
c	$\frac{c}{s}$
t	$\frac{1}{s^2}$
t^n	$\frac{n!}{s^{n+1}}$
$t^{\frac{-1}{2}}$	$\sqrt{\frac{\pi}{s}}$
e^{-at}	$\frac{1}{s+a}$
e^{at}	$\frac{1}{s-a}$
te^{-at}	$\frac{1}{(s+a)^2}$
$t^n e^{-at}$	$\frac{n!}{(s+a)^{n+1}}$

PROPERTIES OF LAPLACE TRANSFORM

Function		Laplace transform $L(f(t)) = F(s)$
Linearity:	$af(t) + bg(t)$	$aL(f(t)) + bL(g(t)) = aF(s) + bG(s)$
change of scale:	$f(at)$	$\frac{1}{a} F(s/a)$
first shifting property	$e^{-at} f(t)$	$F(s + a)$
derivative	$f'(t)$	$sF(s) - f(0)$
	$f^n(t)$	$s^n F(s) - s^{n-1} f(0) - s^{n-2} f'(0) \ldots . f^{n-1}(0)$
	$\int_{-\infty}^{t} f(t)\,dt$	$\frac{F(s)}{s}(s)$
Convolution theorem:		
$f(t) * g(t) = \int_{-\infty}^{t} f(\tau)\, g(t-\tau)\, d\tau$		$L(f(t)) \cdot L(g(t))$
		$-\frac{d}{ds} F(s)$

(*Contd.*)

(*Contd.*)

Function		Laplace transform $L(f(t)) = F(s)$
$t^n f(t)$, n, $tf(t)$ a positive integer		$(-1)^n \dfrac{d^n}{ds^n} F(s)$
	$\dfrac{f(t)}{t}$	$\int_s^\infty F(s)ds$
Initial value theorem	$\underset{t \to 0}{\mathrm{Lt}}\ f(t) =$	$\underset{s \to \infty}{\mathrm{Lt}}\ sF(s)$
Final value theorem	$\underset{t \to 0}{\mathrm{Lt}}\ f(t) =$	$\underset{s \to \infty}{\mathrm{Lt}}\ sF(s)$

Appendix C

FOURIER TRANSFORM

Fourier transform of x(t) is defined by:

$$x(w) = \int_{-\infty}^{\infty} x(t)\, e^{-iw\tau}\, dt$$

Inverse Fourier transform of X(w) is: $x(t) = \dfrac{1}{2\pi} \displaystyle\int_{-\infty}^{\infty} X(w)\, e^{iwt}\, dw$

TABLE 1
Continuous-Time Fourier Transform

Property	Fourier Transform $X(w)$	Inverse Fourier Transform $x(t)$
Linearity	$a_1X_1(w) + a_2X_2(w)$	$a_1x_1(t) + a_2\, X_2(t)$
Time Shifting	$e^{-iwt_0}\, X(w)$	$x(t - t_0)$
Frequency shifting	$X(w - w_0)$	$e^{iw_0t}\, x(t)$
Change of	$\dfrac{1}{\lvert a\rvert} \times \left(\dfrac{w}{a}\right)$	$x(at)$
Time reversal	$X(-w)$	$x(-t)$
Duality	$X(t)$	$2\pi\, x(-w)$
Time differentiation	$iw\, X(w)$	$\dfrac{d}{dt}\, x(t)$
Frequency	$\dfrac{d}{dw}\,(X(w))$	$(-\,it)\, x(t)$
Integration	$\pi X(0)\, \delta(w) + \dfrac{1}{iw}\, X(w)$	$\int_{-\infty}^{t} x(\tau)\, d\tau$
Convolution	$X_1(w)\, X_2(w)$	$x_1(t) * x_2(t) = \int_{-\infty}^{\infty} x_1(t)\, x_2(t-t)\, dt$
Multiplication	$\dfrac{1}{2\pi}\,(X_1(w) * X_2(w))$ $= \dfrac{1}{2\pi}\, X_1(\lambda)\, X_2(w-\lambda)\, d\lambda$	$x_1(t)\, x_2(t)$

(Contd.)

(*Contd.*)

Property	Fourier Transform $X(w)$	Inverse Fourier Transform $x(t)$
Real signal	$X(w) = A(w) + iB(w)$ $A(w)$ = Real parts of $X(w)$ $B(w)$ = Imaginary part of $X(w)$	$x(t) = x_e(t) + x_0(t)$ $x_e(t)$ = even component $x_0(t)$ = odd component
Parse Val 's Theorem	$\int_{-\infty}^{\infty} \lvert X(w)\rvert^2\, dw = 2\pi \int_{-\infty}^{\infty} \lvert x(t)\rvert^2\, dt$	
	1	$\delta(t)$
	e^{-iwt_0}	$\delta(t - t_0)$
	$2\pi\delta(w)$	1
	$2\pi\delta(w - w_0)$	$e^{+iw_0 t}$
	$\pi[\delta(w - w_0) + \delta(w + w_0)]$	$\cos w_0(t)$
	$-i\pi[\delta(w - w_0) - \delta(w + w_0)]$	$\sin w_0 t$
	$\pi\delta(w) + \dfrac{1}{iw}$	$u(t) = \begin{cases} 1 & t > 0 \\ 0 & t < 0 \end{cases}$
	$\dfrac{1}{a - iw}$	$e^{at}\, u(-t),\ a > 0$
	$\dfrac{1}{a + iw}$	$e^{-at}\, u(t),\ a > 0$
	$\dfrac{1}{(a + iw)^2}$	$t\, e^{-at}\, u(t),\ a > 0$
	$\dfrac{1}{(a + iw)^n}$	$\dfrac{t^{n-1}}{n!}\, e^{-at}\, u(t),\ a > 0$
	$\dfrac{2a}{a^2 + w^2}$	$e^{-a\lvert t\rvert}\ a > 0$
	$e^{-a\lvert w\rvert}$	$\dfrac{1}{a^2 + t^2}$
	$e^{-w^2/4a}$	$\sqrt{a/\pi} \cdot e^{-at^2}$
	$\dfrac{\sin(wa)}{wa}$	$\begin{cases} 2a & \lvert t\rvert < a \\ 0 & \lvert t\rvert > a \end{cases}$
	$\dfrac{2}{iw}$	$Sgn\ t = \begin{cases} 1 & t > 0 \\ -1 & t < 0 \end{cases}$
	$w_0 \sum_{k=-\infty}^{\infty} \delta(w - kw_0)$, $p = -\infty\ w_0 = \dfrac{2\pi}{T}$	$\sum_{k=-\infty}^{\infty} \delta(t - kT)$

DISCRETE FOURIER TRANSFORM

$$X(\Omega) = \sum_{n=-\infty}^{\infty} x(n)\, e^{-i\Omega n}$$

$$x(n) = \frac{1}{2\pi} \int_{-\pi}^{\pi} X(\Omega)\, e^{i\Omega n}\, d\Omega$$

TABLE 2

Property	Fourier Transform $X(\Omega)$	Inverse Fourier transform (sequence) $x(n)$
Periodicity	$X(\Omega + 2\pi) = X(\Omega)$	$x(n)$
Linearity	$a_1 X_1(\Omega) + a_2 X_2(\Omega)$	$a_1 x_1(n) + a_2 x_2(n)$
Time shifting	$e^{-i\Omega n_0} X(\Omega)$	$x(n - n_0)$
Frequency shifting	$X(\Omega - \Omega_0)$	$e^{i\Omega_0 n}\, x(n)$
Time reversal	$X(-\Omega)$	$x(-n)$
Frequency	$\frac{d}{d\Omega}(X(\Omega))$	$\frac{n}{i}\, x(n)$
Accumulation	$\pi X(0)\, \delta(\Omega) + \frac{1}{1 - e^{-i\Omega}} X(\Omega)$	$\sum_{k=-\infty}^{\infty} x(n)$
Convolution	$X_1(\Omega)\, X_2(\Omega)$	$x_1(n) * x_2(n) = \sum_{k=-\infty}^{\infty} x_1(k)\, x_2(n-k)$
Real sequence	$X(\Omega) = A(\Omega) + iB(\Omega)$	$x(n) = x_e(n) + x_0(n)$
Parseval's theorem	$\sum_{n=-\infty}^{\infty} \lvert x(n) \rvert^2 =$	$\frac{1}{2\pi} \int_{-\pi}^{\pi} \lvert X(W) \rvert^2\, dW.$
	1	$\delta(n) = \begin{cases} 1 & n = 0 \\ 0 & n \neq 0 \end{cases}$
	$e^{-\Omega n_0}$	$\delta(n - n_0)$
	$2\pi\delta(\Omega)$	$x(n) = 1$
	$2\pi\delta(\Omega - \Omega_0)$	$e^{i\Omega n_0}$
	$\pi[\delta(\Omega - \Omega_0) + \delta(\Omega + \Omega_0)]$	$\cos \Omega_0 n$
	$-i\pi[\delta(\Omega - \Omega_0) - \delta(\Omega + \Omega_0)]$	$\sin \Omega_0 n$
	$\pi\delta(\Omega) + \frac{1}{1 - e^{-i\Omega}}$	$u(n) = \begin{cases} 1 & n \geq 0 \\ 0 & n < 0 \end{cases}$
	$\frac{1}{1 - ae^{-i\Omega}}$	$a^n u(n), \quad \lvert a \rvert < 1$
	$\frac{1}{(1 - a e^{-i\Omega})^2}$	$(n+1)a^n u(n) \cdot \lvert a \rvert < 1$

(*Contd.*)

(*Contd.*)

Property	Fourier Transform $X(\Omega)$	Inverse Fourier transform (sequence) $x(n)$
	$\dfrac{1-a^2}{1-2a\cos\Omega+a^2}$	$a^{\lvert n\rvert} \quad \lvert a\rvert<1$
	$\dfrac{\sin\left(\Omega\left(N+\frac{1}{2}\right)\right)}{\sin(\Omega/2)}$	$x(n)=\begin{cases}1 & \lvert n\rvert\le N\\ 0 & \lvert n\rvert>N\end{cases}$
	$X(\Omega)=\begin{cases}1 & 0\le\lvert\Omega\rvert\le w\\ 0 & \text{elsewhere}\end{cases}$	$\dfrac{\sin wn}{\pi n},\ 0<w<\pi$
	$\Omega_0\sum_{k=-\infty}^{\infty}\delta(\Omega-k\Omega_0)$	$\sum_{k=-\infty}^{\infty}\delta(n-kN_0)$
	$\Omega_0=\dfrac{2\pi}{N_0}$	

Appendix D

Based on "Tables of the Binomial Probability Distribution", National Bureau of standards Applied Mathematics series No.6, Washington D.C., Government Printing Office 1950.

Binomial Distribution Function

$$B(x; n, p) = \sum_{k=0}^{x} \binom{n}{k} p^k (1-p)^{n-k}$$

n	x	0.05	0.10	0.15	0.20	0.25	0.30	0.35	0.40	0.45	0.50
2	2	0.9025	0.8100	0.7225	0.6400	0.5625	0.4900	0.4225	0.3600	0.3025	0.2500
	1	0.9975	0.9900	0.9775	0.9600	0.9375	0.9100	0.8755	0.8400	0.7975	0.7500
3	0	0.8574	0.7290	0.6141	0.5120	0.4219	0.3430	0.2746	0.2160	0.1664	0.1250
	1	0.9928	0.9720	0.9392	0.8960	0.8438	0.7840	0.7840	0.6480	0.5748	0.5000
	2	0.9999	0.9990	0.9966	0.9920	0.9844	0.9730	0.9571	0.9360	0.9089	0.8750
4	0	0.8145	0.6561	0.5220	0.4096	0.3164	0.2401	0.1785	0.1296	0.0915	0.625
	1	0.9860	0.9477	0.8905	0.8192	0.7383	0.6517	0.5630	0.4752	0.3910	0.3125
	2	0.9995	0.9963	0.9880	0.9728	0.9492	0.9163	0.8735	0.8208	0.7585	0.6875
	3	1.0000	0.9999	0.9995	0.9984	0.9961	0.9919	0.9850	0.9744	0.9590	0.9375
5	0	0.7738	0.5905	0.4427	0.3277	0.2373	0.1681	0.1160	0.0778	0.0503	0.0312
	1	0.9774	0.9185	0.8352	0.7373	0.6328	0.5282	0.4284	0.3370	0.2562	0.1875
	2	0.9988	0.9914	0.9734	0.9421	0.8965	0.8369	0.7648	0.6826	0.5931	0.5000
	3	1.0000	0.9995	0.0078	0.9933	0.9844	0.9642	0.9460	0.9130	0.8688	0.8125
	4	1.0000	1.000	0.9999	0.9997	0.9990	0.9976	0.9947	0.9898	0.9815	0.9688
6	0	0.7351	0.5314	0.3771	0.2621	0.1780	0.1176	0.0754	0.0467	0.0277	0.0156
	1	0.9672	0.8857	0.7765	0.6554	0.5339	0.4202	0.3191	0.2333	0.1636	0.1094
	2	0.9978	0.9842	0.9527	0.9011	0.8306	0.7443	0.6471	0.5443	0.4415	0.3438
	3	0.9999	0.9987	0.9941	0.9830	0.9624	0.9295	0.8826	0.8208	0.7447	0.6562
	4	1.0000	0.9999	0.9996	0.9984	0.9954	0.9891	0.9777	0.9590	0.9308	0.8906
	5	1.0000	1.0000	1.0000	0.9999	0.9998	0.9993	0.9982	0.9959	0.9917	0.9844
7	0	0.6983	0.4783	0.3206	0.2097	0.1335	0.0824	0.0490	0.0280	0.0152	0.0078
	1	0.9556	0.8503	0.7166	0.5767	0.4449	0.3294	0.2338	0.1586	0.1024	0.0625
	2	0.9962	0.9743	0.9262	0.8520	0.7564	0.6471	0.5323	0.4199	0.3164	0.2266
	3	0.9998	0.9974	0.9879	0.9667	0.9294	0.8740	0.8002	0.7102	0.6083	0.5000
	4	1.0000	0.9998	0.9988	0.9953	0.9871	0.9712	0.9444	0.9037	0.8471	0.7734
	5	1.0000	1.0000	0.9999	0.9996	0.9987	0.9962	0.9910	0.9812	0.09643	0.9375
	6	1.0000	1.0000	1.0000	1.0000	0.9999	0.9998	0.9994	0.9984	0.9963	0.9922
8	0	0.6634	0.4305	0.2725	0.1678	0.1001	0.0576	0.0319	0.0168	0.0084	0.0039
	1	0.9428	0.8131	0.6572	0.5033	0.3671	0.2553	0.1691	0.1064	0.0632	0.0352
	2	0.9942	0.9619	0.8948	0.7968	0.6785	0.5518	0.4278	0.3154	0.2201	0.1445
	3	0.9996	0.9950	0.9786	0.9437	0.8862	0.8059	0.7064	0.5941	0.4770	0.3633
	4	1.0000	0.9965	0.9971	0.9896	09727	0.9420	0.8939	0.8263	0.7396	0.6367
	5	1.0000	1.0000	0.9998	0.9988	0.9958	0.9887	0.9747	0.9502	0.9115	0.8555
	6	1.0000	1.0000	1.0000	0.9999	0.9996	0.9987	0.9964	0.9915	0.9819	0.9648
	7	1.0000	1.0000	1.0000	1.0000	1.0000	0.9999	0.9998	0.9993	0.9983	0.9961
9	0	0.6302	0.3874	0.2316	0.1342	0.0751	0.0404	0.0207	0.0101	0.0046	0.0020
	1	0.9288	0.7748	0.5995	0.4362	0.3003	0.1960	0.1211	0.0705	0.0385	0.0195
	2	0.9916	0.9470	0.8591	0.7382	0.6007	0.4628	0.3373	0.2318	0.1495	0.0898
	3	0.9994	0.9917	0.9661	0.9144	0.8343	0.7297	0.6089	0.4826	0.3614	0.2539
	4	1.0000	0.9991	0.9944	0.9804	0.9511	0.9012	0.8283	0.7334	0.6214	0.5000

(Contd.)

(*Contd.*)

n	*x*	0.05	0.10	0.15	0.20	0.25	0.30	0.35	0.40	0.45	0.50
	5	1.0000	0.9999	0.9994	0.9969	0.9900	0.9747	0.9464	0.9006	0.8342	0.7461
	6	1.0000	1.0000	1.0000	0.9997	0.9987	0.9957	0.9888	0.9750	0.9502	0.9120
	7	1.0000	1.0000	1.0000	1.0000	0.9999	0.9996	0.9986	0.9962	0.9909	0.9805
	8	1.0000	1.0000	1.0000	1.0000	1.0000	1.0000	0.9999	0.9997	0.9992	0.9980
10	0	0.5987	0.3487	0.1969	0.1074	0.0563	0.0282	0.0135	0.0060	0.0025	0.0010
	1	0.9139	0.7361	0.5443	0.3758	0.2440	0.1493	0.0860	0.0464	0.0232	0.0107
	2	0.9885	0.9298	0.8202	0.6778	0.5256	0.3828	0.2616	0.1673	0.0996	0.0547
	3	0.9990	0.9872	0.9500	0.8791	0.7759	0.6496	0.5138	0.3823	0.2660	0.1719
	4	0.9999	0.9984	0.9901	0.9672	0.9219	0.8497	0.7515	0.6331	0.5044	0.3770
	5	1.0000	0.9999	0.9986	0.9936	0.9803	0.9527	0.9051	0.8338	0.7384	0.6230
	6	1.0000	1.0000	0.9999	0.9991	0.9965	0.9894	0.9740	0.9452	0.8980	0.8281
	7	1.0000	1.0000	1.0000	0.9999	0.9996	0.9984	0.9952	0.9877	0.9726	0.9453
	8	1.0000	1.0000	1.0000	1.0000	1.0000	0.9999	0.9995	0.9983	0.9955	0.9893
	9	1.0000	1.0000	1.0000	1.0000	1.0000	1.0000	1.0000	0.9999	0.9997	0.9990
11	0	0.5688	0.3138	0.1673	0.0859	0.0422	0.0198	0.0088	0.0036	0.0014	0.0005
	1	0.8981	0.6974	0.4922	0.3221	0.1971	0.1130	0.0606	0.0302	0.0139	0.0059
	2	0.9848	0.9104	0.7788	0.6174	0.4552	0.3127	0.2001	0.1189	0.0652	0.0327
	3	0.9984	0.9815	0.9306	0.8389	0.7133	0.5696	0.4236	0.2963	0.1911	0.1133
	4	0.9999	0.9972	0.9841	0.9496	0.8854	0.7897	0.6683	0.5328	0.3971	0.2744
	5	1.0000	0.9997	0.9973	0.9883	0.9657	0.9218	0.8513	0.7535	0.6331	0.5000
	6	1.0000	1.0000	0.9997	0.9980	0.9924	0.9784	0.9499	0.9006	0.8262	0.7256
	7	1.0000	1.0000	1.0000	0.9998	0.9988	0.9957	0.9878	0.9707	0.9390	0.8867
	8	1.0000	1.0000	1.0000	1.0000	0.9999	0.9994	0.9980	0.9941	0.9852	0.9673
	9	1.0000	1.0000	1.0000	1.0000	1.0000	1.0000	0.9998	0.9993	0.9978	0.9941
	10	1.0000	1.0000	1.0000	1.0000	1.0000	1.0000	1.0000	1.0000	0.9998	0.9995
12	0	0.5404	0.2824	0.1422	0.0687	0.0317	0.0138	0.0057	0.0022	0.0008	0.0002
	1	0.8816	0.6590	0.4435	0.2749	0.1584	0.0850	0.0424	0.0196	0.0083	0.0032
	2	0.9804	0.8891	0.7358	0.5583	0.3907	0.2528	0.1513	0.0834	0.0421	0.0193
	3	0.9978	0.9744	0.9078	0.7946	0.6488	0.4925	0.3467	0.2253	0.1345	0.0730
	4	0.9998	0.9957	0.9761	0.9274	0.8424	0.7237	0.5833	0.4382	0.3044	0.1938
	5	1.0000	0.9995	0.9954	0.9806	0.9456	0.8822	0.7873	0.6652	0.5269	0.3872
	6	1.0000	0.9999	0.9993	0.9961	0.9857	0.9614	0.9154	0.8418	0.7393	0.6128
	7	1.0000	1.0000	0.9999	0.9994	0.9972	0.9905	0.9745	0.9427	0.8883	0.8062
	8	1.0000	1.0000	1.0000	0.9999	0.9996	0.9983	0.9944	0.9847	0.9644	0.9270
	9	1.0000	1.0000	1.0000	1.0000	1.0000	0.9998	0.9992	0.9972	0.9921	0.9807
	10	1.0000	1.0000	1.0000	1.0000	1.0000	1.0000	0.9999	0.9997	0.9989	0.9968
	11	1.0000	1.0000	1.0000	1.0000	1.0000	1.0000	1.0000	1.0000	0.9999	0.9998
13	0	0.5133	0.2542	0.1209	0.0550	0.0238	0.0097	0.0037	0.0013	0.0004	0.0001
	1	0.8646	0.6213	0.3983	0.2336	0.1267	0.0637	0.0296	0.0126	0.0049	0.0017
	2	0.9755	0.8661	0.6920	0.5017	0.3326	0.2025	0.1132	0.0579	0.0269	0.0112
	3	0.9969	0.9658	0.8820	0.7437	0.5843	0.4206	0.2783	0.1686	0.0929	0.0461
	4	0.9997	0.9935	0.9658	0.9009	0.7940	0.6543	0.5005	0.3530	0.2279	0.1334
	5	1.0000	0.9991	0.9925	0.9700	0.9198	0.8346	0.7159	0.5744	0.4268	0.2905
	6	1.0000	0.9999	0.9987	0.9930	0.9757	0.9376	0.8705	0.7712	0.6437	0.5000
	7	1.0000	1.0000	0.9998	0.9988	0.9944	0.9818	0.9538	0.9023	0.8212	0.7995
	8	1.0000	1.0000	1.0000	0.9998	0.9990	0.9960	0.9874	0.9679	0.9302	0.8666
	9	1.0000	1.0000	1.0000	1.0000	0.9999	0.9993	0.9975	0.9922	0.9797	0.9539
	10	1.0000	1.0000	1.0000	1.0000	1.0000	0.9999	0.9997	0.9987	0.9959	0.9888
	11	1.0000	1.0000	1.0000	1.0000	1.0000	1.0000	1.0000	0.9999	0.9995	0.9983
	12	1.0000	1.0000	1.0000	1.0000	1.0000	1.0000	1.0000	1.0000	1.0000	0.9999
14	0	0.4877	0.2288	0.1028	0.0440	0.0178	0.0068	0.0024	0.0008	0.0002	0.0001
	1	0.8470	0.5864	0.3567	0.1979	0.1010	0.0475	0.0205	0.0081	0.0029	0.0009
	2	0.9699	0.8416	0.6479	0.4481	0.2811	0.1608	0.0839	0.0398	0.0170	0.0065
	3	0.9958	0.9559	0.8535	0.6982	0.5213	0.3552	0.2205	0.1243	0.0632	0.0287
	4	0.9996	0.9908	0.9533	0.8702	0.7415	0.5842	0.4227	0.2793	0.1672	0.0898
	5	1.0000	0.9985	0.9885	0.9561	0.8883	0.7805	0.6405	0.4859	0.3373	0.2120
	6	1.0000	0.9998	0.9978	0.9884	0.9617	0.9067	0.8164	0.6925	0.5461	0.3953
	7	1.0000	1.0000	0.9997	0.9976	0.9897	0.9685	0.9247	0.8499	0.7414	0.6074
	8	1.0000	1.0000	1.0000	0.9996	0.9978	0.9917	0.9757	0.9417	0.8811	0.7880

(*Contd.*)

(*Contd.*)

n	x	0.05	0.10	0.15	0.20	0.25	0.30	0.35	0.40	0.45	0.50
	9	1.0000	1.0000	1.0000	1.0000	0.9997	0.9983	0.9940	0.9825	0.9574	0.9107
	10	1.0000	1.0000	1.0000	1.0000	1.0000	0.9989	0.9989	0.9961	0.9886	0.9713
	11	1.0000	1.0000	1.0000	1.0000	1.0000	1.0000	0.9999	0.9994	0.9978	0.9935
	12	1.0000	1.0000	1.0000	1.0000	1.0000	1.0000	1.0000	0.9999	0.9997	0.9991
	13	1.0000	1.0000	1.0000	1.0000	1.0000	1.0000	1.0000	1.0000	1.0000	0.9999
15	0	0.4633	0.2059	0.0874	0.0352	0.0134	0.0047	0.0016	0.0005	0.0001	0.0000
	1	0.8290	0.5490	0.3186	0.1671	0.0820	0.0353	0.0142	0.0052	0.0017	0.0005
	2	0.9638	0.8159	0.6042	0.3980	0.2361	0.1268	0.0617	0.0271	0.0107	0.0037
	3	0.9945	0.9444	0.8227	0.6482	0.4613	0.2969	0.1727	0.0905	0.0424	0.0176
	4	0.9994	0.9873	0.9383	0.8358	0.6865	0.5155	0.3519	0.2173	0.1204	0.0592
	5	0.9999	0.9978	0.9832	0.9389	0.8516	0.7216	0.5643	0.4032	0.2608	0.1509
	6	1.0000	0.9997	0.9964	0.9819	0.9434	0.8689	0.7548	0.6098	0.4522	0.3036
	7	1.0000	1.0000	0.9996	0.9958	0.9827	0.9500	0.8868	0.7869	0.6535	0.5000
	8	1.0000	1.0000	0.9999	0.9992	0.9958	0.9848	0.9578	0.9050	0.8182	0.6964
	9	1.0000	1.0000	1.0000	0.9999	0.9992	0.9963	0.9876	0.9662	0.9231	0.8491
	10	1.0000	1.0000	1.0000	1.0000	0.9999	0.9993	0.9972	0.9907	0.9745	0.9408
	11	1.0000	1.0000	1.0000	1.0000	1.0000	0.9999	0.9995	0.9981	0.9937	0.9824
	12	1.0000	1.0000	1.0000	1.0000	1.0000	1.0000	0.9999	0.9997	0.9989	0.9963
	13	1.0000	1.0000	1.0000	1.0000	1.0000	1.0000	1.0000	1.0000	0.9999	0.9995
	14	1.0000	1.0000	1.0000	1.0000	1.0000	1.0000	1.0000	1.0000	1.0000	1.0000
16	0	0.4401	0.1853	0.0743	0.0281	0.0100	0.0033	0.0010	0.0003	0.0001	0.0000
	1	0.8108	0.5147	0.2839	0.1407	0.0635	0.0261	0.0098	0.0033	0.0010	0.0003
	2	0.9571	0.7892	0.5614	0.3518	0.1971	0.0994	0.0451	0.0183	0.0066	0.0021
	3	0.9930	0.9316	0.7899	0.5981	0.4050	0.2459	0.1339	0.0651	0.0281	0.0106
	4	0.9991	0.9830	0.9209	0.7982	0.6302	0.4499	0.2892	0.1666	0.0853	0.0384
	5	0.9999	0.9967	0.9765	0.9183	0.8103	0.6598	0.4900	0.3288	0.1976	0.1051
	6	1.0000	0.9995	0.9944	0.9733	0.9204	0.8247	0.6881	0.5272	0.3660	0.2272
	7	1.0000	0.9999	0.9989	0.9930	0.9729	0.9256	0.8406	0.7161	0.5629	0.4018
	8	1.0000	1.0000	0.9998	0.9985	0.9925	0.9743	0.9329	0.8577	0.7441	0.5982
	9	1.0000	1.0000	1.0000	0.9998	0.9984	0.9929	0.9771	0.9417	0.8759	0.7728
	10	1.0000	1.0000	1.0000	1.0000	0.9997	0.9984	0.9938	0.9809	0.9514	0.8949
	11	1.0000	1.0000	1.0000	1.0000	1.0000	0.9997	0.9987	0.9951	0.9851	0.9616
	12	1.0000	1.0000	1.0000	1.0000	1.0000	1.0000	0.9998	0.9991	0.9965	0.9979
	13	1.0000	1.0000	1.0000	1.0000	1.0000	1.0000	1.0000	0.9999	0.9994	0.9979
	14	1.0000	1.0000	1.0000	1.0000	1.0000	1.0000	1.0000	1.0000	1.0000	0.9894
	15	1.0000	1.0000	1.0000	1.0000	1.0000	1.0000	1.0000	1.0000	1.0000	1.0000
17	0	0.4181	0.1668	0.0631	0.0225	0.0075	0.0023	0.0007	0.0002	0.0000	0.0000
	1	0.7922	0.4818	0.2525	0.1182	0.0501	0.0193	0.0067	0.0021	0.0006	0.0001
	2	0.9497	0.7618	0.5198	0.3096	0.1637	0.0774	0.0327	0.0123	0.0041	0.0012
	3	0.9912	0.9174	0.7556	0.5489	0.3530	0.2019	0.1028	0.0464	0.0184	0.0063
	4	0.9988	0.9779	0.9013	0.7582	0.5739	0.3887	0.2348	0.1260	0.0596	0.0245
	5	0.9999	0.9953	0.9681	0.8943	0.7653	0.5968	0.4197	0.2639	0.1471	0.0717
	6	1.0000	0.9992	0.9917	0.9623	0.8929	0.7752	0.6188	0.4478	0.2902	0.1662
	7	1.0000	0.9999	0.9983	0.9891	0.9598	0.8954	0.7872	0.6405	0.4743	0.3145
	8	1.0000	1.0000	0.9997	0.9974	0.9876	0.9597	0.9006	0.8011	0.6626	0.5000
	9	1.0000	1.0000	1.0000	0.9995	0.9969	0.9873	0.9617	0.9081	0.8166	0.6855
	10	1.0000	1.0000	1.0000	0.9999	0.9994	0.9968	0.9880	0.9652	0.9174	0.8338
	11	1.0000	1.0000	1.0000	1.0000	0.9999	0.9993	0.9970	0.9894	0.9699	0.9283
	12	1.0000	1.0000	1.0000	1.0000	1.0000	0.9999	0.9994	0.9975	0.9914	0.9755
	13	1.0000	1.0000	1.0000	1.0000	1.0000	1.0000	0.9999	0.9995	0.9981	0.9936
	14	1.0000	1.0000	1.0000	1.0000	1.0000	1.0000	1.0000	0.9999	0.9997	0.9988
	15	1.0000	1.0000	1.0000	1.0000	1.0000	1.0000	1.0000	1.0000	1.0000	0.9999
	16	1.0000	1.0000	1.0000	1.0000	1.0000	1.0000	1.0000	1.0000	1.0000	1.0000
18	0	0.3972	0.1501	0.0536	0.0180	0.0056	0.0016	0.0004	0.0001	0.0000	0.0000
	1	0.7735	0.4503	0.2241	0.0991	0.0395	0.0142	0.0046	0.0013	0.0003	0.0001
	2	0.9419	0.7338	0.4797	0.2713	0.1353	0.0600	0.0236	0.0082	0.0025	0.0007
	3	0.9891	0.9018	0.7202	0.5010	0.3057	0.1646	0.0783	0.0328	0.0120	0.0038
	4	0.9985	0.9718	0.8794	0.7164	0.5187	0.3327	0.1886	0.0942	0.0411	0.0154
	5	0.9998	0.9936	0.9581	0.8671	0.7175	0.5344	0.3550	0.2088	0.1077	0.0481

(*Contd.*)

(*Contd.*)

n	x	0.05	0.10	0.15	0.20	0.25	0.30	0.35	0.40	0.45	0.50
	6	1.0000	0.9998	0.9882	0.9487	0.8610	0.7217	0.5491	0.3743	0.2258	0.1189
	7	1.0000	0.9998	0.9973	0.9837	0.9431	0.8593	0.7283	0.5634	0.3915	0.2403
	8	1.0000	1.0000	0.9995	0.9957	0.9807	0.9404	0.8609	0.7368	0.5778	0.4073
	9	1.0000	1.0000	0.9999	0.9991	0.9946	0.9790	0.9403	0.8653	0.7473	0.5927
	10	1.0000	1.0000	1.0000	0.9998	0.9988	0.9939	0.9788	0.9424	0.8720	0.7597
	11	1.0000	1.0000	1.0000	1.0000	0.9998	0.9986	0.9938	0.9797	0.9463	0.8811
	12	1.0000	1.0000	1.0000	1.0000	1.0000	0.9997	0.9986	0.9942	0.9817	0.9519
	13	1.0000	1.0000	1.0000	1.0000	1.0000	1.0000	0.9997	0.9987	0.9951	0.9846
	14	1.0000	1.0000	1.0000	1.0000	1.0000	1.0000	1.0000	0.9998	0.9990	0.9962
	15	1.0000	1.0000	1.0000	1.0000	1.0000	1.0000	1.0000	1.0000	0.9999	0.9993
	16	1.0000	1.0000	1.0000	1.0000	1.0000	1.0000	1.0000	1.0000	1.0000	0.9999
19	0	0.3774	0.1351	0.0456	0.0144	0.0042	0.0011	0.0003	0.0001	0.0000	0.0000
	1	0.7547	0.4203	0.1985	0.0829	0.0310	0.0104	0.0031	0.0008	0.0002	0.0000
	2	0.9335	0.7054	0.4413	0.2369	0.1113	0.0462	0.0170	0.0055	0.0015	0.0004
	3	0.9868	0.8850	0.6841	0.4551	0.2630	0.1332	0.0591	0.0230	0.0077	0.0022
	4	0.9930	0.9648	0.8556	0.6733	0.4654	0.2822	0.1500	0.0696	0.0280	0.0096
	5	0.9998	0.9914	0.9463	0.8369	0.6678	0.4739	0.2968	0.1629	0.0777	0.0318
	6	1.0000	0.9983	0.9837	0.9324	0.8251	0.6655	0.4812	0.3081	0.1727	0.0835
	7	1.0000	0.9997	0.9959	0.9767	0.9225	0.8180	0.6656	0.4878	0.3169	0.1796
	8	1.0000	1.0000	0.9992	0.9933	0.9713	0.9161	0.8145	0.6675	0.4940	0.3238
	9	1.0000	1.0000	0.9999	0.9984	0.9911	0.9674	0.9125	0.8139	0.6710	0.5000
	10	1.0000	1.0000	1.0000	0.9997	0.9977	0.9895	0.9653	0.9115	0.8159	0.6762
	11	1.0000	1.0000	1.0000	1.0000	0.9995	0.9972	0.9886	0.9648	0.9129	0.8204
	12	1.0000	1.0000	1.0000	1.0000	0.9999	0.9994	0.9969	0.9884	0.9658	0.9165
	13	1.0000	1.0000	1.0000	1.0000	1.0000	0.9999	0.9993	0.9969	0.9891	0.9682
	14	1.0000	1.0000	1.0000	1.0000	1.0000	1.0000	0.9999	0.9994	0.9972	0.9904
	15	1.0000	1.0000	1.0000	1.0000	1.0000	1.0000	1.0000	0.9999	0.9995	0.9978
	16	1.0000	1.0000	1.0000	1.0000	1.0000	1.0000	1.0000	1.0000	0.9999	0.9996
	17	1.0000	1.0000	1.0000	1.0000	1.0000	1.0000	1.0000	1.0000	1.0000	1.0000
20	0	0.3585	0.1216	0.0388	0.0115	0.0032	0.0008	0.0000	0.0000	0.0000	0.0000
	1	0.7358	0.3917	0.1756	0.0692	0.0243	0.0076	0.0021	0.0005	0.0001	0.0000
	2	0.9245	0.4769	0.4049	0.2061	0.0913	0.0355	0.0121	0.0036	0.0009	0.0002
	3	0.9841	0.8670	0.6477	0.4114	0.2252	0.1071	0.0444	0.0160	0.0049	0.0013
	4	0.9974	0.9568	0.8298	0.6296	0.4148	0.2375	0.1182	0.0510	0.0189	0.0059
	5	0.9997	0.9887	0.9327	0.8042	0.6172	0.4164	0.2454	0.1256	0.0553	0.0207
	6	1.0000	0.9976	0.9781	0.9133	0.7858	0.6080	0.4166	0.2500	0.1299	0.0577
	7	1.0000	0.9996	0.9941	0.9679	0.8982	0.7723	0.6010	0.4159	0.2520	0.1316
	8	1.0000	0.9999	0.9987	0.9900	0.9591	0.8867	0.7624	0.5956	0.4143	0.2517
	9	1.0000	1.0000	0.9998	0.9974	0.9861	0.9520	0.8782	0.7553	0.5914	0.4119
	10	1.0000	1.0000	1.0000	0.9994	0.9961	0.9829	0.9460	0.8725	0.7507	0.5881
	11	1.0000	1.0000	1.0000	0.9999	0.9991	0.9949	0.9804	0.9435	0.8692	0.7483
	12	1.0000	1.0000	1.0000	1.0000	0.9998	0.9987	0.9940	0.9790	0.9420	0.8684
	13	1.0000	1.0000	1.0000	1.0000	1.0000	0.9997	0.9985	0.9935	0.9786	0.9423
	14	1.0000	1.0000	1.0000	1.0000	1.0000	1.0000	0.9997	0.9984	0.9936	0.9793
	15	1.0000	1.0000	1.0000	1.0000	1.0000	1.0000	1.0000	0.9997	0.9985	0.9941
	16	1.0000	1.0000	1.0000	1.0000	1.0000	1.0000	1.0000	1.0000	0.9997	0.9987
	17	1.0000	1.0000	1.0000	1.0000	1.0000	1.0000	1.0000	1.0000	1.0000	0.9998
	18	1.0000	1.0000	1.0000	1.0000	1.0000	1.0000	1.0000	1.0000	1.0000	1.0000

Normal Distribution

$$F_Z(z) = \frac{1}{\sqrt{2\pi}} \int_{-\infty}^{z} e^{-\frac{z^2}{2}} dz$$

z	0	1	2	3	4	5	6	7	8	9
.0	.5000	.5040	.5080	.5120	.5160	.5199	.5239	.5279	.5319	.5359
.1	.5398	.5478	.5478	.5517	.5557	.5596	.5363	.5675	.5714	.5753

(*Contd.*)

z	0	1	2	3	4	5	6	7	8	9
.2	.5793	.5832	.5871	.5910	.5948	.5987	.6026	.6064	.6103	.6141
.3	.6179	.6217	.6255	.6293	.6331	.6368	.6406	.6443	.6480	.6517
.4	.6554	.6591	.6628	.6664	.6700	.6736	.6772	.6808	.6844	.6879
.5	.6915	.6950	.6985	.7019	.7054	.7088	.7123	.7157	.7190	.7224
.6	.7257	.7291	.7324	.7357	.7389	.7422	.7454	.7486	.7517	.7549
.7	.7580	.7611	.7642	.7673	.7703	.7734	.7974	.7764	.7823	.7852
.8	.7881	.7910	.7939	.7967	.7995	.8023	.8051	.8078	.8106	.8133
.9	.8059	.8186	.8212	.8238	.8264	.8289	.8315	.8340	.8365	.8389
1.0	.8413	.8438	.8461	.8485	.8508	.8531	.8554	.8577	.8599	.8621
1.1	.8643	.8665	.8686	.8708	.8729	.8749	.8770	.8790	.8810	.8830
1.2	.8849	.8869	.8888	.8907	.8925	.8925	.8944	.8980	.8997	.9015
1.3	.9032	.9049	.9066	.9082	.9099	.9115	.9131	.9147	.9162	.9177
1.4	.9192	.9207	.9222	.9236	.9251	.9265	.9278	.9292	.9306	.9319
1.5	.9332	.9345	.9357	.9370	.9382	.9394	.9406	.9418	.9430	.9441
1.6	.9452	.9463	.9474	.9484	.9495	.9505	.9515	.9525	.9535	.9545
1.7	.9554	.9564	.9573	.9583	.9591	.9599	.9608	.9616	.9625	.9633
1.8	.9641	.9648	.9656	.9664	.9671	.9678	.9686	.9693	.9700	.9706
1.9	.9713	.9719	.9726	.9732	.9738	.9744	.9750	.9756	.9762	.9767
2.0	.9772	.9778	.9783	.9788	.9793	.9798	.9803	.9808	.9812	.9817
2.1	.9821	.9826	.9830	.9834	.9839	.9842	.9846	.9850	.9854	.9857
2.2	.9861	.9864	.9868	.9871	.9874	.9878	.9881	.9884	.9887	.9890
2.3	.9893	.9896	.9898	.9901	.9904	.9906	.9909	.9911	.9913	.9916
2.4	.9918	.9920	.9922	.9925	.9927	.9929	.9931	.9932	.9934	.9936
2.5	.9938	.9940	.9941	.9943	.9945	.9946	.9948	.9949	.9951	.9952
2.6	.9953	.9955	.9956	.9957	.9959	.9960	.9961	.9962	.9963	.9964
2.7	.9965	.9966	.9967	.9968	.9969	.9970	.9971	.9972	.9973	.9974
2.8	.9974	.9975	.9976	.9977	.9977	.9978	.9979	.9979	.9980	.9981
2.9	.9981	.9982	.9982	.9983	.9984	.9984	.9985	.9985	.9986	.9986
3.	.9987	.9990	.9993	.9995	.9997	.9998	.9998	.9999	.9999	1.0000

Note 1: If a normal variable X is not "standard," its value must be standardized: $Z = (X - \mu)/\sigma$. Then $F_X(x) = F_Z\left(\dfrac{x-\mu}{\sigma}\right)$.

Note 2: For $z \geq 4$, use $F_z(z) = 1$ to four decimal places; for $z \leq -4$, $F_z(z) = 0$ to four decimal places.

Note 3: The entries opposite $z = 3$ are for 3.0, 3.1, 3.2, etc.

Note 4: For $z < 0$ use $F_z(z) = 1 - F_z(-z)$.

Source: Probability & Statistics with Reliability, Queueing and Computer Science Applications, by K.S. Trivedi, Published by Prentice Hall of India Private Limited, 1988.

Poisson Distribution

$$F(r) = \sum_{r=0}^{x} \frac{\lambda^r e^{-\lambda}}{r!}$$

x	λ = .1	.2	.3	.4	.5	.6	.7	.8	.9	1.0
0	.9048	.8187	.7408	.6703	.6065	.5488	.4966	.4493	.4066	.3679
1	.9953	.9825	.9631	.9384	.9098	.8781	.8442	.8088	.7725	.7358
2	.9998	.9989	.9964	.9921	.9856	.9769	.9659	.9526	.9371	.9197
3	1.0000	.9999	.9997	.9992	.9982	.9966	.9942	.9909	.9865	.9810
4	1.0000	1.0000	1.0000	.9999	.9998	.9996	.9992	.9986	.9977	.9963
5	1.0000	1.0000	1.0000	1.0000	1.0000	1.0000	.9999	.9998	.9997	.9994
6	1.0000	1.0000	1.0000	1.0000	1.0000	1.0000	1.0000	1.0000	1.0000	.9999
7	1.0000	1.0000	1.0000	1.0000	1.0000	1.0000	1.0000	1.0000	1.0000	1.0000

x	λ = 1.1	1.2	1.3	1.4	1.5	1.6	1.7	1.8	1.9	2.0
0	.3329	.3012	.2725	.2466	.2231	.2019	.1827	.1653	.1496	.1353
1	.6990	.6626	.6268	.5918	.5578	.5249	.4932	.4628	.4337	.4060
2	.9004	.8796	.8571	.8335	.8088	.7043	.7572	.7306	.7037	.6767
3	.9743	.9662	.9569	.9463	.9344	.9212	.9068	.8913	.8747	.8571
4	.9946	.9923	.9893	.9857	.9814	.9763	.9704	.9636	.9559	.9473
5	.9990	.9985	.9978	.9968	.9955	.9940	.9920	.9868	.9868	.9834
6	.9999	.9997	.9996	.9994	.9991	.9987	.9981	.9974	.9966	.9955
7	1.0000	1.0000	.9999	.9999	.9998	.9997	.9996	.9994	.9992	.9989
8	1.0000	1.0000	1.0000	1.0000	1.0000	1.0000	.9999	.9999	.9998	.9998
9	1.0000	1.0000	1.0000	1.0000	1.0000	1.0000	1.0000	1.0000	1.0000	1.0000

x	λ = 2.1	2.2	2.3	2.4	2.5	2.6	2.7	2.8	2.9	3.0
0	.1225	.1108	.1002	.0907	.0821	.0743	.0672	.0608	.0550	.0498
1	.3796	.3546	.3309	.3084	.2873	.2674	.2487	.2311	.2146	.1991
2	.6496	.6227	.5960	.5967	.5438	.5184	.4396	.4695	.4460	.4232
3	.8386	.8194	.7993	.7787	.7576	.7360	.7141	.6919	.6696	.6472
4	.9379	.9275	.9162	.9041	.8912	.8774	.8629	.8477	.8318	.8153
5	.9796	.9751	.9700	.9643	.9580	.9510	.9433	.9349	.9258	.9161
6	.9941	.9925	.9906	.9884	.9858	.9828	.9794	.9756	.9713	.9665
7	.9985	.9980	.9974	.9967	.9958	.9947	.9934	.9919	.9901	.9881
8	.9997	.9995	.9994	.9991	.9989	.9985	.9981	.9976	.9968	.9962
9	.9999	.9999	.9999	.9998	.9997	.9996	.9995	.9993	.9991	.9989
10	1.0000	1.0000	1.0000	1.0000	.9999	.9999	.9999	.9998	.9998	.9997
11	1.0000	1.0000	1.0000	1.0000	1.0000	1.0000	1.0000	1.0000	.9999	.9999
12	1.0000	1.0000	1.0000	1.0000	1.0000	1.0000	1.0000	1.0000	1.0000	1.0000

x	λ = 3.1	3.2	3.3	3.4	3.5	3.6	3.7	3.8	3.9	4.0
0	.0450	.0408	.0369	.0334	.0302	.0273	.0247	.0224	.0202	.0183
1	.1857	.1712	.1586	.1468	.1359	.1257	.1162	.1074	.0992	.0916
2	.4012	.3799	.3594	.3397	.3208	.3027	.2854	.2689	.2531	.2381
3	.6248	.6025	.5803	.5584	.5366	.5152	.4942	.4735	.4532	.4335
4	.7982	.7806	.7626	.7442	.7254	.7064	.6872	.6678	.6484	.6288
5	.9057	.8946	.8829	.8705	.8576	.8441	.8301	.8156	.8006	.7851
6	.9612	.9554	.9490	.9421	.9247	.9267	.9182	.9091	.8995	.8893
7	.9858	.9832	.9802	.9769	.9733	.9692	.9648	.9599	.9546	.9489
8	.9953	.9943	.9931	.9917	.9901	.9883	.9863	.9840	.9815	.9786
9	.9986	.9982	.9978	.9973	.9967	.9960	.9952	.9942	.9931	.9919
10	.9996	.9995	.9994	.9992	.9990	.9987	.9984	.9981	.9977	.9972
11	.9999	.9999	.9998	.9998	.9997	.9996	.9995	.9994	.9993	.9991
12	1.0000	1.0000	1.0000	.9999	.9999	.9999	.9999	.9998	.9998	.9997
13	1.0000	1.0000	1.0000	1.0000	1.0000	1.0000	1.0000	1.0000	.9999	.9999
14	1.0000	1.0000	1.0000	1.0000	1.0000	1.0000	1.0000	1.0000	1.0000	1.0000

x	λ = 4.1	4.2	4.3	4.4	4.5	4.6	4.7	4.8	4.9	5.0
0	.0166	.0150	.0136	.0123	.0111	.0101	.0091	.0082	.0074	.0067
1	.0845	.0780	.0719	.0663	.0611	.0563	.0518	.0477	.0439	.0404
2	.2238	.2102	.1974	.1851	.1736	.1626	.1523	.1425	.1333	.1247
3	.4142	.3954	.3772	.3594	.3423	.3257	.3097	.2942	.2793	.2650
4	.6093	.5898	.5704	.3512	.5321	.5132	.4946	.4763	.4582	.4405
5	.7693	.7531	.7367	.7190	.7029	.6858	.6684	.6510	.6335	.6160
6	.8786	.8675	.8558	.8436	.8311	.8180	.8046	.7908	.7767	.7622

(Contd.)

(*Contd.*)

x	λ = 4.1	4.2	4.3	4.4	4.5	4.6	4.7	4.8	4.9	5.0
7	.9427	.9361	.9290	.9214	.9134	.9049	.8960	.8867	.8769	.8066
8	.9755	.9721	.9683	.9642	.9597	.9549	.9497	.9442	.9382	.9319
9	.9905	.9889	.9871	.9851	.9829	.9805	.9778	.9749	.9717	.9682
10	.9966	.9959	.9952	.9943	.9933	.9922	.9910	.9896	.9880	.9863
11	.9989	.9986	.9983	.9980	.9976	.9971	.9966	.9960	.9953	.9945
12	.9997	.9996	.9995	.9993	.9992	.9990	.9988	.9986	.9983	.9980
13	.9999	.9999	.9998	.9998	.9997	.9997	.9996	.9995	.9994	.9993
14	1.0000	1.0000	1.0000	.9999	.9999	.9999	.9999	.9999	.9998	.9998
15	1.0000	1.0000	1.0000	1.0000	1.0000	1.0000	1.0000	1.0000	.9999	.9999

Poisson Distribution

x	λ = 5.1	5.2	5.3	5.4	5.5	5.6	5.7	5.8	5.9	6.0
0	.0061	.0055	.0050	.0045	.0041	.0037	.0033	.0030	.0027	.0025
1	.0372	.0342	.0314	.0266	.0244	.0244	.0224	.0206	.0189	.0025
2	.1165	.1088	.1016	.0948	.0884	.0824	.0768	.0715	.0666	.0620
3	.2513	.2381	.2254	.2133	.0217	.1906	.1800	.1700	.1604	.1512
4	.4231	.4061	.3895	.3733	.3575	.3422	.3272	.3172	.2987	.2851
5	.5984	.5809	.5635	.5461	.5289	.5119	.4950	.4783	.4619	.4457
6	.7474	.7324	.7171	.7017	.6860	.6703	.6544	.6384	.6224	.6063
7	.8560	.8449	.8335	.8217	.8095	.7970	.7841	.7710	.7576	.7440
8	.9252	.9181	.9106	.9027	.8944	.8857	.8766	.8672	.8574	.8472
9	.9644	.9603	.9559	.9512	.9462	.9409	.9352	.9292	.9228	.9161
10	.9844	.9823	.9800	.9775	.9747	.9718	.9686	.9651	.9614	.9574
11	.9937	.9927	.9916	.9904	.9890	.9875	.9859	.9841	.9821	.9799
12	.9976	.9972	.9967	.9962	.9955	.9949	.9941	.9932	.9922	.9912
13	.9992	.9990	.9988	.9986	.9983	.9980	.9977	.9973	.9969	.9964
14	.9997	.9997	.9996	.9995	.9994	.9993	.9991	.9990	.9998	.9986
15	.9999	.9999	.9999	.9998	.9998	.9998	.9997	.9996	.9996	.9995
16	1.0000	1.0000	1.0000	.9999	.9999	.9999	.9999	.9999	.9999	.9998
17	1.0000	1.0000	1.0000	1.0000	1.0000	1.0000	1.0000	1.0000	1.0000	.9999

x	λ = 6.1	6.2	6.3	6.4	6.5	6.6	6.7	6.8	6.9	7.0
0	.0022	.0020	.0018	.0017	.0015	.0014	.0012	.0011	.0010	.0009
1	.0159	.0146	.0134	.0123	.0113	.0193	.0095	.0087	.0080	.0073
2	.0577	.0536	.0498	.0463	.0430	.0400	.0371	.0344	.0320	.0296
3	.1425	.1342	.1264	.1189	.1118	.1052	.0988	.0928	.0871	.0818
4	.2719	.2592	.2469	.2351	.2237	.2127	.2022	.1920	.1823	.1730
5	.4298	.4141	.3988	.3837	.3690	.3547	.3406	.3270	.3137	.3007
6	.5902	.5742	.5582	.5423	.5265	.5108	.4953	.4799	.4647	.4497
7	.7301	.7160	.7017	.6873	.6723	.6581	.6433	.6285	.6136	.5987
8	.8367	.8259	.8148	.8033	.7916	.7796	.7673	.7548	.7420	.7291
9	.9090	.9016	.8939	.8858	.8774	.8686	.8596	.8502	.8405	.8305
10	.9531	.9486	.9437	.9358	.9332	.9274	.9214	.9151	.9084	.9015
11	.9776	.9750	.9723	.9693	.9661	.9627	.9591	.9552	.9510	.9467
12	.9900	.9887	.9873	.9857	.9840	.9821	.9801	.9779	.9755	.9730
13	.9958	.9952	.9945	.9937	.9929	.9920	.9909	.9898	.9885	.9872
14	.9984	.9981	.9978	.9974	.9970	.9966	.9961	.9956	.9950	.9943
15	.9994	.9993	.9992	.9990	.9988	.9986	.9984	.9982	.9979	.9976
16	.9998	.9997	.9997	.9996	.9996	.9995	.9994	.9993	.9992	.9990
17	.9999	.9999	.9999	.9999	.9998	.9998	.9998	.9997	.9997	.9996
18	1.0000	1.0000	1.0000	1.0000	.9999	.9999	.9999	.9999	.9999	.9999
19	1.0000	1.0000	1.0000	1.0000	1.0000	1.0000	1.0000	1.0000	1.0000	1.0000

x	λ = 7.1	7.2	7.3	7.4	7.5	7.6	7.7	7.8	7.9	8.0
0	.0008	.0007	.0007	.0006	.0006	.0005	.0005	.0004	.0004	.0003
1	.0067	.0061	.0056	.0051	.0047	.0043	.0039	.0036	.0033	.0030
2	.0275	.0255	.0236	.0219	.0203	.0188	.0174	.0161	.0149	.0138
3	.0767	.0719	.0674	.0632	.0591	.0554	.0518	.0485	.0453	.0424
4	.1641	.1555	.1473	.1395	.1321	.1249	.1181	.1117	.1055	.0996
5	.2881	.2759	.2640	.2526	.2414	.2307	.2203	.2103	.2006	.1912
6	.4349	.4204	.4060	.3920	.3782	.3646	.3514	.3384	.3257	.3134
7	.5838	.5689	.5541	.5393	.5246	.5100	.4596	.4812	.4670	.4530
8	.7160	.7027	.6892	.6757	.6620	.6482	.6343	.6204	.6065	.5925
9	.8202	.8096	.7988	.7877	.7764	.7649	.7531	.7411	.7290	.7166
10	.8942	.8867	.8788	.8707	.8622	.8535	.8445	.8352	.8257	.8159
11	.9420	.9371	.9319	.9265	.9208	.9148	.9085	.9020	.8952	.8881
12	.9703	.9673	.9642	.9609	.9573	.9536	.9496	.9454	.9309	.9362
13	.9857	.9841	.9824	.9805	.9784	.9762	.9739	.9714	.9087	.9658
14	.9935	.9927	.9918	.9908	.9897	.9886	.9873	.9859	.9844	.9827
15	.9972	.9969	.9964	.9959	.9954	.9948	.9941	.9934	.9926	.9918
16	.9989	.9987	.9985	.9983	.9980	.9978	.9974	.9971	.9967	.9963
17	.9996	.9995	.9994	.9993	.9992	.9991	.9989	.9988	.9986	.9984
18	.9998	.9998	.9998	.9997	.9997	.9996	.9996	.9995	.9994	.9993
19	.9999	.9999	.9999	.9999	.9999	.9999	.9998	.9998	.9998	.9997
20	1.0000	1.0000	1.0000	1.0000	1.0000	1.0000	.9999	.9999	.9999	.9999
21	1.0000	1.0000	1.0000	1.0000	1.0000	1.0000	1.0000	1.0000	1.0000	1.0000

x	λ = 8.1	8.2	8.3	8.4	8.5	8.6	8.7	8.8	8.9	9.0
0	.0003	.0003	.0002	.0002	.0002	.0002	.0002	.0002	.0001	.0001
1	.0028	.0025	.0023	.0021	.0019	.0018	.0016	.0015	.0014	.0112
2	.0127	.0118	.0109	.0100	.0093	.0086	.0079	.0073	.0068	.0062
3	.0396	.0370	.0346	.0323	.0301	.0281	.0262	.0244	.0228	.0212
4	.0940	.0887	.0837	.0789	.0744	.0701	.0660	.0621	.0584	.0550
5	.1822	.1736	.1653	.1573	.1496	.1422	.1352	.1284	.1219	.1157
6	.3013	.2896	.2781	.2670	.2562	.2457	.2355	.2256	.2160	.2068
7	.4391	.4254	.4119	.3987	.3856	.3728	.3602	.3478	.3357	.3239
8	.5786	.5647	.5507	.5369	.5231	.5094	.4958	.4823	.4689	.4557
9	.7041	.6915	.6788	.6659	.6530	.6400	.6269	.6137	.6006	.5874
10	.8058	.7955	.7850	.7743	.7634	.7522	.7409	.7294	.7178	.7060
11	.8807	.8731	.8652	.8571	.8487	.8400	.8311	.8220	.8126	.8030
12	.9313	.9261	.9207	.9150	.9091	.9029	.8965	.8898	.8829	.8758
13	.9628	.9595	.9561	.9524	.9486	.9445	.9403	.9358	.9311	.9261
14	.9810	.9791	.9771	.9749	.9726	.9701	.9675	.9647	.9617	.9585
15	.9908	.9898	.9887	.9875	.9862	.9848	.9832	.9816	.9798	.9780
16	.9958	.9953	.9947	.9941	.9934	.9926	.9918	.9909	.9899	.9889
17	.9982	.9979	.9977	.9973	.9970	.9966	.9962	.9957	.9952	.9947
18	.9992	.9991	.9990	.9989	.9987	.9985	.9983	.9981	.9978	.9976
19	.9997	.9997	.9996	.9995	.9995	.9994	.9993	.9992	.9991	.9989
20	.9999	.9999	.9998	.9998	.9998	.9998	.9997	.9997	.9996	.9996
21	1.0000	1.0000	.9999	.9999	.9999	.9999	.9999	.9999	.9998	.9998
22	1.0000	1.0000	1.0000	1.0000	1.0000	1.0000	1.0000	1.0000	.9999	.9999

x	λ = 9.1	9.2	9.3	9.4	9.5	9.6	9.7	9.8	9.9	10.0
0	.0001	.0001	.0001	.0001	.0001	.0001	.0001	.0001	.0001	.0000
1	.0011	.0010	.0009	.0009	.0008	.0007	.0007	.0006	.0005	.0005
2	.0058	.0053	.0049	.0045	.0042	.0038	.0035	.0033	.0030	.0028

(*Contd.*)

(*Contd.*)

x	λ = 9.1	9.2	9.3	9.4	9.5	9.6	9.7	9.8	9.9	10.0
3	.0198	.0184	.0172	.0160	.0149	.0138	.0129	.0120	.0111	.0103
4	.0517	.0486	.0456	.0429	.0403	.0378	.0355	.0333	.0312	.0293
5	.1098	.1041	.0986	.0935	.0885	.0838	.0793	.0750	.0710	.0671
6	.1978	.1892	.1808	.1727	.1649	.1574	.1502	.1433	.1366	.1301
7	.3123	.3010	.2900	.2792	.2687	.2584	.2485	.2388	.2294	.2202
8	.4126	.4296	.4168	.4042	.3918	.3798	.3676	.3558	.3442	.3328
9	.5742	.5611	.5479	.5349	.5218	.5089	.4960	.4832	.4705	.4579
10	.6941	.6820	.6699	.6576	.6453	.6329	.6205	.6080	.5955	.5830
11	.7932	.8732	.7730	.7626	.7520	.7412	.7303	.7193	.7081	.6968
12	.8684	.8607	.8529	.8448	.8364	.8279	.8191	.8101	.8009	.7916
13	.9210	.9156	.9100	.9042	.8981	.8919	.8853	.8786	.8716	.8615
14	.9552	.9517	.9480	.9441	.9400	.9357	.9312	.9265	.9216	.9165
15	.9760	.9738	.9715	.9691	.9665	.9638	.9609	.9579	.9546	.9513
16	.9878	.9865	.9852	.9838	.9823	.9806	.9789	.9770	.9751	.9730
17	.9941	.9934	.9927	.9919	.9911	.9902	.9892	.9881	.9870	.9857
18	.9973	.9969	.9966	.9962	.9957	.9952	.9947	.9941	.9935	.9928
19	.9988	.9986	.9985	.9983	.9980	.9978	.9975	.9972	.9969	.9965
20	.9995	.9994	.9993	.9992	.9991	.9990	.9989	.9987	.9986	.9984
21	.9998	.9998	.9997	.9997	.9996	.9996	.9995	.9995	.9994	.9994
22	.9999	.9999	.9999	.9999	.9999	.9998	.9998	.9998	.9997	.9997
23	1.0000	1.0000	1.0000	1.0000	.9999	.9999	.9999	.9999	.9999	.9999
24	1.0000	1.0000	1.0000	1.0000	1.0000	1.0000	1.0000	1.0000	1.0000	1.0000

x	λ = 11	12	13	14	15	16	17	18	19	20
0	.0000	.0000	.0000	.0000	.0000	.0000	.0000	.0000	.0000	.0000
1	.0002	.0001	.0000	.0000	.0000	.0000	.0000	.0000	.0000	.0000
2	.0012	.0005	.0002	.0001	.0000	.0000	.0000	.0000	.0000	.0000
3	.0049	.0023	.0011	.0005	.0002	.0001	.0000	.0000	.0000	.0000
4	.0151	.0076	.0037	.0018	.0009	.0004	.0002	.0001	.0000	.0000
5	.0375	.0203	.0107	.0055	.0028	.0014	.0007	.0003	.0002	.0001
6	.0786	.0458	.0259	.0142	.0076	.0040	.0021	.0010	.0005	.0003
7	.1432	.0895	.0540	.0316	.0180	.0100	.0054	.0029	.0015	.0008
8	.2320	.1550	.0998	.0621	.0374	.0220	.0126	.0071	.0039	.0021
9	.3405	.2424	.1658	.1094	.0699	.0433	.0261	.0154	.0089	.0050
10	.4599	.3472	.2517	.1757	.1185	.0774	.0491	.0304	.0183	.0108
11	.5793	.4616	.3532	.2600	.1848	.1270	.0847	.0549	.0347	.0214
12	.6887	.5760	.4631	.3585	.2676	.1931	.1350	.0917	.0606	.0390
13	.7813	.6815	.5730	.4644	.3632	.2745	.2009	.1426	.0984	.0661
14	.8540	.7720	.6751	.5704	.4657	.3675	.2808	.2081	.1497	.1049
15	.9074	.9444	.7636	.6694	.5681	.4667	.3715	.2867	.2148	.1565
16	.9441	.8987	.8355	.7559	.6641	.5660	.4677	.3751	.2920	.2211
17	.9678	.9370	.8905	.8272	.7489	.6593	.5640	.4686	.3784	.2970
18	.9823	.9626	.9302	.8826	.8195	.7423	.6550	.5622	.4695	.3814
19	.9907	.9787	.9573	.9325	.8752	.8122	.7363	.6509	.5606	.4703
20	.9953	.9884	.9750	.9521	.9170	.8682	.8055	.7307	.6472	.5591
21	.9977	.9939	.9859	.9712	.9469	.9108	.8615	.7991	.7255	.6437
22	.9990	.9970	.9924	.9833	.9673	.9418	.9047	.8551	.7931	.7206
23	.9995	.9985	.9960	.9907	.9805	.9633	.9367	.8989	.8190	.7875
24	.9998	.9993	.9980	.9950	.9888	.9777	.9594	.9317	.8933	.8432
25	.9999	.9997	.9990	.9974	.9938	.9869	.9748	.9554	.9269	.8878
26	1.0000	.9999	.9995	.9987	.9967	.9925	.9848	.9718	.9514	.9221
27	1.0000	.9999	.9998	.9994	.9983	.9959	.9912	.9827	.0687	.0475

(*Contd.*)

(*Contd.*)

x	λ = 11	12	13	14	15	16	17	18	19	20
28	1.0000	1.0000	.9999	.9997	.9991	.9978	.9950	.9897	.9805	.9657
29	1.0000	1.0000	1.0000	.9999	.9996	.9989	.9973	.9941	.9881	.9782
30	1.0000	1.0000	1.0000	.9999	.9998	.9994	.9986	.9967	.9930	.9865
31	1.0000	1.0000	1.0000	1.0000	.9999	.9997	.9993	.9982	.9960	.9919
32	1.0000	1.0000	1.0000	1.0000	1.0000	.9999	.9996	.9990	.9978	.9953
33	1.0000	1.0000	1.0000	1.0000	1.0000	.9999	.9998	.9995	.9988	.9973
34	1.0000	1.0000	1.0000	1.0000	1.0000	1.0000	.9999	.9998	.9994	.9985
35	1.0000	1.0000	1.0000	1.0000	1.0000	1.0000	1.0000	.9999	.9997	.9992
36	1.0000	1.0000	1.0000	1.0000	1.0000	1.0000	1.0000	.9999	.9998	.9996
37	1.0000	1.0000	1.0000	1.0000	1.0000	1.0000	1.0000	1.0000	.9999	.9998
38	1.0000	1.0000	1.0000	1.0000	1.0000	1.0000	1.0000	1.0000	1.0000	.9999
39	1.0000	1.0000	1.0000	1.0000	1.0000	1.0000	1.0000	1.0000	1.0000	.9999

Based on 'Probability & Statistics' by A.M. Mathai & P.N. Rathie, Macmillan India Limited, 1988 (2nd edition).]

Values of t_α*

ν	α = 0.10	α = 0.05	α = 0.025	α = 0.01	α = 0.005	ν
1	3.078	6.314	12.706	31.821	63.657	1
2	1.886	2.920	4.303	6.965	9.925	2
3	1.638	2.353	3.182	4.541	5.841	3
4	1.533	2.132	2.776	3.747	4.604	4
5	1.476	2.015	2.571	3.365	4.032	5
6	1.440	1.943	2.447	3.143	3.707	6
7	1.415	1.895	2.365	2.998	3.499	7
8	1.397	1.860	2.306	2.896	3.355	8
9	1.383	1.833	2.262	2.821	3.250	9
10	1.372	1.812	2.228	2.764	3.169	10
11	1.363	1.796	2.201	2.718	3.106	11
12	1.356	1.782	2.179	2.681	3.055	12
13	1.350	1.771	2.160	2.650	3.012	13
14	1.345	1.761	2.145	2.624	2.977	14
15	1.341	1.753	2.131	2.602	2.947	15
16	1.337	1.746	2.120	2.583	2.921	16
17	1.333	1.740	2.110	2.567	2.898	17
18	1.330	1.734	2.101	2.552	2.878	18
19	1.328	1.729	2.093	2.539	2.861	19
20	1.325	1.725	2.086	2.528	2.845	20
21	1.323	1.721	2.080	2.518	2.831	21
22	1.321	1.717	2.074	2.508	2.819	22
23	1.319	1.714	2.069	2.500	2.807	23
24	1.318	1.711	2.064	2.492	2.797	24
25	1.316	1.708	2.060	2.485	2.787	25
26	1.315	1.706	2.056	2.479	2.779	26
27	1.314	1.703	2.052	2.473	2.771	27
28	1.313	1.701	2.048	2.467	2.763	28
29	1.311	1.699	2.045	2.462	2.756	29
inf.	1.282	1.645	1.960	2.326	2.576	inf.

Values of χ_α^{2*}

α → ν ↓	a = 0.995	a = 0.99	a = 0.975	a = 0.95	a = 0.05	a = 0.025	a = 0.01	a = 0.005	← α ν ↓
1	0.0000393	0.000157	0.000982	0.00393	3.841	5.024	6.635	7.879	1
2	0.0100	0.0201	0.0506	0.103	5.991	7.378	9.210	10.597	2
3	0.0717	0.115	0.216	0.352	7.815	9.348	11.345	12.838	3
4	0.207	0.297	0.484	0.711	9.488	11.143	13.277	14.860	4
5	0.412	0.554	0.831	1.145	11.070	12.832	15.056	16.750	5
6	0.676	0.872	1.237	1.635	12.592	14.449	16.812	18.548	6
7	0.989	1.239	1.690	2.167	14.067	16.013	18.475	20.278	7
8	1.344	1.646	2.180	2.733	15.507	17.535	20.090	21.955	8
9	1.735	2.088	2.700	3.325	16.919	19.023	21.666	23.589	9
10	2.156	2.558	3.247	3.940	18.307	20.483	23.209	25.188	10
11	2.603	3.053	3.816	4.575	19.675	21.920	24.725	26.757	11
12	3.074	3.571	4.404	5.226	21.026	23.337	26.217	28.300	12
13	3.565	4.107	5.009	5.892	22.362	24.736	27.688	29.819	13
14	4.075	4.660	5.629	6.571	23.685	26.119	29.141	31.319	14
15	4.601	5.229	6.262	7.261	24.996	27.488	30.578	32.801	15
16	5.142	5.812	6.908	7.962	26.296	28.845	32.000	34.267	16
17	5.697	6.408	7.564	8.672	27.587	30.191	33.409	35.718	17
18	6.265	7.015	8.231	9.390	28.869	31.526	34.805	37.156	18
19	6.844	7.633	8.907	10.117	30.144	32.852	36.191	38.582	19
20	7.434	8.260	9.591	10.851	31.410	34.170	37.566	39.997	20
21	8.034	8.897	10.283	11.591	32.671	35.479	38.932	41.401	21
22	8.643	9.542	10.982	12.338	33.924	36.781	40.289	42.796	22
23	9.260	10.196	11.689	13.091	35.172	38.076	41.638	44.181	23
24	9.886	10.856	12.401	13.484	36.415	39.364	42.980	45.558	24
25	10.520	11.524	13.120	14.611	37.652	40.646	44.314	46.928	25
26	11.160	12.198	13.844	15.379	38.885	41.923	45.642	48.290	26
27	11.808	12.879	14.573	16.151	40.113	43.194	46.963	49.645	27
28	12.461	13.565	15.308	16.928	41.337	44.461	48.278	50.993	28
29	13.121	14.256	16.047	17.708	42.557	45.772	49.588	52.336	29
30	13.787	14.953	16.791	18.493	43.773	46.979	50.892	53.672	30
40	20.706	22.164	24.433	26.509	55.758	59.342	63.691	66.766	40
50	27.991	29.707	32.357	34.764	67.505	71.420	76.154	79.490	50
60	35.535	37.485	40.482	43.118	79.082	83.298	88.379	91.952	60
70	43.275	45.442	48.758	51.739	90.531	95.023	100.425	104.215	70
80	51.172	53.540	57.153	60.391	101.879	106.629	112.329	116.321	80
90	59.196	61.754	65.646	69.126	113.145	118.136	124.116	128.299	90
100	67.328	70.065	74.222	77.929	124.342	129.561	135.807	140.169	100

Values of $F^*_{0.05}$

v_2 = degrees of freedom for denominator	v_1 = Degrees of freedom for numerator 1	2	3	4	5	6	7	8	9	10	12	15	20	24	30	40	60	120	∞
1	161	200	216	225	230	234	237	239	241	242	244	246	248	249	250	251	252	253	254
2	18.50	19.00	19.20	19.20	19.30	19.30	19.40	19.40	19.40	19.40	19.40	19.40	19.40	19.50	19.50	19.50	19.50	19.50	19.50
3	10.10	9.55	9.28	9.12	9.01	8.94	8.89	8.85	8.81	8.79	8.74	8.70	8.66	8.64	8.62	8.59	8.57	8.55	8.53
4	7.71	6.94	6.59	6.39	6.26	6.16	6.09	6.04	6.00	5.96	5.91	5.86	5.80	5.77	5.75	5.72	5.69	5.66	5.63
5	6.61	5.79	5.41	5.19	5.05	4.95	4.88	4.82	4.77	4.74	4.68	4.62	4.56	4.53	4.50	4.46	4.43	4.40	4.37
6	5.99	5.14	4.76	4.53	4.39	4.28	4.21	4.15	4.10	4.06	4.00	3.94	3.87	3.84	3.81	3.77	3.74	3.70	3.67
7	5.59	4.74	4.35	4.12	3.97	3.87	3.79	3.73	3.68	3.64	3.57	3.51	3.44	3.41	3.38	3.34	3.30	3.27	3.23
8	5.32	4.46	4.07	3.84	3.69	3.58	3.50	3.44	3.39	3.35	3.28	3.22	3.15	3.12	3.08	3.04	3.01	2.97	2.93
9	5.12	4.26	3.86	3.63	3.48	3.37	3.29	3.23	3.18	3.14	3.07	3.01	2.94	2.90	2.86	2.83	2.79	2.75	2.71
10	4.96	4.10	3.71	3.48	3.33	3.22	3.14	3.07	3.02	2.98	2.91	2.85	2.77	2.74	2.70	2.66	2.62	2.58	2.54
11	4.84	3.98	3.59	3.36	3.20	3.09	3.01	2.95	2.90	2.85	2.79	2.72	2.65	2.61	2.57	2.53	2.49	2.45	2.40
12	4.75	3.89	3.49	3.26	3.11	3.00	2.91	2.85	2.80	2.75	2.69	2.62	2.54	2.51	2.47	2.38	2.38	2.30	2.30
13	4.67	3.81	3.41	3.18	3.03	2.92	2.83	2.77	2.71	2.67	2.60	2.53	2.46	2.42	2.38	2.34	2.30	2.25	2.21
14	4.60	3.74	3.34	3.11	2.96	2.85	2.76	2.70	2.65	2.60	2.53	2.46	2.39	2.35	2.31	2.27	2.22	2.18	2.13
15	4.54	3.68	3.29	3.06	2.90	2.79	2.71	2.64	2.59	2.54	2.48	2.40	2.33	2.29	2.25	2.20	2.16	2.11	2.07
16	4.49	3.63	3.24	3.01	2.85	2.74	2.66	2.59	2.54	2.49	2.42	2.35	2.28	2.24	2.19	2.15	2.11	2.06	2.01
17	3.45	3.59	3.20	2.96	2.81	2.70	2.61	2.55	2.49	2.45	2.38	2.31	2.23	2.19	2.15	2.10	2.06	2.01	1.96
18	4.41	3.55	3.16	2.93	2.77	2.66	2.58	2.51	2.46	2.41	2.34	2.27	2.19	2.15	2.11	2.06	2.02	1.97	1.93
19	4.38	3.52	3.13	2.90	2.74	2.63	2.54	2.48	2.42	2.38	2.31	2.23	2.16	2.11	2.07	2.03	1.98	1.93	1.88
20	4.35	3.49	3.10	2.87	2.71	2.60	2.51	2.45	2.39	2.35	2.28	2.20	2.12	2.08	2.04	1.99	1.95	1.90	1.84
21	4.32	3.47	3.07	2.84	2.68	2.57	2.49	2.42	2.37	2.32	2.25	2.18	2.10	2.05	2.01	1.96	1.92	1.87	1.81
22	4.30	3.44	3.05	2.82	2.66	2.55	2.46	2.40	2.34	2.30	2.23	2.15	2.07	2.03	1.98	1.94	1.89	1.84	1.78
23	4.28	3.42	3.03	2.80	2.64	2.53	2.44	2.37	2.32	2.27	2.20	2.13	2.05	2.01	1.96	1.91	1.86	1.81	1.76
24	4.26	3.40	3.01	2.78	2.62	2.51	2.42	2.36	2.30	2.25	2.18	2.11	2.03	1.98	1.94	1.89	1.84	1.79	1.73
25	4.24	3.39	2.99	2.76	2.60	2.49	2.40	2.34	2.28	2.24	2.16	2.09	2.01	1.96	1.92	1.87	1.82	1.77	1.71
30	4.17	3.32	2.92	2.69	2.53	2.42	2.33	2.27	2.21	2.16	2.09	2.01	1.93	1.89	1.84	1.79	1.74	1.68	1.62
40	4.08	3.23	2.84	2.61	2.45	2.34	2.25	2.18	2.12	2.08	2.00	1.92	1.84	1.79	1.74	1.69	1.64	1.58	1.51
60	4.00	3.15	2.76	2.53	2.37	2.25	2.17	2.10	2.04	1.99	1.92	1.84	1.75	1.70	1.65	1.59	1.53	1.47	1.39
120	3.92	3.07	2.68	2.45	2.29	2.18	2.09	2.02	1.96	1.91	1.83	1.75	1.66	1.61	1.55	1.50	1.43	1.35	1.25
∞	3.84	3.00	2.60	2.37	2.21	2.10	2.01	1.94	1.88	1.83	1.75	1.67	1.57	1.52	1.46	1.39	1.32	1.22	1.00

Values of $F^*_{0.01}$

v_2 = degrees of freedom for denominator	v_1 = Degrees of freedom for numerator: 1	2	3	4	5	6	7	8	9	10	12	15	20	24	30	40	60	120	∞
1	4,052	5,000	5,403	5,625	5,764	5,859	5,928	5,982	6,023	6,056	6,106	6,157	6,209	6,235	6,261	6,287	6,313	6,339	6,366
2	98.50	99.00	99.20	99.20	99.30	99.30	99.40	99.40	99.40	99.40	99.40	99.40	99.40	99.50	99.50	99.50	99.50	99.50	99.50
3	34.10	30.80	29.50	28.70	28.20	27.90	27.70	27.50	27.30	27.20	27.10	26.90	26.70	26.60	26.50	26.40	26.30	26.20	26.10
4	21.20	18.00	16.70	16.00	15.50	15.20	15.00	14.80	14.70	14.50	14.50	14.20	14.00	13.90	13.80	13.70	13.70	13.60	13.50
5	16.30	13.30	12.10	11.40	11.00	10.70	10.50	10.30	10.20	10.10	9.89	9.72	9.55	9.47	9.38	9.29	9.20	9.11	9.02
6	13.70	10.90	9.78	9.15	8.75	8.47	8.26	8.10	7.98	7.87	7.72	7.56	7.40	7.31	7.23	7.14	7.06	6.97	6.88
7	12.20	9.55	8.45	7.85	7.46	7.19	6.99	6.84	6.72	6.62	6.47	6.31	6.16	6.07	5.99	5.91	5.82	5.74	5.65
8	11.30	8.65	7.59	7.01	6.63	6.37	6.18	6.03	5.91	5.81	5.67	5.52	5.36	5.28	5.20	5.12	5.03	4.95	4.83
9	10.60	8.02	6.99	6.42	6.06	5.80	5.61	5.47	5.35	5.26	5.11	4.96	4.81	4.73	4.65	4.57	4.48	4.40	4.31
10	10.00	7.56	6.55	5.99	5.64	5.39	5.20	5.06	4.94	4.85	4.71	4.56	4.41	4.33	4.25	4.17	4.08	4.00	3.91
11	9.65	7.21	6.22	5.67	5.32	5.07	4.89	4.74	4.63	4.54	4.40	4.25	4.10	4.02	3.94	3.86	3.78	3.96	3.60
12	9.33	6.93	5.95	5.41	5.06	4.82	4.64	4.50	4.39	4.30	4.16	4.01	3.86	3.78	3.70	3.62	3.54	3.45	3.36
13	9.07	6.70	5.74	5.21	4.86	4.62	4.44	4.30	4.19	4.10	3.96	3.82	3.66	3.59	3.51	3.43	3.34	3.25	3.17
14	8.86	6.51	5.56	5.04	4.70	4.46	4.28	4.14	4.03	3.94	3.80	3.66	3.51	3.43	3.35	3.27	3.18	3.09	3.00
15	8.68	6.36	5.42	4.89	4.56	4.32	4.14	4.00	3.89	3.80	3.67	3.52	3.37	3.29	3.21	3.13	3.05	2.96	2.87
16	8.53	6.23	5.29	4.77	4.44	4.20	4.03	3.89	3.78	3.69	3.55	3.41	3.26	3.18	3.10	3.02	2.93	2.84	2.75
17	8.40	6.11	5.19	4.67	4.34	4.10	3.93	3.79	3.68	3.59	3.46	3.31	3.16	3.08	3.00	2.92	2.83	2.75	2.65
18	8.29	6.01	5.09	4.58	4.25	4.01	3.84	3.71	3.60	3.51	3.37	3.23	3.08	3.00	2.92	2.84	2.75	2.66	2.57
19	8.19	5.93	5.01	4.50	4.17	3.94	3.77	3.63	3.52	3.43	3.30	3.15	3.00	2.92	2.84	2.76	2.67	2.58	2.49
20	8.10	5.85	4.94	4.43	4.10	3.87	3.70	3.56	3.46	3.37	3.23	3.09	2.94	2.86	2.78	2.69	2.61	2.52	2.42
21	8.02	5.78	4.87	4.37	4.04	3.81	3.64	3.51	3.40	3.31	3.17	3.03	2.88	2.80	2.72	2.64	2.55	2.46	2.36
22	7.95	5.72	4.82	4.31	3.99	3.76	3.59	3.45	3.35	3.26	3.12	2.98	2.83	2.75	2.67	2.58	2.50	2.40	2.31
23	7.88	5.66	4.76	4.26	3.94	3.71	3.54	3.41	3.30	3.21	3.07	2.93	2.78	2.70	2.62	2.54	2.45	2.35	2.26
24	7.82	5.61	4.72	4.22	3.90	3.67	3.50	3.36	3.26	3.17	3.03	2.89	2.74	2.66	2.58	2.49	2.40	2.31	2.21
25	7.77	5.57	4.68	4.18	3.86	3.63	3.46	3.32	3.22	3.13	2.99	2.85	2.70	2.62	2.53	2.45	2.36	2.27	2.17
30	7.56	5.39	4.51	4.02	3.70	3.47	3.30	3.17	3.07	2.98	2.84	2.70	2.55	2.47	2.39	2.30	2.21	2.11	2.01
40	7.31	5.18	4.31	3.83	3.51	3.29	3.12	2.99	2.89	2.80	2.66	2.52	2.37	2.29	2.20	2.11	2.02	1.92	1.80
60	7.08	4.98	4.13	3.65	3.34	3.12	2.95	2.82	2.72	2.63	2.50	2.35	2.20	2.12	2.03	1.94	1.84	1.73	1.60
120	6.85	4.79	3.95	3.48	3.17	2.96	2.79	2.66	2.56	2.47	2.34	2.19	2.03	1.95	1.86	1.76	1.66	1.53	1.38
∞	6.63	4.61	3.78	3.32	3.02	2.80	2.64	2.51	2.41	2.32	2.18	2.04	1.88	1.79	1.70	1.59	1.47	1.32	1.00

Index

D

E

F

G

H

I

J

K

L

M

N

O

P

Q

R

S

T

U

V

W

Y